DATE DUE

JUN 0 2 1997		
JUN 2 5 1997		
FEB 2 8 2002		
JUN 2002		
JUN 0 1 2007		
JUN 2009		

Natural History and Evolution
of Paper-Wasps

Natural History and Evolution of Paper-Wasps

EDITED BY

STEFANO TURILLAZZI

Department of Animal Biology and Genetics
University of Florence, Italy

and

MARY JANE WEST-EBERHARD

Smithsonian Tropical Research Institute

Oxford   New York   Tokyo
OXFORD UNIVERSITY PRESS
1996

Oxford University Press, Walton Street, Oxford OX2 6DP

Oxford   New York
Athens   Auckland   Bangkok   Bombay
Calcutta   Cape Town   Dar es Salaam   Delhi
Florence   Hong Kong   Instanbul   Karachi
Kuala Lumpur   Madras   Madrid   Melbourne
Mexico City   Nairobi   Paris   Singapore
Taipei   Tokyo   Toronto
and associated companies in
Berlin   Ibadan

Oxford is a trade mark of Oxford University Press

Published in the United States
by Oxford University Press Inc., New York

A catalogue record for this book is available from the British Library

Library of Congress Cataloging in Publication Data
Natural history and evolution of paper-wasps
edited by Stefano Turillazzi and Mary Jane West-Eberhard.
Includes bibliographical references and index.
1. Polistes–Congresses. 2. Polistes–Behavior–Congresses.
3. Social evolution in animals–Congresses.
I. Turillazzi, Stefano. II. West-Eberhard, Mary Jane.
QL568.V5N38 1995 595.79'8–dc20 95–14968

ISBN 0 19 854947 4

Typeset by Light Technology (Electronic Books), Fife, Scotland
Printed in Great Britain by
Biddles Ltd, Guildford & King's Lynn

Foreword

William D. Hamilton

Social wasps are among the least loved insects. Even dogs dislike them. Sharing an Englishman's jam sandwich half-way to his mouth, massing in hostility to the Amazonian's gliding and bumping canoe, raiding and even predating their angelic and useful social cousins the honeybees, and perhaps above all simply being born with that bold, swaying and ever-returning flight, that remorseless buzz—all these things estrange us. Joylessly most of us read texts stating them to be important predators, that without them flies and caterpillars would be even more abundant. Yet where statistics will not alter a general impression, another approach might. Every schoolchild, perhaps as a part of religious education, ought to sit watching a *Polistes* wasp nest for just one hour. While probably not more than one in a thousand will be converted to become the kind of wasp addict that contributes to this book, I think that few will be unaffected by what they see. It is a world human in its seeming motivations and activities far beyond all that seems reasonable to expect from an insect: constructive activity, duty, rebellion, mother care, violence, cheating, cowardice, unity in the face of threat—all these are there. The ancient Greeks, moved seemingly by parallel observations, frequently presented *Polistes* wasps (as well as perhaps other wasps disproportionately) in varied paintings and ornaments. Latreille, the French systematist who named the genus (Linnaeus had them in *Vespa*) acknowledged both the Greek attention and the social traits by his selection of the plural of the word for a Greek city state. Between Latreille in 1802 and to the present, however, it took an American shopkeeper, Phil Rau, and an Italian biologist, Leo Pardi, to bring these astonishing insects back into the forefront of biological attention—and along with *Polistes*, to some extent, the rest of their papery clan.

Clearly Rau and Pardi watched and were entranced. Many following their writings, including the present writer, have become entranced likewise. From his careful observations Pardi gave us the concept of an animal dominance hierarchy; this predating the establishment of the idea for any other group. The concept of unconditional instinctive behaviour in animals, undifferentiated by individual propensity and as predictable as the bouncing molecules of a gas had from this time to begin its wane. If mere insects could be so human, what of the rest? Far beyond the humble paper nest of a *Polistes* wasp the whole idea of prediction in ecology was, or ought to have been, immediately shaken—although perhaps we should admit here that another concept, also largely home-grown within biology, that of mathematical chaos, was needed before the full consequence of the new individualism could be seen.

From *Polistes*, wasp studies branch out to all the rest, as the chapters in this book show. As social insects, paper wasps are probably a monophyletic group. Certainly they have elements of a common style however much they differ in physique. How different *Vespa mandarinia* is, for example, from a gnat-like *Parischnogaster* whose slim and tiny nest the mighty hornet, largest of all social insects, might hardly deign to devour even if she discovers it! Yet when working a ball of wood pulp or plant hairs into a firm papery wall, or in knowing when to consume a comrade's egg, both of these types, the huge and the tiny, turn out to be much the same.

In this book almost the entire sweep of the activities of wasps is represented, although most of the emphasis in the chapters is on the internal, social side rather than their impact on the rest of ecology, great as this is. Thirty years ago, when I was an unknown admirer of Pardi and Rau, this social side was my interest too and *Polistes* likewise my favorite. Sure, I had known long before this a *V. germanica* clypeus from a *V. vulgaris* (thier owners peering at me, perhaps, equally nervous and displeased, as it might be, from inside their hole in a plum). And soon I would be proud to tell a *Mischocyttarus* from a *Stelopolybia* (now *Agelaia*) even while in flight and even when, in coloration and shape, the gentle *Mischocyttarus* was a near perfect mimic of the ferocious other. But it was to social life that wasps were providing my touchstone puzzles.

Much of that skill as well as some of the intensity of my curiosity has now gone. At the Castiglioncello conference that originated this book I wandered rather like a man in paradise who finds God's memory embarassingly better than his own, listening to a list of loving and conscientious answers, detail, by detail given upon all those questions that were once so desperately pursued, saved up, and then . . . well, simply forgotten. This is not to say that there is not still an endless line of questions concerning this ill-starred under-appreciated group, but this book is a milestone— and not least I hope in giving certain marvelous animals a better image.

Preface

The chapters in this book are based on lectures presented at the international workshop 'Natural History and Evolution of an Animal Society: The Paper-Wasp Case', held in Castiglioncello, Italy, from 4–7 October 1993, under the sponsorship and organization of the Centro Fiorentino di Storia e Filosofia della Scienza of Florence.

The workshop was planned to mark the 50th anniversary of Leo Pardi's discovery of dominance/subordinance relationships in *Polistes* wasps (Pardi 1942). The main goal of the meeting was to review the status of research on these social wasps, which for many years have been a prime source of case studies in sociobiology. The enthusiastic responses of some of the best-known researchers in Polistinae biology and social behaviour in accepting our invitation, the great quantity of new data and ideas brought out in the course of the meeting, and especially the stimulating discussions and intellectual exchanges which enriched the official program, were the major stimuli for the publication of this volume. We believe that this book represents an important contribution to the body of knowledge about a group of organisms that occupies a privileged position in the history of that still-developing science, the study of the evolution of social behaviour.

The various chapters that make up this book, while preserving the independence and completeness that characterize original contributions by different authors, can be grouped into several sections.

A previously unpublished lecture by the late Leo Pardi, a prominent figure in ethology, opens the volume with a brief history and analysis of the principal characteristics which have made *Polistes* a key group for modern sociobiology. Following this is a chapter by Jim Carpenter, giving an updated systematics of the genus and then several papers which address the main elements of *Polistes* biology: nest architecture, colonial cycle, social parasitism, mating systems, and mechanisms which regulate behaviour outside the context of the colony and the return to the nest.

The next contributions analyze the physiological and biochemical mechanisms which regulate social functions such as dominance, and nestmate recognition. These chapters are introduced by R. L. Jeanne's paper, which furnishes an anatomical and functional overview of the organs (exocrine glands) which produce the substances that regulate various social activities.

A larger set of mainly theoretical papers explores the origin and maintenance of sociality in *Polistes*. These chapters also take into consideration the contributions that other social wasps with similar characteristics can make to the study of social biology.

The concluding chapter by R. Burian examines the role of organisms such as *Polistes* in the history of science from a philosophical point of view.

The Castiglioncello meeting was sponsored not by scientists but by a group of philosophers and historians of science. So the meeting was of interest on two levels: on one, we reflected upon what has been learned about the study of wasps as specimens; and on another we were specimens ourselves, a 'model system' for reflections about how science works. It has often been said that paper-wasps are 'model organisms' for the study of social behaviour. But Burian, in his lecture on us and our ideas, raised a genteel challenge to that idea. For him, a 'model organism' provides a definitive test of some particular hypothesis or idea. As such, can *Polistes* and its wasp relatives measure up to the standards set by *Drosophila* in genetics, the white rat in comparative psychology, or *Escherichia coli* in molecular biology?

This invites reflection on what constitutes a 'model organism' in biology. Certainly *Polistes* is an important 'test organism' for hypotheses about social behaviour, as illustrated by some of the chapters in this book. Queller uses *Polistes* wasps to test ideas about the influence of demographic parametres on social life; Strassmann uses them to test predictions about kin discrimination; and Jeanne points the way by their use to test ideas about the genetic assimilation of learned traits. Their efforts are reinforced by the efforts of others: while Queller looks at the consequences of demographic pattern, Yamane studies their causes (parasites, predators and climate); while Strassmann examines the acuity of kin recognition, Lorenzi, Bagnères and Clément probe its chemical basis, and Gamboa its behavioural mechanisms. This kind of synergism between progress in theory and fact leads to an ever spiralling accumulation of understanding so that in the area of kin recognition, for example, Gamboa can write that 'we have a more complete understanding of kin recognition in *Polistes* than in any other group of animals'. Thus, if *Polistes* does not yet qualify as a 'model organism' in the strict philosophers' sense, it is rapidly gaining the credentials to do so.

However, there is another quality of model organisms suggested by the paper-wasp case. The chapters in this book show that the wasps are being used not only to test hypotheses, but to formulate them. Wasps are idea generators: Wenzel sees a nest under a laundry basket and has an insight about the origin of geographic variation in nest sites—something Hamilton called 'tradition' in his workshop presentation. Cervo and Dani observe annual migrations and suggest a connection with the evolution of social parasitism; Beani observes patrolling marathons of males and formulates a general hypothesis regarding mate choice; Gervet, Anstett and Kjellberg see a general connection between the dominance hierarchies (vs. territoriality) of wasps and the stability of social groups; Ugolini and Cannicci, with an ingenious experiment of simulated passive transport in which female *Polistes* receive only visible stimulation, shed new light on the route-reversal mechanism of homing; and Hansell sees a general significance for nests as environments for social evolution. Clearly, the wasps are not just 'test organisms'. They are vital sources of new hypotheses, which, therefore—perhaps to their benefit—are not entirely products of abstract reasoning. The idea-generating function should perhaps be added to the hypothesis-testing one in the concept of a 'model organism'.

As hypothesis generators *Polistes* and other wasps have played a special role among social organisms. Wasps inspired the subsocial and trophallaxis hypotheses of insect sociality (Roubaud 1916), the recognition of dominance and hierarchy as a widespread feature of animal societies (Pardi 1942), and many important sub-hypotheses relating to kin selection, some of them discussed in this book. Rau proposed a theory of genetic assimilation of learned traits under the title 'Mind as a Forerunner of Evolution' (Rau 1933), an idea that resurfaced thirty years later in the work of Evans (1966a), and now appears again in the chapter by Jeanne. As C. D. Michener (1990, p. 61) has written, 'Wasp behaviour supports theory-making . . . Persons working on other group usually fall back on theory developed with wasps.'

'Idea generator' means that something special happens at the interface between wasps and the human brain, but why is this so? In part it is because of the suitability of wasps for comparative studies, described in the chapter by Pardi. One aspect of this, not usually mentioned, is that students of paper-wasps and related groups have fortunately included taxonomists (most notably Ducke, Bequaert, Richards, van der Vecht, and Carpenter) interested not only in classification but also in broad issues concerning distribution, phylogeny and other aspects of evolution—systematics in its finest sense. But *Polistes* itself also has a special appeal, a kind of wasp 'mystique' that attracts and inspires students of behaviour. Rau marked them with paint and gave them 'Indian names' ('Red Wing', 'Blue Dot' etc.). Hamilton referred to them, in his Castiglioncello lecture, as 'humanoid' and even experimentally tested the possibility that they might show 'anxiety' or special solicitude at the sudden loss and later return of their offspring (they did not). *Polistes* females are interactive and individualistic. They invite consideration of the role of the individual in society, so much so that the emergent qualities of *Polistes* groups, if they exist, may be a neglected aspect of this research (see Wenzel's chapter in this volume).

In keeping with the spirit of the Castiglioncello workshop we have urged the authors to retain, in the written versions of their lectures, speculations and thoughts that might not be permitted in a concise journal article. We hope that by combining scholarly reviews with new hypotheses this volume will serve as a stimulus to new directions of research as fruitful as those that have characterized the paper-wasp case in the past.

June 1995   S. T.
                                                                          M. J. W.-E.

Acknowledgements

This volume grew out of lectures and discussions at the workshop 'Natural History and Evolution of an Animal Society: The Paper-Wasp Case' (Castiglioncello, Italy, 4–7 October 1993) organized by F. Dessì-Fulgheri, A. Pagnini, P. Rossi, S. Turillazzi, and M. J. West-Eberhard and sponsored by the Florence Centre for the History and Philosophy of Science (Scientific Board: Paolo Rossi, President; L. Amaducci; G. Becattini; S. Califano; E. Casari; R. Conti; M. L. Dalla Chiara; F. Dessì-Fulgheri; A. Fonnesu; P. Galluzzi; E. Giusti; C. Luporini; M. Mugnai; E. Panconesi; P. Parrini; M. Piattelli-Palmarini; S. Poggi; M. Polsinelli; G. Toraldo di Francia; D. Torre; A. Pagnini, Director; P. Minari, Scientific Secretary). Thanks to the experience and approach of these people in their sponsorship of stimulating and well-focused scholarly meetings, this workshop was especially conducive to productive interaction, to the great benefit of all of the participants and the contributions published in this volume.

We acknowledge the financial support from the Ministero dei Beni Culturali e Ambientali, Consiglio Nazionale delle Ricerche, Regione Toscana, Comune di Firenze, and Università degli Studi di Firenze.

We thank the Italian section of the International Union for the Study of Social Insects and the secretariat of the Florence Center for the History and Philosophy of Science (Pierluigi Minari and Anna Eberhard).

Special thanks go to Francesco Dessì-Fulgheri, for having encouraged and supported this project in many ways.

W. D. Hamilton, A. Strambi and F. Dessì-Fulgheri presented valuable contributions at the meeting. They and C. Strambi, M. Keeping, W. Eberhard, J. Field, F. Papi and E. Alleva actively participated in general and informal discussion.

The students of the study group on social wasps of the University of Florence collaborated in the organization of the meeting, and also contributed many ideas and good questions during the lively informal discussions that characterized this workshop.

All the contributions in this book have been subjected to peer review and we would like to thank J. Alcock, B. Alexander, J. Billen, M. Breed, J. H. Brockmann, R. Burian, J. M. Carpenter, R. H Crozier, F. R. Dani, F. Dessì-Fulgheri, W. Eberhard, H. E. Evans, R. Gadagkar, G. J. Gamboa, J. Gervet, W. D. Hamilton, M. H. Hansell, A. Hefetz, R. L. Jeanne, M. Keeping, F. Kjellberg, J. Klotz, M. C. Lorenzi, C. D. Michener, P. Nonacs, S. Pagnini, F. Papi, M. Polak, D. Queller, H. K. Reeve, G. Robinson, P. Röseler, J. P. Spradbery, C. K. Starr, J. Strassmann, W. Wcislo and D. Wheeler for their comments and suggestions to the various chapters.

We are grateful to Elisabetta Francescato, Baldassarre Conti, Anna Eberhard, Cinzia Giuliani and Christina Coster-Longman who helped with translation and editing.

Florence, Italy   S. T. and M. J. W.-E.
S. José, Costa Rica
September 1994

Contents

Contributors

Marie-Charlotte Anstett Centre d'Ecologie Fonctionnelle et Evolutive, 1919 Route de Mende, B.P. 5051, 34033 Montpellier Cedex 1, France [12]

Laura Beani Dipartimento di Biologia Animale e Genetica, Università di Firenze, Via Romana 17, 50125 Florence, Italy [6]

Anne-Geneviève Bagnères CNRS, Laboratoire de Neurobiologie, Communication Chimique, 31 Chemin J. Aiguier, 13402 Marseilles Cedex 20, France [10]

Richard M. Burian Center for the Study of Science in Society, Virginia Polytechnic Institute and State University, Blacksburg, Virginia 24061-0147, USA [18]

Stefano Cannicci Dipartimento di Biologia Animale e Genetica, Università di Firenze, Via Romana 17, 50125 Florence, Italy [7]

James M. Carpenter Department of Entomology, American Museum of Natural History, Central Park West 79th Street, New York, New York 10024, USA [2]

Rita Cervo Dipartimento di Biologia Animale e Genetica, Università di Firenze, Via Romana 17, 50125 Florence, Italy [5]

Jean-Luc Clément CNRS, Laboratoire de Neurobiologie, Communication Chimique, 31 Chemin J. Aiguier, 13402 Marseilles Cedex 20, France [10]

Francesca R. Dani Dipartimento di Biologia Animale e Genetica, Università di Firenze, Via Romana 17, 50125 Florence, Italy [5]

Raghavendra Gadagkar Centre for Ecological Sciences, Indian Institute of Science, Bangalore 560 012, India [15]

George J. Gamboa Department of Biological Sciences, Oakland University, Rochester, Michigan 48309-4401, USA [9]

Jacques Gervet URA 1837 UFR SVT Bat. 4 R3, Université Paul Sabatier, 118 Rte de Narbonne, 31062 Toulouse Cedex, France [12]

Michael H. Hansell Institute of Biomedical and Life Sciences, University of Glasgow, Glasgow G12 8QQ, Scotland, UK [16]

Robert L. Jeanne Department of Entomology, University of Wisconsin-Madison, 237 Russell Laboratories, 1630 Linden Drive, Madison, Wisconsin 53706, USA [8]

Finn Kjellberg Centre d'Ecologie Fonctionelle et Evolutive, 1919 Route de Mende, B.P. 5051, 34033 Montpellier Cedex 1, France [12]

Maria Cristina Lorenzi Dipartimento di Morfofisiologia Veterinaria, Università di Torino, Viale Mattioli 25, 10125 Turin, Italy [10]

David C. Queller Department of Ecology and Evolutionary Biology, Rice University, PO Box 1892, Houston, Texas 77251, USA [13]

Joan E. Strassmann Department of Ecology and Evolutionary Biology, Rice University, PO Box 1892, Houston, Texas 77251, USA [11]

Stefano Turillazzi Dipartimento di Biologia Animale e Genetica, Università di Firenze, Via Romana 17, 50125 Florence, Italy [14]

Alberto Ugolini Dipartimento di Biologia Animale e Genetica, Università di Firenze, Via Romana 17, 50125 Florence, Italy [7]

John W. Wenzel Department of Entomology, Ohio State University, 1735 Neil Avenue, Columbus, Ohio 43210-1220, USA [3]

Mary Jane West-Eberhard Smithsonian Tropical Research Institute, c/o Escuela de Biología, Universidad de Costa Rica, Ciudad Universitaria, Costa Rica, Central America [17]

Sôichi Yamane Faculty of Education, Ibaraki University, 310 Mito, Japan [4]

1

Polistes: analysis of a society

Leo Pardi

This chapter is the translation of an unpublished lecture by Leo Pardi (Fig. 1.1), delivered at the opening session of the Tenth Congress of the Italian Ethological Society in Trieste (1984) and presented at the 1993 workshop by Stefano Turillazzi. It is a fitting first chapter for this volume because it gives a basic introduction to the natural history of paper-wasps, as well as an idea of why *Polistes* has been prominent in studies of insect behaviour and social evolution. Informal in tone, this lecture reveals Pardi's enthusiasm, scholarly virtuosity, and sense of humour. It is still timely, even ten years after it was written. We include it here also as a tribute to a major figure in this field and as a glimpse of his personality. Many of the subsequent chapters in this volume refer extensively to Pardi's work and some bear witness to his deep influence on Italian ethology. We made only small amendments to the original text; a few figures have been added for clarity. The translation into English was done by Anna Eberhard.

[Editors' note]

The reason for my title is that it seems to synthesize quite well the interest, almost insistence, with which this particular genus of wasp is now studied all over the world. Before considering the motives for the development of these studies, here is a brief, general systematic and biological foreword. The genus *Polistes*[1] is nearly completely cosmopolitan and, with 203 species and 106 subspecies recognized (but with the systematics in continual development), practically the only member of the tribe Polistini. The other taxon, *Sulcopolistes*, social parasites of great interest, comprises only three European species and, according to some systematicists, should be considered perhaps only as one of the subgenera of *Polistes*. The tribe Polistini, together with the smaller tribe Ropalidiini and the more widespread Polybiini (both tropical), constitutes the subfamily Polistinae, which comprises the vast majority of the family Vespidae.

In complexity of social organization, the Polistinae are, let us say, at the halfway point between the Stenogastrinae and the more evolved Vespinae. And within the bounds of the Polistinae, the Polistini occupy a similarly central position. To illustrate the biological cycle let us consider, as an example, that of a temperate lowland species, *P. dominulus*.[2] In spring the founding females, all inseminated, leave their winter refuges and found new nests. The new nest can be the work of a single female (we would then be dealing with monogyny or haplometrosis) or of an aggregate of females (polygyny or pleometrosis). With polygynic foundation, a division

Fig. 1.1 Leo Pardi (Photo by F. Papi).

of labour is established among the females: one of these becomes for all practical purposes a true queen, specializing in oviposition, in construction, in work on the nest; the others (called auxiliaries) stop laying and behave as workers, taking on the riskier external work, such as foraging. The offspring derive mainly from the queen, but also in part from the auxiliaries, and consist at first of typically sterile females (workers) that assume the task of caring for the young and permit further reproductive specialization of the queen. With the appearance of the workers, the auxiliaries may disappear, ousted by the queen or the workers, or more rarely they may remain as part of the colony. In the course of the summer, and always predominantly from queen eggs, the future female and male reproductives appear. In temperate climates these late females are lazy and rich in fat reserves. They do nothing on the maternal nest and have been called daughter foundresses or gynes. Fairly precociously the queen dies, the gynes mate generally outside the nest with foreign males, and—when the colony dissolves in autumn—repair to the winter shelters, which they will leave in spring to return to the original maternal nest and start over. Variations in this cycle are numerous and it would be fairly interesting to examine them, but we do not have the time to do so.

Let us return to the initial question. Why are *Polistes* currently so extensively studied? More so, for example, than the equally fascinating Vespini? Certainly there are obvious reasons for such a preference:

1. The widespread occurrence of colonies (*Polistes* are very common).
2. The often aerial position of their nests on plants or on human constructions, more rarely in cavities, such as under roof tiles.
3. An especially obvious reason is gymnodomy, the lack of an envelope around the single comb: Fig. 1.2 is, perhaps, one of the first depictions of a *Polistes* nest, from De animalibus insectis by Ulisse Aldrovandi (1638).
4. The small size of the colony (rarely more than 100 individuals in temperate climates), which renders extremely easy the observation and quantification of behaviour, especially of individual interactions.

The abbot Giuseppe Stefano Disderi, vicar-general and ecclesiastical lawyer in Saluzzo, took advantage of these conditions of easy observation. In his delightful 'Vespae Gallicae Historia' of 1816, he writes, in his succinct and vivacious Latin, about his intention to study a *Polistes* female which had founded a nest in the window of his bedroom:

Quum itaque anno 1811 in vitreis fenestrae cubiculi nostri cancellis, Gallicae Vespae foemina, *Polistes Gallicus* Latreille, consederit; ibique nidum struere caeperit, attente horis subsecivis eam indagare statuimus. Res voto cessit. Nidus ad oculi altitudinem structus commodam observandi praebebat facultatem. (Disderi 1816)

To this convenience of observation (not always in the bedroom, however) can be added the large size of the individuals (there are species of giant *Polistes* with individuals 3 cm long), their relative tolerance of manipulation, the ease with which they can be raised in captivity, and their adaptability even to artificial nests made with independently movable, transparent plastic capsules fastened together (Turillazzi 1981).

There are still more essential reasons for the interest in this group: these derive from the relative primitiveness of its social organization and from its extremely wide geographic distribution. The primitiveness of social organization is evident in two traits: the lack of morphological caste differentiation, and the presence of overt and constant reproductive competition within the colony. One can immediately understand how these two conditions together can provoke interest for the understanding of social evolution.

While in higher hymenopteran societies with marked caste differentiation, reproduction is, as a rule, tightly controlled by the queen caste (often a single female), with a very large population of (usually sterile) workers, in the case of *Polistes* the game is wide open: the frequent contemporaneous presence of more inseminated and fecund females, be they of the same generation (as in polygynic associations) or of successive generations (mothers and daughters), feeds a continuous confrontation that is manifested at the behavioural level. And the varying fertility degrees of the contenders means that they find themselves facing various choices, in which are intermixed elements of antagonism and co-operation, of egoism and altruism. And it is on these problems that modern analysis concentrates, guided by the concepts and predictions of sociobiology and, above all, by kin selection theory which—despite criticisms—is still the dominant paradigm of the social evolution of hymenopterans.

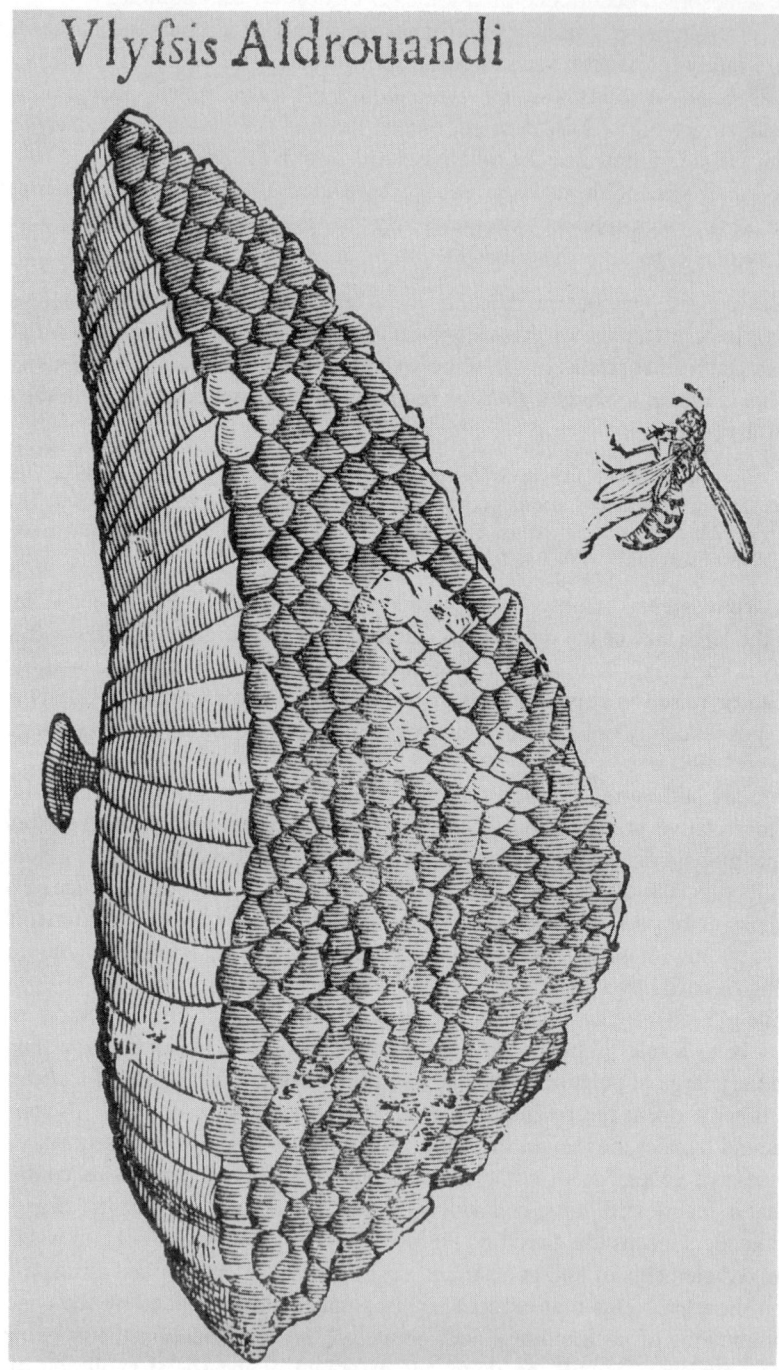

Vlyſsis Aldrouandi

Fig. 1.2 *Polistes* nest, from De animalibus insectis by Ulisse Aldrovandi (1638).

It is also true that in the more primitive Apoidea (Alictinae, bumblebees) there are comparable or parallel phenomena, but these lack—because of their types of nests—that 'commoda observandi facultas' of Disderi. Furthermore, there is an enormous opportunity for comparison between species having different ranges. All ethologists know what this means, as our discipline is in fact described as the 'comparative doctrine of behaviour'.

Polistes, as we have seen, is cosmopolitan, and its range is extensive both latitudinally and altitudinally. This creates enormous possibilities for comparing the multiple adaptations of the cycle and of the social organization itself. In other Vespidae there are no such opportunities.

Before I discuss some of these aspects, permit me to make a brief excursion into history. Without going back to Aristotle, I would like to remark upon the fortunate choice of the generic name. Taking Agassiz's 'Nomenclator Zoologicus' (Agassiz 1845) we find, on page 27 of the Hymenoptera, *Polistes* Latreille 1804, from 'πολιζω', to found a city. Latreille (1804), when he separated *Polistes* from the Linnean genus *Vespa*,[3] could not have made a more felicitous or more foresightful choice of generic name, since even 180 years after that choice the question of colony foundation is still the most current and hotly debated problem in the study of the group.

The enormous work of systematists aside, I would say that the first true ethologist of *Polistes* was the abbot Disderi, who described many things perfectly, from the manoeuvres of construction, to defence and the division of labour. For example the following describes the elimination of superfluous water when the nest is wetted by rain: 'Si pluviae, aut alterius liquoris guttae nidum madefaciant; ore aquam labunt; deglutiunt; deinde guttis ingentibus evomunt ore rejicientes: . . .', ['. . . If raindrops wet the nest they (the workers) lick up and swallow the water, then eliminate it in large drops by vomiting . . .']. And again: '. . . dum in nido haec aguntur, redeunt interim ab agris operariae, quae ad cibos colligendos abierant; reversae ad nidum linguam statim neutris, quae circa nidum laboraverunt, exhibent, ex ore guttam digesti liquoris emittunt, quam istae lambunt . . .' ['. . . meanwhile the workers, who had gone to gather foodstuffs, return from the fields, and immediately—landing on the nest—show to the sisters who had remained at work their tongue emitting from the mouth a drop of the ingested liquor, which the companions lap up . . .'] (Disderi 1816). This is a surprisingly faithful and minute description, one hundred years before Roubaud and Wheeler, of the trophallactic exchanges among adults with which so many of us have been occupied up to the present day: and it is the first allusion, perhaps, to interactions between individuals.

Another extraordinary observer of the habits of *Polistes* was Karl Theodor Ernst von Siebold, the great German zoologist. In his work 'Über die bei *Polistes* wahrzunehemende Partenogenesis' of 1871 there is not only the precise confirmation of the Dzierzon rule for these hymenopterans, but also an impressive number of precise observations on nest building, oviposition, nutrition, defence, male behaviour, and so on, as well as on the anatomy of the reproductive apparatus and on caste differences (Siebold 1871). In addition, von Siebold was the first to point out the polygynic association of foundation, followed a few years later by Pierre Marchal

Fig. 1.3 Phil Rau (by courtesy of M. J. West-Eberhard).

(1896). We have to wait until the 1920s and 1930s for the appearance—in our mi-
nuscule specialists' firmament—of a star of considerable greatness. Phil Rau (Fig.
1.3), a private entomologist of Kirkwood, Missouri, certainly was that. He was—so
to speak—a kind of American Fabre. He resembled him both for his fascinating
prose and for his isolation from academic politics: he was not antievolutionist like
Fabre, but was certainly very Lamarckian. The titles of his works have, at least for
me, the perfume of poetry: 'The duties of a wasp queen, *Polistes pallipes*' (Rau
1935), 'Autumn and spring in the life of the queen *Polistes annularis* and *Polistes
pallipes*' (Rau 1928) and, almost melancholic, 'At the end of the season with
Polistes rubiginosus' (Rau 1929a). But let us not be deceived by the titles. It remains true that Rau, a careful and rigorous field naturalist, set the basis precisely for
the ample comparative work I mentioned on the habits of the numerous American
species, from the temperate to the Mexican and tropical ones. Rau studied and considered everything: nest foundation, nest building, defence, biological cycles, hibernation, migration, homing, caste differentiation, trophallaxis. At his death in
1948 he left unpublished a monograph on the genus.

But the 1930s were no less important, for our little historical review, on this side
of the Atlantic. For in Germany appeared the works of the Swiss Anton Steiner and
of Georg Heldmann, curator and director of the Hessisches Landes Museum of
Darmstadt (Fig. 1.4).

Fig. 1.4 Georg Heldmann (by courtesy of H. Feustel and A. Buschinger).

Steiner is as heavy to read as Rau is entertaining. But from his classic work on thermoregulation, and most of all on the division of labour (Steiner 1930), polistologists have learned precise methods of observation, quantification, and evaluation of behaviour.

Heldmann's work was also fundamental because he was the first to put into evidence the reproductive division of labour and differentiation in social activities which are established in spring among associated *P. dominulus* females with a more fertile and egg-eating queen and other less fertile auxiliary females, which he admitted could be either mated females or over-wintered workers (Heldmann 1936). Rau made similar observations soon thereafter for *P. annularis*, in North America (Rau 1940). Neither one of the two, on the other hand, had paid much attention to inter-individual behaviour. I then had the fortune to be able to show, around the mid-forties (Pardi 1942, 1946, 1951), not only that the polygynic spring community of *P. dominulus* always consists of fertile and inseminated females, but, above all, that among these is formed—after initial interactions—a relatively stable linear hierarchy based on characteristic domination-subordination behaviour and closely correlated with ovary development; that the reproductive division of labour and behaviour are consequences of this; that the ovaries of the subordinates regress; and that the hierarchy also exists among the workers, in which it regulates the appearance of egg-layers (male producers). Nothing like this had before been known in insects or other invertebrates. I will come back to this shortly.

These conclusions were fiercely opposed by Edouard Philippe Deleurance (Deleurance 1952*a*) a worthy student of Grassé, to whom we can credit important studies on construction behaviour and on the colony-cycle of *P. dominulus* (Deleurance 1952*b*, 1955, 1957). But this is an old story that we have gone beyond: the existence and the importance of hierarchy in social organization have been since confirmed in all the other species of *Polistes* studied so far, and its validity, correlations, and consequences have been discovered—though with important variations—even in Polybiini (Marino Piccioli and Pardi 1970; Jeanne 1972) and Ropalidiini (Pardi 1974; Darchen 1976*b*), in Stenogastrinae (Yoshikawa *et al.* 1969) and Vespinae (Montagner 1966), and even in other non-vespid hymenopterans, in some Sphecidae (Matthews 1968) and Halictinae (Ordway 1965) and, above all—as I was able to foresee—in bumblebees (Free 1955; Röseler and Röseler 1977).

It is rather fitting to observe that all the initial research on dominance hierarchy in *Polistes* and in Hymenoptera (by myself and others) was primarily concerned with the immediate causes of the phenomenon, that is with the physiological determination of the behaviour and the effects of the hierarchy on various individuals. The question of adaptive value for single individuals and for the colony was approached fairly timidly, and the evolutionary implications were only barely noted. After all, this was the prevalent trend at the time.

Only after Haldane, Williams and Hamilton and the formulation of the various genetic theories of sociality—only after the advent of sociobiology—do those themes come to the foreground even for our case. And it is undoubtedly Mary Jane West-Eberhard who is to be credited with having fully developed them, first with strong analytical work on the society and the hierarchy of Polistini and Polybiini (West-Eberhard 1969*b*, 1978*a*) then with vast observations and synthetic work (West-Eberhard 1975). It was with data from the study of intragroup competition of these wasps that West-Eberhard brought into a new light both inter-individual competition in its most primitive forms, and the traces of such competition which remain in the organization of the most highly integrated hymenopteran societies. And it is again from this starting point that the American scholar arrived at the proposal of her polygynic-familial hypothesis regarding the paths of the evolution of society in the Hymenoptera, an alternative to mutualistic and matrifilial hypotheses (West-Eberhard 1978*b*).

Now I come—after this historical summary—to some of the problems to which I wish to draw your attention. Since these have to do mostly with hierarchy, I should say immediately that by dominance hierarchy we mean behavioural hierarchy, based on the observation of aggressive and submissive behaviours.

Let us take a look at the behaviours observed in *P. dominulus*. In accordance with a gradient of increasing aggressiveness of the dominant, and schematising a bit, we go from a simple movement of antennae extended toward another individual which evades the encounter, from brief antennations of almost exploratory character followed by immediate separation, to an attitude of imposition with a more vigorous and sustained antennal tapping (Fig. 1.5). In this case the subordinate often responds with a characteristic attitude of deference, crouching, or by regurgitating a drop of liquid, or often even by leaving the nest. Proceeding along the scale of aggressivity, we find widespread nibbling of the subordinate by the dominant, with

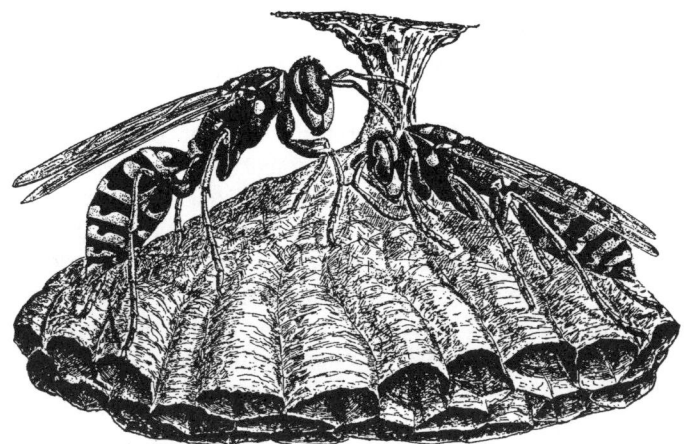

Fig. 1.5 Dominance–subordinance postures in *P. dominulus* foundresses (from Pardi 1946).

akinesis of the subordinate, followed by mounting and grasping with more violent bites, sudden wrenching tugs with extended wings, and finally attempted or actual stinging. In the final case the exit of the subordinate from the nest takes on the appearance of escape.

This is the range of asymmetrical encounters. A symmetrical encounter is a fight in erect position, with blows of legs and antennae. If this does not turn into an asymmetrical encounter and the fight becomes protracted, it progresses to clinching and to a fall while still grappling (the 'falling fight' of American authors).

All these behaviours can be observed in encounters away from the nest and even before nest foundation, but once the nest is founded, high rank in the hierarchy is revealed by other behavioural characteristics as well: by more frequent egg-laying; by egg-eating, especially of others' eggs; and by position on the anterior face of the nest. Very characteristic, or at least more frequent, in females of high rank is abdomen-wagging. This is found both in trophallactic larva-adult interactions and, more often, in interactions among adults: aggressive solicitation of regurgitated drops by a dominant via abdomen-wagging appears to be a continuation of trophallactic solicitation by larvae. American authors have distinguished at least two kinds of abdomen-wagging (tail-wagging and lateral vibrations) (Gamboa and Dew 1981). And it is also thought that the dominant may—during abdomen-wagging—spread on the nest pheromones from the sternal glands, which signal chemically to companions. Downing and Jeanne (1983) have recently demonstrated that there is a good correlation between dominant status and development of the sternal glands.

An etho-morphological problem concerns the inter-specific differences in the encounter ritual. In different groups, obviously, there are considerable differences: this is the case in the Polybiini (*Myschocyttarus* according to Jeanne (1972), and *Belonogaster* according to observations by Maria Teresa Marino Piccioli and myself (Marino Piccioli and Pardi 1970)). Within *Polistes*, the differences sometimes appear small. Perna *et al.* (1978) have forced polygyny in usually monogynyc *P. gallicus* females, and Turillazzi and Ugolini (1978) have created in captivity mixed colonies of *P. dominulus* and *P. gallicus*. The form of the encounters is in these cases almost the same, if anything with an accentuation of the frequency of aggression and a greater frequency of peck-order with respect to peck-right. In other species, as has been shown by West-Eberhard (1982*a*), can be found relevant diversities in forms of encounters. In North America, for example, the behaviour of the temperate-zone *P. fuscatus* is broadly comparable with that of *P. dominulus*, but *P. erythrocephalus* from Colombia has strongly ritualized aggressive behaviour with an extraordinary accentuation of the lateral vibration of the abdomen; *P. major* has special displays called wagging runs and lateral bending; *P. carnifex* shows slow, tense solicitation movements which, when they do not induce appropriate responses, give way to states of fierce aggression. A comparative ethology of *Polistes* from this point of view does not yet exist, but it could be expected to bear interesting fruits.

Now, let us return to the beginning of the biological cycle. Where do the females that in spring associate in the foundation of a single nest come from? Experimental marking in autumn showed many years ago that—in the majority of cases—we are dealing with females from the same colony, which return to nest, like swallows, in the same area.

Since in most cases nest companions come from the same mother, if not from the same father, it is argued that they are sisters or step-sisters, and are in any case closely related. This aspect is crucial for the theory of kin selection, and I will come back to it.

But for recognition to occur is it enough for foundresses to find themselves together at the same site? Certainly this philopatry is important, but there is something else: the discrimination of nestmates, as was first shown by various American authors (Shellmann and Gamboa 1982; Pfennig *et al.* 1983*a*) who put together, in the laboratory away from the nest and the parental site, trios of two companions and one non-companion and quantified the respective tolerance behaviour.

But what are the factors and the mechanism of the discrimination? Recent experiments by Pfennig *et al.* (1983*b*) on *P. fuscatus* and *P. carolina* showed that the key for reciprocal recognition is learned in the adult stage and acquired, soon after emergence, from the nest or the immature offspring that are in it. Since this is a very stable and precociously acquired memory, it has the characteristics of imprinting.

In the American species it seems that only exposure to the nest itself is effective: females isolated from the maternal nest and exposed only to their companions are unable later to detect non-companions.

Another problem that kept us occupied at length can today be said to be essentially resolved: what is the physiological basis of dominant or subordinate status?

Many years ago I recognized the close correlation in *P. dominulus* between rank and ovarian development, not only in polygynic association but also among workers, and I put forth the hypothesis of an influence on behaviour by internal secretions of the ovary and of the corpora allata (Pardi 1942, 1946, 1951). The correlation between hierarchy, ovary development, and size (and thus presumably activity—Röseler *et al.* 1980) of the corpora allata has subsequently been

P. dominulus

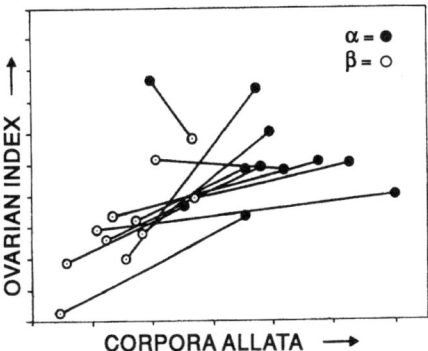

Fig. 1.6 Rank, ovary and C. A. development in *P. dominulus* (from Pardi 1980, based on unpublished data by Marino Piccioli and Pardi).

documented for *P. dominulus*, by Marino Piccioli and myself, in bigynic associations (in Pardi 1980) (Fig. 1.6) and by Turillazzi *et al.* (1982) in polygynic associations. Contemporaneously and independently, by measurements and by determining in vitro the synthetic activity of the juvenile hormone by the corpora allata, Röseler *et al.* (1980, 1984) arrived at identical conclusions: the corpora allata are larger and more active in dominants than in subordinates.

It is worth also noting here the correlation between body size, rank, and ovary development, documented in various species (Turillazzi and Pardi 1977; Dropkin and Gamboa 1981; Strassmann 1983; Sullivan and Strassmann 1984).

All these data refer to already-established bi- or polygynic associations. One recent study by Röseler *et al.* (1984) besides setting up the problem experimentally, has the merit of tackling it—let us say—at the root, before nest foundation. Females were taken from hibernation and either treated or not with juvenile hormone or beta-ecdysone and, in pairs, provoked to see which dominated. Immediately afterward, they were dissected, and their ovaries, corpora allata, and wings measured. From the results it was evident that the individuals treated, for example, with juvenile hormone increase considerably their probability of dominance with respect to the controls. Beta-ecdysone, the hormone produced by the ovaries, has an entirely similar, and perhaps even stronger effect. Injected together the two hormones did not produce a synergistic effect. The authors' conclusion was that differences in the endocrine activity of the corpora allata and of the ovaries during the first days after hibernation are responsible for the position of the females in the dominance hierarchy.

'Among the females that have over-wintered' I wrote in 1946, 'there exist, even before nest foundation, small differences in the ovarian index, and it is in accordance with these differences—or rather with the diversity of physiological classes of which they are the measurable expression [and I was referring here to the corpora allata]—that a dominance hierarchy is established' (Pardi 1946). Today I would like to modify a point in this sentence: the differences are sometimes small but quite relevant.

But once the hierarchy is established, what is the mechanism that determines the ovarian regression of the subordinates and, often, the further development of the ovaries of the dominants? This point can also be said to be resolved:

1. Subordination leads to a trophic disadvantage: the exchanges of liquids, whether from larvae or from the exterior (as in the descriptions of Disderi), are bidirectional in the hierarchy, but there is an evident asymmetry in favour of the dominants. This is demonstrated by observation and by experiments with radioactive marking of honey. If a certain quantity of honey marked with J^{311} is given to the alpha female, she retains high radioactivity for a long time, while the beta, generally, relinquishes it at once (Marino Piccioli and Pardi 1980) (Fig. 1.7). In *P. fuscatus* and other species there is also—according to West-Eberhard (1969*b*)—a preferential passing of solid food from low to high ranks.
2. Low rank in the hierarchy is correlated with a greater expenditure of energy: the alpha remains on the nest and specializes in housework; the subordinates, after an act of domination, are often induced to go out, and specialize in external service. For example, Gamboa *et al.* (1978) report a higher frequency of absences from the nest for the subordinates with respect to the alpha in *P. metricus* from Kansas.
3. The presence of the alpha inhibits the oviposition of the subordinates. An experiment by myself and Cavalcanti (Pardi and Cavalcanti 1951):

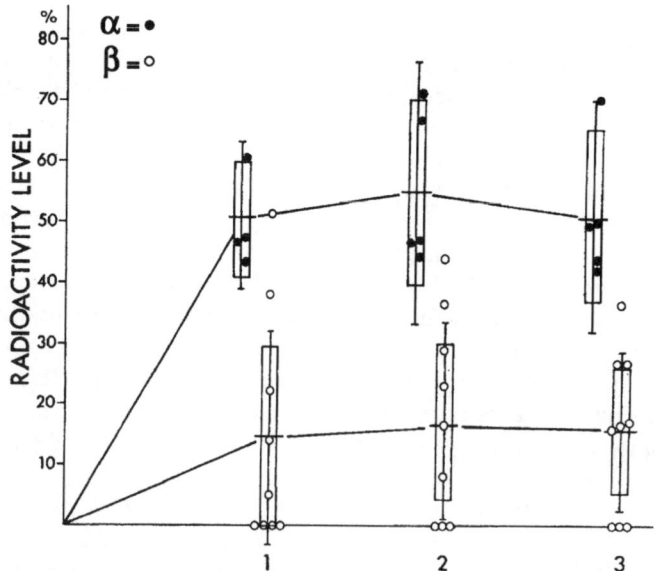

Fig. 1.7 Level of radioactivity reached in the partners of tracer-fed *P. dominulus* in the first three measurements (from Marino Piccioli and Pardi 1980).

On bigynic and trigynic nests toward the end of the polygynic association we found, upon dissection, this type of situation: the ovipositing alpha had an ovary index greater than 2 mm, the ovaries of the beta and the gamma that had not oviposited for 15 or 20 days showed moderate regression for the beta and a still more marked reduction for the gamma. In two trigynic nests in which the subordinates had not oviposited for 15–17 days, one subordinate female was isolated from the other two and kept on the nest for 12 hours, while the other two were kept on the same nest for the other 12 hours, in an inverse light rhythm. After 10 days of this treatment, dissection showed that the isolated gamma female had ovaries considerably more developed than those of the gamma that remained with her alpha, and the isolated beta had more developed ovaries than the beta that also remained with her alpha— the ovaries of this last beta were at the level of those of the gamma that remained with the alpha. Experiments fundamentally comparable to this, by Röseler *et al.* (1984), yielded results for the regression and the reactivating of the corpora allata that are in complete conformity with our findings.

If this is how things are, there should be predictable differences in the ovarian and corpora allata regressions for the associated females according to the level of polygyny. In recent work (Turillazzi *et al.* 1982) we have shown this by measuring ovaries and corpora allata of females of various ranks from mono-, bi-, tri-, or tetra-gynic nests, both immediately after nest foundation (end of April) and at the end of May (in the advanced polygynic association).

Figure 1.8 shows the results for the ovaries, in which each circle represents the mean ovarian index of the alphas, betas, etc. in the nests with different polygyny in April and May. First of all, note the wider variability in May, at the end of polygyny, with respect to April, among females of different rank: this is precisely the reproductive differentiation which I have mentioned many times. It is not a surprise that, in the monogynic nests, the single female present should have smaller ovaries in May— she is alone and has to do everything by herself. But in the bigynic nests as well we find at the end of May that the alphas have, on average, smaller ovaries than the alphas in April. The betas of the bigynic nests, however, present a considerably

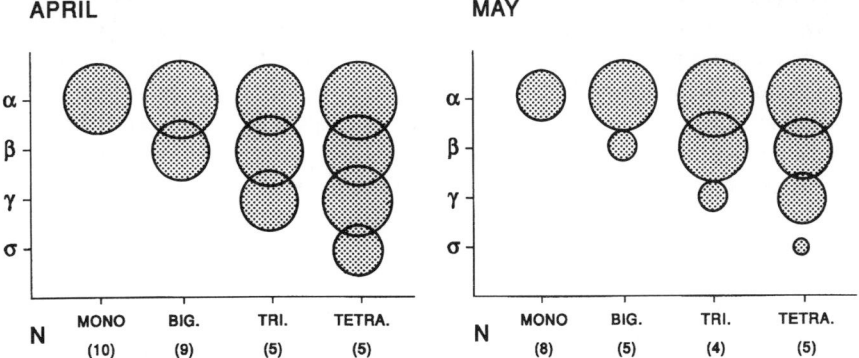

Fig. 1.8 Average ovarian index of females of different ranks on pre-emergence colonies of *P. dominulus* (data from Turillazzi *et al.* 1982).

greater regression than their alphas. In the trigynic nests of May, the alphas, instead of losing have gained a little, while the betas have only barely lost and considerably less than in the bigynics: it is the gammas that reveal a regression comparable to that of the betas in the bigynic nests. In the tetragynic nests, finally, the gain of the alphas appears still more accentuated, the betas have lost—it is true—more than might have been expected, but the gammas (which here are not omega) have lost more than in the trigynic nests, while the maximum regression is found in the delta female. A nearly identical situation is found measuring the corpora allata. It can be concluded that regression of the ovaries and corpora allata depends not only on the absolute rank of a given individual, but also on the number of individuals subordinate to her.

Two other important events in the biology of *Polistes* (other than that of polygynic association) can be related to the system of behavioural dominance:

(1) the appearance of fertile workers;
(2) the appearance of foundress daughters (or gynes), according to American authors, in nests of tropical polistines.

Finally, as is known, the usurpation of *Polistes* colonies by social parasites of the genus *Sulcopolistes* has been interpreted by Scheven (1958) as a special case of dominance hierarchy. But I cannot linger on this and it is time to outline some more strictly sociobiological aspects and draw to a conclusion.

It is well known how kin selection theory faces the problem of the origin of the sterile caste and of eusociality by also basing its affirmations on those famous asymmetries of relationships that in hymenopterans derive from the haplodiploid determination of sex (Hamilton 1964*a* and *b*). The added 'also' is important because in Isoptera there is no such asymmetry. A special problem in the area of kin selection, focused by what has been observed in polygynic associations: I will say rather that the solution offered is one of the kin selection's war horses.

We know that polygynic nests, compared with monogynic ones, have an undoubtedly greater final productivity and longer survival. But here we have not a question of advantages for the colony, but of advantages for single individuals.

At colony foundation, there is frequently a situation that we need to discuss first: that the foundress allows the association of other females subordinate to her—let us say, for the sake of simplicity, a beta. The advantages for the foundress, which remains alpha, are evident: the auxiliary contributes to the care of her offspring and reduces her risks, taking on the dangerous activities away from the nest. But for the beta the probabilities of succumbing increase and her direct reproductive success is reduced, if not cancelled. Why then this altruistic choice?

The explanation given by upholders of kin selection theory is based on two propositions:

1. The associates are as a rule sisters or at least fairly close relatives.
2. The beta that associates herself had from the beginning a reduced reproductive potential: the cost of her sacrifice in direct reproduction is low.

Do these propositions hold? Let us see. The kinship between nest companions is diluted, with respect to the theoretical level, by 3/4 (among true sisters) by factors such as multiple mating by the mother, and presence on the maternal nest of more ovipositing females, and can in the latter case, according to recent estimates, reach a level below that required by the altruistic choice. Nevertheless, it can be asserted that in the majority of cases the first proposition corresponds fairly well to reality.

But this is not enough. Indeed—as Hamilton first observed—a subordinate, even if she were a true sister of a dominant, would have with the daughters of the latter a kinship coefficient of 3/8, while with her own offspring she has an r=1/2. It would seem more convenient for her to found on her own account. Here then is the second consideration, credited to West-Eberhard (1975): let us say that a beta has a reduced reproductive potential—to simplify, half that of an alpha, as in this outline: even allowing for her sacrificing entirely her direct reproduction, if with this sacrifice she doubles her sister's direct reproduction, the beta's total fitness increases (3/8 + 3/8 > 1/2).

Domination, correlated as it is with the state of the ovaries, is the mechanism which establishes immediately at the first encounter which of two females stands to profit from the alternative of indirect reproduction through the greater fertility of her companion, instead of taking a chance with a less productive and more dangerous independent foundation. Furthermore, if we examine the state of ovaries early in the colony cycle, we also find quite a few nests in which the difference between alpha and beta is very small and the beta's renunciation of solitary foundation thus is not immediately comprehensible. The theory has been adjusted with an additional hypothesis that still needs to be verified: where there is a strong difference between alpha and beta, the foundress was joined by the associate. Nests where there is a small difference correspond to cases in which the foundress, though quite fertile, has been overpowered by another more fertile and dominant: so-called usurpation. Then it is understandable that the original foundress should adapt herself to the rank of beta because, all things considered, it is not worth losing the investment already made and starting over. It must then be taken into account—and this again is valid especially for the beta—that the adoption of the subordinate alternative can also be favoured by the probability of succession in the case of the death of the alpha.

Aside from the data, this may appear to be just an ingenious conjecture. Concrete estimates of the total fitness of the various females are rare and complex for many reasons, and yet they appear to agree (though not always) in the admission of a conclusion favourable to the mechanisms postulated by kin selection. A fairly clear situation in this sense has been calculated by Strassmann (1981*d*) for polygynic *P. annularis* nests giving the gain in offspring equivalent of females of varying rank, in nests with different levels of polygyny.

For now, then, accounts on the whole agree. Strassmann (1981*a*) has also recently furnished another quite convincing confirmation of the applicability of the theory in studying the cycle of *P. exclamans* and the association and foundation behaviours in so-called satellite nests.

But it is time to conclude. I have wanted to show, most of all, how a fairly long road has been travelled since von Siebold first noted the polygyny of *Polistes*, and how the recognition of intra-group reproductive competition based on the behaviour of these wasps has led to developments interesting from a physiological as well as an evolutionary point of view—developments that are gradually being extended to all the social vespids and even to other hymenopterans.

I feel obliged to end with a reflection on sociobiological explanations, and in particular on kin selection theory, which, I repeat, is still the weight-bearing axis in this field. No one can deny—even for our case, to which I would like to limit myself—the value of such explanations for the prediction and coherence of interpretation on the selective bases of many behaviours which are hard to understand at first glance. Nevertheless it is difficult to escape some critical remarks which are, it seems, shared by many. I will briefly summarize them as follows:

1. The sometimes obsessive insistence on searching for explanations solely in Hamiltonian arithmetic, to the exclusion of other possible selective mechanisms (reciprocal altruism, imposed altruism, or kin control). This is probably not only valid for the Polistini. All theory on sociality in insects, as West-Eberhard observes, seems still to be too narrowly focused on altruism by choice.
2. The extremely understandable, but in the long run almost annoying approximation in the theme of genetic control of behaviour isolated from observation (see, on this subject, a significant critical analysis by Pratte and Gervet (1980)).
3. A certain tendency, perhaps currently diminishing, to pass over the field of ethophysiology and, more precisely, lack of concern about the sensory basis of the observed or postulated individual choices.

I allow myself to close in a light vein. You all know very well how kin selection readily explains the general social laziness of Hymenoptera males, which seem to have only *one* fixation. I challenge that—their kinship with the maternal offspring is only 1/4, while that with their own is 1! (Hamilton 1972).

It is true that such good-for-nothing-ness is not really absolute: even in *Polistes* there have been sporadic observations of the participation of males in various social duties (Hunt and Noonan 1979) and in the African polybiine *Belonogaster* I myself have observed that the feeding of solid food to larvae by males has a frequency not far from that of the foragers themselves (Pardi 1977). But there it is. We can strike out the rule of universal social laziness of the male.

During one of the many conferences on genetic mechanisms of social evolution, some participants composed some playful verses on the subject.

One of these struck me as funny and an appropriate closing for this essay. I have therefore adapted it to wasps in the following manner:

My virgin worker, you I respect:
Helping momma make sisters
Is your great effect!
But since there are many such lovely wasps,
I prefer instead to be direct!

Notes

1. See the chapter of J.M. Carpenter (this volume) for recent systematics.
2. In Pardi's manuscript this species is still indicated as *P. gallicus* (as in his published papers) though Day changed the name in 1979 (Day 1979)
3. Actually Latreille separated *Polistes* from *Vespa* in 1802 (Carpenter this volume).
4. Deleurance (1952a) saw a possible mechanism for queen control of nestmate reproduction in the fact that the queen removes oviposition stimuli by filling empty cells with her own eggs. Unfortunately, he phrased this as a vigorous attack on Pardi's earlier work which showed the importance of dominance. Pardi (Pardi and Cavalcanti 1951; Pardi 1952) responded with further proof of the important role of dominance. In retrospect, the controversy, like so many in science, was pointless—a case of false dichotomy. As has been pointed out by other authors (Brian 1956; West-Eberhard 1969b) both oviposition and dominance by the queen could contribute simultaneously to queen control.
5. This would more accurately replaced with 'dominance', since linear hierarchy is not ubiquitous but dominance is.

Phylogeny and biogeography of *Polistes*

James M. Carpenter

Polistes is the most widely distributed social wasp taxon, is quite speciose, and its relatively small colonies are readily observed and manipulated, hence its popularity with ethologists. But the species are notorious for morphological and behavioural uniformity, compared with other social wasps. Perhaps because of this fact, there exists nothing like a phylogenetic system for the group. There have been occasional attempts to subdivide the species of the genus, and one worldwide subgeneric classification has been proposed, but that has not been applied to all of the described species. I present here a cladistic analysis of the subgenera and species groups of *Polistes*, which I will use as the basis for a new, phylogenetic classification of the species in the genus. And I will use the results of the cladistic analysis in a study of the historical biogeography of the genus, which has received even less attention than the phylogenetic relationships of the species. There is a long-standing generalization of a century or more that the species of the north temperate zone are derived from ancestral, tropical species. This has been more recently elaborated as the notion that the genus arose in the Oriental tropics, thence dispersing to colonize the world. The analysis I will present accords with the first generalization, but not the second.

Taxonomic history

Subgenera and species groups

The only comprehensive attempt at a subgeneric classification of *Polistes* is that of Richards (1973). Prior to that, a few generic and subgeneric names had come into use, but only on a regional, limited basis. For the most part, workers on the genus did not apply formal nomenclature, dividing the genus instead into informal species groups.

Polistes was separated from *Vespa* as a genus by Latreille in 1802; he included six species. In 1826, Kirby and Spence described a monotypic genus, *Cyclostoma*, for their new species *gigas*; the name *Cyclostoma* being preoccupied, they proposed the name *Gyrostoma* for this taxon in 1828. Saussure (1853–1858) placed this species in *Polistes*, in an appendix to the first revision of that genus. Saussure did not recognize any formal subordinate taxa in *Polistes*, merely arranging it in three divisions. One division was *Gyrostoma*, one comprised *subsericeus* only, and the other included all the remaining species; the last two divisions were separated solely by the shape of the first metasomal segment. Saussure (1857) subdivided the American species into three groups, again based entirely on the shape of the first metasomal segment. Matters then stood until the twentieth century.

Brèthes (1903) pointed out transitions in shape among Saussure's groups, and first called attention to the structure of the mesepisternum as a source of characters to subdivide the New World species of the genus. Brèthes used presence or absence of an epicnemial carina, and Ducke (1904) followed this by using the development of the various sulci on the mesepisternum as key characters. Bequaert (1918) began the use of these characters as the basis for subdivision of the genus worldwide. Bequaert (1937) explicitly stated that the groups formed by this subdivision were not natural, but Bequaert (1938, 1940) continued to use these characters in reviews of species from the Ethiopian and Nearctic Regions, respectively.

Dalla Torre (1904) had proposed names for Saussure's divisions, but these were ignored. Formal taxonomic subdivision really began with the European fauna. Zimmermann (1930) divided the European species into the '*semenowi*' and '*gallica*' species groups. The three species separated as the '*semenowi*' group were later found to be social parasites, with the females living as inquilines on the nests of other European species, and Weyrauch (1937) proposed the new genus *Pseudopolistes* for these species. He also subdivided the remaining European species into two species groups, which he later treated as genera (Weyrauch 1938), proposing the name *Polistula* for the species not placed in *Polistes*. Weyrauch did not designate type species for his new genera until 1939 (Weyrauch 1939), and in the meantime, Blüthgen (1938) proposed a valid subgeneric name for the inquilines, *Sulcopolistes*. Blüthgen (1943) rejected Weyrauch's genera *Polistula* and *Polistes* on the grounds that they were not morphologically distinct, but treated *Sulcopolistes* as a genus, and subdivided *Polistes* into subgenera of his own, *Polistes sensu stricto* and *Leptopolistes*. This classification was generally followed by European authors, until Richards (1973) rejected Blüthgen's subgenera. American workers have tended to criticize the recognition of *Sulcopolistes* as a distinct genus (Evans and West-Eberhard 1970; Wilson 1971). As has been pointed out elsewhere (Carpenter *et al.* 1993) misidentification of the type species of *Polistes* (Day 1979) means that Blüthgen's subgenera are synonyms of *Polistes s. str.* whatever one's view of their taxonomic merits. I have also presented evidence that recognition of *Sulcopolistes* renders *Polistes s. str.* paraphyletic, and synonymized it (Carpenter 1990; van der Vecht and Carpenter 1990; Carpenter et al. 1993).

In the last papers on this subject prior to Richards' work, van der Vecht (1968a, 1972) began formal subgeneric division of the Indo-Pacific species, first validating a manuscript name of Bequaert's, *Megapolistes*, and then establishing the taxon *Stenopolistes*. Van der Vecht called attention to several new characters in justifying these descriptions. Van der Vecht (1968a) treated *Gyrostoma* as a subgenus, and van der Vecht (1972) also recognized *Polistella* as a subgenus. Ashmead (1904) had described *Polistella* as a genus, with *Polistes manillensis* as the type species. Bequaert (1930) pointed out that *Polistella* was described from misidentified specimens of *Protopolybia sedula* (Saussure) (= *Protopolybia exigua* (Saussure)). As noted by van der Vecht and Carpenter (1990), in the case of a misidentified type species under current rules of nomenclature, the International Commission on

Zoological Nomenclature is to designate as type species whichever nominal species will best serve nomenclatural stability. Clearly, that would be *Polistes manillensis*, as has been concluded by all previous workers.

Richards (1973) proposed the first formal subdivision of *Polistes* applied on a worldwide basis. He adduced several new characters in addition to using the traditional features of the mesepisternum. Richards retained *Sulcopolistes* as a separate genus, but as mentioned before placed the remaining European species along with a few Ethiopian ones into *Polistes s. str.* Along with van der Vecht's subgenera, Richards described the new taxon *Nygmopolistes* for two species. Richards proposed five new subgenera for the New World species: *Polistarchus*, *Epicnemius*, *Onerarius*, *Palisotius*, and *Fuscopolistes*. Richards additionally divided *Polistarchus* into six species groups, and *Epicnemius* into three species groups.

Changes to this system during the past twenty years have been limited. Richards (1978a) replaced the name *Polistarchus* with *Aphanilopterus* Meunier after the synonymy was pointed out to him by M. C. Day. Van der Vecht (1984) described a species intermediate in characters between *Polistella* and *Stenopolistes*, and left its placement open. Kojima and Kojima (1988) described several more species having some but not all of the diagnostic characters of *Stenopolistes*. They observed that *Polistella* and *Stenopolistes* are phylogenetically closely related as shown by several synapomorphies, and concluded that whereas *Stenopolistes* could be diagnosed by one apomorphic character, *Polistella* could be differentiated only by the absence of that character, suggesting that *Polistella* is paraphyletic. Nozawa and Itô (1989) published electromorph data on seven species of Japanese *Polistes* and two other paper-wasp species, and phenetic analyses from which they concluded that they had demonstrated closer phylogenetic relationship of *Megapolistes* and *Polistella* to each other than to *Polistes s. str.* Although these authors overlooked it, their phenograms suggested that the genus *Polistes* itself is not monophyletic. I (Carpenter 1990) reanalyzed the data of Nozawa and Itô, and showed that their conclusions rested upon faulty analysis, and that when analyzed correctly their data were uninformative on phylogenetic relationships. In that paper I also listed synapomorphies establishing:

(1) monophyly of the genus *Polistes* itself;
(2) close relationship between *Polistella* and *Stenopolistes*;
(3) close relationship between *Megapolistes* and *Gyrostoma*;
(4) close relationship between *Sulcopolistes* and *Polistes s. str.*

Das and Gupta (1989) divided *Polistella* into three species groups, but made no attempt to include all of the species Richards (1973, 1978b) had placed in the subgenus, limiting their study to Indian species. Their table 3 compared characters of the subgenera occurring in India, and although they attempted no analysis, they did introduce a few new characters. Finally, Carpenter *et al.* (1993) presented electromorph data on the polistine social parasites and their four host species, as well as representatives of two American subgenera, and cladistic analyses that demonstrate paraphyly of *Polistes s. str.* in terms of *Sulcopolistes*. This result was also obtained from analysis of mtDNA data by Chondhary *et al.* (1994).

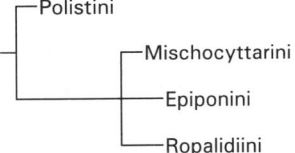

Fig. 2.1 Cladogram of the tribes of Polistinae after Carpenter (1993). The composition of the tribes is: Polistini: *Polistes*; Mischocyttarini: *Mischocyttarus*; Ropalidiini: *Ropalidia, Parapolybia, Polybioides, Belonogaster*; Epiponini: *Apoica, Agelaia, Angiopolybia, Pseudopolybia, Parachartergus, Leipomeles, Marimbonda, Chartergellus, Nectarinella, Protopolybia, Polybia, Protonectarina, Charterginus, Chartergus, Brachygastra, Synoecoides, Epipona, Synoeca, Asteloeca, Clypearia, Occipitalia, Metapolybia*.

Higher classification

The distinctiveness of *Polistes* in relation to other paper-wasps has long been recognized. Bequaert (1918) established a separate subfamily for *Polistes* (and *Gyrostoma*, treated as a genus), contending that the paper-wasps had evolved polyphyletically, following Ducke (1910). Bequaert retained this view of polyphyly, but Richards and Richards (1951) suggested that it was unwarranted, and Richards (1962) reduced Bequaert's subfamily to tribal status within a broader concept of Polistinae, which included Ropalidiini and Polybiini as the other tribes. Charnley (1973) and Jeanne (1980) questioned the distinction of the tribes and I (Carpenter 1981) argued that Polybiini in Richards' sense are paraphyletic, and sank the polistine tribes. I have (Carpenter 1991) presented an analysis of phylogenetic relationships among the polistine genera, and used this as the basis of a new tribal classification for Polistinae (Carpenter 1993). This system recognizes the tribe Polistini for *Polistes* itself, expands Ropalidiini to include the four Old World endemic genera, establishes a new tribe Mischocyttarini for *Mischocyttarus* itself, and places the 22 genera of New World swarm-founding species in the tribe Epiponini. The relationships and composition of the tribes are shown in Fig. 2.1.

Materials and methods

A list of the species of *Polistes* examined is given in Appendix 1. Morphological terminology follows Carpenter (1981) and Carpenter and Rasnitsyn (1990). The male genitalia were dissected, cleared slightly in lactophenol, and examined in glycerin. Cladistic analysis (Hennig 1966) was implemented using the Hennig86 computer program (Farris 1988), results were checked with the NONA program (Goloboff 1993), and character optimization and diagnoses performed with the CLADOS program (Nixon 1992). Two taxa were included as outgroups: the Vespinae, sister-group of the Polistinae (Carpenter 1981), and Mischocyttarini (i.e. *Mischocyttarus*). *Polistes* is the sister-group of the remaining Polistinae (Carpenter 1991, 1993; Fig. 2.1), and Mischocyttarini are a relatively basal lineage within these remaining paper-wasps. Character state codings are presented in the next section.

Table 2.1 Character matrix for subgenera of *Polistes*, *Sulcopolistes*, and two outgroups.

Vespinae	0000000000 0000000000 0000000000 000
Mischocyttarus	0000100000 0000000000 0001000000 000
Nygmopolistes	1100100100 0001000001 1002000000 011
Gyrostoma	3100101101 0101011001 1102000012 011
Megapolistes	2100100000 0001010001 1002000011 011
Polistella	0000110000 0001101001 1002000000 011
Stenopolistes	0010110010 0011101001 1013001000 011
Polistes s. str.	0001100000 0012010111 1002100000 111
Sulcopolistes	0001200000 1012010111 1002100000 111
Aphanilopterus	0000100000 0013000202 2003000000 111
Epicnemius	0000100000 0003010302 2002000000 111
Onerarius	2000100101 0013000302 2002000100 111
Palisotius	0000100100 00130*0302 3002000000 111
Fuscopolistes	0000100100 0013000302 3002001010 111

Asterisk denotes within-taxon variation.

Table 2.2 Character matrix from Table 2.1, with addition of taxa to represent species groups in *Megapolistes*, *Polistella*, *Aphanilopterus*, and *Epicnemius* (see text for explanation).

Vespinae	0000000000 0000000000 0000000000 000
Mischocyttarus	0000100000 0000000000 0001000000 000
Nygmopolistes	1100100100 0001000001 1002000000 011
Gyrostoma	3100101101 0101011001 1102000013 011
M. olivaceus group	2100100000 0001010001 1002010012 011
M. stenopus group	2100100000 0001010001 1002000011 011
P. stigmus group	0000110000 0001101001 1002000000 011
P. adustus group	0000110000 0011101001 1012001000 011
Stenopolistes	0010110010 0011101001 1013001000 011
Polistes s. str.	0001100000 0012010111 1002100000 111
Sulcopolistes	0001200000 1012010111 1002100000 111
A. lanio group	0000100000 0013000202 2003000000 111
A. crinitus group	0000100000 0013000302 3003000000 111
E. bicolor group	0000100000 0003010302 3002000000 111
E. thoracicus group	0000100100 0003010302 2002000000 111
Onerarius	2000100101 0013000302 2002000100 111
Palisotius	0000100100 00130*0302 3002000000 111
Fuscopolistes	0000100100 0013000302 3002001010 111

Asterisk denotes within-taxon variation.

The outstanding autapomorphy of *Polistes* is the conical shape of the first metasomal segment; other, homoplastic autapomorphies include the dorsally narrowed propodeal orifice and convex larval maxilla (Carpenter 1990; and see Results). Within the genus, characters examined were primarily those used by previous authors, particularly Richards (1973). Table 2.1 shows the data matrix used in the initial analysis, which treated subgenera. Table 2.2 shows the data matrix from

Table 2.3 Character matrix from Tables 2.1 and 2.2, reduced to the outgroups and four monophyletic subgenera as reclassified.

Vespinae	0000000000 0000000000 0000000000 000
Mischocyttarus	0000100000 0000000000 0001000000 000
Gyrostoma	1100100100 0001000001 1002000000 011
Polistella	0000110000 0001101001 1002000000 011
Polistes s. str.	0001100000 0012010111 1002100000 111
Aphanilopterus	0000100000 0013000202 2003000000 111

a subsequent analysis, in which certain subgenera were broken up into species groups, as explained below. Table 2.3 shows the data matrix from an analysis of the subgenera as reclassified.

Historical biogeography was investigated using component analysis, as implemented in the COMPONENT program (Page 1989*b*) and TAS program (Nelson and Ladiges 1992). The areas treated (Table 2.4) are those listed by Das and Gupta (1984, 1989), with some modifications, as discussed further below.

Characters

In this section, the coding adopted for each potentially informative character is described. For multistate characters the states are ordered linearly except character #24, which is treated as nonadditive. It will be seen that some taxa are variable for the states of a few of the characters. Intrataxon variability raises the question of coding (Nixon and Davis 1991); the approach taken here is twofold. In the first analysis, taxa were scored with an inferred groundplan (ancestral) state where possible, rather than as missing. A second analysis was also performed, in which subgenera variable for informative characters were divided into species groups in which the pertinent states were of constant distribution. The division is explained in the results section.

1. **Prestigma**: no longer than half the length of the pterostigma, measured along ventral part, 0; more than half the length of the pterostigma, 1; about equal to the length of the pterostigma, 2; longer than pterostigma, 3.

Length of the prestigma was used as a key character by van der Vecht (1972) and Richards (1973), who referred to it as parastigma. They distinguished *Megapolistes* and *Gyrostoma* as having it elongate; Richards including *Nygmopolistes* among the subgenera with a 'short' prestigma. Das and Gupta (1989, table 3) observed however that *Nygmopolistes* has a slightly longer prestigma than the remaining Old World subgenera. Elongation of the prestigma, from a groundplan state of shorter than half the length of the pterostigma, measured along its posterior part, has been repeatedly derived within Vespidae (Carpenter and Cumming 1985; Carpenter 1987). *Nygmopolistes* has the prestigma longer than half the length of the pterostigma (Fig. 2.2(a)), *Megapolistes* and *Onerarius* have it about equal in length (Fig. 2.2(b)), and *Gyrostoma* has it longer than the pterostigma (Fig. 2.2(c)). Some species of *Polistella* (e.g. *sagittarius*) and many *Aphanilopterus* also have the prestigma slightly lengthened, approaching the condition in *Nygmopolistes*.

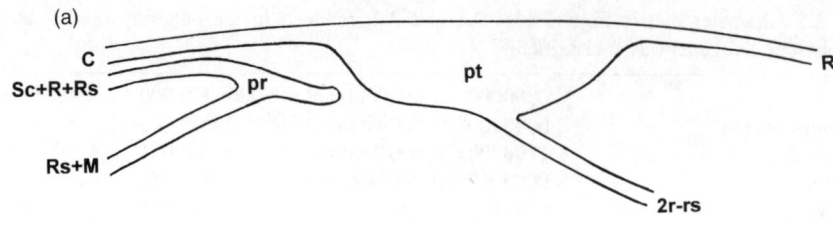

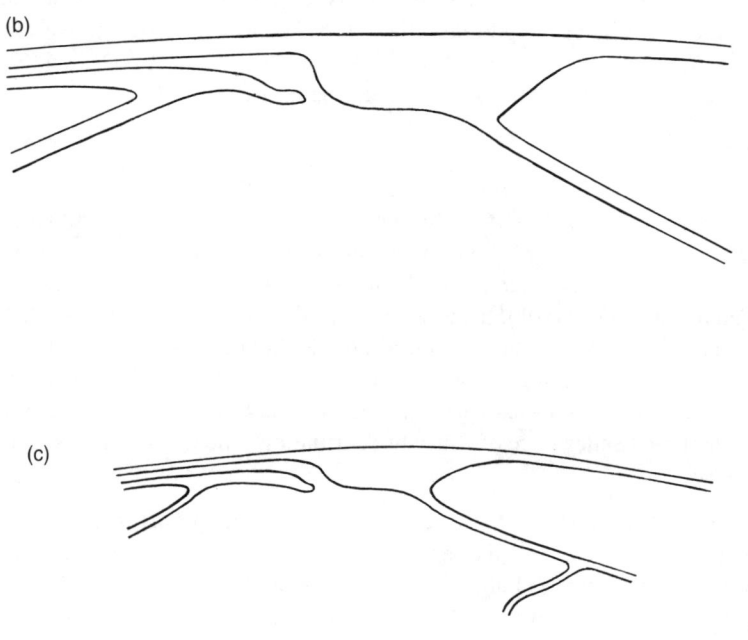

Fig. 2.2 Forewing, veins labelled. (a) *Polistes sulcatus*. (b) *P. jadwigae*. (c) *P. gigas*. pr: prestigma; pt: pterostigma. Scale = 10 mm.

2. **Prestigma apex**: straight, 0; recurved, 1.

The general condition in Vespidae is to have the free abscissa of the prestigma straight, with the R+RS and M veins running smoothly to the apex (e.g. van der Vecht 1972, fig. 33). In *Nygmopolistes*, *Megapolistes*, and *Gyrostoma* the prestigma is slightly recurved just before the tip (Fig. 2.2(a)–(c); van der Vecht 1972, figs. 9, 23; approached in *Onerarius*, but less tapered).

3. **Jugal lobe**: large, preaxillary excision narrow, 0; reduced, preaxillary excision wide, 1.

Van der Vecht (1972) pointed out that the jugal lobe, referred to by him as anal lobe, is small in *Stenopolistes*, and Richards (1973, 1978*b*) used it as a key character to distinguish *Stenopolistes* from *Polistella*. Reduction of the jugal lobe, resulting in a broad preaxillary excision (*cf.* Richards 1973, figs. 4 and 5–6), is derived within Vespidae (Carpenter 1981), and has occurred several times (e.g. Vespinae, Polistinae: *Apoica*). Das and Gupta (1989, table 3) characterized the lobe as smaller in *Polistella* than the remaining subgenera (*cf.* Das and Gupta 1989, figs. 14b and 14c), but differing from *Stenopolistes* by a narrower preaxillary excision. In fact, most species of *Polistella* do not have the jugal lobe reduced in size as shown in Das and Gupta's fig. 14b (not fig. 6 as stated in their table 3) — nor do most *Stenopolistes* have it essentially absent, as shown in their fig. 14a. The two subgenera overlap in size of the lobe *per se*, and the critical difference is the width of the preaxillary excision. Kojima and Kojima (1988) described three new species, all of which share with *Stenopolistes* a wide hindwing preaxillary excision, but two of which lack the other features diagnosing *Stenopolistes*: the hind ocelli are separated by more than an ocellus diametre, metasomal segment I is not truncate in profile, and there is no tubercle on the terminal male metasomal segment. Kojima and Kojima (1988) concluded that a broad excision is the only character that reliably diagnoses *Stenopolistes* apart from *Polistella*; all other previously used characters vary within *Polistella*.

4. **Male antennae**: tapering apically, 0; hooked, 1.

Das and Gupta (1989) used coiling of the apical antennal article of the male (e.g. Bequaert 1918, figs. 263, 266; Das and Gupta 1989, fig. 13k) as a key character to separate *Polistes* from the other Indian subgenera. This condition also occurs in *Sulcopolistes*, while the New World subgenera have the male antennae tapering, as in the other Old World subgenera (e.g. Bequaert 1918, fig. 254; van der Vecht 1972, figs. 20, 37). Even when the apex is modified (e.g. van der Vecht 1972, fig. 43), coiling does not occur.

5. **Clypeal apex**: truncate, 0; pointed, 1; pointed and depressed, 2.

An apically pointed clypeus is the groundplan condition of Polistinae (Carpenter 1991). Depression of the clypeal apex (Fig. 2.3(c)) is diagnostic of *Sulcopolistes* (Guiglia 1971). The state is most clearly apomorphic in the female; most male *Polistes* have the apex somewhat depressed.

6. **Clypeal dorsum**: straight, 0; produced above tentorial pits, 1.

Van der Vecht (1972) characterized *Stenopolistes* as having the clypeus longer than wide, and noted that this is true of some species of *Polistella*. Das and Gupta (1989, table 3) pointed out that in *Polistella* and *Stenopolistes* the clypeus is produced above the anterior tentorial pits (Fig. 2.3(a)). The form of this production gives the dorsal margin of the clypeus a bisinuate appearance (*cf.* Fig. 2.3(a)–(d), a derived condition in Vespidae (Carpenter 1981, 1989*b*), although the details differ in such groups as *Eumenes* (Eumeninae) and Gayellini.

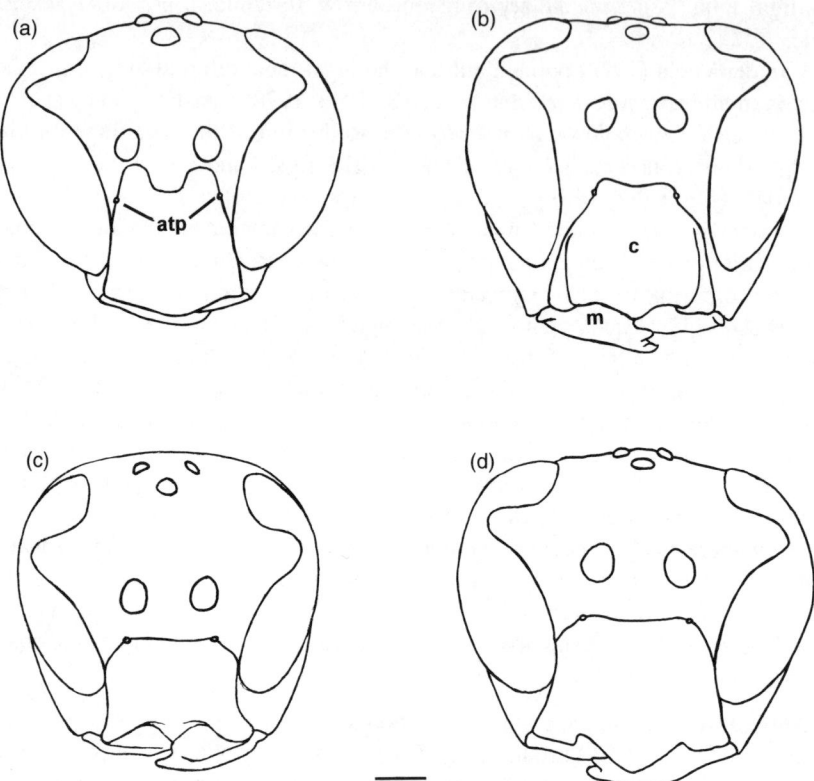

Fig. 2.3 Head, frontal view. (a) *Polistes stigmus*, male. (b) *P. fuscatus*, male. (c) *Sulcopolistes atrimandibularis*, female. 8, *P. wattii*, female. atp: anterior tentorial pits; c: clypeus; m: mandible. Scale = 0.5 mm.

7. **Interantennal carina**: blunt, 0; toothed, 1.

Saussure (1853–1858) pointed out that in *Gyrostoma* the interantennal space is produced into a tooth; it is blunt in other *Polistes* (*cf.* Starr 1992, figs. 41a–b and 41c).

8. **Male eyes**: contacting clypeus, 0; separated, 1.

Whether the eyes contact the clypeus in the male (Fig. 2.3(a)) or are separated (Fig. 2.3(b)) has been used as a key character by Richards (1973, 1978*a*), separating *Fuscopolistes* and *Palisotius* from *Aphanilopterus*. Richards also noted that the females do not always have the same state as the males, and that the state varies within *Epicnemius*, where it is separated in his group 2 (*thoracicus*). The general condition in Vespidae is to have the eyes in contact with the clypeus. The separated condition also occurs in *Gyrostoma*, *Nygmopolistes*, and *Onerarius*, and some species of *Megapolistes* (*jadwigae*, *schach*).

9. **Ocelli**: separated by more than an ocellus diametre, 0; closer, 1.

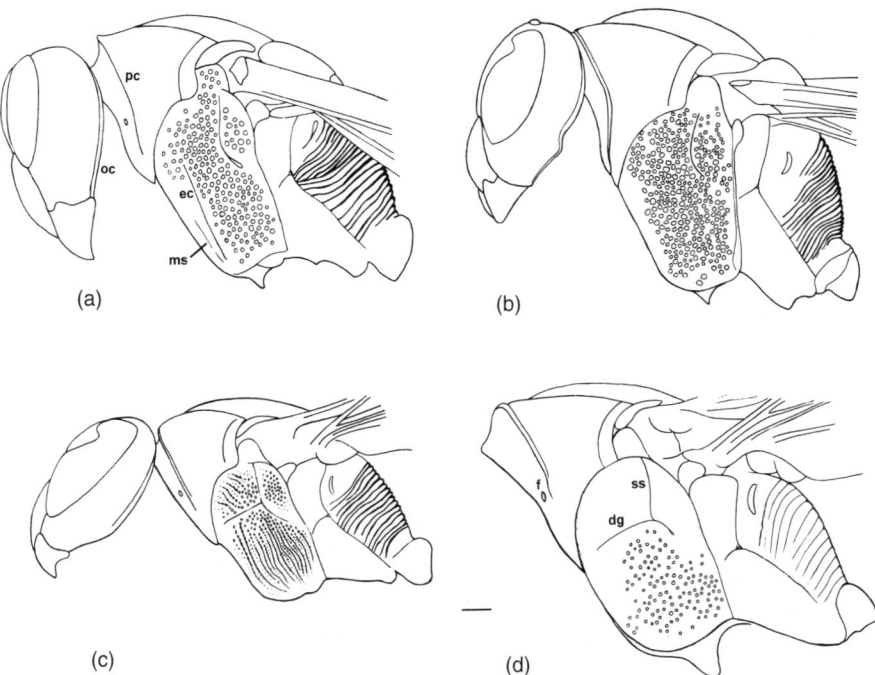

Fig. 2.4 Lateral view. (a)–(c), head and mesosoma. (a) *P. gigas*. (b) *P. tullgreni*. (c) *P. dominulus*. (d) mesosoma, *P. infuscatus anduzei*. Punctation is shown on the mesepisternum; striae on the propodeum. dg: dorsal groove; f: pronotal fovea; ec: epicnemial carina; ms: mesopleural signum; oc: occipital carina; pc: pronotal carina; ss: scrobal sulcus. Scale = 1 mm.

Approximation of the ocelli was used as a distinguishing character for *Stenopolistes* by van der Vecht (1972), and following him Richards (1973, 1978*b*) and Das and Gupta (1989, *cf.* figs. 13e and 13f). However, Kojima and Kojima (1988, figs. 38, 60, 79) illustrated continuous variation in this trait and concluded (p. 79) 'the states with respect to the ocelli are quite variable and do not seem to be of taxonomic importance more than at the species level.' The general condition in Vespidae is to have the ocelli separated by more than an ocellus diametre.

10. **Malar space**: shorter than wide, 0; longer, 1.

Elongation of the malar space, so that it is longer than broad, was used as a key character by Richards (1973). The general condition in Vespidae is to have the malar space quite short, but elongation has occurred in, e.g. Vespinae (Carpenter 1987). The elongate condition occurs in *Gyrostoma* (as noted by van der Vecht 1968*a*; *cf.* Das and Gupta 1989, figs. 13a–c, 15a–e, g–k, and 13d, 15f) and *Onerarius*.

11. **Mandible**: flat, 0; excavated, 1.

Excavation of the external surface of the mandible is diagnostic of

Sulcopolistes (Guiglia 1971), more pronounced in females (*cf.* sexes in Guiglia 1971, fig. XIV).

12. **Male mandibular teeth**: quadridentate, 0; tridentate, 1.

Saussure (1853–1858) pointed out that in male *Gyrostoma* there are only three teeth, with two apical ones well separated from the third, which is a basal flange (e.g. Starr 1992, fig. 42). In other *Polistes* the male mandible is quadridentate (e.g. van der Vecht 1972, figs. 27, 29), although the fourth is somewhat reduced in *Stenopolistes* (Richards 1973; e.g. van der Vecht 1972, figs. 38, 44). Quadridentate mandibles are the groundplan condition in many vespid taxa, including Gayellini, Eumeninae and social wasps (Carpenter 1981, 1988*a*, 1989*b*; Carpenter and Cumming 1985), but reduction in number has occurred (e.g. Stenogastrinae, within Gayellini) as well as increase (within Eumeninae).

13. **Occipital carina**: complete to mandibular bases, 0; evanescent toward mandibular bases, 1.

Richards (1973, p. 87) mentioned this character, termed by him genal keel, as 'often of specific or species group value.' Whether the carina reaches the mandibular bases in the female (Figs. 2.4(a)–(b)) or is found only dorsally (Fig. 2.4(c)) has traditionally been used as a key character (e.g. Ducke 1904; Bequaert 1937, 1938; van der Vecht 1972). Das and Gupta (1989, table 3) pointed out that some subgenera are characterized by one state, but that *Megapolistes* is variable. The phylogenetic variation in this feature within Vespidae has been discussed by Carpenter (1981, 1987, 1988*a*, 1991) and Carpenter and Cumming (1985). A complete carina is the groundplan condition in Polistinae, and there has been reduction all the way to complete loss of the carina. A complete carina is found in *Nygmopolistes*, *Gyrostoma*, most *Megapolistes* and most *Epicnemius*, and a few Ethiopian species of *Polistella* (*aquilinus, fastidiosus, madecassus, tullgreni*).

14. **Pronotal carina**: two, 0; one, 1; not ventrally lamellate, 2; nor laterally lamellate, 3.

Richards (1973) called attention to the relative development of the pronotal carina, pointing out that in New World species the carina generally does not extend to the pronotal fovea (Fig. 2.4(d)), whereas it does in *Polistes s. str.* (Fig. 2.4(c); this is also the case in *Sulcopolistes*), and in the other Old World subgenera it extends below the fovea (Figs. 2.4(a)–(b)), usually all the way to the ventral corner (Fig. 2.4(b)). Although there is some overlap between species of *Polistes s. str.* and some New World species, the carina is not sharply lamellate ventrad of the fovea, and it is usually lowered dorsad to the fovea in the New World species (still sharp in some species of *Epicnemius*). From the vespid groundplan of two carinae, *Polistes* is derived in having only the dorsal one present (Carpenter 1989*b*), and there has evidently been a transformation series in reduction of the carina from the state where it extends to the ventral corner (see Results).

15. **Pronotal fovea**: present, 0; absent, 1.

Presence (Figs. 2.4(a), (c), (d)) or absence (Fig. 2.4(b)) of the pronotal fovea came into use as a key character in *Polistes* with Cheesman (1951), and absence of the fovea was a major character separating *Polistella* and *Stenopolistes* from the other subgenera (Richards 1973). It is also absent in various Ethiopian species of *Polistes s. str.* The fovea is present in the groundplan of Polistinae, including *Polistes* (Carpenter 1989*b*, 1991, 1993).

16. **Epicnemial carina**: absent, 0; present, 1.

As mentioned above, presence (Figs. 2.4(a), (c)) or absence (Figs. 2.4(b), (d)) of the epicnemial carina (prepectal of Bequaert) has been a major character in *Polistes* taxonomy for almost a century. Presence of an epicnemial carina is a groundplan feature of Eumeninae (Carpenter and Cumming 1985), but although I previously treated this as the groundplan condition of social wasps (Carpenter 1987, 1991, 1993), it is clear from the results in Carpenter (1991, 1993) that that interpretation is untenable. The carina is present only in *Polistes* and *Ropalidia* among social wasps; it is absent in Vespinae and Stenogastrinae. *Polistes* and *Ropalidia* are not closely related, and separate origination in these two genera is a more parsimonious interpretation to origination in the polistine groundplan, loss in most genera, and subsequent reacquisition in *Ropalidia*. Further, the carina has evidently been derived multiple times within *Polistes* (see Results). The carina is present in *Megapolistes*, *Gyrostoma*, and *Epicnemius*, is present but weak in *Polistes s. str.* and *Sulcopolistes*, and is variable within a species in *Palisotius* (*major*).

17. **Dorsal groove**: present, 0; absent, 1.

As mentioned above, presence (Figs. 2.4(c)–(d)) or absence (Figs. 2.4(a)–(b)) of the dorsal groove (episternal of Bequaert, epipleural of van der Vecht) has been a major character in *Polistes* taxonomy since Ducke (1904; Richards 1973, incorrectly cited von Ihering 1904, who did not use that character). The groove is present in the polistine groundplan (Carpenter 1991, 1993). It is absent in *Gyrostoma*, *Polistella*, and *Stenopolistes*, and is incomplete or absent in some species of *Epicnemius* and one *Fuscopolistes* (*poeyi*).

18. **Mesepisternal punctation**: coarse, 0; fine, 1; fine and well separated, 2; reduced, 3.

Richards (1973) used this trait as the primary key character to separate *Polistes s. str.* and the New World subgenera (Figs. 2.4(c)–(d)) from the remaining Old World subgenera (Figs. 2.4(a)–(b)). Large punctures are found in other polistines and vespines, and the smaller punctures are evidently the derived state. Fine punctation occurs in a few species of *Megapolistes* (*olivaceus*, *wattii*) and *Stenopolistes* (*meadeanus*, *xantholeucus*), but the difference is otherwise clear. As Richards pointed out, in addition the punctures are reduced in number and not close in the New World subgenera (Fig. 2.4(d)), compared with *Polistes s. str.* (and *Sulcopolistes*; Fig. 2.4(c)). Indeed, among the New World taxa, only some species of *Aphanilopterus* (approximately Group 1 of Richards 1973) show even fine punctation; the usual condition is to have the punctation essentially absent.

19. **Punctation clathrate**: absent, 0; present, 1.

Richards (1973) characterized the thoracic punctation in *Polistes s. str.* as clathrate, brought about by the edges of the punctures running together in lines, giving the appearance of fine striae. This is most obvious on the pronotum and mesepisternum (Fig. 2.4(c)), and the condition is also found in *Sulcopolistes*.

20. **Propodeal orifice**: dorsally rounded, 0; dorsally narrowed, 1; acute and elongate, 2.

A dorsally acute propodeal orifice is a traditional character to separate *Polistes* from other polistines at the subfamily (Bequaert 1918) or tribal (Richards 1962) level. A dorsally broadly rounded orifice is the groundplan state in Vespidae (Carpenter 1981). Richards (1973, p. 89) stated that 'in general the Old World species tend to have a shorter and broader slit than the New World forms' (*cf.* Richards 1973, figs. 2 and 3). The orifice may be quite acute dorsally in Old World species (e.g. *Polistes s. str.*), but is also generally more elongate in the New World species. The elongate condition is evidently further derived.

21. **Propodeal striae**: absent, 0; present, 1; fine, 2; fine and laterally evanescent, 3.

Richards (1973, 1978*a*) used the development of the transverse striae on the propodeum as a key character, with striae not extending to the propodeal border characterizing *Palisotius* and *Fuscopolistes*. He also noted that some of the species of *Aphanilopterus* and *Epicnemius* have this condition, but that others have the striae extending to the propodeal border (Fig. 2.4(d)), as in *Onerarius*. The Old World subgenera have the striae extending to the propodeal border (Figs. 2.4(a)–(c)), and coarser than in the New World species (but fine in a few species, e.g. *meadeanus*, and reduced in a few, e.g. *xantholeucus*). Presence of the striae is evidently an autapomorphy for *Polistes*, and as with other sculptural characters, the New World species are derived in having the striae more finely developed.

22. **Hindtrochanter**: smooth, 0; toothed, 1.

Van der Vecht (1968*a*) pointed out that the hindtrochanter in male *Gyrostoma* is toothed; it is also rather sharply projecting in the female. It is smooth in other *Polistes*.

23. **Claws**: symmetrical, 0; asymmetrical, 1.

This character came into use in *Polistes* taxonomy with Buysson (1905; not 1895 as misprinted by Bequaert 1918). Bequaert (1918, 1938) used it as a key character to distinguish among Ethiopian species. In the asymmetrical condition, the claws of the middle and hind legs have the inner claw much longer and heavier than the outer one (e.g. Bequaert 1938, figs. 1g, 4c–d). This condition is found in *Stenopolistes* and *Polistella*, but varies in the latter, with most species having the claws asymmetrical to some degree, but hardly so in some species (e.g. *smithii*; Bequaert 1938, fig. 7c), and symmetrical in others (e.g. *diakonovi, sagittarius, philippinensis, strigosus, tullgreni*; Bequaert 1938, fig. 9f). The claws are symmetrical or very slightly asymmetrical in other *Polistes* (Richards 1973, p. 99, incorrectly characterized *Megapolistes* as having asymmetrical claws, but in his key on p. 92 correctly stated that they were symmetrical).

24. **Metasomal Segment I shape**: transversely truncate, 0; petiolate, 1; conical, as wide or wider than long, 2; conical, longer than wide, 3. Nonadditive.

As discussed above, the shape of the first metasomal segment has played an important role in *Polistes* taxonomy since the time of Saussure, continuing to Richards (1973, 1978a). Although a conical segment is an autapomorphy for *Polistes*, the variation in shape beyond this is of value primarily at the specific level, because of intermediate forms as discussed by Brèthes (1903) and Bequaert (1918). Richards (1973, 1978a) made several distinctions in shape based upon relative length and width, but the differences are slight at best. However, a clearly longer segment is found in *Stenopolistes, Aphanilopterus*, and some species of *Epicnemius*. Van der Vecht (1972) and Richards (1973, 1978a, b) also paid attention to the relative slope of the first tergum, viewed in profile, especially using an anteriorly truncate tergum to characterize *Stenopolistes* (e.g. van der Vecht 1972, fig. 32). However, that is not perfectly correlated with other characters diagnosing *Stenopolistes* (van der Vecht 1984; Kojima and Kojima 1988), and again, there are intermediates. The character is ordered nonadditively because the states in the outgroups are dissimilar, and evidently autapomorphic at this level for each taxon.

25. **Metasomal Sternum I**: ecarinate, 0; transversely carinate, 1.

Das and Gupta (1989) called attention to this character. A well-developed margin at the base of the first sternum occurs in *Polistes s. str.* (and *Sulcopolistes; cf.* Das and Gupta 1989, figs. 13l and 13n), and some *Polistella*; however it does not occur in many species of the latter, contrary to Das and Gupta's statement. Das and Gupta (1989, table 3) stated that *Gyrostoma* 'usually' has the carina, but on p. 47 stated that it does not. The first sternum has strong transverse striae across its surface, which perhaps may give the appearance of a carinate margin.

26. **Lateral processes of male metasomal Sternum VII**: absent, 0; present, 1.

Van der Vecht (1968a) characterized *Megapolistes* as having two lateral tubercles or processes ('apophyses') on the terminal metasomal sternum (e.g. Starr 1992, fig. 39). These are variably developed; for example, van der Vecht stated that the tubercles were 'blunt and low' in *tenebricosus* (placed by Richards in *Nygmopolistes*). Male *Polistes* have Sternum VII flattened apically, and the lateral margins are often somewhat raised. The lateral margins appear as blunt ridges or weak tubercles in various species of *Polistella, Polistes s. str., Fuscopolistes,* and this condition is pronounced in *Onerarius* (e.g. Richards 1973, fig. 12). Well-developed, elongate processes are unique to *Megapolistes*. Richards (1973) characterized the subgenus by this state, however, processes are not found in all *Megapolistes sensu* Richards. Van der Vecht (1972, fig. 25) described and figured a new species placed in *Megapolistes, stenopus*, which lacks processes on the terminal sternum: it has blunt tubercles, no better developed than *Nygmopolistes*. Other species of *Megapolistes* (e.g. *diabolicus*) have the tubercles only slightly more pronounced. This character thus shows continuous variation in *Megapolistes*, and is apparently not part of the groundplan. It is scored as absent in Table 2.1 (hence invariant), but present in one of the species groups into which *Megapolistes* is divided in Table 2.2.

27. **Disc of male metasomal Sternum VII**: medially slightly depressed, 0; tuberculate, 1.

Van der Vecht (1972, fig. 45) used presence of a median tubercle on the terminal metasomal sternum of the male to distinguish *Stenopolistes* from *Polistella*, which he characterized as lacking a tubercle. Richards (1973) pointed out that *adustus*, which he placed in *Polistella*, has such a tubercle, and he later (1978*b*) described *bambusae*, which likewise although placed in *Polistella* has a tubercle (e.g. Kojima and Kojima 1988, fig. 12). Kojima and Kojima (1988) described two new species of unspecified subgeneric placement, which share with *Stenopolistes* a wide hindwing preaxillary excision, but which lack the metasomal tubercle. Das and Gupta (1989) placed six species in their '*adustus* group', however, not all the species have a tubercle (lacking in *lepcha*). A tubercle is apomorphic, but also characterizes *Fuscopolistes* (Richards 1973).

28. **Base of male metasomal Sternum VII**: without anterior lobes, 0; lobed, 1.

Richards (1973, fig. 12) used presence of concealed anterior lobes on the terminal metasomal sternum of the male to distinguish *Onerarius*. This condition is unique in *Polistes*.

29. **Ventral margin of digitus**: ventrally curved, 0; widened, 1.

Richards (1973) studied the male genitalia of *Polistes*, and called attention to the enlarged, almost saccate digitus of *Megapolistes* and *Gyrostoma* (e.g. Richards 1973, figs. 8a, 9a). This is an outstanding synapomorphy, yet Richards (1973, p. 92) stated that *Fuscopolistes* and *Megapolistes* are similar in having 'the basal portion of the *digitus* very large and the distal membranous process very short', and on p. 100 suggested that *Fuscopolistes* was 'possibly derived' from *Megapolistes*. In fact, the digitus of *Fuscopolistes* is no more similar to that of *Megapolistes* than *Gyrostoma*, as shown in Richards' own figures (8a, 9a, 11a). The digitus in *Fuscopolistes* is widened basally compared with the condition found in most other *Polistes* (*cf.* Richards 1973, figs. 7a and 11a), but in *Megapolistes* and *Gyrostoma* the apical part is also expanded.

30. **Apex of digitus**: narrow, 0; membranous, 1; membranous and saccate, 2; saccate and with process, 3.

As noted above, Richards (1973) characterized *Fuscopolistes* and *Megapolistes* as similar in having the apical part of the digitus short. *Fuscopolistes* has the digitus apically appearing narrow, due to the ventral widening of the digitus, but in *Megapolistes* the apex of the digitus is membranous, and expanded as a sac in most species (e.g. Kojima and Kojima 1988, figs. 4–5). The species where the apex is membranous but not much expanded (*stenopus*, *diabolicus*) are those species without well-developed lateral processes on Sternum VII (character #26). The saccate expansion is also found in *Gyrostoma* (e.g. Richards 1973, figs. 8, 8a). *Gyrostoma* has in addition a fingerlike process, which is elongate, curved, and slightly knobbed distally. This process is an autapomorphy.

31. **Larval teeth**: three, 0; two, 1.

Larvae have been inadequately studied in *Polistes*, but Richards (1978*a*, *b*) keyed seven of the subgenera, and Kojima and Kojima (1988) described larvae of *Stenopolistes*. Kojima (in prep.) provides detailed descriptions of the larvae of

Polistinae, and has kindly allowed me to make use of his data. Tridentate mandibles are the usual condition in the Vespidae, and occur in *Polistes* (e.g. Kojima and Kojima 1988, figs. 24, 52–55, 74, 97). Larvae of *Polistes s. str.*, *Sulcopolistes*, and the New World subgenera have bidentate mandibles (e.g. Nelson 1982, figs. 12–20, 22–28).

32. **Larval maxilla**: not expanded basally, 0; strongly convex, 1.

A strongly basally 'convex' larval maxilla (projecting further laterad than the mandibles; e.g. Nelson 1982, figs. 12–30) was used as a key character by Richards (1978*a*) to separate *Polistes* and *Ropalidia* from other polistines. The trait is evidently an autapomorphy of *Polistes* (Yamane and Okazawa 1981), but varies within *Ropalidia* (Richards 1973, p. 14), and is present in *Belonogaster* (e.g. Kojima and Keeping 1988, figs. 1–2).

33. **Larval tenth abdominal tergum**: flat, 0; with median tubercle, 1.

Richards (1978*a*, p. 438) pointed out that the New World species of *Polistes* are characterized by having a tuberculate tenth abdominal tergum (except *exclamans*), and stated that *Polistes s. str.* lacks such a process. However, Kojima (*in litt.*) informs me that Old World species usually have a tubercle. Nelson (1982, p. 16) referred to the structure as the 'dorsal median process' or 'anal papilla', and stated 'Extreme variation in size and shape prevent its use as a diagnostic character'; he (fig. 32) illustrated considerable variation within one nest of *metricus*. Such a tubercle is lacking in other polistine genera (Richards 1978*a*; Kojima, *in litt.*), and although size and shape vary within species, presence of the tubercle is evidently an autapomorphy of *Polistes*.

Results

Analysis of Subgenera

Exact analysis by implicit enumeration (the 'ie' command of Hennig86, which finds all most-parsimonious solutions) of the data in Table 2.1 resulted in two cladograms (length 58, consistency index 0.77, retention index 0.83). Successive weighting (Farris 1969) was applied as a check of the reliability of the results, as in Carpenter *et al.* (1993). As implemented in Hennig86, successive weighting uses the rescaled unit consistency index (Farris 1989), scaled between 0 and 10 for each character, as a weighting function. The technique resulted in the same two cladograms, which may therefore be considered consistent, in the sense that they imply character weights that do not imply some other cladograms.

Figs. 2.5 and 2.6 show the extremes possible for optimization of the characters on the first cladogram. Fig. 2.5 maps the characters under the fast transformation procedure of CLADOS; Fig. 2.6 maps them under slow transformation. Fig. 2.7 shows the second cladogram, with characters mapped under fast transformation. The cladograms differ only in alternative arrangements of the lineage consisting of *Polistella* + *Stenopolistes*. In Figs. 2.5 and 2.6, it is part of an unresolved trifurcation comprising two other clades, one formed by *Nygmopolistes* + (*Gyrostoma* + *Megapolistes*), and one formed by (*Polistes s. str.* + *Sulcopolistes*) +

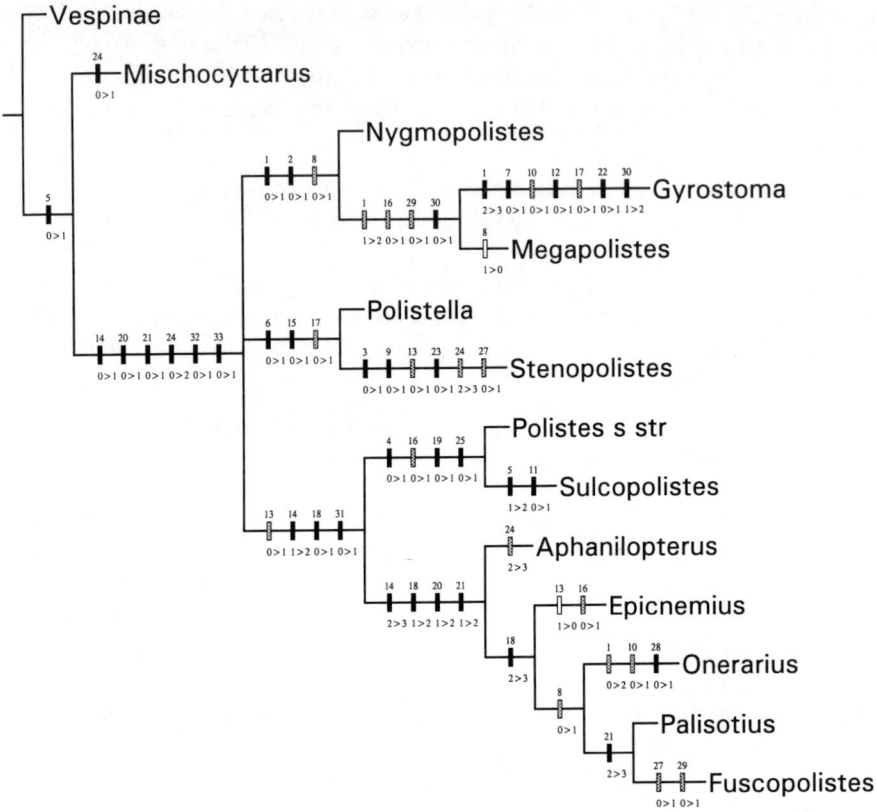

Fig. 2.5 First cladogram reported after exact analysis of the character matrix in Table 2.1. Characters of the ingroup have been optimized by fast transformation as implemented in CLADOS. Character numbers are above the hashmarks; state changes are shown below, with the respective primitive and derived conditions separated by a '>'. Filled hashmarks denote unique origins, greyscaled hashmarks indicate convergent changes, and open hashmarks are reversals.

the New World subgenera. In Fig. 2.7 it is the sister-group to the clade comprising (*Polistes s. str.* + *Sulcopolistes*) + the New World subgenera.

Fig. 2.7 is fully resolved, but the position of the clade comprising *Polistella* + *Stenopolistes* is supported only by optimization of one character, #13, state 1, reduction of the occipital carina, under fast transformation. Character #13 is one of the least consistent in the data matrix: on Fig. 2.7 it originated once but reversed twice, separately in *Polistella* and *Epicnemius* (nor does this reflect the fact that it originated convergently in some species of *Megapolistes*). Under the slow trans-formation optimization, the cladogram collapses to that of Figs. 2.5–2.6. Current parsimony programs treat resolutions of multifurcations as distinct cladograms if they are supported by even one possible optimization. A more stringent require-ment for support would be to accept only those branches supported under all

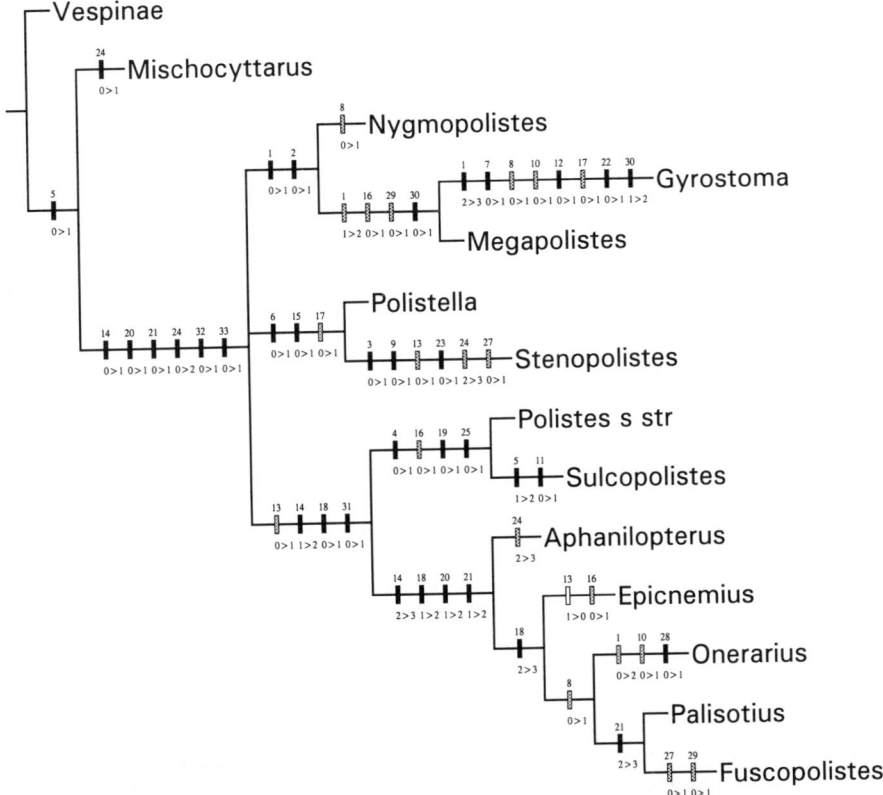

Fig. 2.6 Cladogram of Fig. 2.2, with characters of the ingroup optimized by slow transformation as implemented in CLADOS. Plotting conventions are as in Fig. 2.5.

possible optimizations. Here, that would lead to discarding the cladogram of Fig. 2.7.

The suggestion that *Polistes s. str.* is paraphyletic in terms of *Sulcopolistes* (Carpenter 1990; Carpenter *et al.* 1993) accords with either result; under no optimization is there any character supporting *Polistes s. str.* as monophyletic. *Palisotius* is not supported either, and *Polistella* is supported only by a reversal of character #13 under one optimization on Fig. 2.7, while *Nygmopolistes* and *Megapolistes* are supported only under alternative optimizations of one character (#8, separation of male eyes from clypeus, *cf.* Figs. 2.5 and 2.6), which is another character with the worst fit of the entire matrix (unit character consistency index 0.33). The question of paraphyly thus arises with respect to these subgenera as well, a question which has already been raised regarding *Polistella* (Kojima and Kojima 1988) and *Megapolistes* (Carpenter 1990). The situation with *Palisotius* and *Nygmopolistes* is perhaps less clearcut than for *Megapolistes* and *Polistella*, in that the former two each consist of a very few species of disputed status (*Nygmopolistes* was treated as

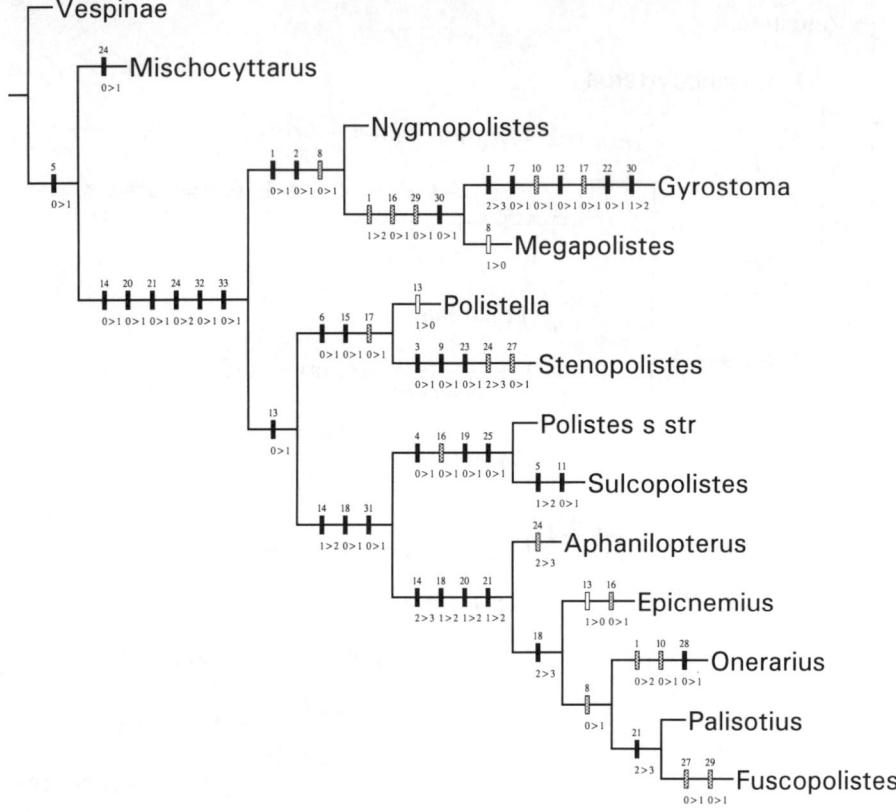

Fig. 2.7 Second cladogram reported after exact analysis of the character matrix in Table 1, with characters of the ingroup optimized by fast transformation as implemented in CLADOS. Plotting conventions are as in Fig. 2.5.

comprising two species by Richards 1973, but one with several subspecies by Das and Gupta 1989; *Palisotius* was treated as comprising two species by Richards 1978*a*, but three species by Snelling 1983). In the next section the issue of paraphyly is explored further.

Analysis of subgenera and species groups

As discussed in the section on characters, several subgenera are variable for some informative characters. The subgenera supported only under alternative optimizations may be paraphyletic. Moreover, two subgenera, *Aphanilopterus* and *Epicnemius*, are variable for characters that establish the resolution among the New World taxa shown in Figs. 2.5–2.7 (*viz.* #18, mesepisternal punctation, and #21, propodeal striae), and each is supported only by homoplastic characters. The question of the naturalness of these subgenera thus also arises. *Aphanilopterus* and *Epicnemius* had been subdivided into species groups (Richards 1973) that are based in part on precisely these characters (Richards 1978*a*). In the data matrix in Table

2.2, these subgenera are subdivided, corresponding to the pertinent characters. Specifically, of the six groups in *Aphanilopterus* of Richards (1973, 1978a), Group 1 has the mesepisternal punctation fine and well separated (state 2 of character #18), but the other species have the punctation reduced (state 3). *Aphanilopterus* has therefore been divided into a group corresponding to Group 1 (which includes the type species of the subgenus, *lanio*) and a group including the remaining species, termed the *crinitus* group. These groups also vary in the extent of the propodeal striae (character #21). Similarly, of the three groups in *Epicnemius* of Richards (1973, 1978a) the *thoracicus* group has the propodeal striae present and fine (state 2 of character #21), but the other species have the striae also laterally evanescent (state 3). *Epicnemius* has therefore been divided into a *thoracicus* group and a group including the remaining species (including the type species of the subgenus, *bicolor*); in this case, the groups also vary in character #8, male eye-clypeal contact.

Polistella had also been subdivided into species groups (Das and Gupta 1989), but they do not correspond to states of the characters analyzed here. Unlike the species groups of Richards (1973, 1978b), Das and Gupta's groups do not include all of the described species assignable to the subgenus. For example, Richards (1973) listed six African species included in *Polistella*, but Das and Gupta (1989, table 4) did not record this subgenus from the Ethiopian Region (they likewise overlooked that *Polistes s. str.* occurs there). Das and Gupta treated only species occurring in India, dividing them into an *adustus* group, *stigma* (properly *stigmus*) group, and *maculipennis* group. The *adustus* group included the species cited by Richards (1973) as having one of the diagnostic characters of *Stenopolistes*, viz. a tuberculate male Sternum VII (state 1 of character #27), but also a species, *lepcha*, lacking that state. No morphological feature characterized their *adustus* group; rather, it was based on colour, namely a black metasoma. The *stigmus* group included two species, diagnosed principally by having the forewing marginal cell with a pigmented spot. The propodeal concavity was said to be shallower and the striae finer than in the third group, the *maculipennis* group, however these characters do not allow separation of the *adustus* group. Three species were included in the *maculipennis* group but one of these, *maculipennis* itself, has the forewing spotted apically—as stated by Das and Gupta (1989, p. 84)! Petersen (1987) cited by Das and Gupta, even treated *maculipennis* as a subspecies of *stigmus*.

Das and Gupta's groups as constituted are thus unsuitable for present purposes, and have been modified as follows. The *adustus* group comprises those species having a tuberculate male Sternum VII. The *stigmus* group includes the species with an apically spotted forewing, while the *sagittarius* group includes all remaining species (*sagittarius* and *strigosus* were the other two species included by Das and Gupta in their *maculipennis* group). These last two groups are lumped into one group in Table 2.2.

Megapolistes is also subdivided into species groups, corresponding to presence or absence of lateral processes on the terminal male sternum (#26), and whether the digitus is apically membranous or membranous and saccate (#30). The *stenopus* group is scored as absent for the processes and digital sac in Table 2.2, while the remaining species (including the type species, *olivaceus*) are scored as present.

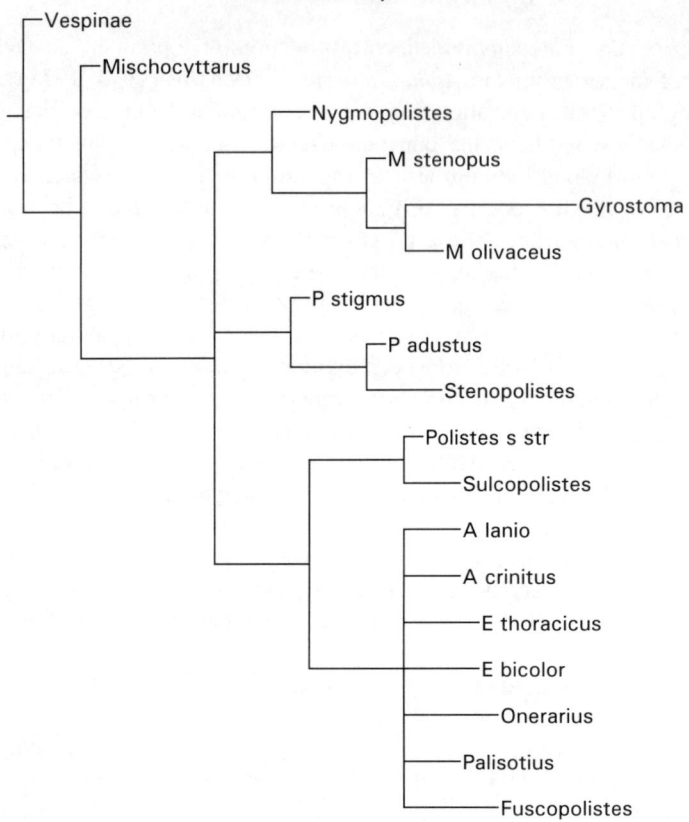

Fig. 2.8 Strict consensus tree for 93 cladograms reported after exact analysis of the character matrix in Table 2.2, and also for the 19 cladograms reported by NONA as meeting the more stringent support requirement explained in the text.

Polistes s. str. is not subdivided into species groups in Table 2.2, as closer relationship between some of its included species and *Sulcopolistes* has already been demonstrated (Carpenter *et al.* 1993). *Palisotius* is also not subdivided. It is not variable for the characters analyzed here, other than at the level of individual species (*viz.* presence of an epicnemial carina, state 1 of character #16).

Exact analysis of the data in Table 2.2 resulted in a report of 93 cladograms (length 64, consistency index 0.73, retention index 0.84). Fig. 2.8 is the strict consensus tree. As before, the cladograms differ in alternative placements of the lineage consisting of *Polistella + Stenopolistes*. The two arrangements presented above were permuted with alternative arrangements within the lineage comprising New World subgenera and species groups. Successive weighting resulted in a report of six cladograms (weighted length 388); these were among the initial 93 cladograms. The technique thus selected the cladograms based upon the most reliable characters, as judged by consistency with the set of all characters (Carpenter 1988*b*). Fig. 2.9 shows the strict consensus tree for these six cladograms.

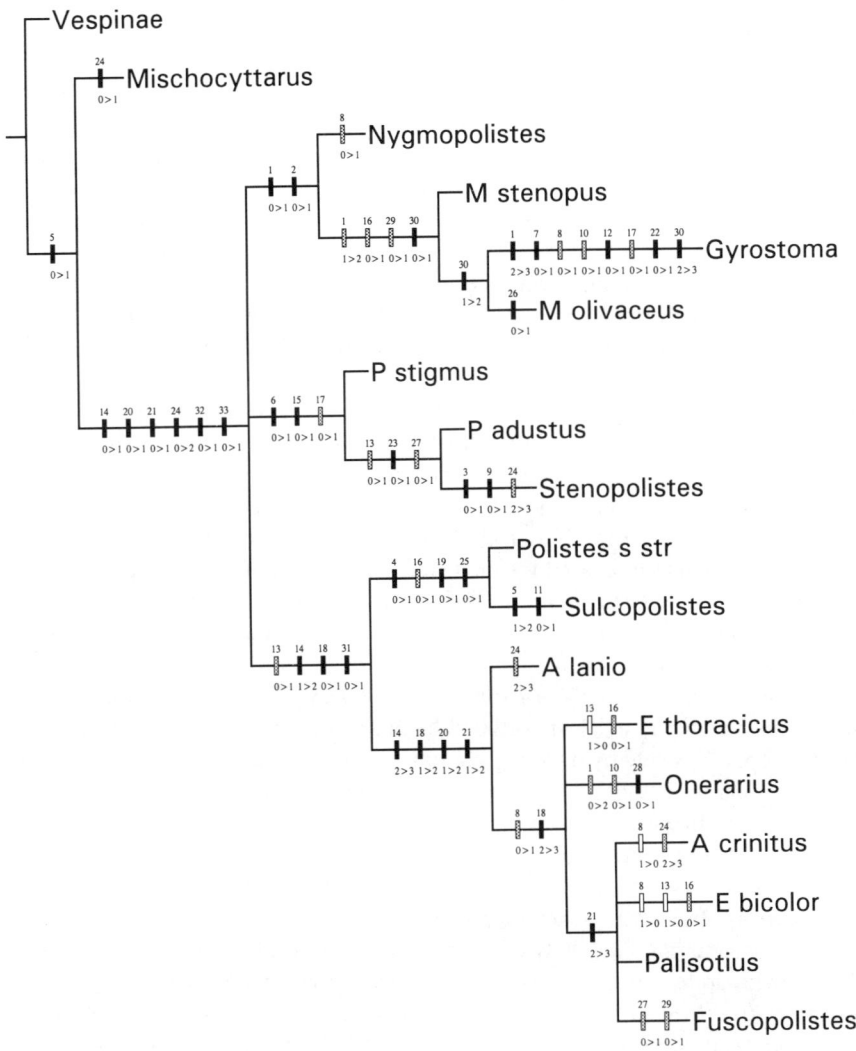

Fig. 2.9 Strict consensus tree for four cladograms (six reported) resulting from successive weighting of the character matrix in Table 2.2; these were among the initial 93 cladograms reported. This is also the consensus tree for the two cladograms reported by NONA as meeting the more stringent support requirement explained in the text. (The 'unique' command of NONA in fact reports this as a third distinct cladogram, to which it assigns the length 64, but that is because NONA does not optimize characters directly on multifurcations, rather on their dichotomous resolutions.) Optimization of the characters of the ingroup is the same under either fast or slow transformation as implemented in CLADOS, with length 65. Plotting conventions are as in Fig. 2.5.

The report of six cladograms, however, is misleading. The cladograms differ by combinations of: (1) the placement of the *Polistella* + *Stenopolistes* clade, based only on one optimization of character #13 as discussed above, and (2) three resolutions within the New World taxa, all supported solely by character #8, male eye-clypeal contact. The three resolutions consist of various permutations recognizing the following sister-group pairings: *Onerarius* + the *thoracicus* group, *Palisotius* + *Fuscopolistes*, and the *crinitus* group + the *bicolor* group. Two cladograms include simultaneous recognition of all three sister-groups, but that is not possible under any one optimization of character #8. It is possible to support either both of the first two sister-group pairings, or the third — not all three. The report of six cladograms is an artifact, similar to that identified in the case of missing values by Platnick *et al.* (1991) for both Hennig86 and the program PAUP (Swofford 1993). The programs report as distinct resolutions those supported by potential optimizations, whether or not the optimizations are simultaneously possible. This is a deficiency of current parsimony programs, introduced by earlier versions of programs such as PAUP that could process only fully bifurcating cladograms, and reported as distinct cladograms that were in reality unsupported resolutions of multifurcations (see Platnick 1987, and Platnick *et al.* 1991). There are in fact only four distinct cladograms. And even these four cladograms do not all meet the more stringent support requirement discussed for the results of analysis of Table 2.1, that is, supported under all possible optimizations. Only the two cladograms with a basal trifurcation as in Figs. 2.5–2.6 do so. The 'unique' command of the program NONA can be used as a check for the more stringent requirement; here, of the initial 93 cladograms reported, NONA discards all but 19, and among the six resulting from successive weighting, it discards all but the two mentioned. Choice among arrangements supported solely by differing placements of a character as inconsistent as #8 does not seem well advised, nor is it necessary for present purposes. Fig. 2.9 is also the consensus tree for these two cladograms, and the discussion that follows is therefore based upon it.

As with analysis of Table 2.1, *Polistes s. str.* is paraphyletic in terms of *Sulcopolistes*, *Palisotius* is not supported either, and *Nygmopolistes* is supported only by character #8. Paraphyly of *Polistella* in terms of *Stenopolistes* is shown by the sister-group relationship between the *adustus* group and *Stenopolistes*; the residue of *Polistella* is also evidently paraphyletic, being unsupported. Paraphyly of *Megapolistes* in terms of *Gyrostoma* is shown by the sister-group relationship between *Megapolistes* excluding the *stenopus* group and *Gyrostoma*; the *stenopus* group is also unsupported. And neither *Aphanilopterus* nor *Epicnemius* appear as monophyletic: the groups into which they were divided in Table 2.2 do not fall out together.

The results of both analyses are discussed further in the next section, in the context of a reclassification of the subgenera. A final analysis, of strictly monophyletic taxa only, is also presented.

Reclassification and final analysis

The present classification is not phylogenetic, and must be revised to make it a phylogenetic system (Hennig 1966). Only monophyletic taxa are useful at information storage and retrieval, because only such taxa correspond to character distribution (Farris 1979)—that is, to the information that is to be stored and retrieved. Revision of the present classification is most readily accomplished by synonymy to eliminate paraphyletic taxa. This is so for two reasons. First, several groups remain unsupported by any character analyzed (see Fig. 2.9), hence elevating them to formal recognition cannot be justified under any circumstances. Second, some groups supported by characters do not correspond to significant ones. The New World taxa exemplify these points, where, as discussed above and shown by the character mapping on Fig. 2.9, the groups and relationships established are supported mostly by homoplastic characters, that vary among the Old World taxa. In view of the marked changes in relationships resulting when the subgenera were subdivided (*cf.* Figs. 2.5–2.7 and 2.9), further subdivision may well alter the relationships yet again, and formal subdivision is thus at best premature.

The synonymies are straightforward. The New World species are a clade supported by four synapomorphies: character #14, state 3, lateral reduction of the pronotal carina; #18, state 2, thoracic punctation fine and well separated; #20, state 2, elongation of the propodeal orifice; and #21, state 2, fine propodeal striae. Within this clade, *Onerarius* has one uncontroverted autapomorphy of the male genitalia, as well as several other, homoplastic apomorphies. But *Fuscopolistes* has only homoplastic apomorphies, while *Palisotius* does not have any, and the groups making up *Aphanilopterus* and *Epicnemius* are separated on Fig. 2.9, with each component species group supported only by homoplastic apomorphies—which are each assigned to the other respective species group. The scattered distribution of *Aphanilopterus* and *Epicnemius* on Fig. 2.9 leaves synonymy of all five of the New World subgenera as the only feasible course.

The sister-group of the New World lineage is the clade comprising *Sulcopolistes* and *Polistes s. str.* This relationship is supported by three uncontroverted synapomorphies: character #14, state 2, ventral reduction of the pronotal carina; #18, state 1, thoracic punctation fine; and #31, state 1, larval mandibles bidentate. Character 13, state 1, occipital carina reduced, also supports this clade, but is homoplastic. *Sulcopolistes* and *Polistes s. str.* are supported as a clade by four synapomorphies: character #4, state 1, male antennae apically hooked; #16, state 1, epicnemial carina present (convergent elsewhere); #19, state 1, thoracic punctation clathrate; and #25, state 1, Sternum I carinate (also in some species of *Polistella*). *Sulcopolistes* is itself evidently monophyletic as shown by characters of the clypeus and mandibles (#5 and 11; the socially parasitic behaviour of the included species may also be treated as a synapomorphy, *cf.* Carpenter *et al.* 1993), but *Polistes s. str.* is paraphyletic in terms of *Sulcopolistes*. *Sulcopolistes* has already been synonymized with *Polistes s. str.* (van der Vecht and Carpenter 1990), and recognition of any of the other previously proposed European subgenera now treated as synonyms of *Polistes s. str.* (Richards 1973; van der Vecht and Carpenter 1990) would still result in paraphyly. *Polistula* and *Leptopolistes* were proposed without any consideration of the world fauna, especially the species of *Polistes s. str.* endemic to the Ethiopian Region. That sort of limited, regional approach to classification has rendered the generic taxonomy of solitary Vespidae 'chaotic' (Parker 1966, p. 153), and it has fared no better in other social wasps (Carpenter 1987), nor in *Polistes*. The synonymy of these taxa is well justified.

Stenopolistes and *Polistella* are supported as a clade by three synapomorphies: character #6, state 1, clypeus dorsally produced; #15, state 1, loss of the pronotal fovea (convergent in a few species of *Polistes s. str.*); #17, state 1, loss of the dorsal groove (also found in *Gyrostoma*, and approached in some species of *Epicnemius* and one of *Fuscopolistes*). *Stenopolistes* is itself monophyletic, as shown by the reduced jugal lobe (character #3), approximated ocelli (#9), and narrowed Segment I (#24), but *Polistella* is paraphyletic in terms of *Stenopolistes*. Kojima and Kojima (1988) have previously suggested that these subgenera be synonymized for this reason. Van der Vecht (1972) separated *Polistella* from *Stenopolistes* entirely by the absence of the characters defining *Stenopolistes*. Richards (1973) discarded one of these characters (forecoxal pubescence in the female) because it occurs in some species of *Polistella*, pointed out that another (Tergum I angular in profile) did not occur in two species he included in the subgenus (*meadeanus, nigritarsis*), and also pointed out another species of *Polistella*, *adustus*, that has another defining trait (tuberculate male Sternum VII). Richards (1978*b*) described another species of *Polistella*, *bambusae*, that also has this trait. Van der Vecht (1984) described a new species, *xantholeucus*, whose subgeneric placement he left open, because it had one of the defining features of *Stenopolistes* (approximated ocelli) but lacked others (Tergum I angular in profile, tuberculate male Sternum VII). Kojima and Kojima (1988) described three new species, and argued that the states of the ocelli were too variable to be of use other than at the specific level, and that only the reduced jugal lobe diagnosed *Stenopolistes*. And as Das and Gupta (1989) pointed out, some (un-specified) species of *Polistella* have the jugal lobe reduced in size, and so there is evidently a transformation series in reduction. The '*adustus* group' of Das and Gupta (1989) apparently contains the sister-group of *Stenopolistes*, but the compo-sition of that species group is unclear (see above). The characters supporting a sis-ter-group relationship between the *adustus* group as construed here either vary within the group as defined by Das and Gupta (1989; character #27, state 1, male Sternum VII tuberculate) or also occur in the residue of *Polistella* (#23, state 1, claws asymmetrical). *Stenopolistes* is thus poorly differentiated from *Polistella*, because of the classical problem of 'intermediates': species showing combinations of the characters originally thought to separate the two subgenera. For both classical and cladistic reasons, *Stenopolistes* should be synonymized with *Polistella*.

The remaining lineage comprises *Nygmopolistes*, *Megapolistes*, and *Gyrostoma*. This clade is supported by two synapomorphies: character #1, state 1, prestigma slightly lengthened; and #2, state 1, prestigma tip recurved. *Megapolistes* and *Gyrostoma* are a clade based on four characters, three of which (#1, state 2, pre-stigma elongate; #16, state 1, epicnemial carina present; #29, state 1, digitus ventrally widened) are convergent, but one of which (#30, state 1, digitus apically membranous) is uncontroverted. But the digitus is variably developed in *Mega-*

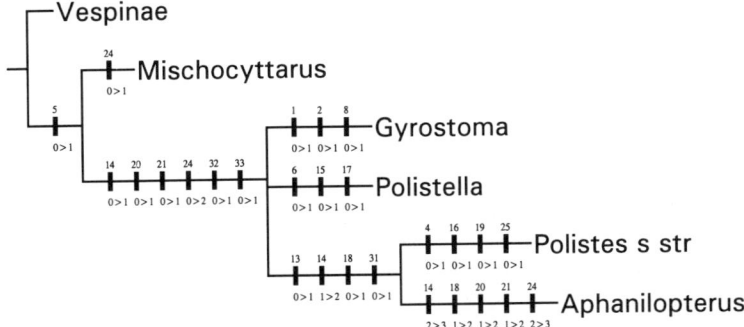

Fig. 2.10 Cladogram resulting from exact analysis of the character matrix in Table 2.3. Optimization of the characters of the ingroup is the same under either fast or slow transformation as implemented in CLADOS. Plotting conventions are as in Fig. 2.5.

polistes, and the species aside from the *stenopus* group are more closely related to *Gyrostoma*, based upon state 2 of character #30, digitus saccate. *Megapolistes sensu* Richards is clearly paraphyletic, and the *stenopus* group is not supported by any character. Van der Vecht (1968a) had included in *Megapolistes* the species, *tenebricosus*, later placed by Richards (1973) in *Nygmopolistes*. Van der Vecht distinguished *Megapolistes* (including *tenebricosus*) from *Gyrostoma* primarily by the absence of the autapomorphies of the latter, and *Megapolistes sensu* van der Vecht is thus also paraphyletic. *Megapolistes* must therefore fall as a synonym of *Gyrostoma*. Recognition of *Nygmopolistes* is at best questionably justified: it is assigned only the character of male eye-clypeal contact (state 1 of #8), a homoplastic character that varies in *Gyrostoma* (including *Megapolistes*). In view of the equivocal nature of *Nygmopolistes*, it is best not to recognize that taxon. Accordingly, it in turn must be synonymized with *Gyrostoma*.

Cladistic reclassification thus results in recognition of four monophyletic subgenera: *Gyrostoma, Polistella, Polistes s. str.*, and *Aphanilopterus*. When the character data of Tables 1 and 2 are reduced to just states assignable to these four subgenera and the outgroups (Table 2.3), exact analysis results in one cladogram (length 27, consistency index 1.0, retention index 1.0), stable to successive weighting. The cladogram (Fig. 2.10) is not completely resolved, but the groups and relationships shown are all supported by multiple characters, which are not homoplastic at this level. A formal reclassification and synonymy is given in Appendix 2. The list follows the conventions proposed by Wiley (1979). *Polistes s. str.* is sister-group of the succeeding subgenus in the list, *Aphanilopterus. Incertae sedis* indicates that it is the position of *Polistella* that is unresolved. A key to the subgenera as reclassified is given in Appendix 3.

Historical biogeography

Background: centre of origin and component analysis

The historical biogeography of *Polistes* has been little studied. Such work as has been done on the subject has been concerned exclusively with the quest for the centre of origin of the genus, in this respect being ingrained with the dominating concept traditional in biogeography (Croizat *et al.* 1974). The general notion that the genus is of tropical origin has been justified with reference to the tropics as the centre of diversity, whether phrased in terms of greater species diversity (e.g. Wheeler 1922; Yoshikawa 1962*a*) or higher-level diversity among Polistinae as a whole (e.g. van der Vecht 1965; Richards 1971). A tropical centre of origin has been recognized both by presence of 'primitive' forms (of nest architecture, West-Eberhard 1969*b*) and by presence of 'later' (van der Vecht 1965) or 'advanced' (van der Vecht 1968*b*) forms. Van der Vecht (1965) and Richards (1971) further suggested that the centre of origin of paper-wasps was south-east Asia, whence they dispersed across Beringea to colonize the Americas. This particular centre of origin was recognized on the basis that all three social wasp subfamilies (*viz.*, Stenogastrinae, Vespinae, and Polistinae) are found together only there, that is, that the Oriental tropics are the centre of diversity for social wasps. Van der Vecht (1965) was inspired by Darlington (1957), who considered that the recently evolved, 'dominant' groups replaced the older groups, which were forced to 're-treat' as the newer groups spread. Van der Vecht (1965, 1968*b*) considered *Polistes* among the advanced, later immigrants to the New World; this was based on his interpretation of the evolutionary polarity of the gland on female metasomal Sternum VI. Richards (1971) argued for the opposite polarity of this character, hence the opposite concept of 'more ancestral' taxa, but retained and elaborated van der Vecht's biogeographic hypothesis. Reeve (1991) suggested that the 'secondary radiation' of *Polistes* in the New World corresponded to an increase in chromosome number, which he related to genetic recombination rates and environmental novelty.

The very concept of centre of origin has been criticized by vicariance biogeographers (e.g. Croizat *et al.* 1974) because of its reliance on *a priori* criteria for the recognition of the centre, criteria that often conflict. That conflict is exemplified in the case of *Polistes*. I previously (Carpenter 1981) criticized the equation of a centre of diversity with a centre of origin in paper-wasps, and suggested that their distribution pattern is broadly gondwanian. I have recently (Carpenter 1993) presented a component analysis of the polistine tribes that indicates a basic Old World/ New World split, with the position of Europe uncertain; Australia, the Oriental tropics, and the Afrotropics formed a group with palearctic Asia. I pointed out that this is consistent with a South American-African vicariant event, during a late stage of the breakup of Gondwana, after Australia had rifted away. Richards (1971) had rejected a connection between South America and Africa, because of a presumed late Cretaceous origin of Vespidae and a Tertiary origin of social wasps. Yet fossils assignable to an extant subfamily of Vespidae have now been found in Lower Cretaceous (Carpenter and Rasnitsyn 1990) and middle Cretaceous (Brothers 1992)

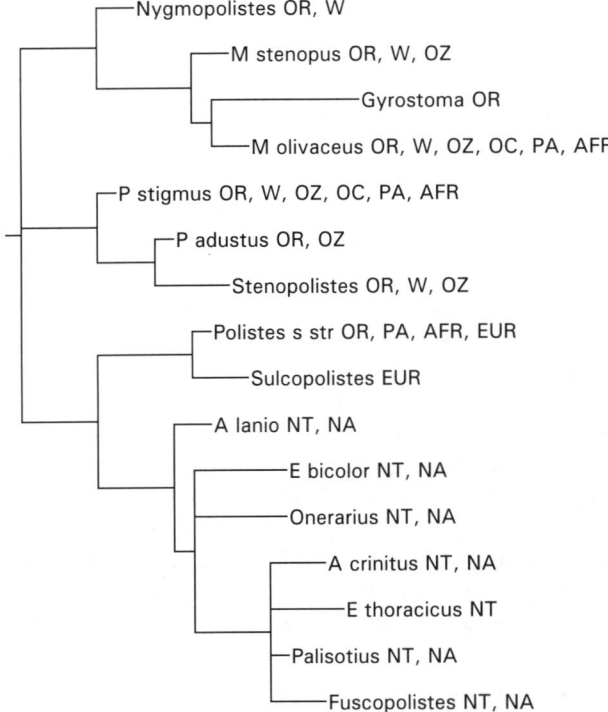

Fig. 2.11 Taxon—area cladogram for *Polistes* based on the consensus tree in Fig. 2.9 and the distributions in Table 2.4.

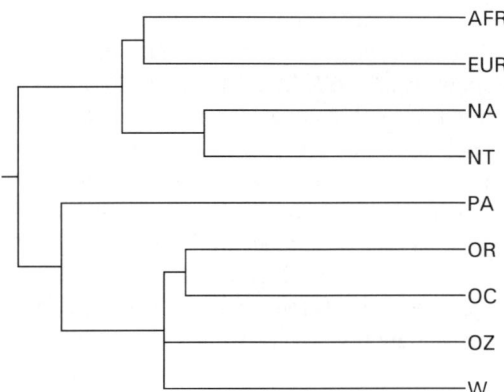

Fig. 2.12 The 'general cladogram' (Nelson, 1979, as modified by Page, 1989a) consensus tree for 42 area cladograms for *Polistes* obtained by component analysis under Assumption 2.

strata, and a fossil social wasp nest is known from the Upper Cretaceous (Wenzel 1990). Predicating biogeographic interpretation upon age of origin as determined by (failure to find) fossils is unsound: fossils indicate the age of first appearance, not the age of origin (Hennig 1966). Biogeographic interpretation is best predicated on distributional data themselves, seeking patterns that may be interpreted by comparison with independent geological evidence adduced after the pattern analysis. Component analysis (Nelson and Platnick 1981) is designed to do just that.

In this study I apply component analysis under Assumption 2 of Nelson and Platnick (1981), as I have done in Carpenter (1993). Component analysis has been controversial, but its justification has also been clearly explained by Page (1990) and I have discussed the subject in detail elsewhere (Carpenter 1992, 1993; see also Morrone and Carpenter 1994). Assumption 2 is the procedure best suited to cope with dispersal, as explained below.

Briefly summarized, in component analysis distributional data are treated as characters and mapped onto a cladogram, producing a taxon-area cladogram (e.g. Fig. 2.11). The taxon-area cladogram is then analyzed under one of several procedures to produce a general hypothesis of the connection among the areas, that is, an area cladogram (e.g. Fig. 2.12). These procedures are methods for dealing with the three problems of widespread taxa, missing areas, and redundancy. If each area contained just one endemic taxon, the construction of area cladograms would be a matter of simply replacing the taxon names with the areas. Widespread taxa occur in more than one area (e.g. the *olivaceus* group in Fig. 2.11, found in six areas). Missing areas are found in one clade but not another (*cf. Polistes s. str.* and *Polistella s. ampl.* on Fig. 2.11). Redundant areas are found in two or more separated clades on the cladogram (most areas on Fig. 2.11).

The procedures that have been formalized, termed 'standard assumptions' by Nelson and Ladiges (1991a, b), are Assumptions 1 and 2 of Nelson and Platnick (1981) and Assumption 0 of Zandee and Roos (1987). Parsimony analysis of area cladograms ('Brooks Parsimony Analysis', Wiley 1987) is similar to Assumption 0 (Page 1990), while 'ancestral areas' approaches (Bremer 1992; and embodied in the controversy over African origin of human mtDNA among Cann *et al.* 1987; Maddison 1991; Vigilant *et al.* 1991; Maddison *et al.* 1992) are similar to Assumption 1. The differences among these procedures may be easily understood with reference to widespread taxa. Under Assumption 0, widespread taxa become synapomorphies for the areas inhabited by such taxa. Assumption 1 allows the area cladogram to be paraphyletic in terms of the areas inhabited by the widespread taxon. Assumption 2 treats each occurrence of the widespread taxon separately, and so the areas can 'float' on the area cladogram. That is, the relationship between the areas inhabited by the widespread taxon can be polyphyletic. A taxon can therefore occur in several areas that are resolved as widely separated on the area cladogram, something the other methods do not readily allow. Assumption 2 attempts to avoid possibly confounding proximity with relationship, and thus allows for dispersal: if the areas on the resulting area cladogram have no close historical connection, the possibility is that the widespread taxon dispersed.

Assumption 2 is therefore the most realistic procedure, but can permit many possibilities in the case of complex data. Its implementation in COMPONENT version 1.5 suffers from memory limitations of the program (*cf.* Morrone and Carpenter 1994). Nelson and Ladiges (1991*b*, p. 481) also suggested that the implementation of Assumption 2 in COMPONENT was deficient, by obscuring 'possibly real complexity', and suggested that the set of Assumption 2 area cladograms produced by COMPONENT could be further resolved by evaluation in terms of three-area analysis of 'nodes' (Nelson and Ladiges 1991*a*). That suggestion has been explored here, by generating area cladograms under Assumption 2 with COMPONENT, and under 'A1' (analysis of nodes) with the program TAS (Nelson and Ladiges 1992), implemented as detailed in the results.

Component analysis: area delimitation

The areas defined for this analysis are modified from those used to list the distributions of the subgenera by Das and Gupta (1984, 1989, table 4). Das and Gupta employed the six classical zoogeographical Regions (Wallace 1876), with the addition of two others: Wallacea, the transition zone between the Oriental and Australian Regions (between Wallace's Line and Aru), and the Oceanic Islands of the Pacific. All these Regions are not only geological composites, they are also composites of many areas of biotic endemism, which are interrelated in ways that do not necessarily correspond to the classical Regions. Lumping the disparate areas into the Regions may lead to results that are artifactual, in that they may differ from a direct analysis of the areas of endemism. Such a direct analysis entails information on cladistic structure at the level of the taxa occupying the areas of endemism, which is not yet possible at the species level for *Polistes*. However, the classical Regions are essentially the basis for the previous work on the genus. I use these Regions heuristically, to investigate how well supported the conclusions of previous work are in a cladistic context.

Table 2.4 shows the distributions of the taxa, and the areas defined for this analysis. The modifications to Das and Gupta's (1989, table 4) distributional listings are as follows. Two corrections have already been alluded to: *Polistes s. str.* and *Polistella* (*stigmus* group) occur in the Ethiopian Region, but the only subgenus listed by Das and Gupta from there was *Megapolistes*. *Polistella* was the only subgenus listed from Wallacea, but *Nygmopolistes* (*tenebricosus*, recorded from Lombok and Flores, Bequaert 1934), *Megapolistes* (*diabolicus*, part of the *stenopus* group, described from Timor; several subspecies of *tepidus* from the Moluccas, see Richards 1978*b*) and *Stenopolistes* (several species from the Moluccas, see Petersen 1990) also occur in that zone. *Megapolistes* was the only subgenus listed from the Oceanic Islands. It is not precisely clear how Oceania was delimited. Das and Gupta (1984, p. 396) gave only 'Hawaii and adjacent islands. Easter islands.' as a definition, although *Polistes* is considered adventive in Hawaii and Easter Island (Richards 1978*a*). However, Das and Gupta's (1984, p. 407) distribution for *olivaceus* listed the Marianas, Fiji, Tonga, Samoa, the Society Islands and the Marquesas as Oceanic Islands. The *stigmus* group thus also occurs in the Oceanic Islands; it has been recorded from Fiji (*stigmus nebulosus*, Yamane and Kusigemati 1985). *Nygmopolistes* was also listed in the Palearctic Region by Das and Gupta

Table 2.4 Distributions of subgenera and species groups of *Polistes* and *Sulcopolistes*.

Taxon	Label	Distribution
Nygmopolistes	OR, W	India to Philippines, Flores
Gyrostoma	OR	India to Taiwan
M. olivaceus group	OR, W, OZ, OC, PA, AFR	Circum-Indian Ocean, East Asia, Solomons to Marquesas
M. stenopus group	OR, W, OZ	Java and Wallacea to Solomons
P. stigmus group	OR, W, OZ, OC, PA, AFR	South Africa to East Asia, Australia, Solomons to Fiji
P. adustus group	OR, OZ	India, New Guinea
Stenopolistes	OR, W, OZ	India to Australia
Polistes s. str.	OR, PA, AFR, EUR	Europe, Africa to East Asia
Sulcopolistes	EUR	Europe
A. lanio group	NT, NA	USA to Argentina
A. canadensis group	NT, NA	USA to Argentina
E. bicolor group	NT, NA	USA. to Argentina
E. thoracicus group	NT	Brazil
Onerarius	NT, NA	USA to Argentina
Palisotius	NT, NA	USA to Argentina
Fuscopolistes	NT, NA	Canada to Mexico

Labels for areas are as follows: OR = Oriental Region; W = Wallacea; OZ = Australian Region; OC = Oceanic Islands; PA = East and Central Asia; AFR = Ethiopian Region; EUR = Europe and Mediterranean Region; NT = Neotropical Region; NA = Nearctic Region.

(1984, 1989), but that appears to be erroneous. Das and Gupta recorded *sulcatus* from northern China and Japan. They (1984, p. 407) cited Dalla Torre (1904) for the record from Japan (of *rugifrons*, treated by Das and Gupta as a synonym of *sulcatus*), but Bequaert (1934, p. 10) had already stated that Dalla Torre's catalogue was in error; *rugifrons* was described from India. Das and Gupta (1984, p. 407) cited Yamane and Yamane (1979) for the record from northern China, but Yamane and Yamane (1979, p. 27) listed *sulcatus* from Nepal, northern India, and southern China. This distribution is in the Oriental Region.

The final modification is an addition. The Palearctic Region has been divided into two areas, separating Europe and the Mediterranean Region from East and Central Asia. The social parasites (*Sulcopolistes*) are endemic to Europe and the Mediterranean Region, while the *stigmus* group and *olivaceus* group do not occur there (the record of *sagittarius* from Greece cited by Das and Gupta 1984, 1989, is doubtless erroneous), but only in eastern Asia. Only *Polistes s. str.* occurs in both Europe and eastern Asia, hence these parts of the 'Palearctic' are related in disparate ways, as has been shown for paper-wasps as a whole (Carpenter 1993).

Component analysis: results

The cladogram for subgenera and species groups (Fig. 2.9) and the distributions of each taxon (Table 2.4) were used as input to construct area cladograms. Fig. 2.11

shows the taxon-area cladogram. Note particularly the extensive redundancy in area distributions. This cladogram contains some paraphyletic taxa, as discussed above, but has the finest resolution of distributions possible. I also attempted an analysis using only the four monophyletic subgenera as reclassified here (i.e. Fig. 2.10). In this analysis, the memory limits of COMPONENT were exceeded, with the number of area cladograms overflowing at more than 1000. The program does not handle redundancy very well, because of the numerous possibilities allowed. The paraphyletic taxa must be borne in mind as a potential artifact—which, of course, applies more forcefully to all previous work. COMPONENT is also limited in that it accepts only fully bifurcating trees. This is irrelevant for the New World taxa: the results would be the same for any resolution of their relationships. But alternative placements of the *Polistella s. ampl.* clade lead to different solutions. Therefore, each of the three possible resolutions of the basal trifurcation of Fig. 2.11 was input into the program, while the New World taxa were arbitrarily resolved.

Component analysis under Assumption 2 results in 14 area cladograms for each of the three resolutions of Fig. 2.11, for a total of 42 distinct area cladograms. Fig. 2.12 shows the 'general cladogram' (Nelson 1979, as modified by Page 1989*a*) for all 42 area cladograms. This is a type of consensus tree that maximizes resolution. Page (1988) has argued that this particular type of consensus may be more appropriate than cladistic parsimony for biogeography, when the areas are related in more than one way. Under parsimony, the areas can be related in only one way, namely, by descent (strict vicariance). By contrast, using consensus techniques to produce general area cladograms combines incongruent area relationships, by permitting them to be in different cliques (Page 1987). The components from incongruent area cladograms are sorted into different cliques, representing the different area relationships, which are summarized by the consensus tree.

TAS does not implement Assumption 2 directly; in the documentation Nelson and Ladiges (1992) suggest an approximation in which three-area statement analysis of nodes (their 'A1') is applied to a taxon-area cladogram in which widespread ranges are reduced in favour of endemic regions. In this case, the widespread taxa could not be reduced to endemic regions, because of the extensive redundancy (Fig. 2.11). Nelson and Ladiges (1992) recommend separate analysis of each clade to avoid combining repeated areas, which they analogize to confounding paralogy and orthology in molecular data. Attempting that here did not resolve all the area relationships. Three-area analysis was therefore applied to the taxon-area cladogram after redundant, widespread distributions were removed by COMPONENT (Fig. 2.13), to minimize as much as possible the impact of both widespread and redundant ranges. Fig. 2.13 was input into TAS and the resulting three-area matrix processed with Hennig86. This resulted in two area cladograms. Fig. 2.14 shows the 'general cladogram' consensus for these two area cladograms.

Area cladograms show the relative order of historical events affecting the areas. Figs. 2.12 and 14 differ primarily in placement of palearctic Asia relative to two main components, one comprising the Ethiopian Region, Europe, and the Nearctic + Neotropics, and the other comprising Australia, Wallacea, the Oceanic Islands, and the Oriental Region. Within each of these main components, the relative order

James M. Carpenter

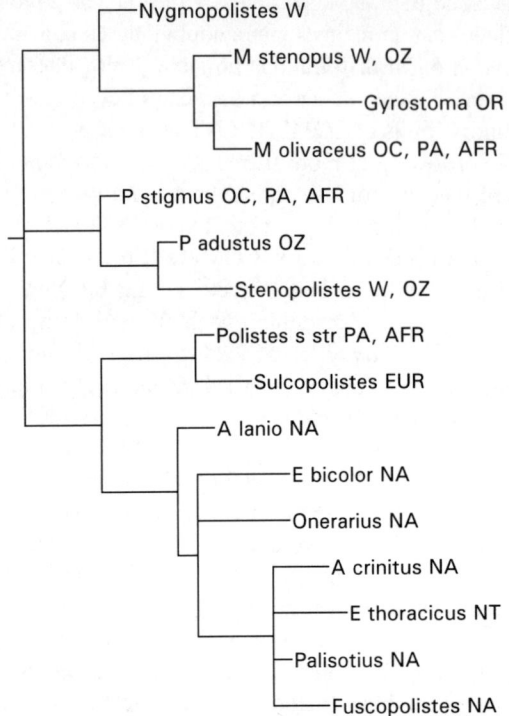

Fig. 2.13 Taxon—area cladogram for *Polistes* based on the consensus tree in Fig. 2.9 and the distributions in Table 2.4, after removal of redundant, widespread areas by COMPONENT (the program performs this as a routine, preliminary step before calculating the possibilities for component analysis under Assumption 2).

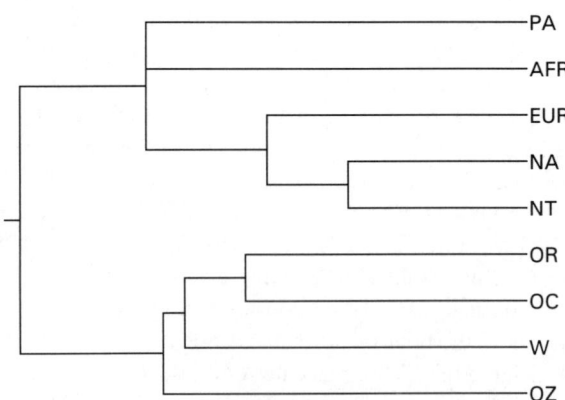

Fig. 2.14 The 'general cladogram' (Nelson 1979, as modified by Page 1989*a*) consensus tree for two area cladograms for *Polistes* obtained by three-area analysis.

of events differs slightly. The primary conclusion to be drawn from either of Figs. 2.12 or 2.14 is that the hypothesis of an Oriental centre of origin, and Beringean dispersal to the New World, is not supported in a phylogenetic context. On none of the area cladograms is the Oriental Region most basal (early event='primitive'), as one might expect of a centre of origin under one criterion. On both Figs. 2.12 and 2.14, the Oriental Region is relatively apical (later event='advanced'), as one might expect of a centre of origin under a second criterion, but that later event also involves the Oceanic Islands, which is therefore just as good a candidate for a centre of origin. On Fig. 2.14, the event involving the Nearctic and Neotropical Regions is just as apical, and so either Region is just as good a candidate for a centre of origin following that hypothesis. Any argument for an Oriental centre of origin would thus be *ad hoc*. As for Beringean dispersal, on both Figs. 2.12 and 2.14 Europe is more closely associated with the New World than is palearctic Asia, and on Fig. 2.12 this is true of the Ethiopian Region as well.

These results accord with the conclusions reached for Polistinae as a whole by Carpenter (1993). In terms of current geological understanding, under either of the hypotheses of Figs. 2.12 or 2.14 the connection between the Old World and New World is better explained by Gondwana than Beringea, via tropical Africa-South America during a late stage of Gondwanian breakup. The general idea of tropical origin of *Polistes* accords with this conclusion. The most basal biogeographic event in the genus separates regions extending to the east of the Indian Ocean from regions to the west for either Fig. 2.12 or 2.14, whatever the sequence of events within the two main components and regardless of the disagreement on the placement of palearctic Asia. If a majority of biogeographic connections in *Polistes* are circum-Indian Ocean, then perhaps the best candidate for the 'centre' of origin of *Polistes* is the present-day Indian Ocean.

Acknowledgements

My initial study of polistine subgenera was done while I held a Smithsonian Postdoctoral Fellowship, and I thank Karl Krombein and Arnold Menke for encouragement. I thank the organizers of this workshop for providing the impetus to finish the work and their editorial efforts, Junichi Kojima for suggestions and specimens, John Wenzel, Chris Starr, Byron Alexander, Mary Jane West-Eberhard, and Stefano Turillazzi for critical reading of the manuscript, and Amy Davidson for assistance with the illustrations.

Appendix 1

List of *Polistes* species examined. Specimens are deposited in the American Museum of Natural History, New York; US National Museum of Natural History, Washington; and Museum of Comparative Zoology, Cambridge. Species nomenclature follows a checklist currently in preparation; subgeneric assignment is based on pers. obs. if not previously published. An asterisk indicates that males were dissected.

Subgenus *Aphanilopterus*

Group 1 of Richards (1973)
annularis (L.) *
aterrimus Saussure *
brevifissus Richards *
buyssoni Brèthes *
canadensis (L.) *
cavapyta Saussure *
comanchus comanchus Saussure *
comanchus navajoe Cresson *
dominicus (Vallot) *
erythrocephalus Latreille *
ferreri Saussure *
huacapistana Richards
infuscatus anduzei Bequaert
infuscatus ecuadorius Richards
infuscatus infuscatus Lepeletier
kaibabensis Hayward *
lanio lanio (F.) *
lanio weberi Bequaert
mexicanus Bequaert *
satan Bequaert

Group 2 of Richards (1973)
binotatus Saussure *
consobrinus Saussure *
maranonensis Willink
myersi Bequaert *
peruvianus Bequaert *
rufidens Saussure
simillimus Zikán *
veracrucis Richards
versicolor flavoguttatus Bequaert *
versicolor kaieteurensis Bequaert
versicolor versicolor (Olivier) *
weyrauchorum Willink *
xanthogaster willei Bequaert
xanthogaster xanthogaster Bequaert*

Group 3 of Richards (1973)
apicalis Saussure *
instabilis Saussure *
stabilinus Richards *

Group 4 of Richards (1973)
oculatus Smith *

Group 5 of Richards (1973)
arizonensis Snelling *
bahamensis Bequaert and Salt *
bequaertellus Snelling
crinitus americanus (F.) *
crinitus crinitus (Felton) *
crinitus multicolor (Olivier)
cubensis Lepeletier
exclamans Viereck *
incertus Cresson *
lineonotus Bohart
minor Palisot de Beauvois *

Group 6 of Richards (1973)
goeldii Ducke *

Subgenus *Epicnemius*

Group 1 of Richards (1973)
bicolor Lepeletier *

Group 2 of Richards (1973)
thoracicus Fox *

Group 3 of Richards (1973)
actaeon Haliday *
billardieri biglumoides Ducke *
billardieri ruficornis Saussure *
cinerascens Saussure *
claripennis Ducke
deceptor Schulz *
geminatus Fox *
melanosoma Saussure *
niger Brèthes
obscurus Saussure
occipitalis Ducke *
pacificus (F.) *
subsericeus Saussure *
testaceicolor Bequaert *

Subgenus *Fuscopolistes*

apachus Saussure *
aurifer Saussure *
bellicosus Cresson *
carolina (L.) *

dorsalis californicus Bohart
dorsalis clarionensis Bohart
dorsalis dorsalis (F.) *
dorsalis neotropicus Bequaert *
flavus Cresson *
fuscatus (F.) *
metricus Say *
perplexus Cresson *
poeyi haitiensis Bequaert and Salt *
poeyi poeyi Lepeletier *
rossi Bohart *

Subgenus *Gyrostoma*

gigas Kirby and Spence *

Subgenus *Megapolistes*

'*stenopus* group'
diabolicus Saussure *
stenopus brandti van der Vecht
stenopus manni van der Vecht
stenopus stenopus van der Vecht *

'*olivaceus* group'
fordi van der Vecht *
jadwigae Dalla Torre *
olivaceus (DeGeer) *
penthicus van der Vecht *
rothneyi carletoni van der Vecht
rothneyi grahami van der Vecht *
rothneyi koreanus van der Vecht
rothneyi tibetanus van der Vecht
schach (F.) *
semiflavus Holmgren
tepidus malayanus Cameron *
tepidus picteti Saussure
tepidus speciosus Buysson
tepidus tepidus (F.)
wattii Cameron *

Subgenus *Nygmopolistes*

sulcatus Smith *
tenebricosus hoplitus Saussure *

tenebricosus leopoldi Bequaert
tenebricosus nigrosericans Bequaert*
tenebricosus sibuyanensis Bequaert
tenebricosus tenebricosus Lepeletier

Subgenus *Onerarius*

carnifex boliviensis Bequaert
carnifex carnifex (F.) *
carnifex rufipennis Latreille

Subgenus *Palisotius*

major castaneicolor Bequaert *
major major Palisot de Beauvois *
major weyrauchi Bequaert
palmarum Bequaert *
paraguayensis Bertoni *

Subgenus *Polistella*

'*adustus* group'
bambusae bambusae Richards *
bambusae humboldti Richards *

'*stigmus* group'
stigmus bernardii Le Guillou *
stigmus dubius Saussure
stigmus maculipennis Saussure
stigmus manillensis Saussure *
stigmus stigmus (F.) *
stigmus tamulus (F.) *
stigmus townsvillensis Soika *

'*sagittarius* group'
albocalcaratus Buysson *
aquilinus Buysson *
assamensis Bingham
bequaerti Schulthess *
bituberculatus Buysson *
defectivus Gerstaecker *
diakonovi Kostylev
fastidiosus Saussure *
haugi Buysson *
humilis synoecus Saussure *
japonicus Saussure *

madecassus Saussure *
madiburensis Schulthess *
mandarinus Saussure *
medius Kojima
philippinensis funebris Bequaert *
philippinensis philippinensis
　　Saussure*
sagittarius indonesicus Bequaert
sagittarius sagittarius Saussure *
saussurei Dalla Torre *
sikorae Saussure *
smithii neavei Schulthess *
smithii smithii Saussure *
snelleni Saussure *
strigosus mimus Bequaert
strigosus strigosus Bequaert *
tullgreni Schulz
variabilis (F.) *
watutus Kojima *

Subgenus *Polistes*

africanus Palisot de Beauvois *
albicinctus Saussure *
associus Kohl *
badius Gerstaecker *
biglumis bimaculatus (Geoffroy) *
bischoffi Weyrauch *
chinensis (F.) *
dominulus (Christ) *

gallicus (L.) *
indicus Stolfa
iranus Guiglia
loveridgei Bequaert
macrocephalus Bequaert
marginalis (F.) *
nimpha (Christ) *
ornatus Lepeletier *
riparius Yamane and Yamane *
tenellus Buysson *

Subgenus *Sulcopolistes*

atrimandibularis Zimmermann *
semenowii Morawitz *
sulcifer Zimmermann *

Subgenus *Stenopolistes*

capnodes capnodes van der Vecht
capnodes incomptus van der Vecht
clavicornis van der Vecht *
lateritius Smith
lycus Cameron
meadeanus (Schulthess)
nigrifrons nigrifrons Smith *
nigrifrons obdurus Cheesman
occultus Kojima
williamsi Petersen *
xantholeucus van der Vecht

Appendix 2

Cladistic classification of *Polistes*. The classification is phyletically sequenced (see Wiley 1979).

Genus *Polistes* Latreille

Subgenus *Gyrostoma* Kirby and Spence

Gyrostoma Kirby and Spence, 1828, Introd. Ent. Ed. 5 (of whole work), Ed. 2 (of vol. III): 36, 631, replacement name for *Cyclostoma* Kirby and Spence, 1826, *non* Lamarck, 1799.

　　Cyclostoma Kirby and Spence, 1826, Introd. Ent. Ed. 1, 3: 36, 633, genus. *Non Cyclostoma* Lamarck, 1799 (Pisces).

Type species: *Cyclostoma gigas* Kirby and Spence, 1826, by monotypy.

Megapolistes van der Vecht, 1968, Bijdr. Dierk. 38: 97, subgenus of *Polistes* Latreille. **NEW SYNONYMY**.

Type species: *Vespa olivacea* DeGeer, 1773, by original designation.

Nygmopolistes Richards, 1973, Rev. Bras. Ent. 17: 91, 93, 98, subgenus of *Polistes* Latreille. **NEW SYNONYMY**.

Type species: '*Polistes sulcatus* Smith, 1852 (= *P. rugifrons* Cameron, 1900)', by original designation.

Subgenus *Polistella* Ashmead, *incertae sedis*

Polistella Ashmead, 1904, Proc. U. S. Natl. Mus. 28: 133, genus.

Type species: *Polistes manillensis* Saussure, 1853, by original designation.

[Misidentification of *Protopolybia sedula* (Saussure) = *P. exigua* (Saussure).]

Stenopolistes van der Vecht, 1972 (1971), Ent. Essays to commemorate the retirement of Prof. K. Yasumatsu: 101, subgenus of *Polistes* Latreille. **NEW SYNONYMY**.

Type species: *Polistes lateritius* Smith, 1857, by original designation.

Subgenus *Polistes* Latreille

Polistes Latreille, 1802, Hist. Nat. Crust. Ins. 3: 363, genus (6 species).

Type species: *Vespa gallica* Linnaeus, 1767, by subsequent designation of Blanchard, 1840, Hist. Nat. Anim. Artic. 3: 397.

Polystes (!) Palisot de Beauvois, 1818, Ins. Recueill. Afrique Amérique: pl. 8; Buysson, 1892, Ann. Soc. Ent. Fr. 61: 59; H. von Ihering, 1896, Zool. Anz. 19: 452.

Eupolistes Dalla Torre, 1904, Genera Insect. 19: 68, name for 'Première division' of *Polistes* Latreille in Saussure, 1853, Ét. Fam. Vesp. 2: 45 (61 species).

Type species: *Vespa gallica* Linnaeus, 1767, by subsequent designation of Richards, 1973, Rev. Bras. Ent. 17 (13): 86.

Pseudopolistes Weyrauch, 1937, Zool. Jahrb. (Abt. Syst. Ökol. Geogr. Tiere) 70: 266, 274, genus (3 species). Unavailable; no type species designated.

Sulcopolistes Blüthgen, 1938 (1937), Konowia 16: 273, subgenus of *Polistes* Latreille.

Type species: *Polistes semenowi* Morawitz, 1889 [= *semenowii*], by original designation.

Polistula Weyrauch, 1938, Arb. Physiol. Angewand. Ent. 5 (3): 273, genus (5 species). Unavailable; no type species designated.

Polistula Weyrauch, 1939, Arch. Naturgesch., N. F. 8 (2): 148, genus.

Type species: *Polistes kohli* Dalla Torre, 1904 [= *Polistes biglumis bimaculatus* Geoffroy, 1785], by original designation.

Pseudopolistes Weyrauch, 1939, Arch. Naturgesch., N. F. 8 (2): 195, validation by type selection of *Pseudopolistes* Weyrauch, 1937.

Type species: *Polistes sulcifer* Zimmermann, 1930, by original designation.

Leptopolistes Blüthgen, 1943, Arch. Naturgesch., N. F. 12 (1): 99, 121, subgenus of *Polistes* Latreille.

Type species: *Polistes associus* Kohl, 1898, by original designation.

Subgenus *Aphanilopterus* Meunier

Aphanilopterus Meunier, 1888, Naturalista Sicil. 7: 302, genus.

Type species: *Aphanilopterus vagabundus* Meunier, 1888 [= *Vespa lanio* Fabricius, 1775], by monotypy.

Polistoides Dalla Torre, 1904, Genera Insect. 19: 68, name for division II of genus *Polistes* Latreille in Saussure, 1854, Ét. Fam. Vesp. 2: 100. *Non Polistoides* Loew, 1873 (Diptera).

Type species: *Polistes subsericeus* Saussure, 1854, by monotypy.

Onerarius Richards, 1973, Rev. Bras. Ent. 17: 91, 94, 101, subgenus of *Polistes* Latreille. **NEW SYNONYMY**.

Type species: *Vespa carnifex* Fabricius, 1775, by original designation.

Fuscopolistes Richards, 1973, Rev. Bras. Ent. 17: 91, 95, 100, subgenus of *Polistes* Latreille.

Type species: *Vespa fuscata* Fabricius, 1793, by original designation. **NEW SYNONYMY**.

Polistarchus Richards, 1973, Rev. Bras. Ent. 17: 94, 95, 101, subgenus of *Polistes* Latreille.

Type species: *Vespa canadensis* Linnaeus, 1758, by original designation.

Palisotius Richards, 1973, Rev. Bras. Ent. 17: 95, 96, 100, subgenus of *Polistes* Latreille.

Type species: *Polistes major* Palisot de Beauvois, 1818, by original designation. **NEW SYNONYMY**.

Epicnemius Richards, 1973, Rev. Bras. Ent. 17: 96, 102, subgenus of *Polistes* Latreille.

Type species: *Polistes bicolor* Lepeletier, 1836, by original designation. **NEW SYNONYMY**.

Appendix 3

Key to the Subgenera of *Polistes* as reclassified

1. Thorax with pronotal fovea, epicnemial carina and dorsal groove all absent (Fig. 2.4(b)); clypeus dorsally produced far above the anterior tentorial pits (Fig. 2.3(a))
                                                                Polistella Ashmead

– Thorax usually with pronotal fovea present (Figs. 2.4(a), (c–d)), if absent the epicnemial carina and dorsal groove are present; clypeus not produced above the anterior tentorial pits (Figs. 2.3(b–d))

                                                                                    2

2. Mesepisternal punctation usually coarse, larger than interspaces (Figs. 2.4(a–b)); pronotal carina lamellate ventrad to fovea (Figs. 2.4(a–b)); forewing with prestigma length >0.5x length of pterostigma (measured along ventral part) and tip slightly recurved (Figs. 2.2(a–c))
                                                        Gyrostoma Kirby and Spence

– Mesepisternal punctation fine, clathrate (Fig. 2.4(c)) or smaller than interspaces

(Fig. 2.4(d)), or absent; pronotal carina shortened or blunt, not lamellate below fovea (Figs. 2.4(c–d)); forewing with prestigma usually short (<0.5x length of pterostigma) and tip straight

3

3. Thoracic punctation clathrate (Fig. 2.4(c)); propodeal striae coarse (Fig. 2.4(c)); male antennae apically hooked; metasomal Sternum I transversely carinate basally; pronotal carina lamellate to fovea (Fig. 2.4(c)); epicnemial carina present (Fig. 2.4(c))

Polistes Latreille

– Thoracic punctation smaller than interspaces (Fig. 2.4(d)) or absent; propodeal striae fine (Fig. 2.4(d)) or absent; male antennae apically tapering; metasomal Sternum I ecarinate basally; pronotal carina usually blunt before reaching fovea (Fig. 2.4(d)); epicnemial carina present or absent

Aphanilopterus Meunier

3

Learning, behaviour programs, and higher-level rules in nest construction of *Polistes*

John W. Wenzel

Many students of social insects view the nest as little more than a substrate upon which social life unfolds. Nest architecture is generally neglected in favour of studies that ask the perennially more fashionable questions, often variations and derivatives of the query 'who laid the eggs?' In *Polistes*, specific issues of genetical relatedness and reproduction capture the greatest attention (Reeve 1991). These studies have inspired several general theories of social evolution and provided tests of others, but studies of nest architecture of *Polistes* remain relatively obscure. Yet, characteristics of the paper nest address many fundamental topics, such as the evolution of complex chains of behaviours (Downing and Jeanne 1987, 1988, 1990; Wenzel 1993), the division of labour (Jeanne 1986*a*), the self-organization of construction behaviour (Deneubourg *et al.* 1992; Karsai and Penzés 1993), and kin recognition (Gamboa *et al.* 1986*b*; Espelie *et al.* 1990; Singer and Espelie 1992).

In social wasps, research seems to have proceeded in a manner opposite to that seen in ants and honey bees: pioneer studies (Deleurance 1950; Janet 1895; Jeanne 1972; West-Eberhard 1969*b*), and the bulk of recent work (Downing and Jeanne 1987; Karsai and Penzés 1993; O'Donnell and Jeanne 1990) focus on individual behaviour patterns, whereas general views of colony-level phenomena are few and recent (Deneubourg *et al.* 1992; Jeanne 1986*a*; Wenzel 1989, 1993). The traditional focus upon individuals is clear from Wilson's appraisal of wasp research: 'Four of the basic discoveries of insect sociology—nutritional control of caste (Marchal), the use of behavioral characters in studies of taxonomy and phylogeny (Ducke), trophallaxis (Roubaud), and dominance behaviour (Heldmann, Pardi)—either originated in wasp studies or were based primarily on them' (Wilson 1971, p. 7). Of these, only Ducke's phylogenetic study is measured above the level of the individual. By contrast, studies directed toward individual behaviour in army ants (Deneubourg *et al.* 1989) or honey bees (Camazine 1991) are recent and depart from a traditional emphasis upon levels of colony or caste.

To some, it may seem that studies at the level of the colony are somehow less 'behavioural' than are those that centre on the actions of individual, marked animals. However, 'natural' units of behaviour are elusive (Miller 1988) and the units that are useful vary according to the level of the problem (Wenzel 1992*a*). Behaviours that are manifested *only* at the level of the colony are meaningless at the level of the individual. For example, 'swarming' is an emergent property of a colony, a behaviour that is not perfectly described by reducing it to the movement pattern of an individual and then multiplying by colony size. A swarming colony of honey bees is something more than 5000 bees flying around: there is no other circumstance under which they soon arrive at the same distant location without benefit of reward, and certainly not when there are various scouts competing for the attention of the departing bees. Such colony-level behaviours mark an organization above that of the individual behaviour pattern. Some studies of self-organized behaviours mark impressive successes in reducing epiphenomena to simple rules (Camazine 1991; Deneubourg *et al.* 1989), but these all rely nonetheless on measurements made at the level of the colony. The dynamics of an army ant swarm raid were already well known (Schneirla 1949) before they could be modelled as simple rules and contingencies of individual behaviour. Although deconstructing a phenomenon into various contingencies differs from mere description of it, this does not eliminate the phenomenon from the model. Whether included directly in a model or not, colony-level data are used to gauge the performance of the model, and these data are therefore critical to our general understanding. The traditional focus upon individual animals in wasp research seems to have left the field with comparatively little interest in colony-level phenomena in paper-wasps. Yet, complete understanding will be available only from a combination of individual and colony-level perspectives.

This chapter is organized into two parts. The first is a brief overview of the sort of variation found among *Polistes,* and this section emphasizes that *Polistes* architecture is surprisingly monotonous in comparison with other genera. I propose that colony size may be related to such uniformity. The second part examines individual behavioural programs and colony-level organization that regulate construction behaviour and produce the stable and regular design. There, I hope to show how such fundamental architectural traits as shape and size are controlled in fashions more complex than popular intuition and simple models would suppose.

Conservative variation in *Polistes* architecture

The dearth of study of nest architecture is not due to difficulty, for *Polistes* is one of the most uniform of the genera, with little variation in the gross architecture of the roughly 200 species world-wide. Indeed, the main difficulty with examining architectural evolution in *Polistes* is that there appears to have been very little of it. By contrast, the comparable primitively eusocial species of *Mischocyttarus* are about as numerous, but display far greater variation in architecture. Less speciose *Ropalidia* displays a still greater and perplexing diversity of nest forms (Wenzel 1991). Even more extreme are the Stenogastrinae (Pardi and Turillazzi 1982; Turillazzi 1991; Wenzel 1991), whose nests vary dramatically between sister species. Nonetheless, *Polistes* remains the single best-studied genus.

Primitive condition

One problem with evolutionary discussions of *Polistes* nests is that the genus is a common model for the ancestral condition, and there is little information about what social wasp nests looked like before *Polistes*. West-Eberhard (1969*b*, p. 92) imagined that nests of ancestors preceding *Polistes* would have had cylindrical cells, each on its own pedicel. Economic considerations would lead to cells suspended from each other in a string, like neotropical *P. goeldii*. She argued that because such forms are found in polistine genera *Mischocyttarus, Ropalidia*, and *Parapolybia* as well as Stenogastrinae, because these are all tropical, and because *Polistes* was assumed to have a tropical origin, such a vertical string should be regarded as primitive, and the radially symmetrical, disc-like, horizontal combs of temperate species should be derived. Based upon a different economic argument, Jeanne (1975) proposed a hypothetical ancestor of modern Polistinae that had a nest different from *Polistes* in that it was broadly attached to the substrate and lacked a pedicel (rather like *Liostenogaster flavolineata*) (Wenzel 1991). The pedicellate comb of *Polistes* is then a derived feature, and one that Jeanne felt might be evolutionarily distinct from other derivations of a suspended comb. He saw the radially symmetrical, horizontal combs as primitive relative to asymmetrical or eccentric designs (Jeanne 1975, fig. 5).

I have argued (along lines similar to those of West-Eberhard, above) that study of other polistine genera suggests that inclined, eccentric combs should be considered primitive relative to horizontal, radially symmetrical combs (Wenzel 1989). Looking to Vespinae, the sister-group of Polistinae, it appears that some kind of disc-like, pendant comb was most likely the ancestral condition preceding *Polistes* because that seems to be the condition for the mutual ancestor of the two subfamilies (Wenzel 1991). Although there is no preserved pedicel itself, good fossil evidence also suggests a suspended comb with paper cell bottoms, very much like modern *Polistes,* was already developed before about 65 million years ago (Wenzel 1990). If *Celliforma favosites* does indeed represent a good preservation of a wasp nest, it may not be alone, for certain 'bee nests' (Nessov 1988) of the Cretaceous of Kazakhstan seem now best placed along with *Celliforma* (Nessov *in litt.*) as some part of the lineage now represented by Polistinae or Vespinae. I have voiced doubt about the homology of radial symmetry across the two subfamilies (Wenzel 1989, p. 694), but evidence for my opinion remains inconclusive. Today, the best we can say is that the primitive state preceding Vespinae and Polistinae was probably a pendent comb.

Modern form and variation

The basic form of the nest is simple and well known. A single, one-sided, paper comb of hexagonal cells is suspended from the back or side by a narrow, rod-like pedicel such that the open ends of the cells point somewhat downward. There are no special structures to provide storage space for food reserves or resting space for the work force. Cells may vary slightly in size, but they are not specialized to serve different functions, such as rearing males versus females, or queens versus workers. There is no protective envelope of any sort. Nest carton is composed of long fibres, usually chewed from a sound, dead, woody source. The ball of pulp is mixed with water and saliva before being put in place and worked into a flatter strip with chewing motions of the mandibles. Silk from vacated cocoons is often used as building material when the nest is mature, but otherwise the carton is usually grey or occasionally brownish in colour. The pedicel is usually reinforced with additional pulp and pure salivary secretion, often becoming shiny, dark, and tough. Temperate species may later build additional, secondary pedicels for added support. The nest may grow by addition of cells in concentric circles centred on the pedicel, or it may grow eccentrically. Larvae are reared one per cell, spinning a completely closed cocoon upon pupation. After the adult emerges from the cocoon, the cell may be reused. In temperate zones, adults generally hibernate away from the nest and nests are rarely used again by wasps returning the following spring.

These generalizations are accurate for the vast majority of species, but a few exceptions warrant mention. Some modifications appear rarely in species that generally have ordinary architecture, such as the multiple combs of *P. exclamans* (Strassmann 1981*a*), *P. metricus* (Gamboa 1981), or *P. fuscatus* (Downing and Jeanne 1986). The few variations found regularly in certain species of *Polistes* are mostly modest changes in the fundamental blueprint (Table 3.1). These innovations may appear substantial, but they are slight in comparison with those found among the Stenogastrinae, *Mischocyttarus* and *Ropalidia* (Wenzel 1991), and include only a few species in the genus. For example, some novelties in *Mischocyttarus* include: long pedicels; wide pedicels; many pedicels with one or few cells each; asymmetrical cells (one side flat); round cells; combs composed of single, horizontal rows of cells; combs of vertical rows of cells; cells in tandem arrangement, each from the lip of the preceding one; combs of cells arranged back-to-back; rows of cells successively displaced distally from one another; cells with one lip much longer than the other; non-cellular structures that surround the comb or join parts of it together; heavy use of salivary secretion; and fusion of combs initiated separately (Wenzel 1991). The variation found in *Mischocyttarus* and in *Ropalidia* begs a question: If the design of *Polistes* is so old, and if the great diversity of nests of recent Polistinae evolved from something like *Polistes*, and if other primitively social taxa have greater diversity of forms despite more limited range, then why is *Polistes* so uniform?

One possibility is that the modern species of *Polistes* are only recently diverged from some common plesiomorphic ancestor of the genus rather than that the species represent ancient lineages. Unfortunately, such an explanation runs into trouble when confronting the cosmopolitan distribution of the genus in the face of the late Cretaceous biogeography of the subfamily (Carpenter this volume). The '*Polistes* species are recent' explanation would require the species to have exploded out across the globe long after vicariance had determined the biogeography of all more apical taxa, *Polistes* going boldly where no other single genus had gone before, producing a quarter of the described species, carrying along mostly plesiomorphies. That seems unlikely to me. Rather, it appears that we must accept that *Polistes* architecture simply hasn't evolved as much as its relatives, and demonstrates impressive stasis sometimes seen in other groups, such as social bees of *Trigona* (Michener and Grimaldi 1988).

Table 3.1 Eighteen exceptional species of *Polistes*, based on elements regularly included in nest architecture.

Architectural Modification	Inferred function	Species	Reference
Multiple independent combs	Parasite control	*P. canadensis*	Jeanne 1979*a*
Long, empty cells around periphery	Protection from rain	*P. biglumis bimaculatus*	Lorenzi and Turillazzi 1986
	Heat conservation	*P. riparius* (cited as *P. biglumis*)	Yamane and Kawamichi 1975
Silk cocoons green or yellow	Crypsis	*P. saussurei*	Wenzel, unpubl.
		P. sagittarius	Wenzel, unpubl.
		P. takasagonus	Wenzel, unpubl.
		P. stenopus	Vecht 1972
Pulp made of short chips	?	*P. stenopus*	Vecht 1972
Comb as long string of cells	Economy	*P. goeldii*	West-Eberhard 1969*b*
	Crypsis	*P. stenopus*	Vecht 1972
Comb bowl- or ball-shaped	?	*P. annularis*	Wenzel 1989
	Consequence of divergent cell walls	*P. takasagonus*	Yamane and Okazawa 1977
		P. snelleni	``
		P. shirakii	``
		P. mandarinus	``
Conical comb	?	*P. carnifex*	Wenzel, unpubl.
		P. tepidus	Yamane and Okazawa 1977
		P. tenebricosus	``
Eggs laid on silk caps producing tandem brood	Efficient cell use	*P. tepidus*	Yamane and Okazawa 1977
		P. tenebricosus	``
		P. jadwigae	``
		P. rothneyi	``
		P. olivaceus	Wenzel 1991

Colony size as a restrictive influence

Mature colony size is one factor that may help explain why *Polistes* nests are more uniform than those of some other primitively eusocial genera. *Polistes* colonies are larger than most representatives of the other primitively social genera. Including both active nests and museum specimens, I am familiar with 18 of 50 Stenogastrinae species, 63 of 200 *Polistes*, 79 of 200 *Mischocyttarus*, 15 of 80

Belonogaster, all 3 *Parapolybia*, and 62 of 150 *Ropalidia* species (estimates of total species per genus are approximate). Of the largest combs I know from a primitively social wasp, most are from various species of *Polistes*. One from Guyana in the Muséum National d'Histoire Naturelle, Paris, has an estimated 2667 cells, some of which produced three pupae; I have seen several *Polistes annularis* nests with over 1000 cells, and one from Rosser, Texas, USA, had 1572 cells (Pierce 1909, p. 17); and *P. olivaceus* nests can grow to over 1400 cells (Alam 1959). Certainly other examples will be found where *Polistes* is common and the growing season is long. By contrast, the highly variable nests of *Mischocyttarus* usually number less than 100 cells. Despite collectors' biases toward bringing home large, impressive specimens for museums, nests of only a few primitively social species of *Ropalidia* commonly attain sizes as large or larger than those of many *Polistes*. *Parapolybia varia* achieves nest sizes even larger than *Polistes* (Yamane 1984), but these nests are composed of many combs supported separately.

Large colony size may require uniformity because the nest is built by a committee of wasps that must have some common goal (below). Whereas Hansell (1987*a*) has argued that the size of some social wasp societies is limited by the nature of the nest, I propose that the converse relation may also hold: that the nature of the nest may be determined by the size of the society. Individual idiosyncracies among a handful of builders may cause no problem in a comb of 100 cells, but any negative aspects of uneven construction might be exaggerated as the comb grows. If there were no restrictions on innovation in design, the integrity of the structure would be sacrificed, and the distribution of resources would be less than ideal. Attaining large colony size in *Polistes* may rely upon more strict adherence to a given groundplan to ensure that critical mechanical requirements of the simple design are not compromised.

My proposal assumes that the nests that are not of a pure form (but rather represent a combination of grossly different building programs) will be suppressed by natural selection. Hansell (this volume) makes a similar argument, suggesting that such variation may lead to speciation because of disruptive selection favouring alternative forms and suppressing the intermediate. In contrast, my scheme suppresses the innovation when it first appears and does not allow it to develop to the point where prezygotic isolating mechanisms could develop. These different views are not mutually exclusive, but it seems apparent that the 200 or so species of *Polistes* do not owe their origins to architectural variation.

Two problems with my proposal are clear. First, and most critical, if innovation is suppressed because of large colony size, then why are species of *Polistes* that typically have small colony size no more innovative than the others? I have no reply to this objection, and I hope that future study of *Polistes* will find either innovation that was overlooked or a more satisfactory explanation for uniformity. Second, why is the greatest diversity of forms found among the swarming genera that also include largest colony sizes? This objection is less damaging because so many aspects of the biology of swarm-founding species of *Ropalidia*, *Polybioides* or Neotropical genera are not strictly comparable to *Polistes* (caste dimorphism (Jeanne 1975, Yamane et al. 1983*b*), organization of work (Jeanne 1986*a*), control of individual reproduction (West-Eberhard 1978*b*), nest cycles, communication, and other features (Jeanne 1991*b*)). It is hardly a surprise that a generalization aimed at *Polistes* does not apply to the more highly social genera. However, it is my impression that even among these genera the species that have very large nests generally build more uniform combs, or show more uniform methods of expansion, than their close relatives that live in smaller colonies.

Several lines of evidence are needed to demonstrate that the price *Polistes* pays for the ability to build large, long-lived, single combs is a reduced freedom to experiment with the design of the comb. Do individual wasps vary in their building programs? Can they recognize departures from an ideal form, and if so do they try to correct errors? Do large combs of a given species vary proportionally more than small combs, and if so does the variation relate to fitness? Is there evidence to suggest that individual behaviour patterns are organized at the level of the colony? These questions are the subject of the next section, and most of them will be answered 'Yes'.

Regulation of construction behaviour in *Polistes*

Site selection: learning and tradition

The first step in building a nest is site selection. *Polistes* choose substrates that include twigs, leaves, dense shrubs and grass, hollow trees and elevated natural cavities, rodent burrows (Hungerford and Williams 1912), and man-made structures. There certainly must be innate components to site selection, but some species display apparent learning also. *Polistes annularis* can build up dense populations, and many researchers have reported highly stereotypical site choices. Interestingly, different populations chose different sites and they share little with other populations other than proximity to permanent open water. Rau (1929*b*) recorded a population nesting almost exclusively in the tops of tall trees in the Mississippi floodplain near St. Louis, Missouri. In contrast, Balduf's (1961) Illinois population was in low vegetation about a metre above the water in the Salt Fork river floodplain. Strassmann (1991) found many hundreds of nests near Austin, Texas, mostly on the west-facing wall of a cliff beside Lake Travis. In the Sabine River region of East Texas, I found *P. annularis* mostly under bridges spanning creeks (Wenzel unpubl.). Hermann's and Dirk's (1975) population in Athens, Georgia, was mostly on buildings, whereas the population of over 500 nests I found near Lawrence, Kansas (Wenzel 1989), was exclusively in prairie hedgerows, generally facing east, and never on farm buildings. Each of these populations chose a specific situation and ignored others that were chosen by other populations. Fidelity to natal sites is well known (Rau 1929*b*, West-Eberhard 1969*b*, Strassmann 1983, Hirose and Yamasaki 1984) and one might infer that site preferences represent local adaptation, but such a proposal is tautological unless advantageous sites are defined by criteria other than the presence of wasps. Furthermore, genetical explanations would have to demonstrate that the variation between populations is heritable, and that different selective regimes have favoured different sites for each of the different populations.

Aspects of persistent variation can be quite specific between lineages within populations. Contemporary colonies of my Kansas *Polistes annularis* nesting only metres apart varied with respect to tree species and growth form, nest height above ground, and placement in the crown (twigs versus limbs), but daughters remained faithful to these characteristics of their mothers' choice when they founded their own nests in other trees the next year (Wenzel 1988). A serendipitous observation strongly suggests that site preference is learned when daughters emerge on their mother's nest, rather than controlled genetically. In the course of other experiments (Wenzel 1989), I transferred 18 natural and ordinary nests to the inside of inverted laundry baskets on posts two metres high and within a few metres of the original nesting site. The next year, females that emerged from two experimental nests chose to initiate nests inside the laundry baskets despite the availability of the same natural sites their mothers had originally chosen. It appears that if there is any preference for 'advantageous' sites, it must be learned according to the rule 'What was good enough for mother is good enough for me' rather than by strict genetic control of preference. In any case, not all choices are advantageous, such as the Kansas population's preference for twigs and small branches. Although one twig 3 mm in diameter was adequate for a nest of 1369 cells, more often such supports broke under the weight of large, successful nests. Despite the fact that a stouter support would be better, numerous reproductive offspring would have already learned to look for thin twigs by the time their mothers' choice caused the termination of a still thriving colony. Differential success of nests for any reason may lead to a local majority favouring a specific nesting situation regardless of whether the salient features of that situation have anything to do with success. Thus, differential site preference may represent tradition more than it does adaptation. There can be considerable variability in site selection, and some poorly chosen individual sites for *Polistes* include the dorsum of a live beetle (Verstraeten 1976) or that of a subordinate wasp (Ishay and Perna 1979, Karsai and Wenzel 1995).

Individual construction programs

Empirical studies

After choosing a site, the wasps begin construction with the pedicel and the first few cells. West-Eberhard (1969*b*) described the motions of the head, body, and antennae of *Polistes* when building. The pulp is collected by walking backward and gathering a ball of fibre between the front legs with the mandibles. The wasp arrives at the nest and makes a rapid inspection of it. The pulp is usually added either to a cell wall that appears shorter than its neighbours, or to the nest margin to make a new cell. In the first case, the antennae sweep across the internal walls of the two cells as the wasp, walking backwards again, lays down the entire pulp mass on the margin of the wall shared by the two cells. The mass is then masticated to make it thin and more uniform, and it becomes a straight line. The wasp seems to maintain a *rule of symmetry*, making sure that the new addition intersects its neighbours at 120 degrees and is equidistant from parallel walls. If the pulp is added to the nest margin, it is placed in the furrow between two cells and is shaped into a small arch as one antenna feels the neighbouring walls and the other waves in the air or across the back of the nest. This arch is later enlarged to become a cup and eventually a cell.

Downing and Jeanne (1986, 1987, 1988, 1990) have provided the most detailed and cohesive view of individual construction behaviour available to date for *Polistes*. This body of work addresses issues of flexibility in the behavioural program of construction, sign stimuli for controlling the program, and similarities and differences between several species regarding the program and the resultant gross architecture. Interspecific comparisons include demonstrations of the potential utility of behavioural-architectural variation in taxonomy, and also adaptive explanations for traits that seem to vary according to latitude.

Downing and Jeanne (1986) found little intraspecific variation in the length of the pedicel or the angle that it forms with the supporting substrate. In a tropical (*P. instabilis*) and a temperate (*P. fuscatus*) New World species, the linear (stepwise and deterministic) sequence of acts was highly rigid until the completion of the first brood cell (Downing and Jeanne 1987), but there was substantial intraspecific variation in the angle of the substrate selected for nesting, and the angle of the first cell with respect to the pedicel or to vertical. After the first cell is built there are many places to add pulp and the program is increasingly flexible. Such behavioural flexibility is believed to be important to account for changes in the comb that inevitably develop as the nest ages, and ultimately to be critical to success of the colony (Downing and Jeanne 1986). Pedicels built by several temperate species tend to be wider, reinforced with pulp, and are sometimes multiple, whereas those of several tropical species are narrow, almost exclusively made of secretion, and are single (Downing and Jeanne 1986). This difference is believed to reflect the great importance of defending the nest against ants (Jeanne 1975) and the known decrease in ant predation pressure as latitude increases (Jeanne 1979*b*).

Downing and Jeanne (1988) explored the degree to which stigmergy may satisfy requirements of *Polistes'* building behaviour, manipulating nests to change the structures the wasp examines prior to the next phase of construction. Stigmergy theory (Grassé 1959) holds that architectural structures are both products and cues, and that builders may change behavioural patterns after they sense the results of their own labour. Originally proposed to explain apparent organization in termite construction behaviour, this idea is useful for devising ways that a complex sequence of behaviours by one or several animals can be stimulated and controlled partly externally without the complete specification of a 'blueprint' in the mind of each animal. Downing and Jeanne (1988) presented a linear model of five major steps in construction up to the point where the nest has two cells. Certain aspects of construction follow the predictions of stigmergy theory. For example, when the pedicel is finished, a flat sheet of paper is added distally to serve as the mutual wall between the first two cells. Downing and Jeanne (1988) found that if the sheet was moved so that it was not centred on the pedicel, the wasp would not correct this, but rather simply proceeded to build the first cell. Some wasps made an effort to centre the cell on the pedicel anyway, demonstrating that the flat sheet is not used alone to determine the placement of the first cell. Nonetheless, the results show that the stimulus provided by the flat sheet operates even when a blueprint for the nest would not be satisfied. Certain cues are evaluated repeatedly and simultaneously, and in some cases improper form is recognized and corrected, indicating more complex system of cues and assessments than a simple version of stigmergy would imply.

Expanding this work, Downing and Jeanne (1990) manipulated *P. fuscatus* nests to find what rules govern a more complicated, branching behavioural program of cell construction. They found evidence for redundant cues, and proposed that a hierarchical use of these multiple sources of information allows the wasp to accommodate for a missing cue, an abnormal situation, and a variety of conditions on the nest. Behaviours associated with repair of holes showed better performance after practice, suggesting learning. The authors stressed the same simultaneous evaluation of multiple cues as was indicated in the earlier paper, and suggested a more complex hierarchical behavioural program than can be presently modelled.

Computer simulation

Whereas Downing and Jeanne (1986, 1987, 1988, 1990) measured and manipulated nests to find rules of construction that might reveal an underlying behavioural program, an alternative approach is to use computer simulation to devise a program adequate to generate the structures observed. Rules are judged according to their performance in the simulation. This approach has had good success in producing simple models that reproduce more complex group behaviours, such as swarm raids of army ants (Deneubourg *et al.* 1989) and brood pattern formation in the combs of honey bees (Camazine 1991). Recently, this approach has been applied to paper-wasp architecture as a first attempt to define how the range of possible structures is affected by incorporating stigmergic responses into the behavioural script (Deneubourg *et al.* 1992). More detailed simulations can produce the simple architecture of a given species (Karsai and Pénzes 1993).

The simulation most relevant to *Polistes* is that by Karsai and Pénzes (1993). They wrote a program for a hierarchical process modelling the initiation of a new cell and extention of a wall of an old one. The model had 11 separate parametres, some of which had several states (e.g. the likelihood of depositing pulp on a wall varied discretely depending on how many of the neighbouring walls were higher than the wall in question). Some values were arbitrary and others derived empirically. All information was strictly local; the model did not include any way for the simulated wasp to know where it was or what the design of the nest was beyond adjacent cell walls. The model produced life-like nests including such features as radial symmetry and an appropriate distribution of pulp (4.3 times as much for extending walls as for initiating cells, roughly that measured in Downing and Jeanne (1987)). These results confirm an earlier (Wenzel 1989, p. 684) assertion that initiating cells randomly about the margin of the comb (without reference to previous efforts) would result in radial symmetry (circular uniformity) and produce the ideal comb *sensu* Saussure (1853–1858) or Karsai and Pénzes (1993).

Higher-level rules

The empirical and model-based studies summarized above have helped make clear the degree to which construction behaviour in *Polistes* can be governed by a few, simple rules operating at the level of the individual. However, reductionist demon-strations of the adequacy of simple rules does not establish that complex or higher-level rules do not exist. As stated in the Introduction, few studies have tried to reveal higher-level phenomena. Indeed, Downing and Jeanne's (1988, 1990) em-pirical studies (above) were designed to test simple rules, and the results suggested that complex and hierarchical rules do apply. This section will examine two funda-mental aspects of *Polistes* nests, eccentricity and size, to demonstrate that there are indeed such higher-level rules.

Eccentricity

Downing and Jeanne (1987) observed that both *P. instabilis* and *P. fuscatus* prefer to add cells to the lower side of an inclined comb. They felt that the eccentricity of *P. instabilis* was therefore a simple result of the fact that the comb was inclined. *Polistes annularis* also produces distinctive, inclined, and eccentric combs that are among the largest for the genus. Using 48 natural nests of *P. annularis*, I found that the relationship between inclination and eccentricity is not adequate to account for the fact that virtually all new nests are highly eccentric including those built at a very shallow angle (Wenzel 1989). Furthermore, the stereotypical eccentric con-struction was limited to the first few weeks of building, and neither the same foundresses building later nor new workers building for the first time showed the same eccentricity nor any relationship between inclination and eccentricity (Wenzel 1989). Downing and Jeanne (1986, fig. 1) may have seen the distinctly different construction by workers in their 'secondary angle of comb back' (Wenzel 1989, p. 692), but they did not pursue the origin of it. Such changes during ontogeny suggest that uniformitarian assumptions are false. If the caste of the builder and the age of the nest affect the placement of new cells, then the program necessary to produce natural nests must be more complicated than can be revealed by studying the actions of an individual builder as representative of the whole colony. The natu-ral program evidently includes hierarchical contingencies that distinguish the nest's age and builder's caste prior to the application of the pulp, and simple rules such as those regarding inclination *per se* are inadequate.

Non-hexagonal cells

Polistes brood cells are usually wider at their mouths than at their bases. If the cell walls are divergent rather than parallel along most of their length, then the combs become curved. Curvature of a comb causes problems for builders because a field of hexagons can not accommodate a curve in two dimensions. Species of *Belono-gaster* (Marino Piccioli and Pardi 1978) compensate for globose nest form by re-moving most of the oldest cells and allowing torsion to crush and distort these regions so that they no longer contribute to the total curvature. In contrast, *Polistes annularis* inserts non-hexagonal cells into the comb (Wenzel 1989), deleting a row by building a five-sided cell (allowing curvature), or inserting a row by building a seven-sided cell (preventing curvature). Each of these cells represent an event when many different builders override the *rule of symmetry* (Construction Programs, above) that generally produces hexagons, and that these builders 'agree' to the

Table 3.2 Placement of pentagonal and heptagonal cells in worker-built comb of *Polistes annularis.*

Cells	Above midline	Below midline	P	Frequency per 1000 worker cells
Pentagons	2	81	<0.001	6.5
Heptagons	13	3	<0.025	1.3

P values by Chi-squared, compared with equal expectation above and below the horizontal midline of the nest.
Pentagons allow curvature by deletion from the hexagonal pattern.
Heptagons counteract curvature by addition to the pattern.

exception throughout the construction of the cell in question and its neighbours. These phenomena were rare, but they were not random (Table 3.2). Deletions were highly significantly associated with the lower region of the nest (permitting the comb to curve so that cells point more directly downward), while insertions were in the upper region (preventing cells from pointing upward). It is easy to imagine adaptive explanations for controlling curvature to allow cells to point downward and not upward, but supporting data are lacking. The steps of construction differ because during deletion the focal cell is surrounded with five neighbours, whereas during insertion it is surrounded with seven. Although building a five-sided cell differs in detail from building a seven-sided cell, it is possible that a single rule governs construction in both situations because (relative to gravity) the geometric relationship between rows is the same when a downward-growing row is deleted as when an upward-growing row is inserted. Non-hexagonal elements are found only in the portion of the comb built by workers regardless of total nest size (Wenzel 1988, 1989, below), hence two *P. annularis* combs of the same size will differ in geometry if one was built mostly by foundresses and the other mostly by workers. The governing rules must be contingent upon the ontogeny of the colony (reflected by caste of builder) in contrast to the simple steps of cell construction *per se.*

Ontogenetic constraints

An interesting generalization that emerges from the studies summarized above is that construction early in the colony cycle appears to be more highly stereotyped than later construction. Downing and Jeanne (1986) found very rigid behaviour patterns during pedicel construction, but stereotypical behaviour is found more broadly than just in linear sequences of acts. Of about 30 000 cells built by *P. annularis* foundresses, only one was not hexagonal, meaning that workers are about 250 times as likely to build these elements as were their mothers (Table 3.2, above). The hallmark eccentricity of *P. annularis,* mentioned specifically in general treatments by Saussure (1853–1858, pl. 8, fig. 4) and Jeanne (1975, fig. 5f), is established in the first few weeks and is followed by random (or uniform) construction later (above, Wenzel 1989). Similarly, the evolutionarily conservative nature of 'early' structures is evident from studies comparing all genera of Vespinae and Polistinae (Wenzel 1991, 1993). For example, pedicels and envelopes built by the foundress in Vespinae differ from homologous structures built by her worker daughters later in the season (Wenzel 1991). The design of the structures built by the foundresses is generally more primitive in a phylogenetic sense, such as the more smoothly linear and laminate construction of envelopes by foundresses compared with the tightly arched, imbricate form of worker envelope construction (but see an important exception in Jeanne 1977*a*). In Polistinae, the design of the pedicel (found to be so rigidly regulated in Downing and Jeanne's experiments) distinguishes the difference between major groups of genera. In the primitive state (shared with Vespinae), the pedicel is flattened distally to provide the common wall between the first two cells, as in *Polistes* and *Mischocyttarus*. A derived condition is found in Ropalidiini where the pedicel flares into a cone that becomes the base of the first cell. The genera of Epiponini are differently derived and more variable (Wenzel 1991). In Epiponini, comb and envelope structures that are built to completion within a day or so of nest foundation (as in *Polybia*, *Brachygastra*, *Epipona*, and closely related genera) are generally less variable from nest to nest than are the homologous structures built over a more protracted period of time by other genera (such as the close relatives of *Synoeca* or *Parachartergus*). Furthermore, changes in architectural characters that are early in a sequence tend to map deeper in the phylogeny of the subfamily than do later characters (Wenzel 1993). Such differences between homologous characters that appear early versus late in the colony cycle are outside the scope of simple models that do not account for higher-level phenomena. Incorporating different probabilities of certain events based on the age of a colony constitutes description of the higher-level order (an ontogeny), not elimination of it.

Expected size

One of the central paradoxes of insect sociobiology is that there is a general inverse relationship between group size and per capita productivity (Michener 1964). Michener's paradox is peculiar because it seems unlikely that evolution would favour formation of larger groups if per capita productivity were higher in smaller groups. Michener himself thought that perhaps the relationship might be an artifact because many failed small colonies disappear before they are measured. West-Eberhard (1975) proposed that perhaps individuals with low productivities are more likely to join others to become the supernumerary females of large groups, thereby decreasing the per capita productivity for the group. Wenzel and Pickering (1991) argued that the relationship is precisely what is expected for progressive provisioning of brood that requires food that is not stored. If each forager operates independently, the expected variation of daily food intake (averaged over the colony as a whole) will be much greater for small groups than for large groups because of the central limit theorem (Wenzel and Pickering 1991). As a result, an optimistic production schedule (one in which there is enough brood to consume all food on good days) will lead to higher per capita productivity in smaller groups. Larger groups will not appear to achieve their full potential when compared with smaller groups because the daily variance is lower for larger groups. In times of hardship, however, small groups will suffer most because of their higher variance.

Brood abortion rates and development times in colonies of different sizes in two species of *Polistes* seem to match expectations of the central limit theorem (Wenzel and Pickering 1991), offering corroborative evidence.

The application of the central limit theorem (above) does not by itself indicate any higher-level rule because it is an inevitable result of the sum of independent efforts by workers that may be assumed to be equal. However, compelling evidence of a higher rule comes from a simple experiment on construction rate, and hence nest size. Nest size (cell number) and nest age would seem to be one of the simplest relationships, likely explainable by the daily iteration of the steps 'initiate cell' and 'lay egg'. It has long been known that prior to the emergence of workers, the founding female initiates cells in order to lay her eggs. Unfortunately, although egg-laying has been long thought to be associated with cell initiation, the relationship of these two activities is not entirely clear. Deleurance (1950) reported that empty cells stimulate oviposition in subordinates as well as queens if the cells are numerous enough. Morimoto (1954c) claimed that the empty cells did not stimulate oviposition, but rather that newly hatched eggs retarded it. Pratte (1990b) said his wasps did not react to hatching larvae, but rather to large larvae, and that activity of the adults was stimulated by the needs of the brood (greatest for large larvae). These studies are contradictory, but one experimental manipulation seems relevant. In the absence of any higher function, removal of both new eggs (which may take 10 days to hatch) and the cells they occupy should not affect the construction rate on the nest. The egglayer still needs to initiate cells only as fast as she needs to lay eggs (presumably the same rate as before). Such a removal of both eggs and marginal cells would inflict a traumatic loss upon nests, and the nest would remain forever somewhat smaller than unmanipulated nests built by foundress cohorts of the same size. However, this proves not to be the case (below).

To challenge the hypothesis of oviposition rate controlling construction, I first measured the growth rate of pre-emergence nests in nature for 33 foundress groups of different size in *Polistes annularis* (Wenzel 1988). A strong linear relationship between foundress number and growth rate was similar to that found in other studies (Jeanne 1972; West-Eberhard 1969b), illustrating the conservative nature of the inherent growth rate. Ten additional natural colonies were then chosen for experimental manipulation, picking colonies that did not differ from those that provide the baseline for the study (Fig. 3.1). I excised the margin of the comb, eliminating new cells and leaving the nest with the same shape and cell number as on a previous day. Between one third and one half of the cells were eliminated, these containing primarily eggs or small larvae. The construction rate on these nests doubled immediately, far surpassing the rate of natural colonies of the same size (Fig. 3.1). Even more surprising is the fact that on the last day prior to the emergence of the first worker, all but one of the ten experimental colonies had not only replaced the lost cells, but even achieved the same nest size as the natural colonies that suffered no traumatic loss of cells (Fig. 3.2).

If the rate of cell initiation is controlled only by the inherent rate of oogenesis or a simple behavioural program, no change should have occurred after manipulation. Certainly, there would be no mechanism to allow the experimental nests to

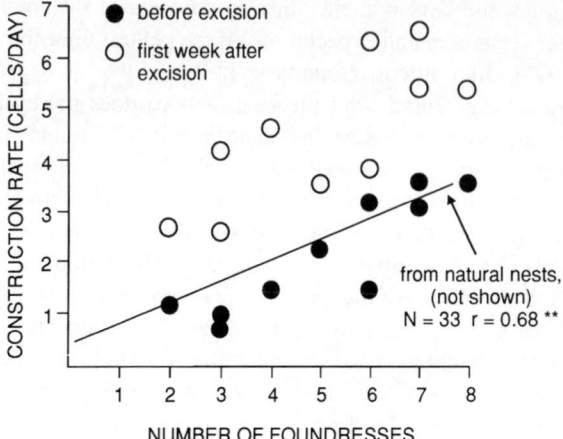

Fig. 3.1 Construction rate on experimental nests (N=10) before (black dots) and after (white dots) excision of comb margin, as compared with 33 unmanipulated nests (regression line). Colonies roughly doubled construction rate after excision, all exceeding the expectation derived from unmanipulated nests. Using the expectation that the rate for the 10 experimental nests would not differ from the regression (half would be above and half below) and comparing to the observation that all 10 are above, Chi-squared has a value of 10.0, (df=1, P<0.002).

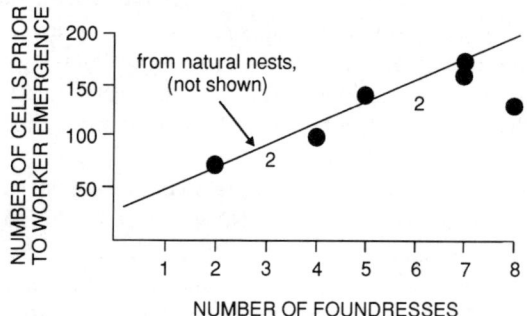

Fig. 3.2 Size of experimental nests of Fig. 3.1 just prior to worker emergence (black dots), as compared with 33 unmanipulated nests of Fig 3.1 (regression line). Numeral 2 signifies two nests have same value. Experimental colonies have overcome the deficit of cells caused by excision of comb and present an excellent fit to the natural expectation.

overcome their deficit and achieve the size of natural nests. Furthermore, release from care of the eliminated brood cannot explain the increased productivity. First, neither the eliminated eggs nor small larvae require much care. Second, the causal relationship, if any, is generally inferred to be that queen oogenesis affects workers, not that worker burdens affect queen oogenesis (Pratte 1990b). Third, even if inhibition through hatching larvae is real (Morimoto 1954c), it seems a little far-fetched to think that the ratio of eliminated brood on nine of ten nests was precisely right to allow colonies of different sizes to achieve natural size just before appearance of workers. It appears that the wasps themselves have some 'idea' about how big the nest should be for a founding cohort of given size, and when the nest was reduced they increased their effort to achieve that size. The optimal nest size must be somewhat less than the greatest possible size, otherwise groups that suffered loss would not be able to catch up with those that did not lose cells. It seems logical that the greater part of effort in rearing brood is in provisioning larvae rather than initiating cells, and that ultimate cell number is determined more by the group's ability to feed the larvae than by its ability to build cells. Evidently, the natural nests did not build as many cells as they could have.

The results of this experiment show that the effects of the central limit theorem proposed by Wenzel and Pickering (1991) do not go far enough to characterize the size of *Polistes* nests in the foundress period. Wenzel and Pickering relied on the assumption that brood production would be on an optimistic schedule, but the reconstruction of the cut nests (above) to arrive at natural size shows that the natural nests are somewhat smaller than their colonies could have built. It appears that the wasps themselves account for expected productivity by building a nest of given size, and that the size of the nest (and therefore the number of brood) is controlled by a property more complex than a wasp's simple repetition of 'initiate cell' and 'lay egg'. The size of the nest is itself a trait that has evolved, and therefore may be subject to some form of homeostasis.

Conclusions

Students of social wasps have kept a narrow focus to their work compared with the traditions of myrmecology and melittology. Intense study of relatedness and reproduction has relegated to marginal status such issues as the organization of work, mechanisms of kin recognition, and role of learning. Indeed, topics of reproduction have overshadowed virtually all mechanisms by which a mass of animals of limited individual capacity displays what appears to be an integrated, purposeful intelligence. This is ironic because the enormous success of social insects is probably more a result of the mechanisms of behavioural integration than of the fact that few females reproduce directly. The details of nest construction provide the greatest single resource for studying the rules that transform a number of wasps into a cooperative colony. Whereas some aspects of colony life are controlled directly by the queen and her interest, construction ultimately becomes a task performed by many individuals without reference to a commanding authority. The way in which such a committee works to produce the nest represents the distinction between an anarchy of solitary individuals and a colony of social individuals.

This discussion of *Polistes* architecture has focussed on two main issues: why are *Polistes* nests not more variable, and how is construction behaviour controlled? I propose that architectural uniformity (relative to comparable genera) may be a result of large comb size. To support this idea, I offer the argument that mechanical demands and those of the organization of work should constrain innovation if single combs are to achieve large size. I offer the opinion that most of the variation in other primitively social genera is in species that typically have small nest sizes, and that *Polistes* typically have larger colonies. Of course, this line of reasoning does not explain why *Polistes* with small colony size apparently do not vary much.

Mechanisms of control of behavioural programs range from rigidly innate patterns to learned cues. Different species will probably be found to vary in mechanism of control itself, but we do not have enough information to know how general our conclusions are. Work by various researchers shows that individuals do vary in their selection of sites, and that conspecific foundresses vary in their behaviour patterns after the first cell is built. Individuals do recognize improper form and may try to correct it, so simple stepwise models are not sufficient to explain control of construction behaviour. There is a positive maintenance of 'correct' form through a complex hierarchy of contingencies and multiple cues.

Stereotypical patterns appear at the level of the individual, colony and species. Early building efforts are very rigidly controlled, as shown by studies of pedicel construction or the determinants of eccentric comb form. Later efforts vary more, including modified elements (such as nonhexagonal cells) virtually absent from earlier efforts. The differential variation between structures built early versus those built late in the colony cycle appears to extend to phylogenetic variation as well. If true, this principle suggests that behavioural novelty arises most often in later steps of a sequence or in features typical of mature colonies and may represent an important fundamental pattern of behavioural evolution.

Both theoretical and empirical studies show that nest size itself is not just the result of repetitious cycling through a simple stepwise program. Larger colonies have larger nests, but the relationship is not due to simple additivity of individual efforts. There is a kind of homeostasis in the architecture, one that operates at a level broader than the fundamental behaviours of adding pulp to the nest. Wasps appear to correct for a nest that is too small and refrain from building one that is too large.

Certain aspects of construction behaviour in *Polistes* are explainable as the output of identical, independent actors, each under rigid control via local cues. Other aspects indicate organization of work at a level higher than the behavioural output of a stereotyped individual. The lesson in these results is that we must be willing to incorporate studies of colony-level phenomena if we are to achieve complete understanding of wasp biology. As we move ahead with the study of individuals, keeping true to our tradition, we should not overlook the answers and inspiration available from examinations of the integrated product of all individuals, the phenomena of the colony as a whole.

Acknowledgements

I thank the Centro Fiorentino di Storia e Filosofia della Scienza for the invitation to join this symposium. J. M. Carpenter, I. Karsai, S. Turillazzi, and M. J. West-Eberhard improved the manuscript, which was prepared in part under a Kalbfleisch Fellowship, American Museum of Natural History.

4

Ecological factors influencing the colony cycle of *Polistes* wasps

Sôichi Yamane

The genus *Polistes* is a large cosmopolitan group of the subfamily Polistinae comprising over 200 known species (Reeve 1991; Carpenter this volume). The species have adapted to a wide range of environments from tropical to cold temperate climates. Wasps of this genus build uncovered, single-combed nests which are suspended by a narrow petiole. The nests are small with generally fewer than 250 cells (Reeve 1991). The social biology of this group is one of the major lines of sociobiological research in insects. Various aspects of their social phenomena, as well as systematics, geographical distribution and colony cycle, have been comprehensively reviewed and discussed by Reeve (1991).

Polistes colony cycles share several common features:

1. Colonies are founded independently by a single foundress or an association of several foundresses.
2. Colony cycles are, as a rule, annual, and termination of colonial activity is usually associated with the loss of the foundress.
3. Males and gynes are produced in a discrete mass or masses toward the end of the nesting period after the all workers have emerged (West-Eberhard 1969*b*; Akre 1982; Reeve 1991).

In the temperate and probably also the subtropical zones, climatic seasonality demands an annual colony cycle unless there is to be bivoltinism. We know, however, that this cannot be true in the tropics, because colonies are successfully initiated at any time of the year. So the question arises why equatorial colony cycles do not vary more in length, and why they last about the same length of time as elsewhere. Although colony cycles seem primarily to be regulated by intrinsic and climatic factors, they are also affected to varying degrees by ecological factors peculiar to the areas the wasps inhabit. *Polistes* colonies often suffer serious damage from brood predation by various kinds of insects and vertebrate animals, from infestations by parasitoid wasps and flies, and scavenging moths that feed upon the meconia. Some of these predators and parasitoids are fatal to colonies, whereas others only hinder their development or hasten their disintegration (Jeanne 1975; Miyano 1980; Strassmann 1981*c*; Matsuura 1984; Strassmann *et al.* 1988). The impact of these animals on wasp colonies, however, has not yet been thoroughly assessed.

This paper compares the colony cycles of *Polistes* wasps in major climatic zones, and summarizes environmental and ecological factors which probably influence their nesting periods (= active periods in the colony cycle). It also discusses how the nesting period of *Polistes* is affected by ecological factors, and how they have responded to these.

Comparison of colony cycles in major climatic zones

Distribution of the genus Polistes

Polistes wasps are distributed over all the continents except Antarctica (see Reeve 1991, fig. 1). In the northern hemisphere, apparently thanks to the North Atlantic Ocean Current, the northernmost border of their distribution extends to 60° N in the Scandinavian Peninsula in Europe, while it lies at about 50° N in the Far East, and at 52° N in British Columbia, in North America. In the southern hemisphere, *Polistes* is distributed over three continents (to about 40° S), except for the southernmost area of South America.

At these distribution borders the climate is cold temperate—in Oslo (about 60° N) the annual mean air temperature (henceforth abbreviated to AMT) is 3.7 °C (1951–1980) and there are five winter months during which monthly mean air temperatures are below zero. Vespids in these climatic areas are dominated by *Vespula* and *Dolichovespula* of the subfamily Vespinae, which can reach the polar zones of the Eurasian and the North American Continents (Matsuura and Yamane 1990).

Colony cycles in major climatic zones

Reeve (1991) divided the *Polistes* colony cycle into the following four phases:

(1) the founding phase (= pre-emergence phase), from the founding of the colony to the emergence of the first worker;
(2) the worker phase, from the emergence of the workers until that of the first reproductive form(s);
(3) the reproductive phase, from the appearance of males and/or gynes to disintegration of the colony;
(4) the intermediate phase, between colony disintegration and founding in the next season or cycle.

A combination of the worker and reproductive phases is called the post-emergence phase (West-Eberhard 1969*b*).

Cool temperate zones

Yamane (1969) described the colony cycle of two species, *P. riparius* (cited as *P. biglumis*) and *P. snelleni* in central Hokkaido (42° 15' N, AMT 7.3 °C), northern Japan (see Fig. 4.1 for reference):

1. Colonies of both species are founded by a lone foundress (rate of polygynous foundation was 1.3% for *P. riparius*: Makino and Aoki 1982) in early May. The

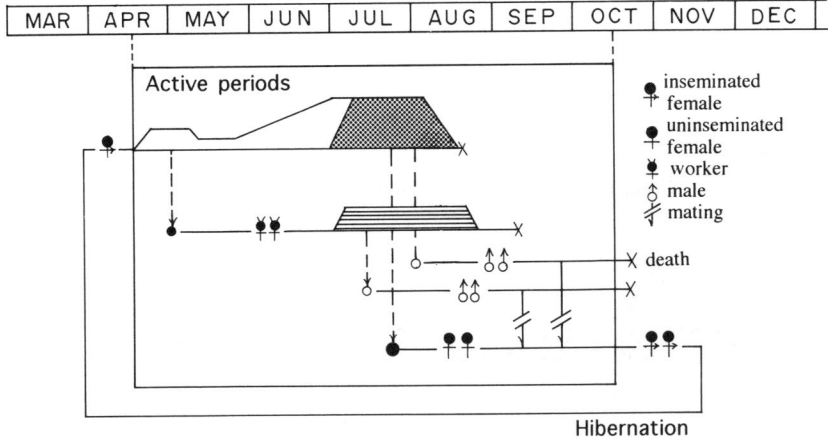

| MAR | APR | MAY | JUN | JUL | AUG | SEP | OCT | NOV | DEC |

Fig. 4.1 Schematic representation of the colony cycle of *Polistes chinensis* in central Japan, a warm temperate district. Colony is initiated by an inseminated foundress in mid-April. She lays worker-producing eggs until the end of June, and then lays eggs producing gynes and a few males. Workers emerge in mid-June and sexuals from early August after all workers have emerged. In this population, a majority of males are produced by some egg-laying workers in a queenright condition. The nest is vacated in mid-October. Changes of oviposition activity are roughly shown for the foundress and workers together with kinds of eggs (open area: worker eggs; dotted area: gynes and males; striped area: males). (Based on Miyano 1980).

foundress is often replaced by another conspecific foundress which has lost its own nest (Makino 1985*a*)

2. Five to 15 (usually less than 10) workers emerge from mid- to late July.
3. Ten to 20 males and 30–50 gynes emerge from early to late August. Both sexes are produced together in *P. riparius*, but males always emerge in a discrete mass prior to females in *P. snelleni*.
4. The colonies disintegrate in early to mid-September, indicating that the nesting period lasts about four months, of which two are occupied by the founding phase. The intermediate phase lasts almost eight months. At Oketo, a sub-montane (43° 40' N, 600 m a.s.l., AMT 2.7–3.8 °C) district in eastern Hokkaido, the nesting period in *P. riparius* is further reduced to 3.5 months. But colonies never lose the worker caste (Yamane and Kawamichi 1975). Similar patterns can also be seen in *P. chinensis antennalis* (Fig. 4.2A) and *P. rothneyi iwatai* along their distribution border on Okushiri Island (42° 10' N, AMT 8.0 °C) of Hokkaido (Yamane 1972).

P. biglumis bimaculatus lives near the tree line in the Alps (45° 00' N, 1600–2350 m a.s.l.), the highest altitude known for *Polistes* distribution in the northern hemisphere, where the climate is typically alpine (Lorenzi and Turillazzi 1986). Colony activity lasts 3.5–4 months from late May to late September, and only a small number (mean 17.5 inds., n = 60) of adults are produced in each colony. Since

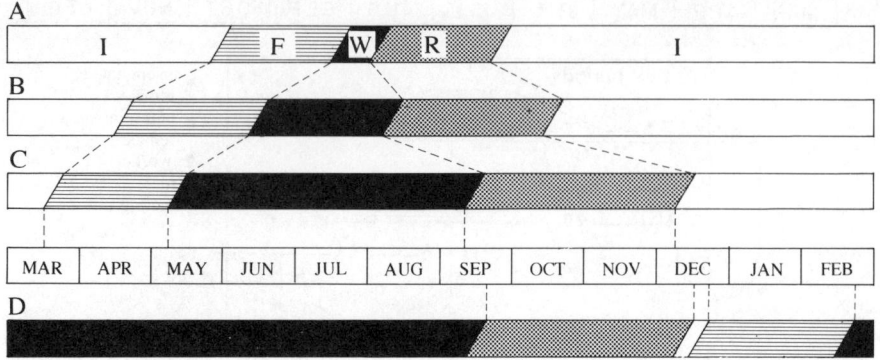

Fig. 4.2 Comparison among colony cycles of cold and warm temperate populations of *Polistes chinensis* ((A) Hokkaido (42° N), northern Japan, and (B) Ibaraki (36° N), central Japan) and subtropic populations of *Polistes gigas* (C) and *Parapolybia varia* (D) (Kenting (22° N), southern Taiwan). F: founding phase, W: worker phase, R: reproductive phase, and I: intermediate phase. (Reconstructed from Yamane 1972; Miyano 1980; Yamane unpubl.; Yamane 1980.)

workers and future gynes are present at the beginning of the post-emergence phase, the worker phase is extremely reduced or virtually absent in this population.

   P. fuscatus in south-eastern Michigan (42° N, AMT 8–10 °C) has a colony cycle essentially similar to that observed in Hokkaido (West-Eberhard 1969*b*). Some differences are: (1) Colonies are initially founded by a lone foundress, but many of them are later joined by multiple females. (2) Duration of the nesting periods is a little longer (4.5 months) and more workers are produced. In one nest with 234 cells 32 workers were produced, though worker number in other nests may be lower, judging from the fewer constructed cells (mean 180, range 117–217, n = 4, calculated from West-Eberhard 1969*b*, fig. 16).

Warm temperate zones

The colony cycle of *Polistes* has been studied most intensively in warm temperate areas, as represented by *P. dominulus* in Europe (Pardi 1942, 1946, cited as *P. gallicus*), *P. chinensis* in Japan (Morimoto 1954*a*, *b*, *c*, 1959; Miyano 1980, 1983, 1986), *P. jadwigae* (Yoshikawa 1962*c*, 1963*a*, *b*). Here the case of *P. chinensis* is mainly referred to for comparison with that of a cool temperate population.

   The typical colony cycle of *P. chinensis* in central Japan (35° 57' N) was presented by Miyano (1980) as follows (Fig. 4.1; Fig. 4.2B; Table 4.1):

1. Colonies are founded by a lone foundress in mid-April.
2. On two intensely observed colonies, 40 and 90 workers, respectively, emerged from mid-June until early August.
3. 31 and 103 males and 100 and 104 gynes were simultaneously produced after all the workers had emerged.
4. Nests are vacated in mid- to late October, after about six months of colony

Table 4.1 Comparison of the productivity in *Polistes chinensis* colonies at two different latitudes in Japan.

Locality	AMT [1] (°C)	No. cells built	Number of adults produced			
			Workers	Males	Gynes	Total
Ibaraki (36° N; n=2)	13.2	285–575 (430)	49–91 (70)	31–133 (82)	100–104 (102)	171–327 (249)
Okushiri Isl. (42° N; n=7)[2]	8.2	31–99 (59)	1 (0.3)	0–7 (2.4)	1–17 (7.1)	8–58 (28.7)

[1] Annual mean air temperature. Climatic conditions on Okushiri Island were more rigorous than indicated by AMT, since nest sites were exposed to strong sea winds.
[2] Since numbers of adults are those present at collection, real figures, especially workers, are larger. Total number of adults is the count of cells from which adults emerged. (Based on Yamane 1972 and Miyano 1980.)

activity. The active part of the colony cycle in this species is further prolonged in northern Kyushu (33° 35' N) (Morimoto 1954a, b, c, 1959).

5. Some workers lay eggs even in queenright colonies and produce most of the males in their colonies (Miyano 1986).

Subtropical zones

The nesting period of species living in these is longer than that at higher latitudes, yet the relative asynchrony present in the tropics is lacking (Strassmann 1981a). Strassmann and co-workers have intensively studied colony cycles and related phenomena from an evolutionary point of view (Strassmann 1981a, c, 1989b; Strassmann et al. 1988; Hughes and Strassmann 1988). Strassmann et al. (1988) summarized the colony cycle of *Polistes bellicosus* in Texas, USA (c. 30° N) as follows:

1. Colonies are initiated by one to several overwintered foundresses in March.
2. The first offspring emerge in May. They are mostly workers, but a few are males which attempt mating with workers which will be queens on nests where the original queen has disappeared.
3. Replacing workers (second queens) remain on the nest and lay eggs destined to be males and gynes.
4. The males and gynes emerge from late September to October, after which the colonies disintegrate.

Taiwanese species and populations, at latitudes (22°–25° N) lower than Texas, also have synchronous, annual colony cycles. *P. gigas*, the largest *Polistes* species, has the following colony cycle at Kenting, the southernmost tip (22° 30' N; AMT, 25.2 °C) (Fig. 4.2C) (Yamane, unpublished):

1. Colonies are mostly initiated rather sporadically, during March, each by a lone foundress.

2. More than 20 workers emerge from May to September, and thus the worker phase lasts nearly five months.
3. Males and gynes appear with the cessation of worker production.
4. Male gigantism is conspicuous, and colonies produce fewer males than females (5–10 males and >20 females, sex ratio < 0.3).
5. Colonies disintegrate by mid-November, and never overwinter. The total duration of the nesting period reaches eight to nine months.

Colony cycles of other Taiwanese species vary, particularly in the length of the nesting period, although their essential life patterns do not differ. Colonies of *P. takasagonus* and *P. japonicus*, belonging to subgenus *Polistella*, last about six months, while those of *Megapolistes* (= *Gyrostoma*, Carpenter this volume) species from seven to eight months.

Tropical zones

The life history of *Polistes* in this zone is represented by West-Eberhard's (1969*b*) intensive study on *P. erythrocephalus* (cited as *P. canadensis*) at Cali, central Colombia (4° N).

1. Although there were dry and rainy seasons, colonies in all developmental stages can be found throughout the year. So, there is little synchrony of colony founding and disintegration is found.
2. Nests are usually initiated by a lone foundress and thereafter foundress associations are gradually formed.
3 Males appeared on three observed colonies at 11, 11 and 70 days, respectively, after the emergence of the first adult.
4. On one continuously observed nest, both workers and non-worker females emerged, but not in two discrete groups as in temperate and subtropical species.
5. The termination of colonies is usually associated with the disappearance of a reproductively active female (queen). The regeneration of colony activities by a substitute queen occurred in a colony from which the queen was removed, and thus colony longevity could be prolonged.
6. The number of cells counted in vacated nests was usually less than 100, suggesting that colony longevity of this species is usually five to six months. The largest inhabited nest seen in Colombia contained 750 cells, from which the growth time was roughly estimated at 470 days (Table 4.7).

Some fragmentary observations made at Padang (0° 57' S, AMT 27.0 °C, annual rainfall 4700 mm), West Sumatra reveal the following facts (Yamane *et al.* 1989) (Table 4.7):

1. Since the seasons are poorly defined by loosely dry and rainy periods, colony cycles of *P. tenebricosus hoplites* are asynchronous.
2. The largest nest of *P. tenebricosus* had 734 cells, from which colony longevity was estimated at least at 7.5–8.5 months.

3. The worker phase seemed to last more than three months, and in the case of the loss of the foundress from one observed nest in this phase, no substitution of the queen occurred.

4. Three completed nests of *P. stigma* had 44–139 (mean, 97.3) cells, suggesting that their longevity was less than six months. From one active nest, the reproductive forms seemed to emerge towards the end of active period. According to Suzuki and Ramesh (1992), colonies of *P. stigma* seem to be founded more frequently during the hot season (March to June) in South India (11–15° N).

Comparison of colony cycles under various climatic conditions

Seasonal synchrony of colony cycles

In the temperate zones where the seasons are well defined, the colony cycle is primarily subject to cyclic phenomena as in other social vespids (Spradbery 1973; Matsuura and Yamane 1990). A similar trend is seen in the subtropics, where seasonality still exists, though less conspicuously. In colder districts colony founding occurs during a short period on fine and warm days, while in warmer areas it tends to occur sporadically during a prolonged period. In general, colony disintegration also occurs at a fixed period in these climatic zones, but more sporadically than foundation, as factors other than cyclic ones, such as colony size attained, parasitism by social wasps and other insects, are involved.

In the tropics, however, high and constant air temperatures enable wasps to start new colonies at any time of the year, and the synchrony among colony cycles is lost (Richards and Richards 1951; West-Eberhard 1969b; Yamane and Okazawa 1977; Yamane *et al.* 1989, Suzuki and Ramesh 1992). Such a phenomenon is also seen in other social wasps (Richards and Richards 1951; Jeanne 1972; Yamane *et al.* 1990; Matsuura 1991). Seasonality of rainfalls often occurs also in the equatorial zones, such as in Panama, Colombia and West Sumatra, where the rainy season is not clearly separated from the dry season, but usually colony activities are not substantially influenced by this factor unless it is very rigorous (West-Eberhard 1969b; Yamane 1986).

Length of the nesting period

The nesting period may depend on species specific, climatic, and other biological factors. For healthy colonies, regardless of the species, it is naturally reduced in colder districts due to the long winter. Active periods for *P. chinensis* last six months in central Japan, whereas these fall to 3.5 months in Hokkaido (Fig. 4.2A, B). At lower latitudes, however, colony longevity is primarily subject to factors intrinsic to particular species and ecological ones such as predation. In the subtropics of Taiwan, colonies of *P. gigas* last nearly nine months, whereas those of other small- to medium-sized species (e.g. *P. japonicus* and *P. takasagonus*) last only half a year. Certain evidence of the occurrence of perennial colonies, as seen in many swarm-founding Polistinae (Richards and Richards 1951; West-Eberhard 1978b; Jeanne 1991b) has so far not been published for *Polistes*. In most species of this genus, colony longevity may primarily depend on the foundress's life span in areas with weak climatic constraints.

Colony cycle patterns

The length of some phases in colony sequences are climate-influenced. The length of the intermediate phase (hibernation and related periods), if any, is determined by the length and rigorousness of winter. These phases occupy three-quarters of the year in the highest latitudes, but are very short or absent in lower latitudes.

The active period influenced by climate may be the worker phase, which characterizes eusocial life in the Hymenoptera. Its length varies from a few weeks in high latitudes, due to an extreme reduction in worker number, to five months in lower latitudes (Fig. 4.2A–C). Complete absence of the worker caste under climatic extremity is not known, as shown by a worker specimen of *P. riparius* collected in southern Sakhalin (46° N). This suggests that *Polistes* never loses its eusociality. However, a return to solitary life in extremely cold climates is possible in the eusocial halictine bee, *Lasioglossum calceatum*, probably because of its weak caste differentiation (Sakagami and Munakata 1972). The length of the reproductive phase is also affected by climate, because it depends on the number of workers produced in the previous phase. It is also affected by social parasitism (Lorenzi *et al*. 1992). The length of the founding phase, almost identical to the egg-to-adult development time, is relatively constant (50–60 days), though prolonged in some tropical species, e.g. *P. erythrocephalus* (65 days) (West-Eberhard 1969*b*), due to conflicts among foundresses.

Social structure

The social structure of *Polistes* is relatively homogeneous across species and climatic zones if compared with other polistine genera. The most marked difference in structure arises during the pre-emergence phase. Colonies of temperate species or populations are often initiated by a lone foundress or initiated by a single foundress which is later joined by others, whereas those of the subtropics and tropics are more often initiated by multiple foundresses (Reeve 1991). The post-emergence colonies are mostly monogynous and the foundress virtually dominates all the others and usually monopolizes the oviposition. Workers are, as a rule, not inseminated and their ovaries are hardly developed in a queenright condition. However, a few workers of *P. chinensis* normally lay eggs in queenright colonies, producing almost all males (Miyano 1983).

Queen replacement

It is well known that if the foundress is lost for some reason, one or more dominant workers soon develop their ovaries and lay unfertilized eggs which give rise to males. However, workers in orphan colonies of some temperate and subtropical species, such as *P. snelleni* (Suzuki 1985), *P. jadwigae* (Miyano 1991), *P. exclamans* (Strassmann 1981*a*), and *P. bellicosus* (Strassmann *et al*. 1988) often or occasionally mate and lay female-producing eggs. Strassmann (1981*a*) stated that the insemination of workers became possible by early male production, which originated as an adaptation to the frequent loss of queens before the emergence of males and reproductive females in autumn. West-Eberhard (1969*b*) stated that queen substitution probably occurs frequently in tropical *P. erythrocephalus* when the original queen has disappeared. Although she did not determine that the substitute queens had been inseminated, it is noteworthy that males are present at any season. If the associate foundresses are still present in the colony at the time of queen loss, one of these females may replace her. West-Eberhard (1986*a*) dissected a despotic queen and 70 subordinate females in a mature colony of *P. canadensis*, and found only the queen and two subordinates, which were young and unmated, had mature oocytes. One of the two females could take over as substitute queen if the legitimate queen were lost, although she would not lay female-producing eggs. Substitute queens may be more prevalent in subtropical and tropical populations, which have longer nesting periods than temperate ones. In cold temperate areas, such substitution may be of no use, as most eggs which produce reproductives are laid before the first workers emerge (Yamane 1969; Lorenzi and Turillazzi 1986).

Colony size

As shown in *P. chinensis* (Yamane 1972; Miyano 1983), colony size in terms of numbers of cells and adults in a given species shows latitudinal differences within the temperate zone (Table 4.1). Some species or populations in warmer districts construct very large nests, e.g. 1886 cells were reported for *P. annularis* (Nelson 1968). *P. olivaceus* on Fiji Islands also constructs gigantic round combs with diametres of 30–40 cm and containing more than 1500 cells (Yamane, unpublished). However, when mean values of many species from various climatic areas are compared, no significant relationship between mature nest size and latitude is recognized (Reeve 1991). Although the mean number of constructed cells ranges from 56 to 492 cells, that of most species lies around an overall mean of about 135 cells for the genus (Reeve 1991). Reeve, therefore, considers that despite experiencing longer warm periods, subtropical and tropical colonies do not tend to have larger nests than temperate colonies.

Factors influencing the colony cycle

In this section, three external factors—climatic conditions, infestation by nest scavengers and parasitoids, and predation—are examined with respect to possible influences on the colony cycle. Figure 4.5 summarizes relations between selected natural enemies in three categories, i.e. predators, parasitoids and nest scavengers, with chronology of their attacks on *Polistes* colonies together with a rough estimate of the intensity of the damage.

Climatic conditions

Among various climatic factors, seasonal changes of air temperature may be the most effective at limiting the length and pattern of the nesting period at medium and high latitudes as well as high altitudes. The length of the nesting periods in *P. biglumis bimaculatus*, the Eurasian species reaching the highest latitude, is expected to be about three months even on lowlands, judging from a very low annual mean air temperature (3.7 °C). Colony activity of the same species at the highest altitude (45° 00' N, 1600–2350 m a.s.l.) in the Alps last 3.5–4 months (Lorenzi and

Turillazzi 1986; Lorenzi *et al.* 1992). Furthermore, short seasons in these districts may favour social parasitism, which reduces the nesting periods of the host colony cycle (Lorenzi *et al.* 1992). The importance of climatic factors may fall at lower latitudes, and in the equatorial zones they are no longer the limiting factor. Effects of rainfall on colony activity are probably not serious, judging from the fact that colonies of *Polistes*, *Ropalidia*, and other social wasps can continue a normal life in West Sumatra with a maximum monthly rainfall of 500 mm (Yamane 1986; Matsuura 1990).

Infestation by nest scavengers and brood parasitoids

As in many other social wasps (Nelson 1968; Kistner 1982; Matsuura and Yamane 1990), *Polistes* nests are scavenged and parasitized by various insects. Makino's (1985*b*) list of parasitoid insects of polistine colonies includes four lepidopterous families that scavenge meconia left in the bottom of the cells, and ten parasitoid families belonging to three orders (Hymenoptera, Neuroptera and Diptera, of which Neuroptera are not known to parasitize *Polistes*).

Lepidopterans

According to Makino (1985*b*), more than 11 moth species from four families (Pyralidae, Tineidae, Cosmopterygidae and Gelechiidae) are known to infest nests of 16 *Polistes* species (Table 4.2). They scavenge meconia excreted by post-feeding larvae and some of them subsidiarily prey upon pupae and prepupae. Although no such cases have been recorded from Australia and Africa, these moths are common at least in Australia. A few species are purely predaceous (Strassmann 1981*c*).

Colonies of *P. chinensis* very often suffer damage from larvae of *Anatrachyntis* sp., a cosmopterygid moth (Miyano 1980). The moth larvae infest nests in which the wasp larvae have just appeared. The relative number of infested nests was 44.6% in the worker phase, reaching 95% at the end of the reproductive phase. These larvae are primarily scavengers, tunnelling from cell to cell and eating meconia, but they probably also eat sealed broods in declining colonies. Larvae of other pyralid moths, *Chalcoela* spp., seem to be purely predaceous and inflict serious damage on *Polistes* in North America (Nelson 1968; Strassmann 1981*c*).

These lepidopterous larvae notably reduce colony productivity, but are not thought to be a direct cause of colony failure (Miyano 1980). They also cause serious damage to the nest itself by boring through cell walls and bottoms, thus lowering its physical durability. Extensively bored nests cannot persist for long, and such structural damage apparently hastens colony disintegration.

High rates of infestation on the population base suggest that wasps do not have sufficient countermeasures against these moths, especially in the later periods of colony life. Wasps of *Polistes exclamans* try to defend their colony with the 'parasite dance', a kind of alarm response to attacks by *Chalcoela iphitalis* moths (Strassmann 1981*c*). But these moths attack wasp nests more at night, and the darkness seems to hamper the wasps' visual search for the enemies.

Jeanne (1979*a*) in the lower Amazon region (2° 32' S), analyzed the defensive measures of *Polistes canadensis* against attacks by an unidentified tineid moth,

Table 4.2 List of parasitoid moths that scavenge meconia and/or feed upon sealed brood of wasps. After Makino (1985*b*)

Moth species	No. of *Polistes* species infested	Localities
1. Pyralidae (4 spp.)		
Chalcoela iphitalis	7	N. America
Dicymolomia pegasalis	7	N. America
Hypsopygia postflava	1	Japan
H. mauritialis	1	Japan
	(13 spp.)	
2. Tineidae (5 spp.)		
Tinea fusipunctella	1	N. America
T. latebricola	1	C. America
T. carrariella	2	N. America
Taeniodictys servicella	1	C. America
Antipolistes anthracella	2	C. America
Gen. sp.	1	S. America
	(6 spp.)	
3. Cosmopterygidae (1 sp.)		
Anatrachyntis sp.	1	Japan
	(1 sp.)	
4. Gelechiidae (1 sp.)		
Epithectis sphecophila	1	C. America
	(1 sp.)	
Total 11 moth species	16 *Polistes* species	

whose larvae burrow cells to scavenge and occasionally feed upon wasp pupae. This wasp builds nests in open sites, and post-emergence nests have 2-12 combs. Of the colonies examined 56% were infested by moths, and a smaller percentage (30%) of total combs contained moths. Jeanne hypothesized that selection by this moth caused the evolution of an unusual behaviour—the construction of multiple separate combs. In this way broods are divided into age-segregated batches and each is safe from infestation by moths until the first adult emerges from that particular comb. The removal of disused, infested combs may prevent infestation of remaining combs. It seems probable that the colonies would break up earlier than usual if such measures had not evolved.

Hymenopterans

More than 20 species belonging to six families of this order have been known to parasitize *Polistes* broods (Makino 1985*b*). These families are Eulophidae, Chalcididae, Torymidae, Ichneumonidae, Trigonalidae and Mutillidae. More than 26 species of *Polistes* are reported to be attacked by these enemies (Table 4.3).

Latibulus argiolus, an ichneumonid wasp, parasitizes larvae of *P. dominulus* (cited as *P. gallicus*) in Europe (Guiglia 1971) and *Latibulus* sp. (cited as *L. argiolus* by some authors) parasitizes *P. snelleni* (Townes *et al.* 1965) and *P. riparius*

Table 4.3 List of hymenopterous parasitoids on *Polistes* brood. After Makino (1985*b*)

Parasitoids	No. of *Polistes* species parasitized	Localities
1. Trigonalidae (2 spp.)		
Seminota marginata	1	S. America
Seminota sp. nr. *depressa*	1	S. America
	(2 spp.)	
2. Ichneumonidae (14 spp.)		
Sphecophaga vesparum burra	2	N. America
Pachysomoides fulvus	10	N. America
P. stupidus	3	N. America
P. flavescens	1	C. America
P. iheringi	2	S. America
P. vespicola	1	S. America
Toechorychus albimaculatus	1	S. America
Latibulus argiolus	2	Europe
L. siculus	2	Europe
L. sp. (cited as *L. argiolus*)	2	Japan
Arthula flavofasciata	3	Taiwan
Arthula sp.	1	Australia
Mesostenus gladiator	1	Europe
Ephialtes extensor	1	Europe
	(22 spp.)	
3. Eulophidae (6 spp.)		
Elasmus polistes	4	N. America
E. japonicus	3	Japan
E. schmitti	2	Europe
E. biroi	1	Europe
Tetrastichus midulans	2	Europe
Pediobius ropalidae	>1	Africa
	(>12 spp.)	
4. Chalcididae (1 sp.)		
Brachymeria discreta	1	C. America
	(1 sp.)	
5. Torymidae (1 sp.)		
Monodontomerus minor	1	N. America
	(1 sp.)	
6. Mutillidae (3 spp.)		
Dasymutilla castor	1	N. America
Pycnotilla barvara var. *brutia*	1	
Tropidotilla littoralis	1	
	(3 spp.)	
Total 27 parasitoid species	>26 *Polistes* species	

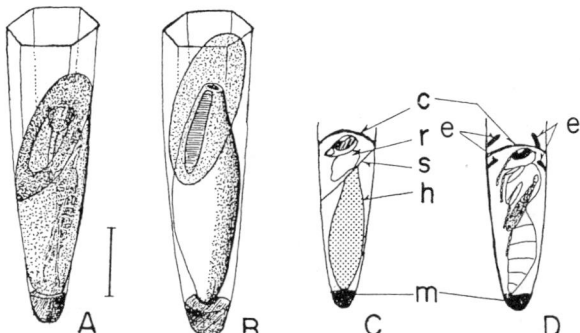

Fig. 4.3 Cells of *Polistes riparius* parasitized by *Latibulus* sp. Cutaway views of cells containing a soft cocoon in July (A) and a hard cocoon which overwinters (B) of *Latibulus* sp. C: section of a cell showing the cocoon cap (c), remnants of host (r), silken sheet (s) and a hard cocoon (h); D: eggs (e) of the parasitoid laid in a cell (redrawn from Makino 1983).

Table 4.4 Estimated reduction in the gyne productivity of *Polistes riparius* colonies caused by parasitism by *Latibulus* sp. in Sapporo, northern Japan. Cited from Makino (1983)

Year and locality	No. parasitized nests/sample size	No. of meconia	Loss in colonial productivity (%) caused by parasitisms of:	
			spring form[1]	summer form[2]
1980 Misumai	16/17 (94.1%)	7–152 (54.5±8.8)	0–9.1% (2.81±0.63)	0–30.3% (7.63±2.33)
1980 Toyotaki	23/23 (100%)	14–70 (33.7±3.3)	0–46.7% (12.09±2.49)	0–41.9% (13.83±2.83)

In the right 3 columns, the range and the mean ± S.E. among nests are given.
[1] (No. soft cocoons/No. meconia) x100.
[2] (No. hard cocoons/No. meconia) x100.

(cited as *P. biglumis*, Makino 1983) in Japan (Fig. 4.3). According to Makino (1983), parasitism of this species among colonies of *P. riparius* was very high (about 80%) in northern Japan, but it was rather sporadic and usually not fatal to the colonies. When the loss of worker broods reared by the foundress was high, however, the production of males and gynes was considerably disturbed (Table 4.4) (Makino 1983) and it hastened the disintegration of the colony.

Elasmus japonicus, an eulophid, is known to infest broods of four species in Japan (Matsuura 1977). It mostly invades late pre-emergence colonies of *P. jadwigae* in southern Japan. Although the rate of parasitism is very low (1–3%) on the population base, infested colonies suffer serious damages to their development and longevity.

Dipterans

More than six species belonging to three families (Tachinidae, Sarcophagidae and Phoridae) of this order have been known to parasitize the broods of more than seven *Polistes* species (Makino 1985*a*).

Litte (1981) observed that attacks by *Megaselia* sp., a phorid fly, cause serious damage on colonies of *Mischocyttarus labiatus*. After parasitism by this fly, a quarter of founded colonies in a tropical rain forest of Colombia were forced to abandon their nests. Colonies of this wasp were founded by lone or multiple foundresses, and 75% of the attacked nests were reconstructed in the case of multi-female foundations. In the temperate regions, such a serious event has not been observed.

Predation by other insects and vertebrates

Wasp nests are very attractive to predaceous insects, birds and mammals because they contain a considerable mass of proteinous sources (pupae and larvae), and the brood can never escape from the nest even when attacked (Jeanne 1975). Many kinds of natural enemies prey upon wasp nests at various developmental stages, and the predation is usually fatal to the colony. Thus, predation greatly affects the ecological longevity of wasp colonies. In response to their pressures, wasps have evolved various forms of defence (Wilson 1971; Jeanne 1975, 1980).

Ants

Ants are typical cursorial hunters searching for food in response to chemical and visual cues (Jeanne 1975). Pre-emergence colonies of social wasps are particularly vulnerable to ant attacks because these colonies are defended by a lone or a few foundresses and are easily invaded by ants in the absence of their owners. Many ant species have been recognized as predators of social wasp colonies, and the single-petiole nest and the smearing of ant repellent over the petiole are thought to have evolved primarily against ants which predominate at lower latitudes (Jeanne 1970, 1975, 1979*b*).

Successful attacks by most ant species on wasp colonies may decrease with an increase in the wasps' defensive abilities during the post-emergence phase. In most temperate regions, developed colonies of *Polistes* can resist ants, though ants are still one of major causes of failure. In *P. chinensis*, 8.6% of a population's total pre-emergence colonies were destroyed by *Lasius niger* and *Pristomyrmex pungens*, whereas only 2.2% of total post-emergence colonies failed after their attacks (Table 4.5) (Miyano 1980). Predaceous ants may rank high among the failure factors of post-emergence colonies in the lower latitudes because of their extremely high population densities (Jeanne 1979*b*). In contrast, *P. biglumis bimaculatus* colonies at high altitudes in the Alps are never attacked by ants and the occurrence of smearing ant repellent over the petiole is significantly reduced (Lorenzi and Turillazzi 1986).

Army ants of the subfamily Ecitoninae may have a special status as predators of social wasp colonies in South and Central America, because their ways of life are completely different from those of other ant groups. Their distinguishing characteristics include massive predatory raids against other insects in particular,

Table 4.5 Life table for colonies of *Polistes chinensis* in central Japan in 1977. Modified from Miyano (1980)

Phases	*lx*	Mortality factors	*dx*	*qx*(%)	Moth(%)*
Pre-emergence	162	Unknown	49	30.2	3.1
		Moth larvae	3	1.9	
		Ants	14	8.6	
		Artifacts	3	1.9	
		Total	69	42.6	
Worker	93	Unknown	6	6.4	44.6
		Moth larvae	2	2.2	
		Ants	2	2.2	
		Rain	2	2.2	
		Artifacts	6	6.4	
		Total	18	19.4	
Reproductive	75	Disintegration	75	100.0	96.0

* Per cent colonies infested by larvae of an *Anatrachyntis* moth.

and great nomadic movements of their colonies (Schneirla 1971). They raid more actively during the nomadic phase, when the colony contains several larval stages. Therefore, only colonies located around the migration route or bivouac point of the ant colony encounter the raid. Their attacks are so well organized and intense that most wasp colonies discovered by these ants succumb. The significance of army ants as a threat to wasp colonies has been emphasized by some authors, e.g. Richards and Richards (1951), Wilson (1971) and Jeanne (1975). Jeanne (1975) states from his experiences in the lower Amazon region that these ants cause the failure of more wasp colonies than do birds. It has not yet been fully studied how *Polistes* and other social wasps have responded to such a type of attack with life cycle strategies.

Genera *Aenictus* and *Dorylus* of subfamily Dorylinae, driver ants, living in Africa and tropical Asia (*Aenictus* also in Australia) have broad dietary habits (Hölldobler and Wilson 1990). However, only *Dorylus* has been listed as preying upon polistine colonies (Bequaert 1918), and both genera are believed to have little to do with social wasp colonies (C. K. Starr, pers. comm.).

Hornets

Some species of the genus *Vespa* have been major pests of other social wasps in the Old World (Ohgushi *et al.* 1990; Matsuura and Yamane 1990). *V. tropica* and *V. ducalis* are outstanding among the genus *Vespa* for their extremely confined prey preferences and specialized hunting behaviour. Reports on their biology is mainly based on the Japanese population of *V. ducalis* (Sakagami and Fukushima 1957*a*; Matsuura and Yamane 1990). *V. tropica* is distributed in most areas of South-east Asia, India, Greater and Lesser Sunda Islands, and east to New Guinea, while *V. ducalis* is distributed in Japan, Taiwan and other districts of the Far East (Starr

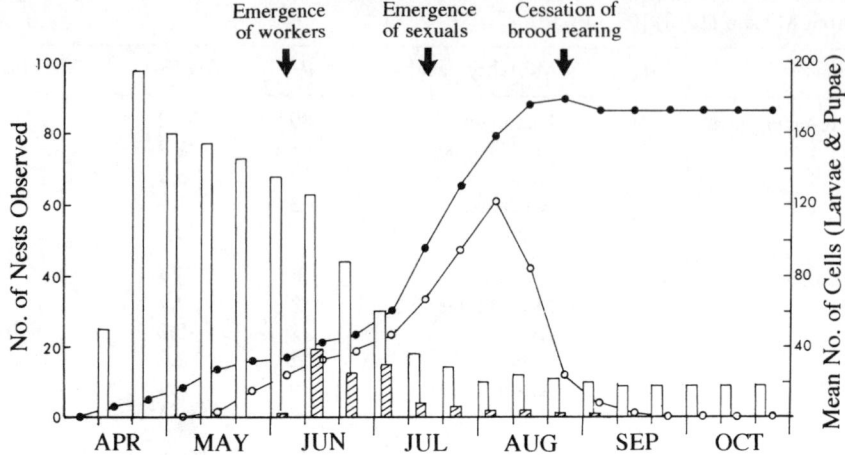

Fig. 4.4 Seasonal changes in the number of active nests of *Polistes jadwigae* (open column) and that of the nests attacked by *Vespa ducalis* (= *V. tropica*, hatched column) in Wakayama, southern Japan in 1971. Solid and open circles: mean numbers of cells and larvae + pupae in *P. jadwigae* colonies (redrawn from Matsuura 1984).

1992). These species prey exclusively upon pupae and larvae of the polistine and stenogastrine wasps. On the prey nest the forager pulls out pupae and prepupae (if not abundant, large larvae too) from the cells after breaking the cocoon caps. It imbibes the victim's body fluids, and then returns to its own nest to feed its own larvae with regurgitated fluid food.

Response of the victim colony to attacks by *V. tropica* or *V. ducalis* varies with the species, but most species almost entirely lack any defensive response (Matsuura 1977; Matsuura and Yamane 1990). Colonies are completely destroyed through repeated attacks by the same hornet. Matsuura (1984) studied the predation by *V. ducalis* (cited as *V. tropica*) on *P. jadwigae* colonies from 1969 to 1975 in southern Honshu, Japan (Fig. 4.4). Prior to the foundation of their own nests, *V. ducalis* queens begin to attack nests of this and other species in early June, when workers on the prey nests have just emerged. This enemy, therefore, causes failure of only post-emergence colonies in the temperate zone. The percentage of victimized nests during the worker phase reached 45.5% in *P. jadwigae* nests in action at the beginning of this phase; 66.7% of raided nests of *P. jadwigae*, or 30.0% in *Parapolybia indica*, were reconstructed near the original sites, but eventually failed after renewed attack (Matsuura 1984; Sekijima *et al.* 1980).

Estimates suggest that the impact of *V. ducalis* on *Polistes* colonies is great. According to Matsuura (1984), one *V. ducalis* colony built 268 cells and produced 71 workers, 61 males and 75 new queens in southern Japan. To produce these hornets, 150–200 polistine colonies must be exploited. In Sumatra *V. tropica* was seen to build huge nests with 5000–6000 cells, and they may produce 500–1000 new queens by the end of the active period (Matsuura 1990). To sustain such huge colonies, thousands of *Polistes* colonies must be sacrificed if the hornets depend solely on them. Matsuura discovered four such active colonies in an area of 1 km². Since, in fact, *V. tropica* also depends heavily on colonies of other social wasps in Sumatra, such as *Ropalidia* and stenogastrines (Williams 1919; Hansell *et al.* 1982; Turillazzi 1985*a*; Matsuura 1990; Ohgushi *et al.* 1990), and probably also *Parapolybia* and *Polybioides*, real predation pressure on the *Polistes* colonies may be less. Both species also hunt honey bees in flight (Starr 1992). The occurrence of many huge colonies of *V. tropica* in the wet tropics is apparently possible because of the abundance of large polygynous colonies of *Ropalidia* and probably *Polybioides*. In any case, extremely high predation pressure of *V. tropica* seems to affect the ecological longevity of *Polistes* colonies.

Some large-sized species, such as *P. gigas* and *P. tenebricosus*, especially the former, seem to be able to resist attacks by *V. ducalis* and *V. tropica*, respectively (Yamane *et al.* 1989). *P. gigas* is smaller than *ducalis*, but in Taiwan I once saw some 10 wasps on a *P. gigas* nest in a joint threat against a *V. ducalis* worker just alighting on the nest. Disturbed by their agitation, the *V. ducalis* worker failed to land and eventually flew away. Workers of *P. tenebricosus* sometimes behave in a similar manner against *V. tropica*, though it is uncertain whether this species is strong enough perfectly to resist the hornets' attacks (Yamane *et al.* 1989).

Among other *Vespa* species, *V. mandarinia*, the largest among the subfamily, attacks not only solitary insects but also colonies of honey bees and other social wasps including *Polistes* (Matsuura 1984). But *Polistes* colonies are not a major item on its menu. *V. affinis* often preys upon stenogastrine and *Ropalidia* colonies in Sumatra, but it also relies on various kinds of non-social insects and its hunting technique is not specialized (Ohgushi *et al.* 1990).

Birds and mammals

Jeanne (1975) listed the predators of social wasp broods, but the number of vertebrate predators he gave was small. These animals are primarily visual hunters, have a keen sense of hearing and smell, and are very sensitive to the approach of humans. So it is not easy to confirm the actual scene of their predation. Possible avian predators on the polistine colonies include honey buzzard (*Pernis apivorus*) (*Polistes* spp., Japan: Matsuura pers. comm.), *Cyanocitta cristata* (*P. exclamans*: Pulich 1969), *Piranga rubra* (*P. pallipes* and *P. variatus*: Rau 1941), and unidentified birds (*P. bellicosus*: Strassmann *et al.* 1988). Of *P. biglumis bimaculatus* nests at a high altitude in the Alps 17.6% (*n* = 34) were probably destroyed by some unidentified birds throughout pre- and post-emergence phases (Lorenzi and Turillazzi 1986). Mammals known to prey upon *Polistes* brood are *Cebus variegatus* (pet animal) (*P. versicolor*: Zikan 1951), foxes (*P. riparius*: Yamane and Kawamichi 1975), racoons and possums (*P. bellicosus*: Strassmann *et al.* 1988). Other vertebrates, such as primates (including human beings (Yamane 1992)), that prey upon broods of vespines and other genera of polistines, must also be potential predators of *Polistes*. These predators remove the entire nest from the substrate, and the wasps lose all they have invested in. Jeanne (1975) stated that birds as predators are considered just as important as ants.

Predation pressure of birds and mammals (probably racoons and possums) on a natural population of *P. bellicosus* was high in Texas (Strassmann *et al.* 1988). From 6 to 46% of nests were victimized by the predators during any of five given periods from 22 March to 11 August in 1985 and 1986. They concluded that group living may be maintained in this population because larger groups have advantages in recovering from the predation.

Social parasitism

In the Polistinae, only *Polistes* is known to have obligate, permanent social parasites belonging to the same genus (*Sulcopolistes* by previous authors, Carpenter 1991) (Cervo and Dani this volume). They have no worker caste and their females depend entirely on the host species for nest construction and brood rearing (Taylor 1939). The distribution of the three parasitic species is limited to areas of South Europe around the Mediterranean Sea (Guiglia 1971). Although their distribution is localized, colonies of certain species of the subgenus *Polistes s. str.* in this area may suffer great damage from their parasitism. Biology of these species and their role of social parasite and predator of the host species has been recently studied intensively (Cervo and Dani this volume).

Cervo *et al.* (1990c) observed the impact of *P. atrimandibularis* on colony productivities of *P. biglumis bimaculatus* at a high altitude (1800 m a.s.l.) in the Alps. Unlike other social parasites, females of this species can control two kinds of host nests, 'nursery' nests solely for reproductive purposes and 'supply' nests for food sources of the parasite brood. In mid-August supply nests have fewer cells than nursery nests (38.2 vs. 52.6 cells, data in 1987 and 1988 combined), because host productivity is reduced by heavy predation of pupae and larvae. About 20% of founded colonies of *P. biglumis bimaculatus* were parasitized by *P. atrimandibularis*. The number of host wasps emerged from these parasitized nursery nests was less than half that of the social parasites. Activities of the parasitized colonies end earlier than intact ones (Lorenzi *et al.* 1992).

Conclusions

Although the colony cycle of *Polistes* is flexible in response to different climatic and ecological factors, it still has some features specific to the genus. These seem to have arisen mainly through phylogenetic constraints, within which given species or populations seem to have responded in a variety of ways to the obstacles they encountered. The relationships among these probable factors are summarized in Figure 4.6.

Phylogenetic constraints on Polistes

Carpenter (1991) tabulated the characters used for cladistic analysis of the Polistinae, of which the following three are related to behaviour:

1. Absence of swarm-founding. Like most Vespinae excluding *Provespa* (Matsuura 1990), and unlike swarm-founding genera of Polistinae (Jeanne

1991*b*), all *Polistes* are believed to initiate colonies independently. Although there are some suggestions of swarm founding (Rau 1941), its occurrence has not yet been demonstrated (Jeanne 1980; Reeve 1991).

2. Absence of meconium extracting behaviour. Like all the other New World polistine genera (Jeanne 1980), and unlike four Old World genera (*Ropalidia, Parapolybia, Polybioides* and *Belonogaster*) (Vecht 1962, 1966; Marino Piccioli 1968; Yamane 1984), wasps of this genus do not remove, through a hole made at the cell bottom, meconia excreted by the post-feeding larvae. The meconium always remains in the bottom of the cell.

3. Absence of nest envelope. *Polistes* never constructs an outer envelope surrounding the comb. Furthermore, *Polistes,* with a few exceptions, builds a single comb, which is always suspended by a petiole. This may limit ultimate nest size (Wenzel 1991).

In addition to the above, some characteristics of the colony cycles are common to the genus *Polistes:*

1. Colony cycles are, as a rule, annual and most colonies have nesting periods shorter than 6–8 months (Fig. 4.2). There are some unpublished records of colonies lasting more than one year, but these are rather exceptional. This is consistent with most independent-founding Polistinae and Vespinae. The short colony cycle may be essentially the result of the longevity of the foundress as a queen being shorter than one year. This short longevity may relate to poorer nutritional conditions compared with honey bee queens during the larval stages, and hard work during the founding stage.

2. Males and gynes are, as a rule, produced in a discrete mass only towards the end of colonial activity, after all the workers have been produced (Fig. 4.1). This type of production schedule, as well as small colony size, may make the occurrence of swarm-founding difficult. In some species, a few males which can inseminate workers are produced early in the cycle. In contrast, males and potential gynes are sporadically produced from an early period of the post-emergence phase in *Mischocyttarus* and *Ropalidia* (Jeanne 1972; Yamane 1986).

3. The foundress, which has become senescent, is usually not replaced by another inseminated, potential egg-layer during the post-emergence phase. So colonies usually terminate or decline with the loss of the foundress. Workers of some species, such as *P. exclamans, P. bellicosus,* and *P. snelleni* sometimes mate with early males and lay female-producing eggs when the colony is orphaned (Strassmann 1981*a;* Suzuki 1985; Strassmann *et al.* 1988), but how this prolongs nesting periods in the tropics, i.e. climatically unrestricted circumstances, is not well known. All *Polistes* species studied have a determinate colony cycle, and the early males do not seem to put those species in a different colony-cycle mode.

  By contrast, the prolongation of colony activity by repeated replacements of the dominant egg-layer (called serial polygyny by West-Eberhard 1978*b*), is a normal process in *Mischocyttarus drewseni* (Jeanne 1972), *Ropalidia variegata* *jacobsoni* (Yamane 1986) and *R. marginata* (Chandrashekara and Gadagkar 1992) among others. Indeed, many *Ropalidia* colonies survive more than one year. In one *R. variegata* colony, the first daughter expelled her mother soon after her emergence and she in turn was driven out by her sister one month later (Yamane 1986). Some potential gynes which fail in succeeding to the colony may initiate their own nests elsewhere, as in some other *Ropalidia* species (Gadagkar 1985).

Ecological constraints

As mentioned in Section 3, some ecological factors, e.g. infestation by other insects and predation by various insects and other animals, have indirect and/or direct effects on the colony cycle. Further, some of these are closely related with the behavioural traits of *Polistes*. Predation and infestation of *Polistes* colonies at various developmental stages by selected enemies are summarized in Figure 4.5.

Infestation by scavenger moth larvae is one of the ecological factors believed to exert the most vigorous selective pressure on colony longevity. This type of moths is cosmopolitan and infests wasp nests at high rates after mid-phase of the nesting periods. Once the nests have been bored, they lose structural strength, and cannot be used for long. Furthermore, reuse of old nests may carry the risk of residual parasitism. Indeed, *Polistes* wasps never use them again, with a few exceptions (Distefano 1971). Easy infestation by moths may be partly a result of the lack of a nest enve-

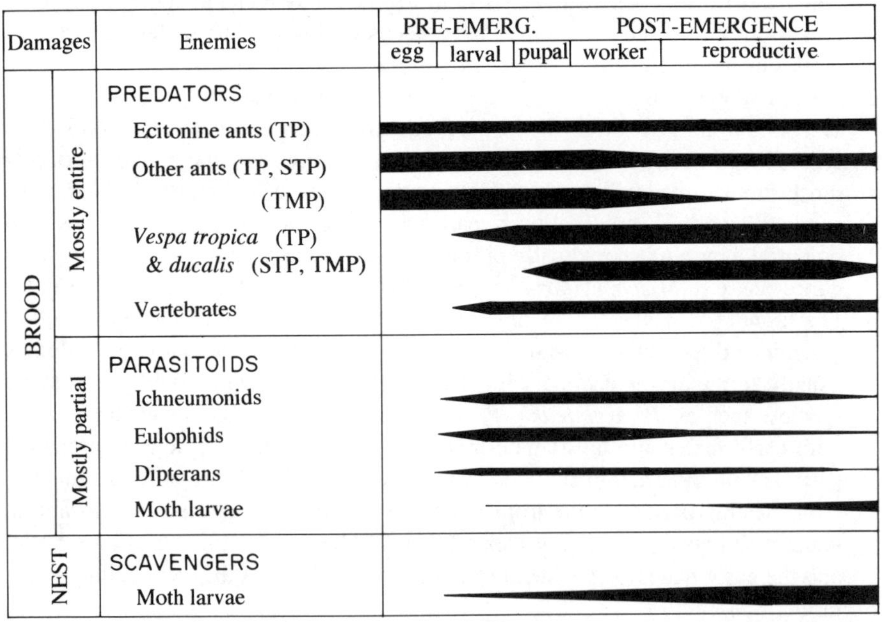

Damages		Enemies	PRE-EMERG.			POST-EMERGENCE	
			egg	larval	pupal	worker	reproductive
BROOD	Mostly entire	PREDATORS Ecitonine ants (TP) Other ants (TP, STP) (TMP) *Vespa tropica* (TP) & *ducalis* (STP, TMP) Vertebrates					
	Mostly partial	PARASITOIDS Ichneumonids Eulophids Dipterans Moth larvae					
	NEST	SCAVENGERS Moth larvae					

Fig. 4.5 Predation and infestation of *Polistes* colonies at various developmental phases by selected natural enemies in three categories. Horizontal bars indicate roughly estimated intensity of damages. TP: tropics, STP: subtropics and TMP: temperate areas.

Table 4.6 Number of recorded polistine species infested by nest-scavenging moths and parasitoids. Reconstructed from Makino (1985*b*)

Enemies	*Polistes*	*Mischocyttarus*	Other Old World genera	Other New World genera
Moths	16	2	>3[2]	0
Hymenopterans	>26[1]	6	>6	3
Neuropterans	0	0	0	4
Dipterans	>19	>4	>5	4

[1] Including 2 *Fuscopolistes* species.
[2] Including *Belonogaster* spp. (Richards 1969) and *Ropalidia* sp. from Australia (Tillyard 1926).

lope. That the envelope effectively prevents the nest from invasion by moths is demonstrated by vespine nests—their nests in Japan and Taiwan are hardly ever infested by moth larvae even at the end of season (Matsuura and Yamane 1990).

Table 4.6 shows the number of species infested by scavengers and parasitoids for five polistine groups classified according to their behavioural traits. The Old World meconium-extracting group includes only an unidentified *Belonogaster* species as parasitized by *Pyroderces orphnographa*, a cosmopterygid moth, recorded by Richards (1969) from Ghana, and a *Ropalidia* species (by *P. anaclastis*) recorded from Queensland, Australia, by Tillyard (1926). *Mischocyttarus* is comparable to *Polistes* on various points, such as independent founding, small exposed nests, leaving meconia in cells. At least two species of this genus are known to suffer moth attacks. However, no species of the swarm-founding genera in the New World are so far known to be attacked by moths, though 11 species are attacked by other parasitoids including neuropterans. The lack of scavenging moths in these groups may have resulted partly from insufficient surveys, and partly from large colony populations and enveloped nests.

I have not yet discovered whether nests of meconium-extracting genera are subject to attacks by scavenger moths in Japan, South-east Asia or Australia, where *Polistes* nests are commonly infested by them. This suggests that these genera have only a few such enemies. I found that over 100 observed colonies of *Parapolybia varia* were never infested by moths over a 12-month prolonged nesting period in southern Taiwan (Fig. 4.2D). Such a prolongation may partly be facilitated by being free from moth attacks. *Ropalidia plebeiana* (Yamane *et al.* 1991) and *R. revolutionalis* (Yamane unpubl.) often reuse old combs built in the previous season, if these have not been heavily weathered. Uncovered combs of some other species, e.g. *R. marginata* and *R. cyathiformis* (Gadagkar, pers. comm.) are also durable and could be used for more than one year.

Predation of wasp broods by various predators, such as *V. tropica* in the Asian tropics, army ants in the New World tropics, many ant species, especially in lower latitudes, and birds and mammals, is without doubt rigorous. *Polistes* wasps have developed various countermeasures against them. For example, high predation pressures of ants at lower latitudes and intraspecific colony usurpation in some

Table 4.7 Length of nesting period and nest size in *Polistes* species with short-term and long-term colony cycles in subtropical and tropical areas.

Type of colony cycle and species	Locality (climate)	Length of nesting period	No. cells/nest	
			Average	Largest
Short-term type				
Subgen. *Polistella*				
P. *japonicus*[1]	Taiwan (ST)	6–7	mostly 40–50	>100
P. *takasagonus*[1]	Taiwan (ST)	6–7	mostly 100	>300
P. *shirakii*[1]	Taiwan (ST)	6–7	mostly 100	>200
P. *sagittarius*[2]	W. Sumatra (T)	5–6?	74(2)	76
P. *stigma*[2]	W. Sumatra (T)	5–6?	97(3)	139
Long-term type				
Subgen. *Gyrostoma*				
P. *gigas*[1]	Taiwan (ST)	9–10	60	126
Subgen. *Nygmopolistes*				
P. *sulcatus*[1]	Taiwan (ST)	7–8	mostly 80–120	150
P. *tenebricosus*[2]	W. Sumatra (T)	7–9	242(11)	734
Subgen. *Megapolistes*				
P. *rothneyi*[1]	Taiwan (ST)	7–8	150–300	
P. *olivaceus*[1]	Fiji Is. (T)	?		>1500
P. *tepidus malayanus*[3]	New Guinea (T)	7–10?	208(10)	662
Subgen. *Aphanilopterus*				
P. *erythrocephalus*[4]	W. Colombia (T)	5–15.6?	95	750

[1] Yamane (unpublished), [2] Yamane *et al.* (1989), [3] Yamane and Okazawa (1977), and [4] West-Eberhard (1969*b*).

populations may, at least in part, be responsible for the evolution of multiple-foundress associations (Gamboa 1978; Jeanne 1979*b*; for more general discussion, Itô 1993). One response in their colony cycle seems to be a tendency to diverge into two types of life cycles: short-term (less than 6–7 months) and long-term (over 6–8 months) (Table 4.7). Species belonging to the former type are small- to medium-sized and have small colony populations (10–30 workers), while species of the latter type are medium- to large-sized with large colony populations (> 30 workers) or have special defensive measures as seen in *P. gigas*, though this distinction is not clearcut. Since the chance of encountering predation is proportional to the duration of the nesting period, species that form defenceless colonies can evade enemies by producing gynes and males in a short cycle.

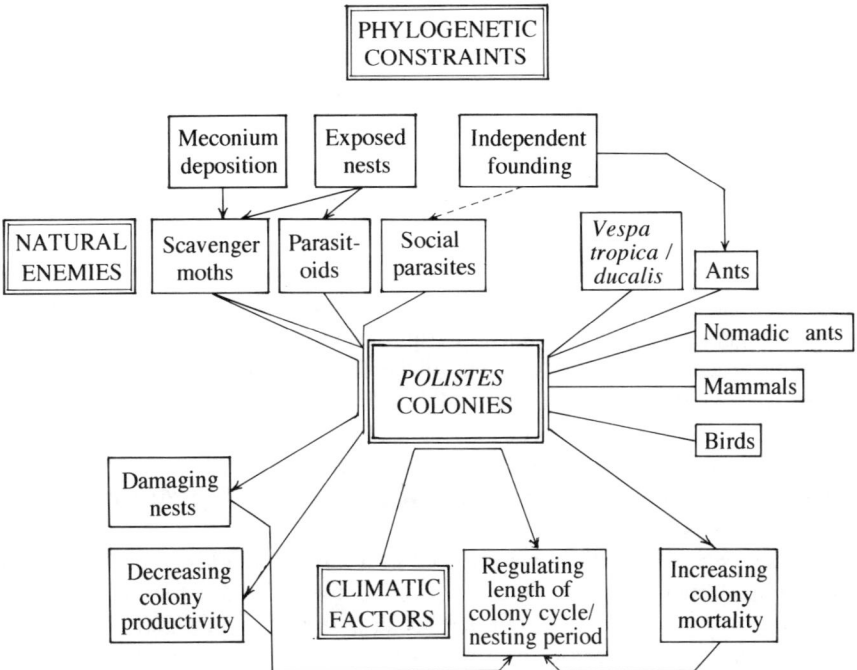

Fig. 4.6 Diagrammatic figure showing how the phylogenetic traits of *Polistes* wasps are related with natural enemies, and how climatic factors and these natural enemies affect the population consequences in *Polistes* wasps.

Reeve (1991) surmised that the longevities of tropical colonies may be entirely determined by chance mortality events, judging from the results of *P. erythrocephalus* studied by West-Eberhard (1969*b*). But, as he also argues, optimum colony longevity would be set at a certain length, e.g. 6–7 months to compromise between fatal predation before gyne production and the number of gynes produced if the colony could survive without interference. My classification of tropical colony cycles into two types is tentative, and presently I cannot show the actual causes of such a divergence, if any. More quantitative, as well as qualitative, data would be needed to clarify this point.

Acknowledgements

I thank J. Kojima, S. Makino, two referees, C. K. Starr and M. C. Lorenzi and editors of this volume, S. Turillazzi and M. J. West-Eberhard for their critical reading of the earlier draft of this paper and comments and stylistic corrections on it. Many persons who attended the workshop held at Castiglioncello gave valuable comments and suggestions on this paper. M. Matsuura, S. Makino and S. Miyano permitted me to use their figures and tables.

5

Social parasitism and its evolution in *Polistes*

Rita Cervo and Francesca R. Dani

Wasps in general, and *Polistes* in particular, constitute for West-Eberhard (1991) a group of organisms especially suited to inspire and inform the study of social evolution. Not by chance was *Polistes* one of the first groups on which kin selection theory was tested (Hamilton 1964*a* and *b*). Though this theory is today the most accredited for explaining the origin of eusociality, it is possible that various selective factors may have operated together in determining its several appearances, making a single explanation unlikely (West-Eberhard 1975; Andersson 1984; Brockmann 1984).

Although social parasitism may be seen as a kind of anti-social behaviour, this formerly neglected topic may hold new insights for the study of social behaviour. For example, it has been proposed that social parasitism might play a role in the evolution and maintenance of sociality in hymenopterans (Lin and Michener 1972; Gamboa 1978; Brockmann 1993). Some behaviours that are at the base of insect sociality, such as kin and nestmate recognition, can be interpreted as adaptive behaviours that function not only to facilitate co-operation between genotypically similar individuals, but also as a defence mechanism to suppress parasitic tendencies (Fisher 1987). Fisher (1987) states that 'the study of parasitism may complement, and contrast with, that of altruism as a way of understanding the evolution of social behaviour.' This aspect of behaviour should not be underestimated, as it is a strategy widespread, if with a varying frequency, in the various taxa of social insects (Wilson 1971) as well as of birds (Payne 1978). No parasitic species have ever been described in termites. In ants, social parasitism is well represented in some subfamilies (e.g. Formicinae and Myrmicinae) while it is not documented at all in others (Wilson 1971; Hölldobler and Wilson 1990); the phenomenon of social parasitism occurs only in the primitively eusocial bees, such as bumblebees and allodapine bees (Michener 1974; Alford 1975). Obligate social parasites are present in the subfamilies of Vespinae and Polistinae (Weyrauch 1937; Beaumont and Matthey 1945; Scheven 1958; Akre 1982) but no cases are known in the Stenogastrinae. Although the obligate parasites among *Polistes* wasps are rare (only three out of 203 species stated by Akre in 1982), their study can reveal new and useful information on the mechanisms that are fundamental to a *Polistes* society: already in 1942 Pardi pointed out a notable resemblance between the role of an alpha female in a polygynic colony and that of a parasite invading a host colony. It can be said that, in a certain sense, a parasite amplifies certain basic behaviours of a *Polistes* colony (Turillazzi 1992).

The earliest attempt, today still widely accepted, to explain the origin and evolution of social parasitism in the wasps was made by Taylor in 1939. This author formulated a scheme, based on four stages, explaining the evolutionary route to social parasitism:

(1) intraspecific, facultative temporary parasitism;
(2) interspecific, facultative temporary parasitism;
(3) interspecific, obligatory temporary parasitism;
(4) interspecific, obligatory permanent parasitism.

Although the third stage of this scheme is a fairly probable intermediate step, it has not been documented in social wasps. In this review, knowledge about the parasitic behaviour of *Polistes* paper-wasps will be presented following Taylor's evolutionary pathway.

Temporary facultative intraspecific parasitism

The simplest form of social parasitism present in *Polistes* wasps is that of temporary facultative intraspecific parasitism, the first stage of the Taylor scheme. This type of parasitism involves foundresses that usurp a colony of the same species by evicting the resident queen or by occupying an unattended comb (take-over: Kasuya 1982). At first, reports documenting this type of parasitism in wasps were rather sporadic because they came from studies designed to investigate other aspects of *Polistes* biology (Yoshikawa 1955; Strassmann 1981a, 1985a; Klahn and Gamboa 1983; Lorenzi and Turillazzi 1986; Hughes and Strassmann 1988). Only recently have studies specifically concerned with intraspecific parasitism been conducted (Gamboa 1978; Makino 1985a, 1989a; Klahn 1988; Makino and Sayama 1991; Dani and Cervo 1992; Gamboa *et al.* 1992; Lorenzi and Cervo 1993, 1995). These studies have investigated the incidence and causes of intraspecific parasitism, the mode of usurpation, and the reproductive success of usurped colonies. At the present, intraspecific usurpation has been reported for 11 species of *Polistes* (Table 5.1), although this number is likely to increase rapidly as other species are studied.

The rate of usurpation in populations of the species analyzed varies from 3% to about 100% (Table 5.1), but it must be kept in mind that the number of colonies on which the percentages were calculated is highly variable (from 17 to 370).

From the available data it is evident that usurpers tend to be females that have lost their own nests (Strassmann 1981a; Makino 1985a, 1989a; Klahn 1988; Makino and Sayama 1991; Dani and Cervo 1992; Lorenzi and Cervo 1995). In fact, independent of the species, all the usurpers whose origin was known had previously lost their nests due to predation (Makino 1985a, 1989a; Klahn 1988; Makino and Sayama 1991; Lorenzi and Cervo 1995). It must also be pointed out that the majority of usurpations in the course of the reproductive season coincide with the period of late pre-emergence, in other words, with the period in which nest predation is at its peak (Dani and Cervo 1992; Lorenzi and Cervo 1995) and nest defence is relatively weak.

Table 5.1 *Polistes* species in which intraspecific usurpation has been observed and percentage of usurped colonies in the studied population.

Species	% usurped nests	No. of checked nests	References
Polistes jadwigae	?	?	Yoshikawa 1955
Polistes riparius	13	281	Makino 1989a
Polistes snelleni	19	100	Makino and Sayama 1991
Polistes biglumis bimaculatus	18	125	Lorenzi and Cervo 1995 Lorenzi and Turillazzi 1986
Polistes gallicus	7	157	Dani and Cervo 1992
Polistes metricus	10–100	74	Gamboa 1978
Polistes exclamans	?	?	Strassmann 1981a
Polistes fuscatus	3 (in Michigan)	288	Noonan 1979
	20 (in Iowa)	370	Klahn 1988
Polistes annularis	7	141	Hughes and Strassmann 1988
Polistes carolinus	5	39	Hughes and Strassmann 1988
Polistes bellicosus	6	17	Hughes and Strassmann 1988

Facultative parasitism may also be intensified by other factors, such as the lack of sites suitable for nesting and belated emergence from hibernation due either to cooler wintering sites or, as Matsuura and Yamane (1990) suggest for Vespinae, as a protection against the high predation to which wasps and colonies are subject at the beginning of the season. At the moment, however, there is no evidence that such factors might constitute an important contribution to the appearance of intraspecific usurpation in Polistinae.

Among the species in which intraspecific usurpation has been noted there are some with strictly solitary foundation and some with associative foundation. Gamboa (1978) and Klahn (1988) found that colonies with multiple foundresses are less subject to intraspecific usurpation than those with a single foundress. The same authors hypothesized that defence against usurpation is one of the principal selective factors that maintain the associative strategy.

Recent observations by Gamboa *et al.* (1992), however, indicate that conspecific usurpation pressures are intense for multiple-foundress colonies in observed population of *P. fuscatus*: on average, each colony experiences one usurpation attempt per day throughout the preworker stage. According to the same authors, the low frequency of successful usurpations in multiple-foundress colonies is because of the decreased probability that nests are left unattended rather than to co-operation between foundresses in repelling invaders.

One criterion used by Klahn (1988) to diagnose nest usurpation is that the usurper replaces only the dominant foundress and not the subordinate ones. This criterion, though applicable to the great majority of usurpations, does not take into account the cases of usurpation in which the original owner is not expelled from the usurped colony, as has been observed in *P. biglumis bimaculatus*, a species with strictly solitary foundation (Lorenzi and Cervo 1995).

It should be emphasized that since the object of usurpation is the nest with its immature offspring, the expulsion of the real owners is of secondary importance in usurpation behaviour. In fact, as we shall see in the following, even facultative and obligatory interspecific parasites of *Polistes* sometimes allow the host foundress to remain on the colony after usurpation.

If we consider as usurpations all of the situations in which a female joins an existing colony and assumes the dominant position without expelling the other foundresses, we necessarily include many colonies of species with associative foundation. According to Field (1992), some behaviours common among nestmates, such as domination and egg-eating, can be seen as parasitism at the individual level. It then appears evident that the line between associative foundation and intraspecific usurpation is not clear from the behavioural point of view, but can be defined on a temporal basis. Usurpation is involved when the substitution of the dominant female is observed in colonies already advanced in their colonial cycle. We have not here taken into consideration kinship among involved females as a factor of distinction between these two strategies. In fact, though it is known that co-foundresses tend to be close kin and usurpers usually are not closely related to the host colony, associations are not always composed exclusively of females from the same original nest (Reeve 1991). In addition, Klahn (1988) has found, in 10 cases of usurpation, a certain degree of kinship between the usurper and the previous owner of the usurped nest.

If reports on the existence of the phenomenon of usurpation are relatively uncommon, even more scarce are studies of the behaviour and success of females that usurp conspecific colonies, of their relations with host workers, and of their reproductive success. Knowledge about these aspects is essential to allow an evaluation in cost-benefit terms of usurpation behaviour as compared with nest-building behaviour.

As already mentioned, data indicate that the usurpation of conspecific colonies is particularly concentrated in the period prior to and shortly after the emergence of the first workers. Some direct observations and the finding of bodies of conspecific foundresses near the colonies indicate that the strategy of intrusion into a foreign colony is usually violent (Makino 1985*a*; Klahn 1988; Gamboa *et al.* 1992; Cervo and Lorenzi, in prep.; Dani and Cervo, pers. obs.).

Notes on the destiny of successfully usurped nests are scarce. A behaviour performed by the intruder during the first days following usurpation, common to all the species studied, is destruction of the immature brood belonging to the previous owner of the nest. This destruction of the offspring is particularly concentrated on eggs and young larvae; however, larger larvae and pupae, which will soon become workers to labour for the reproductive success of the intruder, are spared (Klahn 1988; Makino 1989; Makino and Sayama 1991; Lorenzi and Cervo 1992). This differential destruction of immature offspring also occurs after foundresses are experimentally induced to adopt a foreign nest (Klahn and Gamboa 1983; Cervo and Turillazzi 1989; Lorenzi and Cervo 1992).

Current evidence indicates that the reproductive success of a female which usurps a colony, or associates belatedly with another conspecific female, is rather

Table 5.2 Mean number and standard deviation of adults which emerged from usurped and non-usurped colonies (number of colonies, in brackets) of *P. biglumis bimaculatus* (Cervo and Lorenzi 1995), *P. fuscatus* (Klahn 1988), *P. snelleni* and *P. riparius* (Makino and Sayama 1991).

Species and caste	Usurped colonies Mean ± SD (No. of colonies)	Non-usurped colonies Mean ± SD (No. of colonies)
Polistes biglumis bimaculatus reproductives + workers	13.2 ± 8.6 (5)	27 ± 5.2 (5)
Polistes fuscatus site 1 reproductives	44.4 ± 38.4 (11)	88 ± 74.4 (31)
Polistes fuscatus site 2 reproductives	59.1 ± 71.9 (29)	161 ± 80 (59)
Polistes snelleni reproductives	11.4 ± 14.2 (8)	33.5 ± 19.9 (16)
Polistes riparius reproductives	4.5 ± 4 (4)	39.8 ± 9.8 (4)

In the usurped colonies of *P. biglumis bimaculatus* about half as many adults emerge as in non-usurped colonies: this proportion, taking into consideration only the reproductives, is between 1/2 and 1/3 in *P. fuscatus* colonies, 1/3 in *P. snelleni*, 1/8 in *P. riparius*.

low (Table 5.2). It has been argued that the low reproductive success registered in *P. fuscatus*, *P. riparius* and *P. snelleni* usurpers is probably a result of conflict between the usurper and the first emerging workers, daughters of the previous owner of the nest (Klahn 1988; Makino 1989*a*, 1991; Makino and Sayama 1991). It seems that workers have evolved a series of counter-measures to usurpation, ranging from expulsion of the foreign female, to abandonment of the nest and destruction of the immature offspring of the usurping female. In contrast, observations of usurped *P. biglumis bimaculatus* colonies revealed no greater aggression between workers and usurpers than among queens and workers in control colonies (Cervo and Lorenzi, in prep.).

Counter-measures adopted by workers, observed in *P. fuscatus* and *P. riparius*, seem to indicate that intraspecific usurpers are not always capable of controlling the daughters of the previous foundress, or of 'convincing' them that they are close kin. Similarly, *P. gallicus* workers emerging in the presence of a foreign female experimentally made to occupy an alien nest have well-developed ovaries (Cervo and Turillazzi 1989). This confirms the foreign female's difficulty in controlling the reproductive capacity of unrelated workers.

Klahn (1988) hypothesizes that workers can easily recognize a non-related female they encounter on the nest by her odour, which is different from that of their nest. The workers, as well as future foundresses, learn the colony's scent from the nest itself, not from themselves or their nestmates (Gamboa *et al.* 1986*a*; Gamboa this volume). Klahn therefore tries to explain the workers' acceptance of some foreign females, and the rejection of others, by the ability of the usurper to change the scent of the nest. This hypothesis fits well with the obvious rubbing behaviour that is performed by foreign queen on host comb. During this behaviour the females rub last abdominal sternites against the comb. This behaviour has been described as stroking by West-Eberhard (1982*b*, 1986*a*) in *P. canadensis* and more recently analyzed thoroughly in *P. dominulus* foundresses by Dani *et al.* (1992*b*). Stroking is performed conspicuously by *P. gallicus* females following their experimentally induced arrival on foreign colonies (Cervo and Turillazzi 1989) and by *P. biglumis bimaculatus* females after a natural usurpation of a conspecific colony (Lorenzi and Cervo 1992). The ability of usurper to impose its own scent on the comb would thus increase the probability that the workers that emerge from the nest will accept the invader. Though the abundant sternal glands (see Jeanne this volume) of foundresses seemed to be the most probable source of this secretion (West-Eberhard 1982*a*; Cervo and Turillazzi 1989; Dani *et al.* 1992*a*), recent chemical and behavioural studies on *P. dominulus* show that both the fifth and sixth gastral sternal glands produce substances that work as allomones (Dani *et al.* 1994*a*). Analogous findings were already reported for the sixth sternal glands of *P. fuscatus* (Post *et al.* 1984*b*) and of *P. annularis* (Espelie and Hermann 1990). As chemical analysis indicate that the Dufour's gland of *Polistes* is a source of cuticular hydrocarbons (Dani *et al.* 1994*b*), it is possible that pheromones are released from this gland during the stroking behaviour.

Data on the productivity of usurped colonies (Table 5.2) indicate that females that adopt the strategy of usurping a conspecific nest have lower reproductive success than dominant females in queenright colonies.

Usurping a conspecific nest is probably the best strategy when reproductive success has been annulled by the loss of the female's own nest and by the lack of time to found a new successful one. The wide spread of intraspecific usurpation suggests that, in such cases, even if the reproductive rate of an intraspecific usurper is less than half that of the average non-parasitic female, some offspring are better than none. It is probable that facultative usurpers do not possess behaviours as specialized as those shown by obligatory parasites that cannot reproduce without usurpation.

Temporary facultative interspecific parasitism

A somewhat more complex stage of social parasitism in wasps is that of temporary facultative interspecific parasitism (the second stage in Taylor's scheme), in which females of one species may either found their own colonies or usurp those of another species. In this case we are also dealing with a temporary parasitism, since the foreign female's workers could take the place of the host's workers.

This stage of parasitism has been described in the genus *Vespa* (Sakagami and Fukushima 1957*b*; MacDonald and Matthews 1975), but there exist only few and sporadic observations of it in polistine wasps (Snelling 1952; Hunt and Gamboa

1978; O'Donnell and Jeanne 1991).

We recently observed two cases of usurpation of *P. gallicus* nests by *P. dominulus* (unpublished records). The usurpations occurred shortly after nest foundation and, in one of the two cases, the usurping female came from an undamaged nest located a few centimetres away. Thus the cause of this parasitism seems to be different from that of observed intraspecific usurpations, which have followed usurper nest loss. In both cases the foundresses of the two species coexisted on the colony, and the usurper became dominant over the nest's owner, which assumed typically worker tasks. During the beginning of this cohabitation we observed the performance of stroking behaviour by both females and the frequent offer of drops of salivary fluid by the usurper to the host, a behaviour also observed in the obligate social parasites (as will be explained later).

O'Donnell and Jeanne (1991), reported the extensive disappearance of immature brood in the nest occupied simultaneously by *P. instabilis* (host species) and *P. canadensis* (usurper species), similarly to what happens in intraspecific usurpation.

In the cases we observed, analogous to the findings of O'Donnell and Jeanne (1991), the females of *P. dominulus*, the usurping species, turn out to be of greater size than those of *P. gallicus*, the usurped species. According O'Donnell and Jeanne (1991), such size differences render interspecific usurpation easier than intraspecific usurpation.

The small number of usurpations observed does not allow us to determine the reproductive success of interspecific usurpers.

The scanty records of interspecific parasitism, though it is more easily recognized than intraspecific parasitism, indicate that this strategy is not frequently adopted by *Polistes* wasps. This may be due to the dissimilar behaviours and chemical signals used by the different species, which may make it difficult for a usurper to become integrated into a foreign colony.

Permanent obligatory interspecific parasitism

The last stage in Taylor's scheme is that of obligatory and permanent interspecific parasitism or inquilinism. In this case the worker caste is lost, as is the capacity to found a nest. The parasite must usurp a colony of a host species and depend entirely on the host's workers to raise her own offspring, which consist of reproductives only. Among *Polistes* only three species of obligatory parasites are known: *P. sulcifer*, *P. semenowi* and *P. atrimandibularis*. Prior to Carpenter (1991) revision, these three species were grouped into a separate genus, *Sulcopolistes*.

Cladistic analyses of data from allozyme polymorphisms (Carpenter *et al.* 1993) and from rRNA sequences (Choudhary *et al.* 1994) brought evidence for defining the polistine social parasites as a monophyletic group (see Carpenter, this volume).

Knowledge about obligatory parasites in *Polistes* has been augmented recently through studies conducted especially by our group. Until now, work on these species was scarce (Beaumont and Matthey 1945; Scheven 1958; Distefano 1968; Demolin and Martin 1980), owing mainly to their limited range of distribution, which comprises the zones surrounding the Mediterranean and the Caspian basins

(Guiglia 1971), and to their sparse abundance even in this area. Research by our group has shown that some zones have a moderate abundance of these parasites, particularly areas at the foot of massifs or mountain chains. In these areas the rate of parasitism is rather elevated and fairly constant over the years (around 20% for the *P. atrimandibularis* study area and even higher in some *P. sulcifer* and *P. semenowi* study areas).

Given the greater concentration of polistine obligatory parasites at the foot of mountainous areas, some link between the evolution of obligatory parasitism and this environment can be hypothesized. According to Lorenzi and Turillazzi (1986), it is possible that obligatory parasitism in *Polistes* may have in fact evolved from mountainous species like *P. biglumis bimaculatus* which, because of the compression of the colonial cycle, display a high rate of intraspecific usurpation (the time-constraints hypothesis of Brockmann 1993).

It is possible that selective pressures on species that nest at higher altitudes and latitudes may have favoured the evolution of parasitic behaviour. As Wcislo (1987) observes, the usurpation of another's nest is a strategy usually adopted more by temperate than tropical species. Since colony cycles of tropical species are not syncronized, this reduces the availability of suitable nests in the opportune stage of development.

In this regard, useful information on social parasitism in Polistinae could be deduced from their migratory habits, which have been observed recently by our research group (Turillazzi *et al.* 1993). Many females and males of the three species of parasites—as well as wasps of some of the host species—were observed (Distefano 1968; Guiglia 1971; Borsato 1992; Turillazzi *et al.* 1993; Brandmayr, Luchetti, Osella, Vigna Taglianti, pers. comm.), at the end of summer, at the tops of various Italian mountains, which are non-nesting zones for *Polistes* (except for the mountainous species, *P. biglumis bimaculatus*). These wasps undertake altitudinal migrations from the place of emergence, usually at the base of the same mountains, to the tops of the mountains where, according to preliminary data, copulation seems to take place. It has not yet been established when most of the females perform the inverse migration, but many have been found to winter in these same areas, under fairly harsh climatic conditions. Migrations to the tops of mountains, which constitute particularly salient features of the landscape, could facilitate the meeting and outbreeding of individuals of rare species, as are the obligatory parasites of *Polistes*. These insects, in fact, because of their low density, present a high level of dispersion of the two sexes. The correlation between rarity and the use of hilltops as meeting points has been described for various other species of insects (Thornhill and Alcock 1983). The use of 'salient points of the landscape' as mating places has, in fact, already been shown for *Polistes* (Beani *et al.* 1992). Despite their rarity, *Polistes* social parasites are usually concentrated in valley areas. Here they probably have no difficulty in mate-finding, but they may run a high risk of inbreeding.

Altitudinal migrations may also be related to the abundant trophic opportunities present during the summer in mountain zones. However, it is probable that the trophic and reproductive advantages are associated, since trophic resources can favour meeting and breeding between partners (Turillazzi *et al.* 1993).

PARASITES

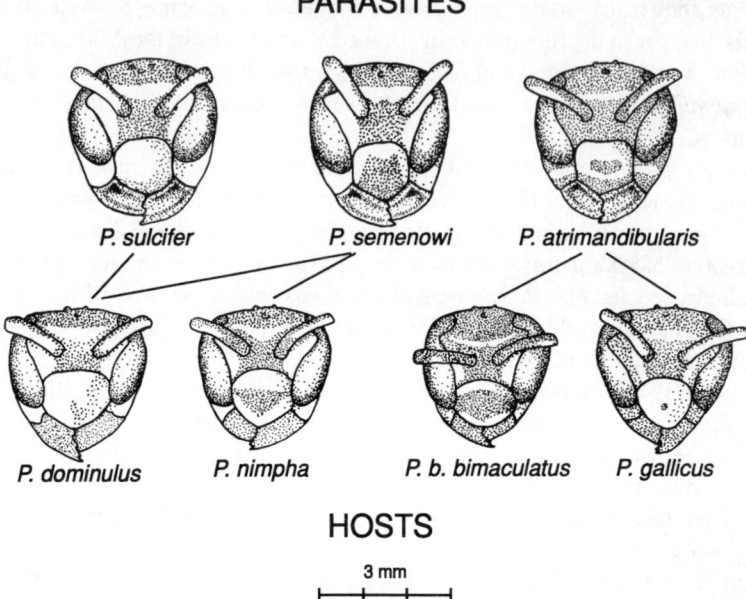

HOSTS

3 mm

Fig. 5.1 Heads of females belonging to the three species of *Polistes* obligatory social parasites (on the top) and the four host species (on the bottom). The lines match each parasite with the respective host species.

Hamilton (1972) points out that the migratory habits observed in some species of wasps and bees (in zones with colder climates) could lead to a decrease in genetic kinship within a population and thus favour the establishment of parasitic behaviour. These migrations could thus be involved in the evolution of parasitic behaviour (see West-Eberhard, this volume) and could help account for the particular geographic distribution of *Polistes'* obligatory parasites. The migratory habits would have then remained in the behavioural repertoire of the parasite species, facilitating the meeting of males and females (Cervo 1990).

All of the most common species of *Polistes* that live in Europe are subject to usurpations by at least one of the parasite species (Fig. 5.1). *P. sulcifer* parasitizes a single species, *P. dominulus*; the other two parasites have more than one host species. In particular, *P. semenowi* parasitizes *P. dominulus* as well as *P. nimpha,* and *P. atrimandibularis* can parasitize *P. biglumis bimaculatus* (mountainous) as well as *P. gallicus* (very common on the Italian plains). Experiments we recently conducted in the laboratory to evaluate the preference of host species, offering nests of different species to the parasites, have shown that *P. nimpha* is potentially a third host species for *P. atrimandibularis*, although this is undocumented in nature (unpublished observations).

Usurpations generally occur at a specific time considerably after colony foundation. This time of usurpation may be tied to the timing of emergence from wintering places, which occurs later in mountainous zones than on the plains (Cervo 1990). Recently collected data do not allow us, for the moment, to draw any conclusions about this hypothesis. A temporal gap between nest initiation by the host and invasion by the parasite has also been reported for various species of bumblebee and Vespinae social parasites (Taylor 1939; Alford 1975; Reed and Akre 1983; Fisher 1983).

This delay in usurpation gives the parasite access to pre-worker colonies of considerable size. These colonies will soon produce workers, which will provide a work force for the parasite and empty cells for oviposition by the parasite once it has removed the less developed immatures (Cervo 1990; Cervo *et al.* 1990*a*). Whereas colonies located in colder zones are usurped in the mid-pre-emergence period (Cervo *et al.* 1990*a*), colonies on the plains are usurped at the end of the pre-emergence period or even at the beginning of the post-emergence period (Cervo 1990). Since the colonial cycle in colder zones is shorter, usurpation must take place earlier for the parasite female to have sufficient time to raise her own offspring (Cervo *et al.* 1990*a*).

Field data indicate that, on average, parasites tend to usurp larger, more advanced nests among those available in the population (Cervo 1990). Binary choice experiments conducted on *P. sulcifer* in the laboratory confirm that parasites preferred larger nests in more advanced stages of development (Cervo *et al.* 1993). In another series of experiments in which nests offered, without adults, were hidden from the parasites' view by a sheet of thin blotting paper, *P. sulcifer* females recognized the host species' nest and chose the one most appropriate for usurpation solely by its scent, before establishing contact with it. The results appear to indicate that it is the scent of the immature brood, not that of the material with which the nest is built, that permits its identification. This ability to recognize the species and the stage of development of a nest to be usurped without alighting is certainly advantageous for a parasite, since it avoids pointless and risky landings on unusable nests (Cervo *et al.* 1993). Once the parasite female identifies the nest to usurp, she takes possession of it. Studies conducted in captivity show that *Polistes* parasites, similarly to Vespinae and bumblebee social parasites (Alford 1975; Jeanne 1977*b*; Greene *et al.* 1978; Akre 1982; Reed and Akre 1983, 1990; Fisher 1984), use both aggressive (Scheven 1958; Turillazzi *et al.* 1990; Cervo *et al.* 1990*b*) and non-aggressive (Demolin and Martin 1980; Cervo *et al.* 1990*a*; Schwammberger 1993) strategies.

Polistes parasites, besides possessing behavioural adaptations for invasion of a host nest, also have morphological traits, such as the development of certain body parts and a greater robustness of the exoskeleton, that give them an advantage during the fights with their adversaries (Cervo 1994). The mandibles, which are used as the principal offensive weapons during struggles with the hosts (Turillazzi *et al.* 1990), are particularly well developed in these parasites. The mandibles possess a longitudinal groove that may add strength, given the increased size of this body part (Cervo 1994). In *Polistes* parasites, in contrast to what is observed in other obligatory wasp parasites (Reed and Akre 1982; Ondricek-Fallscheer 1992), the sting has no morphological modifications, which is consistent with its lack of use in invasion fights (Cervo 1994).

P. sulcifer females always adopt an aggressive strategy for introducing them-selves into the host colony, which generally leads to the expulsion or death of the highest ranking host female (Turillazzi *et al.* 1990). *P. semenowi*, adopts an aggres-sive invasion mode but often permits the dominant female to remain on the nest (Cervo *et al.* 1990*b*). However, from observations of a *P. nimpha* colony, it seems that under certain conditions *P. semenowi* may also adopt a non-violent strategy for entering a host colony (Demolin and Martin 1980; Cervo *et al.* 1990*b*).

The third species, *P. atrimandibularis*, seems to have the most flexible strategy in that it adopts either a violent or a non-aggressive strategy depending on the spe-cies it is usurping and on the period of the season in which the invasion occurs. When parasitizing *P. biglumis bimaculatus*, a mountain species with solitary foun-dation, the *P. atrimandibularis* female uses a temporarily submissive invasion method, initially submitting to continued attacks of the host female before gradu-ally dominating her. Thus the parasite maintains the host female on the nest without injuring her: this permits the use of her labour until the emergence of the workers, which occurs some time after invasion (Cervo *et al.* 1990*a*). When this parasite usurps nests of *P. gallicus*, a plains species with solitary foundation, it adopts a non-violent yet less submissive strategy, and does not maintain the host foundress on the nest (Cervo *et al.* 1992). Recent observations of invasions of *P. nimpha* colonies (unpublished data), and of invasions of *P. biglumis bimaculatus*, but later in the season (Cervo, pers. obs.), indicate that in these cases *P. atrimandibularis* adopts a decidedly more violent strategy, and fights with the host females.

It has been suggested that non-aggressive usurpation followed by cohabitation of parasite and host is characteristic of the most specialized social parasites (Reed and Akre 1983; Fisher 1984, 1985), and that it depends on the specificity of the parasite to its host. A generalist parasite (i.e. one with many host species) should usurp a colony violently, since it does not possess species-specific behaviours and means of communication (Fisher 1984). However, the studies of *Polistes* parasite invasion methods indicate that it is the most generalist species that adopts a non-violent strategy followed by cohabitation with the foundress. This apparent contradiction might be explained by the possibility that *P. biglumis bimaculatus* represents the original host for *P. atrimandibularis* (as suggested by Scheven 1958), for which the parasite has thus evolved more specialized behaviours, and that the other two spe-cies it occasionally usurps represent more recent hosts.

Based on available data, it is quite probable that the parasite changes invasion strategies according to the situation in which the host nest is found; the plasticity of the nest invasion strategies seems, in fact, to be associated with the eco-ethological characteristics of the host species, such as the number of foundresses present on the host nest, the length of the colonial cycle of the host species and the point in the cycle at which the invasion occurs (Cervo *et al.* 1990*b*).

If the host foundress submits to the parasite in lieu of resisting her, she can still attain some reproductive success. In fact, from data available on *P. sulcifer* and *P. atrimandibularis*, a small number of the host's reproductives emerge from many parasitized colonies, demostrating that the host's reproductive success is not completely eliminated (Cervo 1990). The host can also achieve some reproductive success when, as has been observed by Mead (1991) in a *P. dominulus* colony, the *P. semenowi* usurper, after a period of oviposition on the host colony, abandons it for unknown reasons. This results in the production of reproductives of both species (Demolin, pers. comm. quoted by Röseler 1991; Mead 1991).

P. sulcifer and *P. semenowi* females perform an intense and prolonged stroking behaviour immediately after their arrival on the host colony (Turillazzi *et al.* 1990; Cervo *et al.* 1990*b*). This behaviour appears nearly identical to, though more intense than, that observed in intraspecific *P. biglumis bimaculatus* usurpers (Lorenzi and Cervo 1992) and in *P. dominulus* foundresses on normal colonies (Dani *et al.* 1992*a*). Compared with the other two parasites, *P. atrimandibularis* females stroke their abdomen with less intensity and frequency and also display a vigorous lateral vibration of the abdomen (Cervo 1990).

Stroking by females may function to apply a recognition pheromone to the nest. The workers can learn the odour of their nest and then compare it to that of the dominant individual. As already discussed for intraspecific parasitism, it could be a decisive factor in the parasite's success to be able to impose her own scent by masking that of the previous dominant, thus deceiving the workers about their kinship with her (Cervo and Turillazzi 1989; Cervo *et al.* 1990*a*; Turillazzi *et al.* 1990; Dani *et al.* 1992*b*). Alternatively, stroking behaviour may serve the parasite for assuming the host nest odour. Recent findings on *P. atrimandibularis* (see Lorenzi *et al.* this volume) suggest that *P. atrimandibularis* adopts this second strategy while usurping *P. biglumis bimaculatus* nests.

The offering of drops of liquid by the parasite to the hosts (Turillazzi *et al.* 1990; Cervo *et al.* 1990*a*) and the prolonged and intense licking that the parasite sometimes performs on the hosts, as previously observed by Scheven (1958), are probably other behaviours involved in a possible chemical control of the colony.

The most distinctive aspect of the behaviour of *P. atrimandibularis* is its ability simultaneously to control more than one host colony (Cervo *et al.* 1990*c*). *P. sulcifer* and *P. semenowi*, once established on the host nest, behave like typical obligatory parasites, never leaving the nest and depending entirely on the host workers to raise their own reproductives (Cervo 1990). *P. atrimandibularis* females, however, carry out intense extra-colonial activity during which they usurp other nests of the same host species (secondary nests). From these they take larvae and pupae and transfer them to the primary nest, where they are used to feed the immature offspring (Cervo *et al.* 1990*c*).

Thus, to define *P. atrimandibularis* as an obligatory and permanent parasite in Taylor's sense, as has been done until now, seems inappropriate. In fact, this parasite carries out colonial activities such as nest defence and thermoregulation, and helps provide food for raising offspring. Its behaviour differs from that of a *Polistes* foundress only in that it is incapable of founding a nest, although construction activity still seems to be part of its behavioural repertoire (Cervo *et al.* 1990*c*).

The predatory behaviour of *P. atrimandibularis* on secondary nests, described initially for *P. biglumis bimaculatus* colonies (Cervo *et al.* 1990*c*), has also been observed for the plains host species, *P. gallicus* (Cervo *et al.* 1992). It is difficult to ascertain whether *P. atrimandibularis* evolved this behaviour on *P. biglumis* *bimaculatus* in response to the host species' brief colonial cycle and thus to the necessity of collaborating with the host to obtain greater reproductive success, or whether this behaviour evolved independently of ecological conditions.

This kind of behaviour, in which an inquiline is also predator of its host species, has never been described for other social parasites. It differs from the robbing behaviour of some stingless bees in which the robbers invade the nests of other bee species to take from them the stored food and not the host's immature brood. It is different also from predatory behaviour reported in some ants (Hölldobler and Wilson 1990) because it is performed by an inquiline. The behaviour displayed by *P. atrimandibularis* could cast new light on the evolution of social parasitism in *Polistes*. The predation of immature broods of different species could, in fact, represent another pathway in the evolution toward obligatory parasitism different from that defined by Taylor. Alternatively, this behaviour could represent a subsequent adaptation in the stage of obligatory and permanent parasitism. In other words, once specialized in usurpation, and in the maintenance of rigid control over colonies of the host species, a parasite could increase its reproductive success by using this ability to plunder other nests to obtain food for its own offspring (Cervo 1990).

Other factors, such as the level of reproductive success on usurped colonies and the ability to maintain control of the host colony, can help us understand the evolutionary position of this parasite and thus define its level of specialization and dependence on the host species.

The mean number of the parasite's reproductives that emerge from colonies of *P. biglumis bimaculatus* invaded by *P. atrimandibularis* is roughly equal to the total number of individuals (including both workers and reproductives) that emerge from non-parasitized colonies (Lorenzi *et al.* 1992). This indicates that the parasite's success in terms of reproductive offspring is superior to that obtained by a host foundress nesting in the same environment.

The total productivity of parasitized colonies (workers and reproductives of the host, and reproductives of the parasite) is also greater than that of non-parasitized colonies, despite no significant difference in the sizes of the combs (Lorenzi *et al.* 1992). The greater number of individuals produced may be due to use of the ring of peripheral cells (which is rarely used by *Polistes*) for egg-laying by the parasite (Lorenzi *et al.* 1992). Thus, the female parasite manages to obtain a high number of reproductives by emptying the cells containing the host's immature offspring and replacing them with her own eggs; by controlling the host's reproduction; by contributing to the raising of the immature offspring; and finally by using all the space available on the nest (Cervo 1990). The host's productivity, by contrast, is severely limited by the parasite's exploitation: the number of host progeny that emerge from a parasitized colony is about half that of a non-parasitized colony (Lorenzi *et al.* 1992). From these data it is evident that *P. atrimandibularis* achieves high reproductive success when it usurps *P. biglumis bimaculatus* colonies.

Analysis of the ovarian development of host workers taken from *P. biglumis bimaculatus* colonies after a certain period of cohabitation with *P. atrimandibularis* indicates that this parasite exercises complete reproductive control. The control is similar to that of host queens in normal *P. biglumis bimaculatus* colonies (Cervo and Lorenzi 1994).

On colonies invaded by the other two parasitic species, *P. sulcifer* and *P. semenowi*, it is rather common to observe the workers laying eggs undisturbed (Cervo 1990). Dissection of workers from parasitized nests taken from nature or kept in the laboratory indicate, however, that *P. sulcifer* is capable of controlling the reproductive capacity of its host, as a dominant female does with her workers on a normal nest (Turillazzi *et al.* 1991). This control is not complete, as is demonstrated by the presence of fertile workers found by dissecting females taken from colonies parasitized by *P. sulcifer* in nature. This also occurs, however, in non-parasitized colonies (Turillazzi *et al.* 1991).

The presence of a *P. biglumis bimaculatus* foundress on colonies parasitized by *P. atrimandibularis* opens the possibility that it is the host queen, rather than the parasite, that inhibits ovarian development in emergent workers. Cohabitation of this parasite with the host colony's foundress could indicate, as we have seen, a more advanced form of parasitism, involving subjugation of the host queen by chemical and/or behavioural means, or it could be a necessity dictated by the inability of the parasite to maintain reproductive control over the host colony (Cervo 1990; Cervo and Lorenzi 1994). The latter possibility would suggest the presence of a less advanced grade of parasitism. Some preliminary data seem to indicate that the presence of the host foundress is necessary for the success of the parasitized colony (Cervo and Lorenzi 1994).

It is clear that wherever we place the predatory behaviour of *P. atrimandibularis* in the outline of the evolution of social parasitism, the path of temporary parasitism proposed by Taylor is a highly probable route for the evolution of inquilinism in *Polistes*. In fact, as has been illustrated here, the behaviour of facultative parasites seems to be very similar to that of obligatory parasites. Furthermore, the mechanisms with which a foreign female introduces herself into the social system of a colony and takes control of it seem to be similar in intraspecific and interspecific usurpation (Cervo and Turillazzi 1990). The behaviours linked to obligatory social parasitism are thus part of the behavioural repertory of *Polistes* foundresses (Cervo 1990). It is plausible that the various forms of temporary and permanent parasitism are each an adaptation to particular ecological/climatic conditions and do not necessarily have to evolve toward inquilinism (Yamane 1978, quoted by Matsuura and Yamane 1990).

Recently West-Eberhard (1986*b*) proposed that obligate parasite behaviour might evolve through the fixing of a pre-existing intraspecific parasitic behaviour without requiring the existence of behaviourally intermediate species. This appears to be true in *Polistes*, given the resemblance between intra- and interspecific parasitic characteristics.

In conclusion, similar to what emerges from the analysis of the *Polistes* social system (Reeve 1991), recent data obtained on social parasitism in *Polistes* show a notable flexibility in the strategies adopted by facultative and obligatory social parasites, probably dictated by variation of ecological conditions.

Since research in this field is in its early stages, there is useful information yet to come for the expansion of our knowledge about social parasitism in *Polistes*, and for the evaluation of its role in the evolution and maintenance of sociality in these insects.

Acknowledgements

This review is a synthesis of the work carried out by many researchers during the past several years under the guidance of S. Turillazzi, whom we also thank for the useful suggestions offered on this work.

We are also grateful to G. Gamboa, B. Wcislo and M. J. West-Eberhard for their helpful critical reading of the chapter and to A. Eberhard for the revision of the English text.

6

Lek-like courtship in paper-wasps: 'a prolonged, delicate, and troublesome affair'

Laura Beani

Sexual selection and prolonged male displays

Courtship may be 'a very tedious affair, going on hour after hour.' With these words two American experts on spiders, George and Elisabeth Peckham (1889, quoted in Cronin 1991), remarked on the long wait for successful matings that characterizes much research on sexual behaviour. Their study was one of the early detailed attempts to test Darwin's theory of sexual selection by means of a female 'aesthetic' choice. They concluded that 'the males vie with each other in making an elaborate display, not only of their grace and agility, but also of their beauty, and that the females, after attentively watching the dances and tournaments (...), select for their mates the males that they find most pleasing.' One century later, 'dancing' and 'athletic ability' of males in the courtship of fruitflies has been explained as a costly, therefore an honest, signal of mate quality; females mate 'only with those males that are able to keep up' with them during the courtship dance (Maynard Smith 1991).

Although the perspective within the Darwinian framework for sexual selection is changing, the Peckhams' study contains some issues suitable to introduce a review about mating behaviour in paper-wasps. First is the critical role of female choice— for a hundred years a rather neglected topic in discussions of evolution. Maynard Smith recently noted in his foreword to the book of Helena Cronin, *The Ant and the Peacock* (Cronin 1991): 'when, in 1956, I published a paper showing how female fruit flies choose males, I do not remember receiving a single reprint request.' Second, the same 'elaborate display' may be involved in both intra- and inter-sexual selection, without any rigid dichotomy (Halliday 1992). Third, a group of males displaying to each other are really competing as if they were actually fighting (West-Eberhard 1983). Fourth, a main function of courtship is quality advertisement (Trivers 1972). In view of a subtle mate assessment, protracted and 'very tedious' displays seem to provide only redundant information about male conditions and/or attributes.

In this review, I am concerned with mating strategies in paper-wasps, and mainly with the temporal and spatial aspects of male behaviour. Whereas courtship dances may go on for hours, males of several species of *Polistes* occupy contiguous territories which do not contain resources, for days, for weeks, sometimes for months. These aggregation sites are scent-marked, continuously patrolled and defended against intruders, but only sporadically visited by females. Lek-territoriality, implying both struggles and self-advertisements, is involved in intra- and inter-sexual selection, even if female mate choice is more obscure than in courtship dances. This paper has two main goals: first, to review detailed reports of mating behaviour in the genus *Polistes*, focusing attention on lekking at landmarks (the commonest mating system), on intraspecific plasticity, and on female mate choice at arenas; second, to explain why male activity is so prolonged when female presence itself is sporadic, and why matings are observed so infrequently in *Polistes*. About 60 years ago, Phil Rau 'wondered why one sees in the literature no record of the mating of *Polistes*' (1929*a*). Still today, the paucity of such reports is a mystery in the subject of insect leks.

Resource- and female-defence: two neglected strategies in paper-wasps

In a total of 24 *Polistes* species (Table 6.1), male activity is centred on nesting/ hibernation areas in seven species (eight including *P. canadensis* of Colombia); on landmarks (more or less conspicuous, ranging from hilltops to a few stones in the fields) in 14; and on both (landmarks containing hibernacula) in three species. By contrast, foraging areas are rarely patrolled by males. This agrees with the predictions of Emlen and Oring (1977): scattered feeding sites (i.e. flowering patches, caterpillars) and a rather unspecialized foraging behaviour do not favour a resource-defence strategy, which in this survey occurs only as an alternative tactic (by smaller *P. fuscatus* individuals), or sporadically, in relation to an unpredictable mountain environment (in *P. atrimandibularis* and *P. biglumis bimaculatus*).

Male paper-wasps do not appear to monopolize access to females, even when male activity is focused on nesting (*P. carolina, canadensis* of Colombia, *chinensis antennalis, erythrocephalus, tepidus*) and hibernation sites (*P. fuscatus, jadwigae* and *versicolor*). There is no evidence of any direct, aggressive, exclusive control of harems of emerging/hibernating females, despite the fact that the potential for mate monopolization ought to be high where females are spatially and temporally clustered (Emlen and Oring 1977). In paper-wasps, female-defence polygyny appears to be curtailed by a combination of causes (Beani *et al.* 1992): the aggressiveness of workers at nests, which drive 1–2 week old males away from their natal nest, at least in *P. dominulus, P. gallicus* and *P. biglumis bimaculatus* (attacks against males tethered near nests were not observed, however, in *P. canadensis*, Polak 1992); the non-receptivity of females at nests (*P. canadensis*, Polak 1992), where gynes are not inseminated (*P. fuscatus,* Noonan 1978); the presence of already fertilized females at hibernacula (*P. fuscatus,* Noonan 1978; *P. dominulus*, Turillazzi 1980).

Mate-searching efforts are focused in proximity to—rather than directly on— female concentrations. For example, *P. erythrocephalus* males 'were stationed near the flight path of females attending newly founded nests', and *P. fuscatus* males 'often sit near cracks through which females enter hibernacula' (West-Eberhard 1969*b*). There are some reports of sexual activity directly on nests in *P. tepidus*

Table 6.1 Nuptial systems and sexual records in 23 *Polistes* spp.

Polistes species	Focus/and alternative sites of male displays In brackets: activity period	Male behaviour/and alternative tactics	Sexual interactions: c=copulations a=copulatory attempts In brackets: mating season	References
annularis	Landmarks (tall buildings) (Oct–overwintering)	Interspecific leks, hibernant clusters	Not observed; high % of fertilized females (in spring)	Rau and Rau 1918, Hermann and Gerling 1974, Lin 1972
atrimandibuaris	Ground-level landmarks (scattered stones)/resources (flowers) (Aug–mid-Sept)	Leks/resource defence	Not observed	Turillazzi *et al.* 1993, Cervo, pers. comm.
bellicosus	Prominent landmarks (fire and bell towers) later used as hibernacula (mid-Oct–Dec)	Leks and heterospecific swarms	13 pairs (late), which fell to the ground; 12 fertilized	Reed and Landolt 1991
bernardii comis	Prominent landmarks (a line of poles)	Collective patrol routes	Not observed	Richards 1978b
biglumis bimaculatus	Ground-level landmarks /resources (stones, green and flowering patches) (mid-Aug–mid-Sept)	Leks/resource defence	3 **a** on nests (early), 1 **a** on flowers (early), 2 **c+1 a** on lek (late); **c+a** on tied females	Beani and Lorenzi 1992, Beani *et al.* 1992
canadensis (Costa Rica)	Landmarks at hilltops (trees along ridges) (mid-May–mid-Sept)	Leks/patrol routes through territories	1 **c**; **c+a** on tied females	Polak 1992, 1993a, b
(Colombia)	Newly founded nests	Territoriality	Not observed	West-Eberhard 1982b

Table 6.1 (*cont.*)

Polistes species	Focus/and alternative sites of male displays In brackets: activity period	Male behaviour/and alternative tactics	Sexual interactions: c=copulations a=copulatory attempts In brackets: mating season	References
carnifex	Landmarks at hilltops (trees)	Leks	Not observed	Polak 1993b
carolina (=*rubiginosus*)	Nesting sites (a hole in a wall)	'Ball-like mass' of males	c+a on newly-emerged queens (late)	Rau 1929a
chinensis antennalis	Nesting sites (shelters)	Heterosexual, pre-hibernant clusters	Observed; high % of fertilized females (late)	Yoshikawa 1963b, Kojima and Suzuki 1986
commanchus navajoe	Landmarks at hilltops (trees, shrubs, rocks) (mid-Jul–Oct)	Leks	4 c in 100 h of obs. (late) 6 c in 100 h of obs. (mid-Aug)	Matthes-Sears and Alcock 1986
dominulus (cited as *gallicus*)	Prominent landmarks (tall trees, electricity poles (mid-Jul–Sept)	Leks/patrol routes through territories	6 c+13 a (late); c+a on tied and dead females	Beani and Turillazzi 1988a, 1990a
dorsalis	Prominent landmarks (fire and bell towers) used as hibernacula (mid-Oct—Dec)	Leks and heterospecific swarms	a (late)	Reed and Landolt 1991
erythrocephalus (=*canadensis*)	Nesting sites (bushes)	Territoriality at perches	1 c 'on a waning nest'; high % of unfertilized new-foundresses	West-Eberhard 1969b, Hamilton pers. comm.
exclamans	Landmarks (tall buildings)	Leks/non-aggressive	Not observed	Lin 1972

	Nesting and hibernation sites/ landmarks/resources	Territoriality/swarms/ patrol routes at resources		
fuscatus	Nesting and hibernation sites, landmarks (buildings, towers)/resources (flowers) (Aug–Dec)	Territoriality/swarms/ patrol routes at resources	Not observed at nests; c in a crevice; a around towers (late); a+c on tied females	West-Eberhard 1969*b*, Noonan 1978, Post and Jeanne 1983*a*, Reed and Landolt 1991
gallicus (=*foederatus*)	Landmarks (low and tall trees, poles) (Sept–mid-Nov)	Swarms and collective patrol routes	8 **c**+7 **a** (late); **a**+**c** on tied females	Distefano 1972, Beani and Turillazzi 1988*b*, 1990*a*, *b*
gigas	prominent landmarks (tall trees) (Oct–Nov)	Territoriality (exploded leks?)	Not observed	Yamane, pers. comm.
jadwigae	Hibernation sites (crevices in walls, eaves) mid-Oct–mid-Nov	Territoriality	42 **c** at crevices; high % of females inseminated in Nov (late)	Kasuya 1981*a*
major	Landmarks in a gully (trees, bushes)	Individual perches (leks?)	**a** on tied females	Wenzel 1987*a*
metricus	Prominent landmarks (fire and bell towers) used as hibernacula (mid-Sept–Dec)	Leks and heterospecific swarms	**a** (late)	Reed and Landolt 1991
nimpha	Horizontal landmarks (hedges, fences, enclosure walls) (Sept–Oct)	Leks	2 **a** at leks (late)	Turillazzi and Cervo 1982, Beani *et al.* 1992
semenowi	Ground-level landmarks (slabs, green tufts, heaps of stones) (mid-Aug–Sept)	Leks/patrol routes	1 **a** at leks; **c** on tied females	Lorenzi *et al.* 1994*a*, Cervo, pers. comm.
tepidus	On nests (Feb–Mar)	Courtship	Multiple **c**	Hook 1982
versicolor	Hibernation sites (crevices)	Courtship?	Observed (late)	Hamilton, pers. comm.

(Hook 1982) and, more rarely, in *P. erythrocephalus* (one case of 'male and female in copula on a waning nest', West-Eberhard 1969*b*). Matings were also observed in pre-hibernation clusters (*P. carolina, chinensis antennalis, fuscatus* and *versicolor*). However, males are generally absent from both nests and hibernacula during the core of their flight season. For example, in some Italian *Polistes* reviewed here (*P. biglumis bimaculatus, dominulus, gallicus, nimpha*), sporadic copulations and copulatory attempts on nests and at hibernacula might represent marginal tactics (Beani *et al.* 1992).

Lek-like aggregations at landmarks

For the 24 species surveyed here (Table 6.1), the key word is probably 'lek', i.e. an aggregation of males displaying in an arena, where they defend more or less clumped territories (in our study area, a few branches of a tree, some stones, a bush, segments of fences or poles, corners of buildings, and so on). These so-called 'symbolic' territories contain no resources useful to females, but are only display and mating sites, basically stable in space and in time ('traditional nuptial arenas', for example a mulberry-tree occupied by *P. dominulus* males each summer between 1982 and 1987, until it was destroyed). The lek, a Scandinavian word meaning 'play', term was used by Darwin to define the encounter sites of some galliforms without paternal care which mate on particular patches of ground used solely for this purpose. In this bare stage, the only performance is the contemporaneous, repeated, sometimes bizarre exhibition of males: in the case of paper-wasps, struggles, pursuits, threats and self-advertisements (flights, marking, attendance at habitual perches and so on).

Leks at landmarks are observed in 10 species, 12 if we include two uncertain cases: *P. gigas* males defend tall trees in Ken-Ting Park, in Taiwan, which is perhaps an example of 'exploded' lek; *P. major* males are territorial in the laboratory, but less clearly so in the field. Males of a further two species (*P. bernardii comis* and *P. gallicus*) patrol and mark prominent landmarks in swarms, with a site-tenacity comparable to that of a territorial species. Insect swarms of varying configurations have recently been considered to be lek-like mating aggregations (Bradbury and Davies 1987; Svensson and Petersson 1992). In line with this unifying view, the same trees and poles have been adopted both as swarm markers in *P. gallicus* and as territorial arenas in *P. dominulus* (Beani and Turillazzi 1990*a*). The presence of displaying males may reinforce the initial settlement at landmarks through scent-marking (the release of pheromones probably occurs when males rub their abdomens, faces, and legs on habitual perches; see Beani and Calloni 1991*b*).

Lek-territoriality and swarming are described in a further three species (*P. bellicosus, dorsalis* and *metricus*) around bell and fire towers where females overwinter. This intermediate category could perhaps include *P. fuscatus* males, which gather on tall sunlit buildings with nests and hibernacula. Indeed, the classical definition of lek as non-resource based territoriality at arbitrary sites, was refocused in the broader 'hotspot model' (Bradbury and Gibson 1983), which predicts that males settle where the probability of encountering receptive females is high (i.e. near hibernacula). I wonder, along with Bradbury (1985), 'whether the so-called arbitrary sites are truly arbitrary, or whether they are also modulated by resource distribution and female transit routes.' Landmarks themselves may fall within a 'resource proximity strategy', *sensu* Bradbury. At least in the Italian *Polistes* considered here, landmarks are in the middle of waste lands and near abandoned farm buildings containing nests and hibernacula; puddles and flowering patches in the surroundings are potential foraging areas.

Landmarks, besides being strategic mate-encounter points, may represent major orientation cues. The landscape is one of the orienting factors for paper-wasps. In displacement experiments (Ugolini and Cannicci this volume), *Polistes* females were able to orient themselves homewards by using both sun and landmark sightings. If landmarks have orientation value, then territoriality at landmarks is inappropriately defined as merely 'symbolic', as Pardi (pers. comm.) has remarked. Landmarks may occasionally constitute transit nodes and suitable stopping sites. For example, *P. major* males patrol the trees of a gully, a typical bottleneck in the desert environment. *P. canadensis* males defend trees and shrubs growing along prominent ridges, which may serve as orientation guides and natural barriers to wasps travelling across flat plains. Again, *P. semenowi* and *atrimandibularis* males form leks at peaks of Central and South Italy, on light-coloured stones (in heaps for the former species, scattered in the grass for the latter; Cervo, pers. comm.). Peaks represent orientation guides and mate-encounter points in relatively rare species, like these social parasites of other paper-wasps. Finally, several species of insects fly along the gravel-bed where the leks of *P. biglumis bimaculatus* are situated. The apparently insignificant stones of this dry stream combine the conspicuousness of a white, glossy surface against a green background with an advantageous microclimate in a mountain environment—warmer temperatures as a result of their thermal inertia and their position at ground level.

Behavioural plasticity

A long-term study on *P. biglumis bimaculatus* (Beani and Lorenzi 1992) offers a good example of the flexible settlement rules of male clusters. During two successive breeding seasons (August–September 1988 and 1989), classical arenas of tiny, contiguous, scent-marked territories (a few stones each) were located along the above mentioned dry stream, a white 5–10 m wide strip in the green meadows. At the beginning of August 1989, after a dry spring and summer, a small group was also territorial on a nearby flowering tuft of *Cirsium eriophorum*, visited by many insects (*Polistes* females included) owing to the scarcity of nectar sources. But this resource-defence represented only a marginal tactic, flowering plants being perhaps too numerous and scattered in 1988 and too difficult to defend from intruders in 1989. Moreover, there were some copulatory attempts on parental nests early in the season, before lek formation.

In 1990, after another dry season, a summer flood enlarged and excavated the gravel-bed, covering pebbles with mud; consequently the formerly white strip assumed an irregular and less distinctive appearance. In this altered scenario, territories were centred on low green bushes of *Adenostiles alpina* projecting onto the stream. Since summer flowering was extremely scarce, then these plants combined a certain degree of attractiveness, because of the presence of aphid honey and shelter from the wind, and conspicuousness, on account of their location along the borders. One year later, in 1991, the dry stream resembled a moraine, a 10–50 m wide strip running through grassland. Males gathered on green tufts (both with and without flowers or aphids) growing along the gravel-bed and standing out against the clear background.

In 1991, in a tract of stream 30 m long, I searched for territorial males on 30 tufts of *Adenostiles alpina* of similar size (diametre: about 1 m; height: 30–50 cm), all lacking in food sources: 15 tufts were scattered among stones on a light-coloured background, and 15 were in the grass, each at a distance of 3–5 m from a tuft among stones. *P. biglumis bimaculatus* males were present on only 1 out of 15 tufts in the grass, but on 11 out of 15 tufts in the stones (Chi-square=13.86, d.f.=1, P<0.001). Inside a traditional mating area, the river-bed, the focus of male activity ranges from feeding to landmark sites, each of which has a certain conspicuousness as a display site and is compatible with the strict thermal requirements of this mountain species. Even if their behaviour has not been extensively studied, the males of *P. atrimandibularis*, a social parasite of *P. biglumis bimaculatus*, also appear to be territorial at both flowers and stones.

Intra- and inter-season changes suggest an 'adaptive behavioural plasticity (...) in response to environmental cues' (West-Eberhard 1989) and to strong competition for mates. The plasticity of mate-location by males is a recurring trait in paper-wasps. Strict lek-territoriality combines with wide-range patrolling in *P. canadensis*, *dominulus*, *fuscatus*, *semenowi* and perhaps *exclamans* (see Table 6.1); in some species, the territorial option is size-dependent (Post and Jeanne 1983*a*; Beani and Turillazzi 1988*a*; Polak 1993*a*), but an individual may switch between different tactics. Sexual behaviour may be particularly subject to geographic variation (West-Eberhard 1983). 'The same species may have different habits in different countries', Darwin noted (1871) about the occasional 'nuptial assemblages' of some birds. Indeed, *P. canadensis* males in eastern Colombia exhibit territoriality at nests (West-Eberhard 1982*b*), whereas in Costa Rica they perform lekking/patrolling behaviour at landmarks (Polak 1992); similar cases are likely to occur elsewhere but, due to the lack of comparative and long-term studies, the phenomenon has been under appreciated.

Mate choice at male aggregation sites

Despite a certain degree of behavioural plasticity, the behaviour of *Polistes* males follows the rather fixed and general rule of unisexual aggregations at peculiar sites. These assembly points, which are sometimes multi-specific (Lin 1972; Beani and Turillazzi 1988*b*; Reed and Landolt 1991; Turillazzi *et al.* 1993), may be more or less conspicuous, close to resources and crowded with males. The patrol routes of several males from one marking site to another produce a notable visual/chemical signal: a wind-tunnel experiment has recently revealed the presence of male sex attractants in *P. exclamans* (Reed and Landolt 1990). In response to species-specific pheromones released on habitual perches, *P. fuscatus* and *dorsalis* females 'fly upwind in a zigzag pattern to a male perch site' (Reed and Landolt 1991). The only natural copulation observed in *P. canadensis* occurred 'on a leaf in an occupied territory' (Polak 1993*a*), and in *P. major* 'on one of the male's marked perches' (Wenzel 1987*a*). Females of *P. comanchus navajoe* mate mainly with perch holders (Matthes-Sears and Alcock 1986). In *P. jadwigae*, the territorial 'abdomen-bending males' copulated at marking sites near hibernacula (Kasuya 1981*a*).

Data collected on the Italian *Polistes* species (Beani and Turillazzi 1988*a, b*, 1990*b*; Beani *et al.* 1992) confirm a central role of scent-marking and lekking/ swarming behaviour in male mating success. Sexual interactions occur mainly at habitual and marked perches in all the species. Tied females are easily detected only if placed at marking points. In two laboratory experiments, the more active males of *P. dominulus* and *gallicus* paired more often than individuals with low frequencies of flights, fights, marking and so on. In the field, residents interacted with females more than did transients in *P. dominulus*, and routine patrollers did so more often than newcomers in *P. gallicus* (Beani and Turillazzi 1988*a, b*, 1990*a*). The few natural copulations recorded in *P. biglumis bimaculatus*, *nimpha* and *semenowi* were always achieved by territorial males at leks (Beani *et al.* 1992, Lorenzi *et al.* 1994*a*).

These data suggest a possible mate choice by females. Receptive females, crossing areas where males are conspicuous and active, might stop at highly contested and marked perches, where they have a good chance of finding a 'superior' pre-tested mate (that is, a territorial male, a winner during recent fighting, a male larger than average in the case of size-conditional territoriality, or a male that is fit for patrolling activity and scramble-competition). The opportunity for female choice in *Polistes* is enhanced by a general control mechanism over sperm transfer and egg fertilization (see Eberhard 1985); for example, *P. fuscatus* females may be unfertilized even after a long genital linkage (Post and Jeanne 1983*b*), although fertilization could fail because of different factors (e.g. no sperm available, non-functioning genitalia, etc.). In this species, as in *P. major* (Wenzel 1987*a*) and *P. dominulus* (Beani and Turillazzi 1988*a*), there is also some evidence for female ability to dislodge males grasping them (West-Eberhard 1969*b*; Post and Jeanne 1983*a*). A low level of female receptivity was also observed in mating trials with *P. fuscatus* (Larch and Gamboa 1981; Post and Jeanne 1983*b*), *P. exclamans* (Reed and Landolt 1990) and *P. canadensis* (Polak 1993*a*).

According to the hypothesis of Alcock for hilltopping insects (1987), there are different grades of female choice of males gathered at display sites: from a default strategy (the visiting females simply cross nuptial arenas and mate with males able to intercept them, likely the winners of scramble-competition), through the selection of a popular perch site (high probability of mating with territorial residents), to a subtle mate assessment after direct inspections of displaying males, as proposed for the dances of spiders and fruitflies. All these degrees of mate choice are probably represented in paper-wasps. Several males may grapple around a gyne (i.e. *P. gallicus*). Perch holders often achieve a higher success than non-territorial floaters, as described above. An example of sequential approach to possible partners is provided by some cage observations on *P. dominulus*: a female stopped at perches of different residents and mated with two of them (Beani and Turillazzi 1988*a*).

Rarity and timing of sexual interactions

Apart from a more or less evident female choice, the key problem is the low frequency of sexual interactions observed at most of these display sites, which are the focus of male activity and presumably also of mating behaviour. Alcock, in his review on leks and hilltopping insects (1987), reported a rate of one mating every 10–60 hours of observation (11 during 100 hours in *P. comanchus*), but for many species no matings were recorded, in spite of the time spent at hilltops. In two field studies conducted on *P. dominulus* and *gallicus*, the rate was comparable (respectively, one mating about every 40 and 25 hours). Alcock proposed that several factors may combine to yield low observation rates of mating: a single mating per season in many species (but probably not in *Polistes*: see Metcalf 1980; Page and Metcalf 1982; Lester and Selander 1981, in addition to the above-mentioned case of multiple copulations in *P. dominulus*); an adult life span of some weeks; brief duration of copulations, which are thus easily overlooked; the existence of several alternative mating tactics that it is difficult to identify and investigate simultaneously.

Among the possible causes of the paucity of mating reports, I would stress the lengthy male displays. In 18 out of 24 *Polistes* species (Table 6.1), male activity covers a period of at least one month. Focal observations and long-term studies in Italian *Polistes* revealed 3–5 h of patrol/perching behaviour every sunny day, over a mean period of 14 days in *P. dominulus* residents and 13 days in *P. gallicus* routine patrollers (Beani and Turillazzi 1988*a*, 1990*a*). These figures underestimate the length of time males spend at display sites, because a male might explore various landmarks before settling on one site, or simply might not be censused immediately. The maximum site-fidelity was 34 days for the former species and 35 days for the latter. Durations were shorter in *P. nimpha* and *P. biglumis bimaculatus* (respectively, 21 and 17 days), whose male aggregations were less extensively sampled.

Considering the very low daily probability of mating, males could increase their opportunities for sexual interactions just by spending as many days as possible at nuptial arenas. In such a system, male fitness may be a function of the number of days spent displaying. But some temporal features of this strategy remain rather obscure. At least in *P. dominulus* and *P. gallicus*, there is a noticeable discrepancy between the peak of male density at aggregation sites and the peak of female visits to the same sites. At two tall trees occupied by *P. dominulus* males, 28 out of 55 conspecific females (about a fifth of the male population censused there: 284 individuals) were captured in September and October, whereas the maximum male presence was in August; furthermore, 18 of 35 females collected at landmarks around the middle of September were unfertilized. On the same two trees, 615 *P. gallicus* males were censused between September and November. The highest male density was recorded in the first 15 days of October, while 25 of 51 conspecific females were captured there after mid-October (8 out of a sample of 14 were unfertilized).

The timing of the few natural matings observed in paper-wasps agrees with a certain delay in female arrival at male aggregation sites. Sexual interactions occur 'late'—relative to the male flight season—in 9 of the 18 lekking/swarming species (Table 6.1, data in brackets). For example in Italian *Polistes*, copulations and copulatory attempts—i.e. without genital linkage—were recorded between 20 August and 17 September in *dominulus*, between 12 October and 2 November in *gallicus* (except one copulation on 28 September), between 27 and 31 October in *nimpha* and between 23 August and 8 September in *biglumis bimaculatus* (except 3 copulatory attempts at a nest on 18 August). By contrast, several successful experiments with restrained females have shown that males are sexually mature and motivated from their first settling at landmarks.

However, the increase of matings in late-season could be just a side-effect of changes in numbers of flying males: at the end of the activity period, a low male density ought to facilitate the observation of mating. At least for *P. dominulus* this possibility can be rejected: in correspondence with a decreasing number of *P. dominulus* males at landmarks, a massive arrival of *P. gallicus* swarms occurred at the same sites (Beani and Turillazzi 1990*b*); thus, observations in September (when most copulations occurred) were perhaps more difficult than earlier. Again, focal observations of the same territorial individuals at leks confirm a rather delayed mating success (respectively, after 12–23 days from the first appearance at the lek for *P. dominulus* and after 9–14 days for *P. biglumis bimaculatus*).

The 'marathoner' hypothesis

The lengthy nature of courtship was first noted by Darwin (1871), in insects (Chapter XI: 'The courtship of butterflies is a prolonged affair'), as well as in other groups (Chapter XIX: 'the persevering efforts' in producing sounds 'during the season of courtship by many insects, spiders, fishes, amphibians and birds'). Some recent studies indicate that females of different species use at least indirectly the duration of sexual displays and male attendance at nuptial arenas to select a mate (e.g. crickets: Hedrick 1986; frogs: Halliday 1987, Ryan 1988; sage grouse: Gibson and Bradbury 1985; red deer: McComb 1987; see also Burk 1988). As regards paper-wasps, could three rather enigmatic findings—low mating frequency, prolonged male activity, copulations late in the season—be related to one another and be involved in mate choice? A possibility is that females approach displaying males primarily in the second half of the male flight season, when only vigorous males still persist in their displays and the territorial system is well-defined. The paucity of mating reports might be mainly a by-product of the 'waiting strategy' of both females and males, rather than a result of constraints in observations, failure to locate prime male activity sites, rarity of sexual events, and so on. The daily distribution of sexual interactions (nearly always in the afternoon, at least in Italian species) supports the idea of a potential female screening for persevering males.

In line with this 'marathoner' hypothesis, which is only tentative given the scarcity of reports of mating, patrollers could 'parasitize' true marathoners at leks, avoiding the costs of territoriality. This could be the case for *P. dominulus* transient males, which patrol different landmarks and shift from one perch to another, not defending and rarely marking them, but intercepting females at these points in the absence of their legitimate owners (three copulations and five copulatory attempts, typically at usual perches, during field and cage observations). If a certain degree of plasticity exists, such males might act like marathoners only late in the season, when the payoff would be higher and the cost lower, due to the reduced number of rivals. Indeed, in a cage experiment, three individuals switched from wide-patrolling to territorial behaviour when three residents disappeared at the end of August (Beani and Turillazzi 1988a).

If females arrive only at the middle of this 'endurance competition', why don't all males wait to occupy a territory, initially behaving as transients or remaining inactive? It appears that only long-term participants are assured of a place in the final contest. The turnover at territories is limited, at least in temperate species in which the male life span coincides with the male flight season (see Polak 1993a, for a tropical species). When the last *P. nimpha* residents disappeared, the lek on the hedge simply finished (Turillazzi and Cervo 1982). In territorial interactions, 'owners, as a rule, won the contests' (Kasuya 1981a); this is true for *P. jadwigae* as well as for *P. comanchus* ('most male-male encounters (552 chases at all) ended quickly with the departure of the intruder male', Matthes-Sears and Alcock 1986), *P. fuscatus* ('in all but two observed grapples (n=44) the resident male repelled the intruder', Post and Jeanne 1983a), *P. exclamans* and *annularis* ('owners were never observed to yield their perch to aggressive intruders', Lin 1972). Lek structure itself seems to prevent tardy intrusions. In leks of *P. dominulus*, *nimpha* and *biglumis bimaculatus*, territorial males experimentally removed from stable arenas were replaced by neighbouring residents more frequently than by newcomers. In the last species, lower levels of aggression towards neighbours than strangers (more grasps and dorsal contacts, fewer threats; Beani, in prep.) could further reward a prolonged and faithful attendance at leks.

Besides these constraints to late settling in a lek, cheating could be hindered by individual recognition of perch holders by females. Lloyd, who considered this possibility in a paper on identification and sexual selection in a bumblebee and other insects (Lloyd 1981), mentioned dominance hierarchies in *Polistes* as an example of complex recognition. 'Individuality and mate choice go hand in hand', Lloyd remarked, stressing precisely the role of male 'endurance' in sexual selection. 'Successes in holding a prime position in a lek, at a nest hole, and on a territory, and in continuing the regular rounds of a circuit over days, all give a partial measure of endurance. (...) By scoring endurance females have a good and reliable measure of male quality'. If *Polistes* females can recognize and avoid short-term cheaters (those that slip onto perches late in the season), it is reasonable that males would set up territories early in the season; moreover, they might achieve some early matings. The high number of reports of females reluctant to couple at display sites (most of the 'copulatory attempts' of Table 6.1) could be explained in terms of mate selection by the females.

The 'marathoner hypothesis' is similar to a mate choice rule proposed for vertebrates, namely 'an old male is a good male' (Trivers 1972; Halliday 1983).

Male emergence is rather synchronized in many paper-wasps, preceding the emergence of future foundresses. Therefore, an individual displaying at landmarks relatively late in the season is likely to be a mature experienced male. This is supported by the finding that mainly yellow-eyed subjects attend male aggregations (pers. obs.); eye colour, which changes from black, through brown, up to yellow (Strassmann 1981a), is an age-dependent trait. Moreover, persistent males at display sites also could be healthy individuals (i.e. non-parasitized, see Polak 1993b). Finally, from a female's point of view, waiting before copulating would be a parsimonious way of screening for superior mates, since the females have to overwinter before egg-laying. In tropical species of wasps, in which mating behaviour is not restricted to a season as in temperate latitudes, daily temporal features of male displays might also play a role. A preliminary study on Stenogastrinae wasps revealed that the likelihood of sexual interactions and successful matings is higher in the second half of the daily flight period, when only a few males persist in their courtship 'stripes-display' and defend their aerial territories (Beani and Turillazzi 1994).

Regarding the 'curious nuptial assemblages' of birds, Darwin noted (1871, Chapter XIV): 'the stronger males would simply have driven away the weaker, and then at once have taken possession of as many females as possible; but if it is indispensable for the male to excite or please the female, we can understand *the length of the courtship* and the congregation of so many individuals at the same spot (...) *very early in the spring*. (...) The whole affair was evidently considered by the birds as one of the highest importance. (...) we may conclude that *the courtship is often a prolonged, delicate, and troublesome affair.*' The rules of this game—several males displaying together before the arrival of females, the females free to visit many arenas and potential mates, the absence of resources and paternal care, and finally the marked persistence of male 'paradoxical' activity—suggest a more or less unhindered mate choice, probably congruent with the results of male-male competition, in birds as in wasps. In light of a strategy of waiting both for receptive females and for superior males, the 'prolonged, delicate, and troublesome' male activity could acquire a new meaning, which is still to be explored more fully.

Acknowledgements

This review would be not possible without the strict collaboration, in the field and in laboratory experiments, with S. Turillazzi, R. Cervo, M. C. Lorenzi, F. Dani and F. Bertolino. I should also like to express my special thanks to M. J. West-Eberhard, who first discussed the 'marathoner hypothesis' with me, to W. Eberhard for providing ideas and criticisms on earlier drafts, to helpful referees J. Alcock and M. Polak, and to P. Christie, who carefully revised my English.

7

Homing in paper-wasps

Alberto Ugolini and Stefano Cannicci

In flying hymenopterans, the majority of studies have concerned observations or experiments held within the insects' known territory, or their home range (internal homing), and most of this research has focused on the problem of nest location and identification on return from foraging flights (immediate orientation) (see Wehner 1981).

Neither homing ability, following passive displacement and release from a presumably unknown area (external homing), nor the capacity and mechanisms of orientation have been studied in depth in flying hymenopterans, apart from bees, in spite of the early origins of this type of research. Indeed, as Batra (1977) mentions, it seems that the first records on the subject date back to Vedic times and recount the use of *Xylocopa* as messengers (like homing pigeons) because of their ability to return to their nests from distances as far as 4 km away. In this type of experiment, preference has generally been given to pre- or sub-social flying hymenopterans. Far less literature is available on eusocial hymenopterans (cf. Wehner 1981, 1984, 1993), although, in theory at least, they would appear to be choice material for this type of study for the following reasons:

(1) higher spatial-temporal constancy of the nesting site;
(2) the relatively easy source of a good number of individuals, and finally
(3) easier standardization of some biological parametres prior to experimentation.

However, the very fact that these insects fly means they do not lend themselves easily to experimental manipulation.

Therefore a revue of the literature published to date on homing ability following passive displacement in the social vespids, updated where appropriate with new findings, seemed an interesting proposal.

Flight range extension

Obviously it is impossible to discuss homing from unknown territory without knowing the size of the territory with which the animal is, at least partially, familiar. Studies of this type in the vespids are summarized in Table 7.1, from which it may be deduced that paper-wasps (*Polistes*) usually forage within 100 m of their nests. Akre *et al.* (1975) used metal tags and magnets to capture individuals of *Vespula pensylvanica* and found that 80% of the workers foraged within 370 m of their nest.

Table 7.1 Summary of experiments conducted to determine the extent of foraging area and/or flight range distance in different species of wasps.

Species	Foraging distance (m)	References
P. chinensis	70 max.	Suzuki 1978
P. chinensis	20	Suzuki 1978, Kasuya 1980
P. jadwigae	40	Hibino 1980
P. metricus	102	Dew and Michener 1978
P. fuscatus	48	Dew and Michener 1978
V. pensylvanica	380	Akre *et al.* 1975
V. rufa	~ 300	cf. Edwards 1980
V. germanica	~ 300	cf. Edwards 1980
V. vulgaris	~ 300	cf. Edwards 1980
V. mandarinia	1000–8000	Matsuura and Yamane 1990

Studies on *V. rufa, V. germanica,* and *V. vulgaris* are less precise, but it seems that these species normally forage within 250–300 m of their nest, reaching distances as great as 1–4 km in cases of particularly rich supplies. *Vespa mandarinia* most frequently forages within 1 km of its nest but has been observed at a distance of 8 km (Matsuura and Yamane 1990).

This topic was the object of some preliminary experiments we conducted for 3 days in July and August, 1992, in an attempt to evaluate the normal foraging flight range in workers of *Polistes dominulus*. Sheets of paper (30 × 20 cm) coated with glue were used for the purpose, which were placed at various distances from the nest. The experiments were held in the middle of a slightly sloping meadow, where most of the vegetation consisted of Graminaceae and Leguminosae. The 90 traps were set at 50 m intervals as symmetrically as possible round the nest from a radius of 50 m outwards up to a maximum of 300 m, and left in place for 24 hours (the maximum time the glue remained sticky). A total of 38 wasps was caught over the 4 experimental days, 84% of which were within 200 m of the nest (Fig. 7.1A) and only 5% at a distance of 300 m. From the distribution of the traps which had caught the individuals (Fig. 7.1B) it was apparent that the workers followed a preferential direction on their foraging trips, as Suzuki (1978) had previously observed in *P. chinensis antennalis*. This finding is in accordance with the results of previous trials on workers of *P. dominulus* in which the outgoing and incoming directions of individuals from one colony were taken with a compass. From these observations (Fig. 7.1C) it emerged that these individuals leaving the nest also headed in a common direction towards a given objective (in this case a thick hedge of shrubs about 100 m away). By recording consecutive directions of the same wasp (individually marked) in a given interval of time (30 min.) it was also possible to detect individual foraging directions in workers of *P. dominulus* (Fig. 7.2A) and *V. orientalis* (Fig. 7.2B): clearly there is a high concentration of individual vanishing points in one particular direction for each individual.

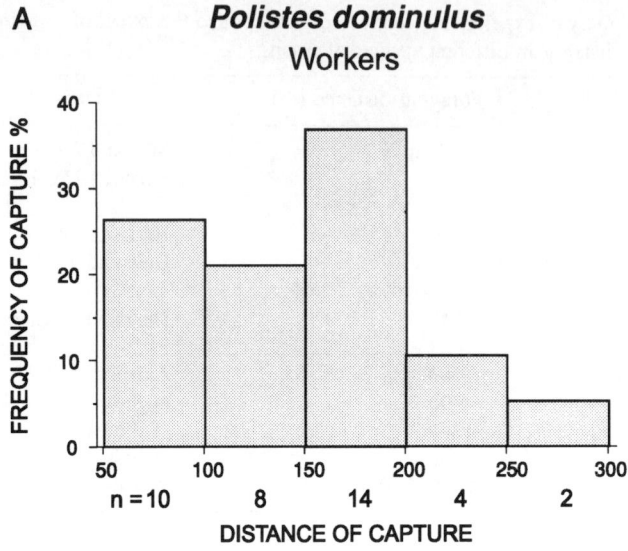

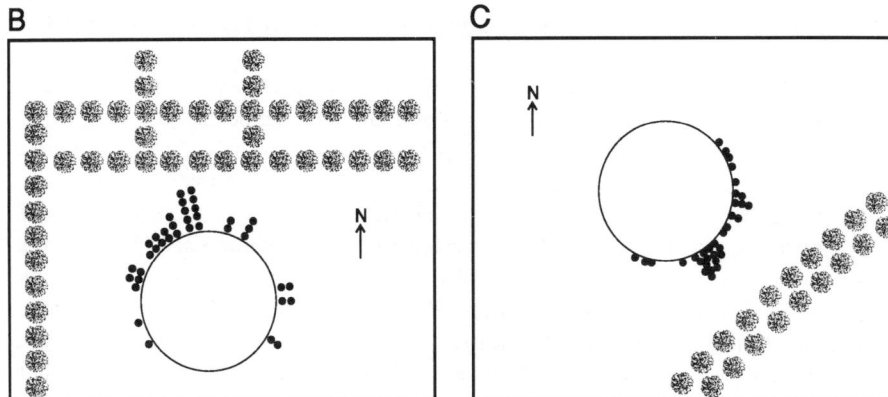

Fig. 7.1 *P. dominulus* workers. Flight range. A: capture frequency at various distances from the nest; B: directions of individuals caught by the traps; C: outgoing directions of individuals from one colony.

Homing ability

Most experiments on homing ability after passive displacement have been conducted on a few species of the genus *Polistes* (Table 7.2). Unfortunately, because of the small number of individuals used in these trials and the rather imprecise accounts of the experimental methods, it is difficult to interpret the results or use them for comparison with other species.

In spite of the pioneering nature of their work, the only authors to have conducted

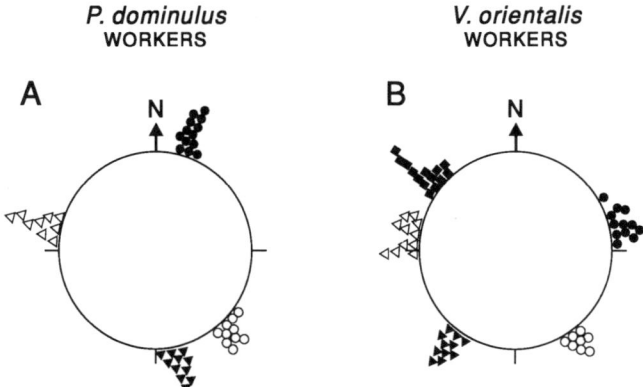

Fig. 7.2 Individual foraging directions. A: *P. dominulus*; B: *V. orientalis*. The different symbols refer to single individuals. Each symbol represents one direction.

Table 7.2 Release experiments after various techniques of passive displacement in different species of vespids.

	Species	Displacement method	Release dist. (m)	n	HP %	Reference
F + W	*P. b. bimaculatus* and/or *P. nimpha*	D ? (with the nest)	200	17	70	Weyrauch 1929
" "	" "	" "	500	21	62	"
" "	" "	" "	800	10	50	"
" "	" "	" "	1000	16	44	"
" "	" "	" "	2500	9	0	"
W	*Polistes* sp.	L	1000	11	9	Molitor 1939
W	*P. hebraeus*	D	50	9	55	Kundu 1965, 1967
W	*V. sylvestris*	D	50	25	92	Rabaud 1924
W	*V. sylvestris*	D	200	23	56	Rabaud 1924
W	*V. vulgaris*	L	150 c.	15	47	Molitor 1937

F = females; W = workers; L = displacement under the sun; D = displacement in the dark. HP = homing performances.

their experiments in such a way as to allow a modern, albeit limited, interpretation of the results are Rau and Rau (1918). Both females (= foundresses and auxiliaries) and workers of *Polistes fuscatus pallipes* (=*Polistes pallipes*) were released various distances from their nests (from 200 m to 3200 m) after displacement in the dark. Homing times were not recorded accurately, but nevertheless the results clearly indicate that whilst homing ability is high in females (Fig. 7.3A), even over long distances (33% were from 3200 m!), it is far lower in the workers (Fig. 7.3B) (falling to 0 at 1300 m) and depends on their age at release: none of the individuals under 10 days old released at 200 m returned home (0%, n=16).

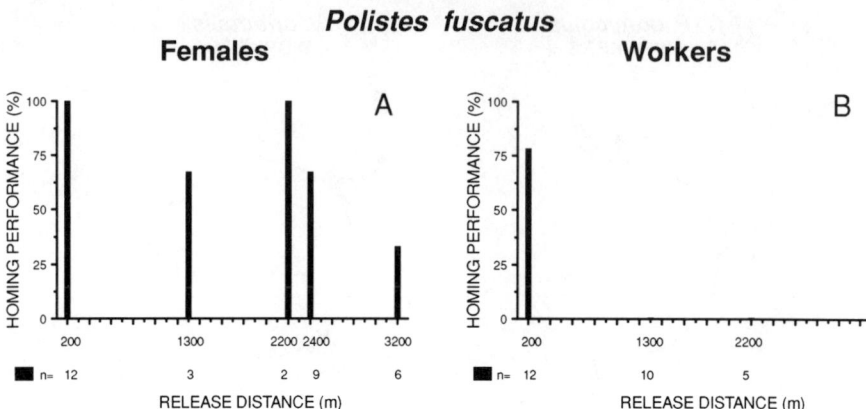

Fig. 7.3 *P. fuscatus pallipes.* Homing ability in females (=foundresses and auxiliaries) (A), and workers over 10 days old (B). n, sample size for each release distance. Data from Rau and Rau (1918).

More recently, the problem of homing ability in paper-wasps (Ugolini 1986*a*; Ugolini and Cannicci 1991) and in flying hymenopterans in general (Tengö *et al.* 1990; Schöne and Tengö 1991; Wehner 1992) has been the object of renewed interest. In our experiments we released *Polistes dominulus* (cited in some papers as *Polistes gallicus*), *Polistes gallicus* (= *Polistes foederatus*), *Polistes nimpha* and *Vespa orientalis* after passive displacement in various directions and from different distances from their nests. Each wasp was caught on its nest (in the case of *V. orientalis* on their return), individually marked and put into a transparent plexiglass test tube which was then hermetically sealed.

Spectrophotometre analysis (Perkin Elmer λ9) of the test tube shows that a good part of the ultraviolet light passes through the material (50% transmission at 340 nm), but the sky light polarization pattern (which the wasps may perceive during their journey from the nest to the release point) is altered.

During displacement, the wasps were either allowed full view of the landscape, sky and sun or else transferred to the release site in the dark. Depending on the release distance (from 50 m to 2000 m), displacement time varied between 2 and 25 minutes. Once at the established release site, the test tubes were opened one at a time to allow each wasp to leave the tube and fly away alone. When possible, homing times and homing performances were recorded within 24 h of release (the night time resting hours, as noted at nests, were not counted).

Displacements were carried out with females in the pre-emergence period taken from haplo- and pleometrotic nests, young workers less than 7 days old (with no flight experience) and workers over 7 days old which had almost certainly already memorized some of the landmarks round their nests during their orientation flights (Pardi 1948*b*, 1951).

Homing in females

Homing performance patterns in females of *P. dominulus* (Fig. 7.4A) confirm previous findings (Ugolini 1986*b*): homing performance is very high within 400 m of the nest, regardless of whether they have been displaced in the dark or with a view of the sun, but falls rapidly from 400 to 600 m to then remain practically stable—around 50%—up to a distance of 2 km. Homing times (Fig. 7.4B) in wasps displaced under the sun increase at about 600 m. In females displaced in the dark (Fig. 7.4B), homing times again reflect the pattern described above, but accentuate the difference between releases from 200 and 400 m.

Homing performance was higher and homing times faster in *Polistes dominulus* foundresses from haplometrotic than from pleometrotic nests (Fig. 7.4C, D).

Similar results were obtained in foundresses of *P. gallicus* (Fig. 7.4E, F), although homing in general was less efficient, especially from distances of over 400 m.

Homing in workers

P. dominulus workers, about 5 days old, were displaced in full view of the sun and surrounding landscape then released 1 km away. Up to the time of their capture, these wasps had never flown outside their nests. When the wasps left the test tubes at the release site, they flew in increasingly large loops (from 1 to 4–8 m approximately), often repeatedly returning in proximity of the test tube and sometimes even landing on it. This behaviour was invariably observed in all wasps under one week old and sometimes in wasps displaced in the dark. As occurred in the trials by Rau and Rau (1918), none of the 10 young workers returned to their nest.

Workers over 7 days of *P. dominulus* (Fig. 7.5A) and *P. gallicus* (Fig. 7.5C) displaced under the sun, both showed a similar relationship between release distance and homing performance: up to 200 m homing performance is high (100%), then falls almost linearly up to about 2 km (Ugolini 1985). However, it should be noted that homing performance for the 1 km releases in *P. dominulus* almost doubled that of *P. gallicus*. The homing performance of individuals of the two species which were displaced in the dark showed a similar trend to that just described (Fig. 7.5A, C). Homing times in both *P. dominulus* (Fig. 7.5B) and *P. gallicus* (Fig. 7.5D) increase sharply in the 400 and 600 m releases, regardless of the method of displacement.

In the 1000 m releases, some *P. nimpha* workers were able to home with homing performances very similar to at least *P. dominulus* (Fig. 7.6).

Workers of *V. orientalis* over 1 week old can return to their nest from a distance of 1 km if they are displaced under the sun (Ugolini *et al.* 1987) (Fig. 7.7A); homing performance in this species also gradually falls as the release distance increases. However, it is interesting to note that displacement in the dark caused a sharp fall in homing performance at 500 m and total failure at 1 km. Similarly, there is a marked increase in homing times (Fig. 7.7B) in wasps displaced in the dark and released at distances of more than 200 m.

Page 132

Alberto Ugolini and Stefano Cannicci

Polistes dominulus females
pleometrotic

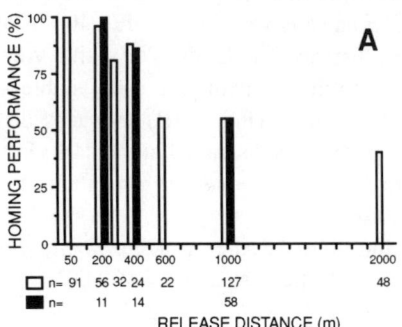 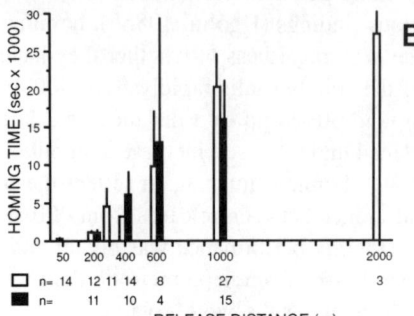

haplometrotic

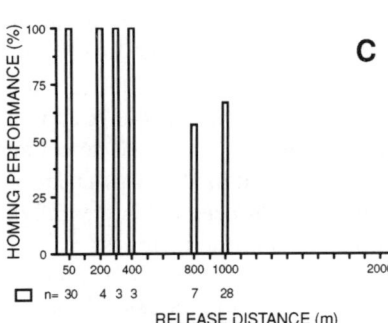 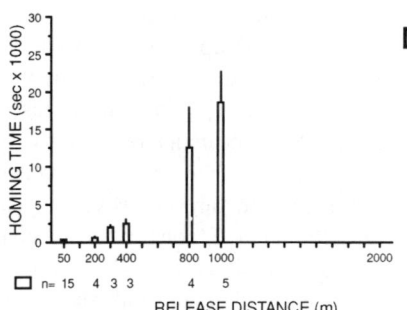

Polistes gallicus females

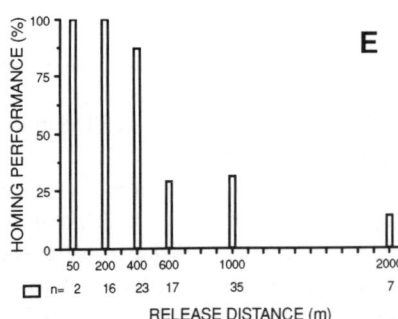 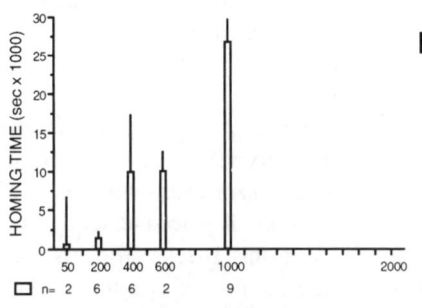

Fig. 7.4 *P. dominulus* (A, B, C, D) and *P. gallicus* (E, F) females. Homing ability (A, C, E) and homing times (B, D, F). White bar, displacement under the sun; black bar, displacement in the dark. Homing times are given by the mean value and standard error. A, B, females from pleometrotic and (C, D) haplometrotic foundations of *P. dominulus*.

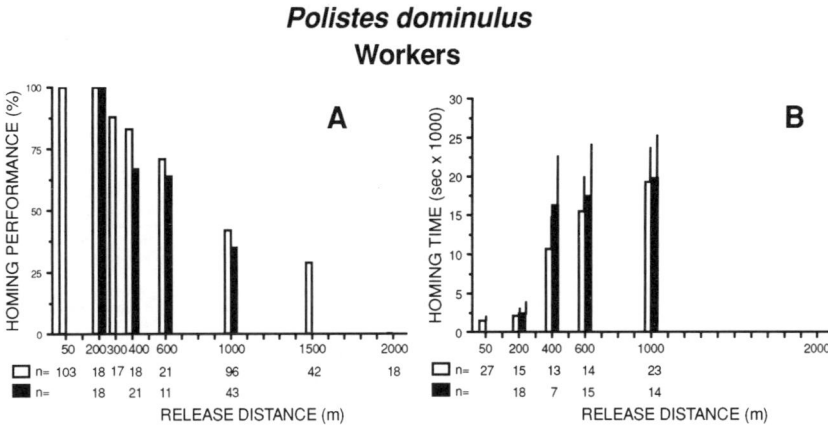

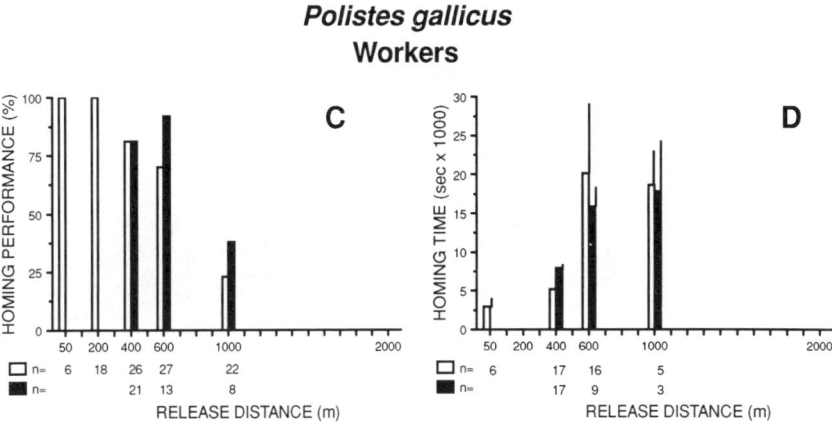

Fig. 7.5 Homing performance (A, C) and homing times (B, D) in workers. For further explanations see Fig. 7.4.

Homing in 'caged' wasps.

To evaluate the real importance of territorial familiarity in homing ability, 'caged' individuals of *P. dominulus* (see also Ugolini 1986a) and *P. gallicus* were also tested: the wasps (females in the pre-emergence period and workers) were taken on their nests to the experimental site 5–10 km away, where they were kept for 7 days in wood and wire mesh cages (18 × 15 × 15 cm), so they could not acquaint themselves with their surroundings. They were regularly provided with building material, water, honey and *Tenebrio molitor* (L.) larvae. At the end of this period, the wasps were

(1) immediately caught and released 1 km away from their cage;

1000 m RELEASES

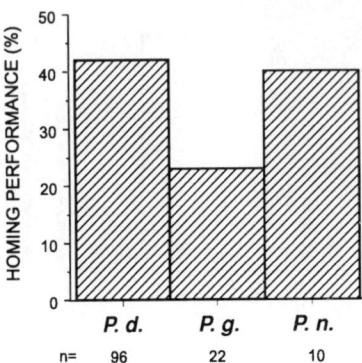

n=   96   22   10

Fig. 7.6 Releases from 1 km under the sun. Homing performances: Pd, *P. dominulus*; Pg, *P. gallicus*; Pn, *P. nimpha*.

Vespa orientalis
Workers

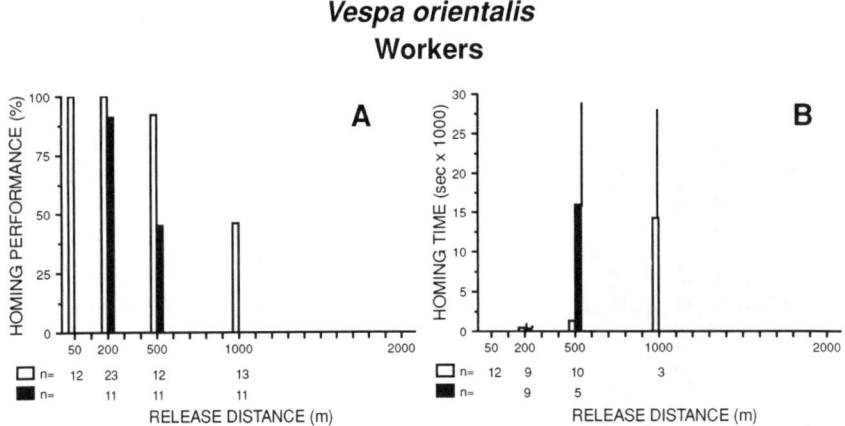

Fig. 7.7 *V. orientalis* workers. A: homing performance; B: homing times. For further explanations see Fig. 7.4.

(2) caught after allowing them a maximum (theoretical) of 2 hours; or,
(3) 4 hours flying time outside their cage.

When possible, the number of flights and flying times outside the nests were also recorded for each wasp.

Females of *P. dominulus*, displaced under the sun and released 1 km away, show that homing ability clearly depends on the amount of time they are allowed to familiarize themselves with their nest surroundings (Fig. 7.8A). In fact, although none of the *P. dominulus* females with zero hours flying time outside their cages returned to their nest (Fig. 7.8A), some wasps come back home after only 1 hour

Polistes dominulus
Females

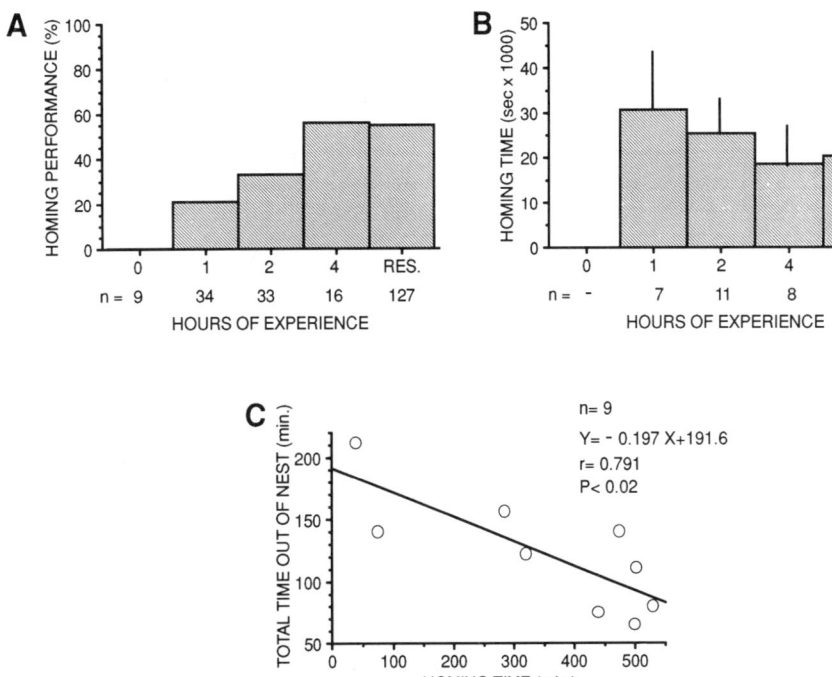

Fig. 7.8 Homing in 'caged' wasps. *P. dominulus* females; A: homing performance, 1 hr—4 hrs indicates the maximum theoretical time the wasps were allowed to fly in their new territory. B: homing times; C: relation between time spent outside nest and homing times. The number of pairs (n), regression line equation and correlation coefficient (r) with probability level (P) are also given.

flying experience (although homing times were considerably higher than those of the residents). Homing performance increases with the maximum theoretical time allowed outside the nest; after only 4 theoretical hours the wasps reach the same values as the residents in both homing performance and homing times (Fig. 7.8A, B).

In some cases, it was possible to record the precise amount of time each wasp spent outside the nest. As previously observed (Ugolini 1986a), and now confirmed (Fig. 7.8C), a significant correlation exists between homing times and the total time spent outside the nest (and probably, therefore, with the degree of territorial familiarity).

Although less evident, results for the 'caged' workers of *P. dominulus* (Fig. 7.9A) confirm the observations on homing performance and homing times in the females (Fig. 7.9B).

Polistes dominulus
Workers

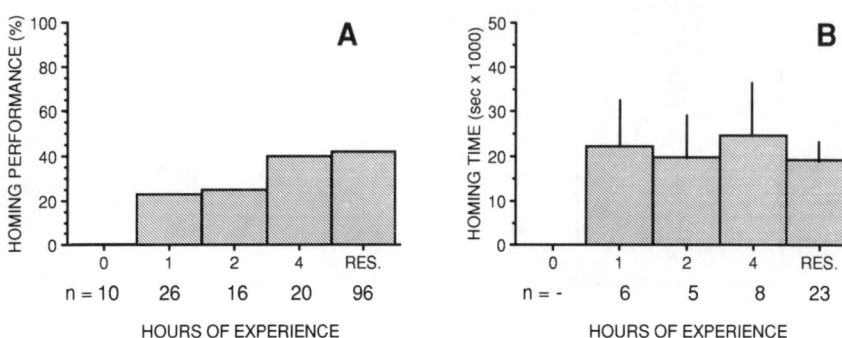

Fig. 7.9 Homing in 'caged' wasps. *P. dominulus* workers. For further details see Fig. 7.8.

However, *P. gallicus*, regardless of caste, seems unable to return home from a distance of 1 km (Table 7.3) even after 4 hrs theoretical flight over the new territory. But resident wasps, kept in open cages and released at the same time as the others, exhibit normal homing ability.

Table 7.3 Homing performance in 'caged' individuals of *P. gallicus* with limited (4 hrs) or full (residents) territorial flight experience.

	4 hours	Residents
Females	n = 21	n = 31
	0 %	35 %
Workers	n = 34	n = 23
	0 %	22 %

Another type of experiment was also conducted (in 1986) to further evaluate the influence of territorial familiarity on homing ability. This type of experiment, using a greater number of individuals and recording the flight direction at vanishing point, should furnish interesting information on the type of search the wasps perform after the vanishing point. These wasps, workers caught on a nest about 4 km from the experimental site, were kept in cages using the same methods described above. After a week, the cages were opened for 4 hours and the wasps were free to fly out and familiarize themselves with their surroundings and with the cage exterior. The cages were then closed and, with the wasps inside, transferred in the dark to a second locality about 1 km away. The wasps were then collected in their cages and displaced in transparent test tubes to a third locality, at a distance of approximately 1 km from both first and second site, where they were released.

On account of the small number of individuals, the results for the 'triangle' experiment can only furnish very preliminary indications on the importance of territorial familiarity and on the type of search for home. The results show that four individuals, out of the six released, returned to the area where they were allowed to fly outside their cages; it should be noted, however, that the other two wasps returned to the site from which they were passively displaced with vision of the sun and landscape flow.

Initial orientation

The capacity for initial homeward orientation after passive displacement, i.e. the direction (in this case taken with a compass) the wasps assume at vanishing point (8–10 m from their release site), has already been amply demonstrated in both females and workers of *P. dominulus* (Ugolini 1981, 1983, 1985, 1986*b*), even when they are deprived of any possible terrestrial orienting cues at the moment of release (arena releases). Landscape flow (natural or artificial) during displacement is one of the factors the wasps use to determine the direction of displacement, and consequently of their nests, at the moment of release (Ugolini 1987; Ugolini and Samoggia 1991). This was demonstrated in two types of experiments: 1. The wasps were first displaced in full view of the sun and landscape, carried back without visual cues (the controls were allowed full vision during both displacements) and finally released (in a plexiglass arena with the landscape screened from view) 50 m away from their nest at a point perpendicular to the direction of the first tract of the displacement. 2. The wasps, in test tubes, were placed between two strips of material with black and white vertical stripes (= the artificial landscape, Fig. 7.10), which were powered by two electric motors to run in the same direction). Thus the wasps could see the artificial landscape flowing past them in the same direction as a nest-to-release-site displacement. The wasps used for this experiment were also released in the arena set at 90° to the direction of the false displacement.

The arena (Fig. 7.11) used for both of these experiments consists of two plexiglass discs (1 cm thick) separated by a 3 cm gap. The transmittance of

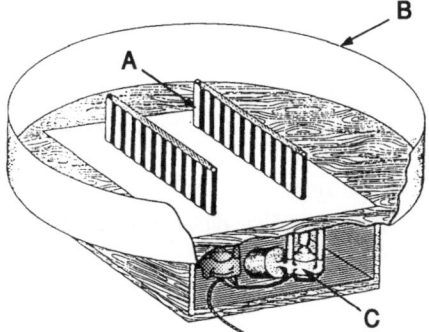

Fig. 7.10 Apparatus used to generate the false displacement. A: striped material (= artificial landscape); B: white plexiglass screen; C: electric motors. The wasps were set in the middle of the two strips of material (from Ugolini 1987, modified).

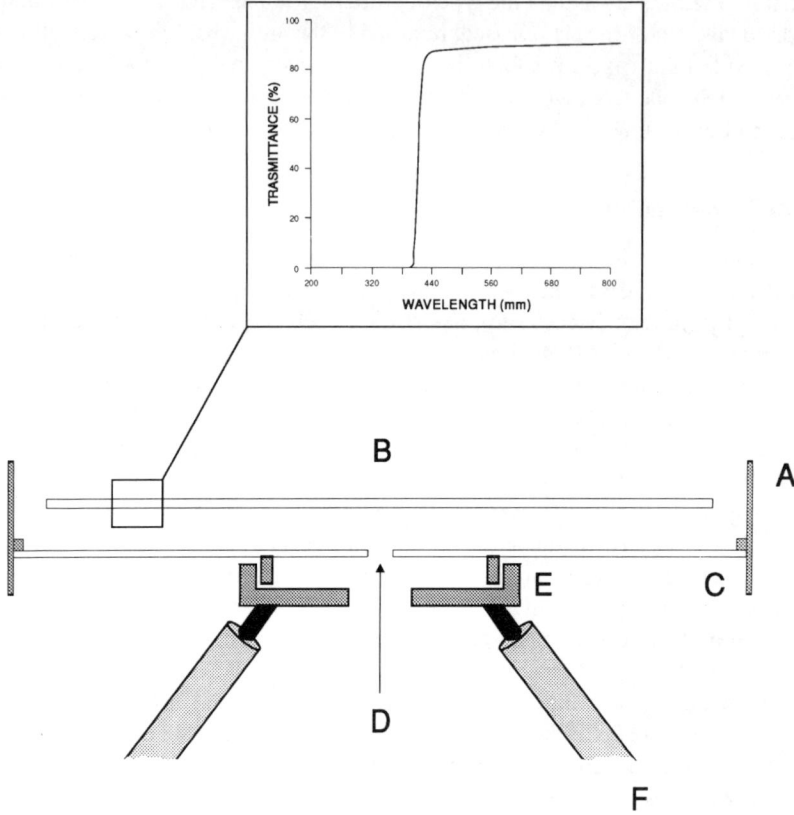

Fig. 7.11 Cross section of the arena. A: cylindrical white plexiglass screen; B, C: transparent plexiglass discs, one on top of the other (diameter of lower disc = 1 m); D, hole for insertion of test tubes; E: apparatus for turning the arena; F: tripod. The transmittance diagram for the upper disc is shown in the inset.

plexiglass (inset Fig. 7.11) clearly shows that very little UV-light reaches the wasps in the arena and consequently they probably cannot perceive any associated orienting cue (see Waterman 1981; Rossel and Wehner 1986; Wehner 1993). Each test tube was inserted into a hole in the lower disc. As the wasp climbs the test tube wall, it finds itself alone inside the arena.

From the results, landscape flow, whether natural (Fig. 7.12A, B) or artificial (Fig. 7.12C, D), clearly influences the capacity for finding the home direction: while the controls (A, C) are well oriented towards the true home direction both in and outside the arena, the mean direction of the experimentals (B, D) corresponds to the expected flight direction (i.e. deviated by 90°). Therefore, landscape is one of the orienting factors perceived during displacement. The above experiments refer to foundresses and auxiliaries (Ugolini 1987), but basically the same applies to workers (Ugolini and Samoggia 1991). To determine the direction of displacement,

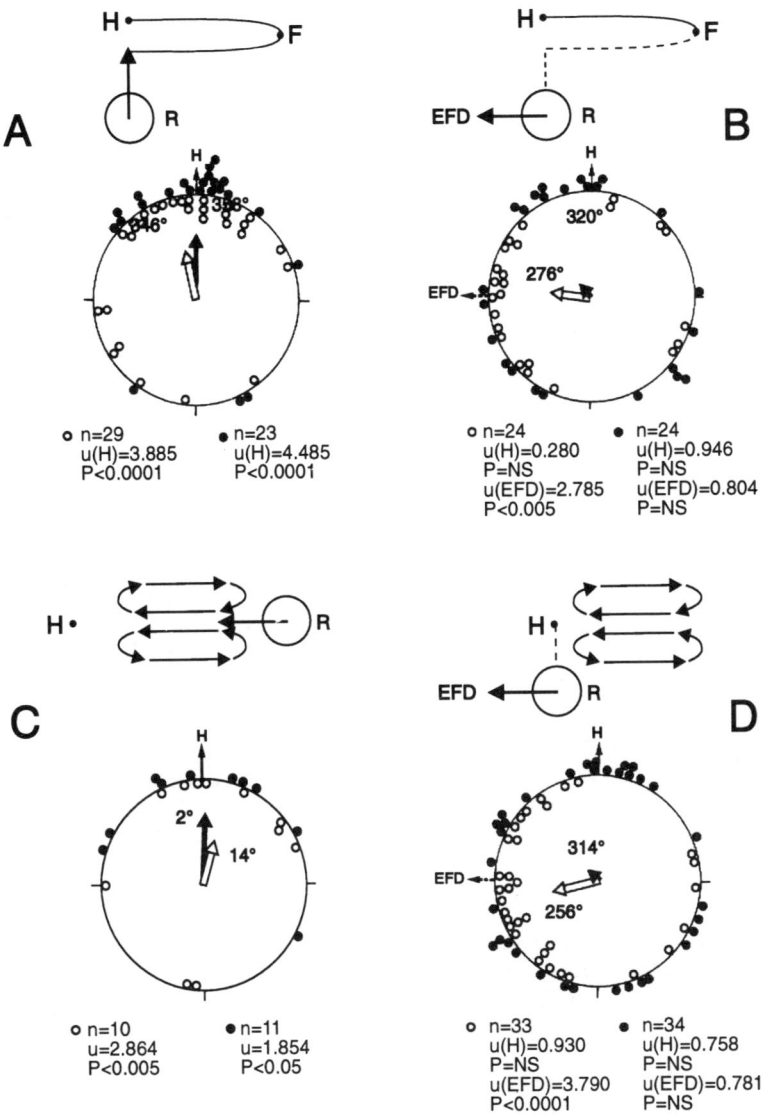

Fig. 7.12 *P. dominulus* females. Initial orientation experiments. A: wasps displaced in view of the sun and natural landscape from H (home) to R, arena release site. Distance H—F = 1 km. B: wasps displaced as in (A) from H to F but surroundings totally screened from view F to R; EFD: expected flight direction. C, D: false displacement using the artificial landscape apparatus. White dots, directions of wasps released inside the arena; black dots; directions outside the arena. The arrows inside the distributions are the mean resultant vectors (max. length = 1 = circle radius); n, sample size; u, V test value with probability level, P (see Batschelet 1981) (from Ugolini 1987, modified).

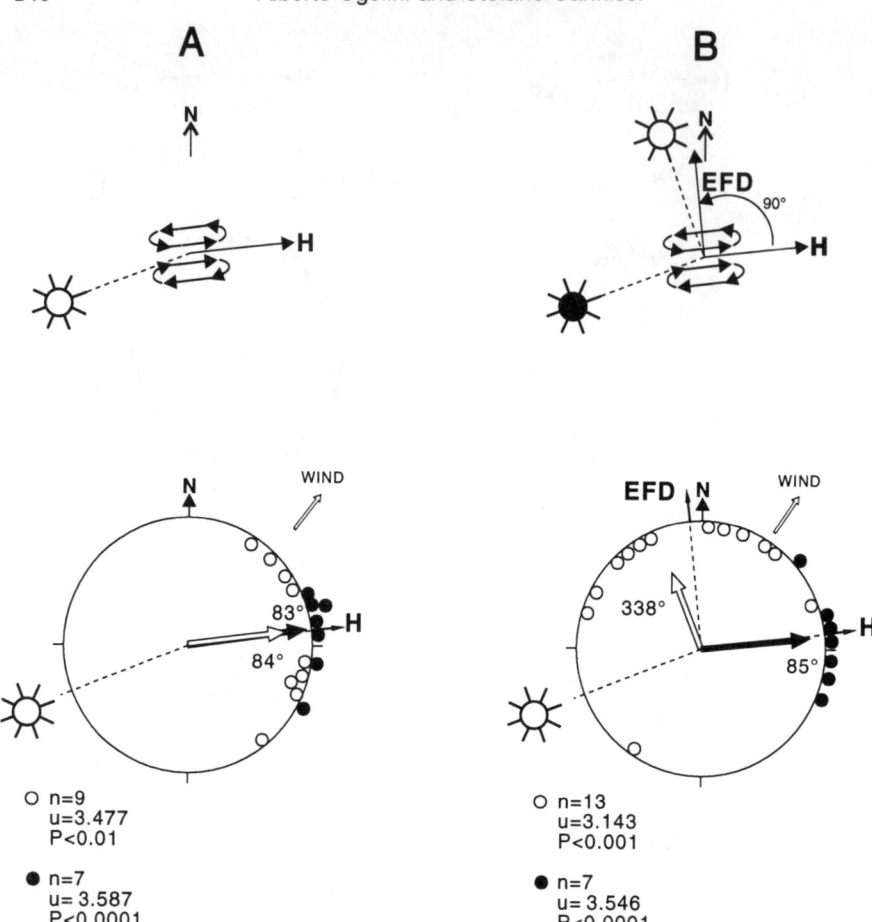

Fig. 7.13 *P. dominulus* workers. Arena releases from 50 m. A: controls with the vision of artificial landscape flow. B: experimentals displaced with the same method as in (A), but with the sun screened from view (black sun) and deflected by a mirror (open sun). For further explanation see Fig. 7.12.

however, the wasps must combine a further orienting factor with the apparent motion of the landscape. We had already hypothesized that this could be the sun (Ugolini 1987; Ugolini and Samoggia 1991): the sun compass is one of the most common orienting factors in nature and furthermore Jacobs-Jensen (1959) had already proved that workers of *Vespula germanica* use the sun to maintain precise foraging directions on leaving the nest. To verify this point, we conducted the classic experiment in which the sun's azimuth is deviated by a pre-established number of degrees (Santschi 1911). As this would have been impossible to execute during the return flights, and the arena with a diametre of 1 m would not have been suitable for the purpose, we deflected the sun's azimuth during the false displacement. Immediately prior to the false displacement, the sun was screened from view and its mirror image reflected onto the wasps, from an azimuth deviated by 90° in a clockwise direction. The striped artificial landscape was run in such a direction as to induce the wasps to keep to an angle of 105° with the sun during their return journey (Fig. 7.13).

In Fig. 7.13 the controls (A) are clearly well oriented in the homeward direction both in and outside the arena, whilst the mean direction of the experimentals (B), where the sun's azimuth is deflected by 90°, is deviated by 106° (in the expected flight direction) with respect to the controls. Once outside the arena, as already observed in previous experiments (Ugolini 1986b, 1987; Ugolini and Samoggia 1991), the wasps can adjust their orientation and assume the correct homeward direction, probably using landmark sightings (difference between experimentals and controls outside the arena = 1°).

Discussion and conclusions

Evaluation of foraging areas and distance from the nest in social vespids presents all the usual difficulties typical of this kind of study, but is further complicated by the great difficulty (and often impossibility) of tracking the individuals in the field. This limitation also prevents reaching any conclusions on homing mechanisms, which will be discussed shortly. An indication of the activity range area in *Polistes* can be gleaned from the few investigations on the subject (Dew and Michener 1978; Suzuki 1978; Kasuya 1980; Hibino 1980); however, as interesting and useful as these results may be, in our opinion they could have underestimated the foraging distance. In fact the experimenters followed the wasps on foot and with the naked eye—it is therefore quite reasonable to imagine that the slower wasps which foraged nearer the nest were the easiest to follow. During our observations, for example, we noticed that on leaving the nest, foragers of *P. dominulus* often reached a height of 3–4 m before heading in any particular direction, presumably towards foraging grounds farther afield. When this occurred, we regularly lost sight of them after 30–50 m. However, as previously reported (Ugolini 1986b), the results for females of *P. dominulus* and *P. gallicus*, allow some conclusions to be drawn regarding foraging distances. Since homing performance can be assumed to reach its maximum in releases within the known territory or in its immediate vicinity, the homing performances from our releases between 200 and 600 m, would indicate that the normal flight range for at least some species of *Polistes* is about 300 m. This estimate is supported by the increase in mean homing times from distances over 400 m, which are higher than expected if they depended on release distance alone.

These points also apply to *V. orientalis*: homing performance after displacement in the dark remains high within 500 m of the nest. This distance is rather short considering the flight ability of this species, but could depend on the fact that some apiaries had been placed 200–400 m from the nests to encourage *V. orientalis* to nest and forage in the vicinity (Ishay, pers. comm.). Nonetheless, these points are subject to various criticism:

(1) flight distance determined by homing experiments may be overestimated, since

the different homing mechanisms can guarantee high homing performance regardless of accuracy, especially in releases not too far from the nests;

(2) the 'mean' distance does not account for any (probable) relevant individual differences depending on A) the various objectives of foraging flights (water, nectar, meat, building material), B) the period in the colonial cycle, C) the different tasks wasps perform during their lifetime (Pardi 1951);

(3) the radius of presumably familiar territory round the nest estimated by homing experiments does not consider the important factor of flight directionality, either in single individuals or in some cases in the majority of individuals from a single colony;

(4) it is extremely difficult to determine the (probable) differences in foraging distances in the different castes.

Nevertheless, our capture-marking-recapture experiments in the field, together with the data in the literature, support the results of the homing experiments. Furthermore, if we consider the normal flight distance found in the vespines (Akre *et al.* 1975; Edwards 1980), within 400 m, and bear in mind the obvious differences in flight ability between the vespines and polistines, it is unlikely that the flight range for species of the genus *Polistes* exceeds 300 m, although there may be interspecific and intercastal differences.

The release experiments after passive displacement conducted by various authors (Table 7.2) undoubtedly reveal good homing ability in the genus *Polistes*, and probably in the vespids in general, in both females and workers. Moreover, this capacity is found, with an evident adaptive value, in all the central place forager hymenopterans (Wehner 1983, 1992). In *Polistes*, however, some differences in homing ability do exist, which could partly depend on a different ecology. This is difficult to justify in the light of our present knowledge on the subject (in the release area, *P. dominulus*, *P. gallicus* and *P. nimpha* are largely sympatric). The most noticeable differences are plainly inter- and intracastal. Among the different castes, homing ability is clearly far less distance-dependent in females than in workers: in females homing performance remains fairly constant from 600 m to 2 km, whilst in the workers it appears to be closely tied to distance: over 200 m it falls almost linearly. It should be emphasized (Rau and Rau 1918) that *P. fuscatus* females return home from a distance of 3200 m, whilst none of the workers returned from 1300 m (Fig. 7.3A, B). The differences found in the females, can, in our opinion, be attributed to the motivation urging the females to return home quickly, thus minimising the risk of predation on the nest (see e.g. Hermann and Blum 1981; Jeanne this volume).

Whilst the results of the 'caged wasps' trials and—although very preliminary—the 'triangle' experiment clearly demonstrate the importance of territorial familiarity in homing of *P. dominulus* and *P. gallicus*, they also show that females of both species can return to their nests from areas which in all probability lie outside their normal flight range. Some differences in homing ability, both interspecific and intercastal, still remain unanswered: for instance, homing performance in females from haplometrotic foundations is higher in *P. dominulus* than in *P. gallicus*. There are some unexplained differences in homing ability not obviously related to what is known about the ecology of the species.

Females and workers seem to employ a common orienting mechanism which, at least in the releases outside the known territory, may be considered as 'route based orientation' following the classification proposed by Papi (1992). However, it should be underlined that homing ability from unknown territories is distinctly lower in workers than in females, and in the workers falls linearly for distances over 200 m. In our opinion, at least two explanations may account for this intercastal difference, common to both species: (1) different motivations for returning home in individuals belonging to different castes; (2) differences in the type and/or efficiency of their homing mechanisms (see also Ugolini and Samoggia 1991). Both points become more meaningful when considering the damage a colony would suffer at the loss of a leader (or one of the females) compared with a worker. It is also possible to imagine an initial phase of homeward orientation, using the sun and landscape flow which is common to all the wasps, followed by a second based on a sort of systematic or random search for the goal (see Jamon 1987). This second capacity could be different between the castes. In fact, it seems highly probable that females and workers of *P. dominulus* can orientate homewards in the correct direction by integrating the angle they assume during displacement (active or passive) minus 180°, but it has not been demonstrated that they do the same with distances. If this is not the case, we can imagine that a change of strategy may come into play during the homeward journey, which not only differs from individual to individual but also from caste to caste. In spite of the small number of wasps tested, and disregarding premature conjectures as to the existence of cognitive maps in wasps (for this argument in *Apis* see Gould 1986; Wehner and Menzel 1990; Dyer 1991; Wehner 1992), the 'triangle' experiment suggests that just such an individual strategy does exist or may be changed during the return journey. This hypothesis could, in particular, explain the high mean and variability in homing times in releases conducted from areas presumably unknown to the wasps.

Desiderata

Little research has been done on the behaviour of social wasps away from the nest. It is to be hoped that in future greater attention will be given to research in the field of behavioural ecology in the social wasps, as this would help bridge this gap in our knowledge of these hymenopterans.

Note added in proof: Geiger *et al.* (1994) tested the influence of landscape flow on orientation during passive displacement in honey bees. Results show that bees do not use landscape flow to return home. However, the discrepancy between bees and wasps could be due to several causes (Geiger *et al.* list four). The differences in methods of displacement and release should also be noted. Moreover, it has recently been demonstrated that bees use optic flow to acquire distance information (Esch and Burns 1995).

8

The evolution of exocrine gland function in wasps

Robert L. Jeanne

Fifty years ago, when Leo Pardi was doing his landmark studies of dominance in *Polistes*, most of the exocrine glands we know of now in wasps had already been described, often in minute anatomical detail (Bordas 1908; Heselhaus 1922). Understanding of gland functions, however, was limited largely to the venom gland and some of the glands producing structural products—silk production by the labial gland of last instar larvae, and nest construction cement by the labial gland of the adult female. The rest of the glands opening in the head were assumed to produce secretions related to feeding, and the function of Dufour's gland was unknown. Virtually nothing was known of the existence of pheromones and allomones and their glandular sources.

At present we know of 16 exocrine glands in adult polistine females (Fig. 8.1), plus a 17th (7th metasomal tergal gland) in vespine queens (Landolt and Akre 1979a). Yet we are reasonably certain of the functions of less than a third of these: poison gland, thoracic (labial) gland, 6th sternal, 5th sternal, plus a possible function of Dufour's gland. The task of understanding the role of glands in social behaviour is made more difficult by the existence of behavioural responses to chemical releasers for which morphologically discrete glandular sources apparently do not exist. Colony odour, kin recognition, and the thermoregulation pheromone are examples of social functions that appear to be mediated by general cuticular odours. On the chemical side, identification of active components of only three alarm pheromones, a thermoregulation pheromone, one queen pheromone, and two defensive allomones have been reported to date for wasps (see below).

In fact, we still know so little about gland function and chemistry in wasps that it is hard to get a feel for how much diversity there is among genera or subfamilies, much less to discern patterns within the diversity. Nevertheless, it is useful to review what we do know about exocrine glands in the wasps in order to detect emerging patterns that may guide future research.

There are several ways to approach such a review. One could take up the glands in turn, from the front to the rear of the body, and examine the functions of each. Another approach would be to organize a review along taxonomic lines. A third approach, and the one I adopt here, is to focus primarily on gland function. I believe this has the advantage of best highlighting gaps in our knowledge.

As a secondary organizing theme, I consider the functions in the order in which they may have evolved, based on the distribution of these functions in wasps and

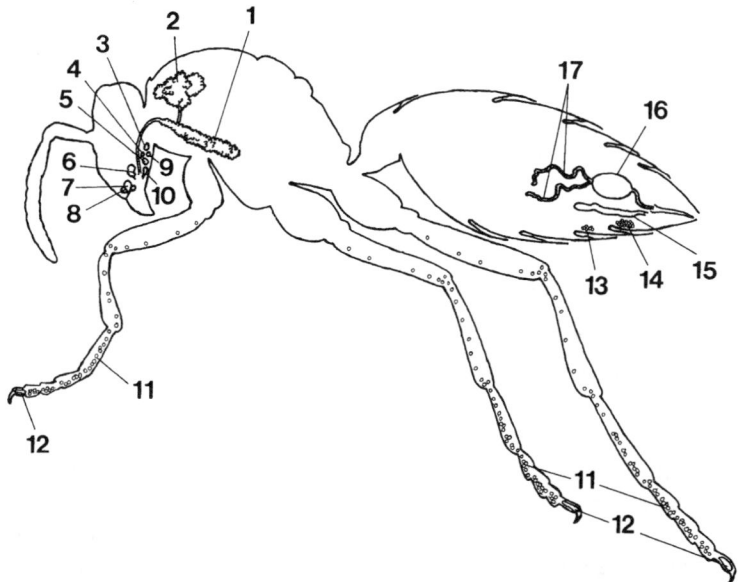

Fig. 8.1 Schematic representation of the exocrine glands of a composite *Polistes* female. Locations and scales are approximate only. Glands of the head and thorax based on *P. dominulus* (Deleurance 1955) and *P. fuscatus* (Landolt and Akre 1979a; Downing and Jeanne 1982). Glands of the legs based on *P. dominulus* (Beani and Calloni 1991a). Glands of the abdomen based on *P. fuscatus* (Landolt and Akre 1979a; Jeanne *et al.* 1983). Tegumental glands (class III cells opening directly through cuticle to the outside) are shown as open circles without ducts. Terminology after Landolt and Akre (1979a) and Beani and Calloni (1991a). All glands are paired except for Dufour's gland. (1) Mesothoracic lobe of thoracic (labial) gland; (2) prothoracic lobe of thoracic gland; (3) sublingual gland; (4) endostipal gland; (5) maxillary-hypopharyngeal gland; (6) hypopharyngeal gland; (7) ectal mandibular gland; (8) mesal mandibular gland; (9) labial palp gland; (10) maxillary-labial gland; (11) leg tegumental glands; (12) tarsal gland; (13) 5th sternal gland; (14) 6th sternal (van der Vecht's) gland; (15) Dufour's gland; (16) venom reservoir (poison sac); (17) venom (poison) gland.

other social Hymenoptera. I will do this by dividing them into three categories: functions that existed in the pre-social condition; those that evolved as, or shortly after the eusociality threshold was crossed, and those that evolved well after eusocial behaviour was established. Although admittedly speculative, such a categorization will, I believe, serve to stimulate thinking about the evolution of exocrine gland function in the social wasps.

It is useful to keep in mind that the functions of exocrine glands can fall into two categories, structural and informational. A product need not have exclusively one or the other function. Products with structural functions include digestive enzymes, adhesives, and the like. Products with informational functions modify the behaviour of organisms perceiving them, whether nestmates, non-nestmates of the same species, or members of other species. Information-bearing products can be further subdivided into cues and signals (Lloyd 1983). Cues are products whose predominant form is not a consequence of selection in an information-carrying context. Kairomones are an example. Signals are products whose predominant form is a consequence of selection in an information-carrying context. Chemical signals include pheromones and allomones.

Functions of exocrine glands

Arising in pre-social groups

Nest construction cement

I would argue that the most fundamental role exocrine glands play in the social organization of wasps is the production of the adhesive for the collected material used in nest construction. It seems clear that this material, whether vegetable fibre or mud, requires some sort of matrix, or glue, to hold it together and to enable the resulting structure to withstand weathering. Although certain bees (Michener 1974) and some solitary wasps (West-Eberhard 1988a) use collected materials for this— resins and other plant products—the majority of wasps appear to utilize exocrine products of some sort. Social sphecids (*Microstigmus comes, Arpactophilus mimi*) use silk secreted from the tip of the abdomen (Matthews and Starr 1984; Matthews and Naumann 1988). The glandular source is not known. Non-social vespids—the eumenines and masarines—apparently use saliva as a cement in their mud nests (Ferton 1921). All the social vespids, as far as we know, also rely on oral secretions.

As to the glandular source(s) of the material, we have made only modest progress since Ormerod (1868) and Janet (1903) surmised that the thoracic (labial) gland (Fig. 8.1: 1, 2) of adult females is the source of the cement, a view subsequently supported on anatomical grounds by Heselhaus (1922). Landolt and Akre (1979b), in an ultrastructural study of the labial gland of queens of *Vespula pensylvanica*, suggested that it produces either a brood food (based on Deleurance's (1955) suggestion for *Polistes*) or the glue used in nest construction. The fact that in *P. fuscatus* this gland is active from the start of nesting, rather than from the appearance of larvae in the nest, supports the latter but not the former hypothesis (Downing and Jeanne 1983). Unpublished work on *Ropalidia opifex* by G. Pfaff at the University of Frankfurt indicates that labial gland volume and cell diametre were significantly larger in building wasps than in foragers (cited by Maschwitz *et al.* 1990), supporting the same conclusion. But all this is based on circumstantial evidence. What is needed is chemical confirmation of the identity of the material in the nest and in the labial gland reservoir, and to date this has not been done.

Considerable progress has been made in recent years concerning the chemical nature of the cement itself. Schremmer *et al.* analyzed the hardened saliva of *Protopolybia chartergoides*, which is used in pure form to enclose the nest, and found it to contain chitin and possibly protein (Schremmer *et al.* 1985). The secretion used to build up the nest petioles of *Polistes annularis* and *P. metricus*, by contrast, was shown not to contain chitin, but instead to be a mixture of carbohydrate and protein (Espelie and Himmelsbach 1990; Singer *et al.* 1992*b*). Similarly, the cellophane-like nest walls of *Ropalidia opifex*, consisting of pure secretion, consist mainly of protein, with no trace of chitin (Maschwitz *et al.* 1990). The adhesive in the nests of *Dolichovespula maculata* also consists of protein (McGovern *et al.* 1988). Either Schremmer *et al.*'s (1985) analysis was flawed or *Protopolybia* belongs to a group that has evolved the use of chitin while other polistines and the vespines use protein. This uncertainty remains to be resolved.

Antibiotic secretions

Most groups of social insects are known to produce secretions with antibiotic properties. The metapleural gland of ants secretes antibiotics (Maschwitz *et al.* 1970; Beattie *et al.* 1986), and the venom of the fire ant also has antibiotic properties (Hölldobler and Wilson 1990). Fungicides and bactericides are produced in the mandibular glands of many bees, and Dufour's gland secretion of some has similar properties (Roubik 1989). These substances are used to disinfect the nest cells before they are sealed.

The nest and brood of wasps are probably just as subject to infection by microbes as are those of bees and ants, making it likely that one or more exocrine glands produce antibiotics. The prolonged licking and/or rubbing with the gaster of the cell walls by females of wasps may serve to spread an antimicrobial secretion that protects the brood and nest from invasion by pathogens. Gambino recently reported that the larval salivas of *Vespula pensylvanica* and *V. vulgaris* have antibiotic properties (Gambino 1993). It is not known how this property is used, but it may relate to the fact that these wasps are carrion-feeders. Perhaps the use of such resources increases the risk of infecting the nest with harmful bacteria. However, the secretion of Dufour's gland in the stenogastrine genus *Parischnogaster* was found not to inhibit growth of bacteria and yeasts (Turillazzi 1985*d*).

It is remarkable that the wasps have received so little attention in regard to antibiotic secretions.

Nestmate recognition

Odour-based nestmate recognition systems are widespread, even among pre-social insects. It seems likely, therefore, that the components of the nestmate recognition system we see in social wasps was already largely in place prior to the evolution of eusociality in the group.

Females of *Mischocyttarus, Polistes,* and other genera frequently stroke the gaster over the cell openings of the nest in a movement resembling the application of ant repellent to the nest petiole (Jeanne 1972; Dani *et al.* 1992*b*). The behaviour, coupled with the presence of sternal glands (Fig. 8.1: 13, 14), suggests that a glandular secretion is being applied to the nest. In a recent analysis of this behaviour in *P. dominulus*, Dani and her colleagues showed that solitary foundresses as well as co-founding females perform this behaviour on pre-emergence nests (Dani *et al.* 1992*b*). Among co-founding females the rate of performance seems to correlate with ovary development: the alpha female performs the behaviour frequently, while the beta female's rate diminishes during the course of the pre-emergence period. If the alpha is removed, the rate of beta's stroking rises by the next day to what the alpha's had been. When, on the third day, the alpha is put back, her rate more than doubles what it was on the first day (Dani *et al.* 1992*b*). Dani *et al.* (1992*b*) argue, rightly in my opinion, that the behaviour has nothing to do with nest defence, despite the resemblance of the behaviour to the application of ant repellent to the nest petiole. The fact that solitary foundresses perform it during the entire pre-emergence period argues against any direct role in dominance interactions. That the behaviour occurs when the nest contains only eggs argues against any role in signalling directly to the larvae or pupae. As an alternative, the authors hypothesize that stroking applies a substance to the nest that either represses future ovary development in the brood or informs the brood which adult female is dominant. In view of the findings of Gamboa and his co-workers that nestmate discrimination in *Polistes* is based on odours picked up from the nest, abdominal stroking may also be the mechanism by which the nestmate discrimination odour is transferred from the bodies of adults to the nest, whence it is learned and adsorbed by young adults in their first hours after eclosion (Pfennig *et al.* 1983*a, b*; Gamboa *et al.* 1986*b*). Whether the compounds presumed to be laid down by stroking have one or more of these functions will require some exacting chemical sleuthing and bioassaying.

The so-called 'footprint pheromone' reported by Butler *et al.* (1969) deserves mention in this context. These authors showed for *Vespula vulgaris* that forager traffic through the nest entrance deposits a chemical that is attractive to returning foragers. Neither the source nor the chemical nature of this substance has been determined. I suggest the parsimonious hypothesis that it is the same as the colony odour.

The topic of recognition odours is dealt with in more detail in the chapter by G. Gamboa in this volume.

Reproductive behaviour

In recent years chemicals have been implicated in reproductive behaviour of both male and female wasps. Short range olfactory or contact chemical cues produced by females release copulatory behaviour in males of vespines (reviewed by Keeping *et al.* 1986), but *Polistes* is the best-studied genus in this regard. Venom has been shown to attract males from distances of a few centimetres in still air and >2 m in moving air and to release male copulatory behaviour in *Polistes fuscatus, P. exclamans*, and *Belonogaster petiolata* (Post and Jeanne 1983*d*, 1984; Keeping *et al.* 1986; Reed and Landolt 1990). Volatiles on the cuticle of females are also attractive to males. In a 2.2 m wind tunnel bioassay, males flew upwind in air passed over all tagmata of unmated females, although the greatest response was to the thorax (Reed and Landolt 1990). Reed and Landolt (1990) list as possible sources of the attractant the prothoracic glands, mesothoracic glands, venom gland, and Dufour's gland (Fig. 8.1). Hexane extracts of female heads and thoraxes also released copulatory behaviour in *B. petiolata* males (Keeping *et al.* 1986). Although both the venom and cuticular volatiles of females attract and release copulatory behaviour in *Polistes* males, species recognition is mediated only by cuticular compounds (Post and Jeanne 1984).

Fig. 8.2 Schematic representation of the sternal glands of metasomal sternites 3–7 of the *Polistes fuscatus* male. Based on Post and Jeanne (1983c). Males, along with females, have leg tegumental glands (*P. dominulus*) (Beani and Calloni 1991a) and ectal manidibular glands (*P. major*) (Wenzel 1987a) (not shown; see Fig. 8.1).

Males also produce chemicals in the context of reproduction. In many polistine species males scent-mark their perches and other objects (reviewed by Beani and Calloni 1991a; Polak 1993a). The most common behaviour is a dragging or wiping of the abdomen over the substrate, although four species of *Polistes* also drag their hind legs and *P. major* drags the abdomen and the clypeus and mandibles (Wenzel 1987a). Males have well-developed mats of individually ducted cells (class III cells of Noirot and Quennedey 1991) on the posterior portion of 1–5 of the metasomal sternites in *Polistes* spp. (Fig. 8.2) and the last three in *Mischocyttarus* (Post and Jeanne 1982b, 1983c; Downing *et al.* 1985). The gland cells in *Polistes* open onto the cuticular surface among scattered setae; on sternite seven, which has the largest gland, they open among large, posteriorly curved setae. The distribution of sternal glands in males of other polistine genera is quite variable, with some lacking glands altogether (Downing *et al.* 1985). In *P. dominulus*, one of the species that drags its hind legs, class III gland cells were found in the femur, tibia, and especially in the tarsi of all three pairs of legs of both sexes of (Beani and Calloni 1991a) (Fig. 8.1: 11). These are distinct from the large gland in the distal tarsomere that is thought to provide the fluid enhancing adhesion of the arolium to smooth surfaces (Billen 1986) (Fig. 8.1: 12). The ectal mandibular gland of males of *P. major* is unusually well developed and opens into a conspicuous brush of applicator hairs (Wenzel 1987a). The other glands of the head have not been investigated in males.

Male mating strategies, including marking behaviour, have been described for several *Polistes* species (Turillazzi and Cervo 1982; Post and Jeanne 1983a; Wenzel 1987a; Beani and Turillazzi 1988a; Beani and Calloni 1991a; Reed and Landolt 1991; Beani *et al.* 1992; Polak 1992, 1993a, b), but much remains to be done. It is not yet clear, for example, how the chemicals are used. The fact that unmated female gynes are attracted to secretions of the mandibular and seventh sternal gland of males of *Polistes fuscatus* supports the hypothesis that male scent marks are used in mate attraction (Reed and Landolt 1990), although this may not be universal in the genus. Virtually nothing is known about the role of chemical signals in reproductive behaviour in other genera. Although the use of chemical attractants undoubtedly preceded the origin of eusociality in the wasps, it is too soon to tell if and how different male mating strategies correlate with differences in social systems within the Vespidae.

Arising with, or soon after, eusociality

The defensive role of the venom gland

The females of all social wasps possess stings and, as far as is known, produce venom (Fig. 8.1: 17) that is used in defence against vertebrates, conspecific usurpers, and social parasites (Fisher 1993). The presumed ancestors of the social vespids employed their stings to paralyse the prey used to provision their developing larvae. The shift to a defensive function probably occurred with or shortly after the crossing of the threshold of eusociality, and it is likely that the chemistry of the venom evolved in response to the new function. Although we know something about the venom chemistry and pharmacology of a few species (Blum 1981), we are far from any kind of 'comparative venom chemistry' for the wasps. A study of species that bridge the eusociality threshold could yield correlations between chemical and functional differences. Social species with very small colony size would be especially interesting in this regard. The stenogastrines and some species of *Mischocyttarus*, for example, cannot be induced to defend their nests by stinging a large intruder, but flee instead (Turillazzi 1989; Jeanne, pers. obs.). It would be interesting to know how the venom chemistries of these differ from those of more advanced social species in which a stinging defence is readily induced. Similarly, the venom chemistry of very small-bodied wasps (*Leipomeles, Protopolybia*) whose stings may be too small to penetrate vertebrate skin, may be different from those of larger species that can mount effective stinging defences against vertebrate predators.

Other defensive allomones

It is now well-established that all genera of independent-founding polistines (*Polistes, Mischocyttarus, Belonogaster, Parapolybia, Ropalidia*) have a well-developed gland of individually ducted (class III) cells arranged either in a narrow band or in two lateral clusters at the base of metasomal sternite six (Jeanne *et al.* 1983) (Fig. 8.1: 14). Associated with the gland opening is a brush of long setae with which females smear the secretion of the gland onto the nest petiole (Jeanne 1970). It has now been shown for representatives of all five genera that the coating is effective in repelling many species of predaceous ants (Jeanne 1970; Turillazzi and Ugolini 1979; Post and Jeanne 1981; Kojima 1982, 1983*a*, *b*, 1992; Keeping 1990*b*) (Table 8.1). An active component of the allomone of *Polistes fuscatus* has been identified as methyl palmitate (Post *et al.* 1984*b*). It is not known whether the glands on one or more of the sternites anterior to the 6th (Fig. 8.1: 13) have the same function as the 6th; the lack of a sternal brush on these segments suggests a different function.

In contrast to the independent-founding polistines, most swarm-founders appear to rely on active vigilance and physical removal to prevent intrusion of ants onto the nest. Two species, however, *Nectarinella championi* and *Leipomeles dorsata*, erect physical barriers in the form of short, sticky-tipped stalks closely spaced around the

Table 8.1 Species for which ant-repelling properties of the 6th sternal gland and/or rubbing behaviour in females have been demonstrated.

Species	Rubbing	Ant-repellent	Reference
Mischocyttarus drewseni	+	+	Jeanne 1970
M. atramentarius	+		Silva 1981
Polistes annularis	+		Hermann and Dirks 1974
P. dominulus (cited as *P. gallicus*)	+	+	Turillazzi and Ugolini 1979
P. gallicus (= *P. foederatus*)	+	+	"
P. nimpha	+	+	"
P. fuscatus	+	+	Post and Jeanne 1981
P. lanio lanio	+		Giannotti 1992
Parapolybia indica	+	+	Kojima 1983*b*, 1992
Ropalidia gregaria	+		Kojima 1982
R. fasciata	+	+	Kojima 1983*a*
Belonogaster petiolata	+	+	Keeping 1990*b*
B. grisea	+		Marino Piccioli and Pardi 1970

perimetre of the nest and on the petiole of the leaf bearing the nest, respectively (Williams 1928; Schremmer 1977, 1983). The stalks, at least in the case of the former species, are of vegetable fibre (Schremmer 1977). The source of the sticky droplets is not known, even as to whether the wasps produce it or collect it at plants. This line of defence is clearly a derived condition within the swarm-founding species and may be correlated with the small body size of adults in these genera. It may be that wasps of such small size are unable to deter some of the larger predaceous ants in one-on-one combat, which could have led to selection favouring the evolution of such a physical barrier.

Some of the stenogastrines have independently evolved a different kind of ant barrier, in the form of one or more rings of repellent material around the filament from which the nest is hung. Turillazzi and his co-workers have shown that the material is produced by Dufour's gland, is collected from the tip of the abdomen by the legs and applied to the filament with the mandibles. The same material is used in brood care (see below).

Brood feeding

In *Apis mellifera* the amount of royal jelly fed to a larva determines whether that larva develops into a worker or a queen. So far there is no evidence to indicate that any social wasp adds glandular products to larval food, as honey bees do. But this may be simply because no one has looked. Wasp larvae are largely fed on prey collected by the foragers. When a prey item is returned to the nest by a *Polistes* forager, it is malaxated for several minutes, during which time it becomes visibly drier as the female imbibes most of the liquid in the bolus (Grinfel'd 1977). She then feeds the pellet to one or more late-instar larvae, following which she regurgitates the liquid to early-instar larvae. It is certainly possible that at least this liquid portion may have some oral glandular secretion added to it by the worker before it is fed to larvae. Indeed, Deleurance suggested that this was going on in *Polistes* (Deleurance 1955), but to date there has been no confirmation of this. The suggestion by Spradbery (1973) that the hypopharyngeal gland in *Vespula* might produce such a substance is not supported by studies of gland activity in *P. fuscatus* (Downing and Jeanne 1983).

Several species of Stenogastrinae secrete a pap-like substance from the tip of the abdomen that serves several functions. It is used to handle newly laid eggs and it is placed in nest cells, where it serves as a protective mass for eggs and young larvae and as a food storage substrate for larvae and adults. It is the same material that is used in the ant barriers (Turillazzi 1991). The secretion has recently been shown by chemical analysis to be produced by the hypertrophied Dufour's gland and to consist of hydrocarbons, largely straight-chain alkenes in the range of C_{21}-C_{33} (Keegans *et al.* 1992*b*, 1993). However, this secretion of Dufour's gland, unique among social vespids, does not in itself constitute a larval food.

Arising well after eusociality

Chemical alarm recruitment

If the sequence of my evolutionary scenario is correct, the alarm pheromones represent the first appearance in a social context of true chemical signals, that is, chemicals whose evolved function is to convey information.

The first experimental evidence for alarm pheromones in social wasps was published by Maschwitz (1964), who demonstrated the presence of an alarm pheromone in queens and workers of *Vespula vulgaris* and *V. germanica*. There is now experimental evidence for alarm pheromones in several highly eusocial genera: *Vespula, Vespa, Provespa, Dolichovespula,* and *Polybia* (Maschwitz 1964, 1984; Jeanne 1981*a*; Landolt and Heath 1987), with circumstantial evidence for *Protopolybia, Apoica,* and *Synoeca* (Naumann 1970; Schremmer 1972; West-Eberhard 1982*b*). In all tested cases the releaser of alarm behaviour is a component of the venom.

Alarm-releasing chemicals have been chemically identified for three species so far: *Vespula vulgaris* (methyl-1,6-dioxaspiro[4.5]decanes) (Francke *et al.* 1978); *Vespa crabro* (2-methyl-3-butene-2-ol) (Veith *et al.* 1984), and *Vespula squamosa* (N-3-methylbutylacetamide) (Heath and Landolt 1988). This compares poorly with the ants, for which 48 alarm pheromones had been chemically identified as of 1990 (Hölldobler and Wilson 1990).

While alarm pheromones may well prove to be universal in the vespines and swarm-founding polistines, the independent-founding polistines are another matter. *Polistes* is a case in point. Freisling (1943) had shown that wing-buzzing by wasps on the nest, but not hovering in front of the nest, elicits alarm among nestmates in *P. nimpha* and *P. dominulus*. Wasps with the head and abdomen excised also elicited alarm when they wing-buzzed while in contact with the comb. So when Maschwitz tried and failed to find any evidence of an alarm pheromone in *P. biglumis*, he was led to conclude that in *Polistes*, with its small colonies, mechanical communication, or even independent response to visual cues, was adequate and therefore no chemical alarm signal was necessary (Maschwitz 1964).

The subsequent discovery of alarm pheromones, also in the venom, in three American species (*P. canadensis, P. fuscatus, P. exclamans*) belies this argument (Jeanne 1982; Post *et al.* 1984*a*). This raises the question of why some species of *Polistes* have evolved alarm pheromones whereas others apparently have not. In species with an alarm pheromone, does it supplement, or supplant, wing buzzing? Unfortunately, no one has yet rigorously tested for signal effects of buzzing in any species with an alarm pheromone. In *Polybia occidentalis*, a species with an alarm pheromone, it has been experimentally demonstrated that wing-buzzing does not communicate alarm (Jeanne 1981*a*), but it would be premature to predict that the same is the case for *Polistes*.

What is going on in the remaining independent-founding genera is even less well known. The only study I know of is Keeping's recent careful test for alarm pheromones in *Belonogaster petiolata*, which came up negative (Keeping 1989). No tests have been done on *Mischocyttarus, Parapolybia* or *Ropalidia*. Clearly, such studies would be worth pursuing, as it looks as though the independent-founding species are variable on this trait.

Chemical queen control

Pardi's early work on dominance in *Polistes* was essentially a study of queen control. When exocrine glands become involved in this role, we are dealing with queen pheromones, or queen substances, the set of pheromones by which the queen suppresses reproductive activities of the workers.

West-Eberhard distinguishes between 'queen control' and 'queen recognition', limiting 'queen control' to the effect of suppression of reproduction by workers (West-Eberhard 1977). When we speak of chemical queen control, we are essentially speaking of suppression of workers' ovaries via the primer effect of a queen-derived pheromone. West-Eberhard (1977) points out that much of what has been adduced as evidence for pheromonal queen control in wasps is merely queen recognition.

For example, the queen pheromone of *Vespa orientalis*, identified as δ-*n*-hexadecalactone (Ikan *et al.* 1969), is attractive to workers, has a tranquillizing effect on them, appears to mediate the 'royal court' wherein workers surround the queen and lick her body, and regulates the initiation of queen-sized cells and the overall rate of cell construction (Ishay *et al.* 1965; Ishay 1973). Yet it is not known whether it suppresses ovary development in workers, that is whether it is involved in queen control (Spradbery 1991).

Circumstantial evidence for chemical queen control comes from *Vespula*. Members of the *Vespula rufa* group produce small colonies of a few hundred workers. Queen control appears to be exercised by a combination of pheromones and physical action (Landolt *et al.* 1977). In an experimental division of a colony of *V. atropilosa* into queenright and queenless portions, workers in the queenless side soon began to behave as though queenless: some underwent ovarial development and within eight days began ovipositing (Landolt *et al.* 1977). The fact that the two sides were separated by a double screen, so that workers on the queenless side could not contact the queen or workers on the other side, suggests that if there is a pheromone responsible for ovary suppression it is non-volatile. But the results could equally well be interpreted as indicating behavioural control.

Similar colony division experiments with members of the *Vespula vulgaris* group suggest that there is an ovary-suppressing pheromone and that it is non-volatile. Akre and Reed (1983) showed that development of worker ovaries on the queenless side was suppressed if the combs between the two sides were switched every 24–48 hours, implicating the comb or brood as mediating factors in the dispersal of a queen pheromone. Spradbery showed experimentally for *V. germanica* that it is the comb and not the brood (Spradbery 1991).

What is interesting and worth following up in the vespines is what appears to be a range of the relative involvement of physical dominance vs. pheromones in queen control. The amount of overt aggression occurring in dominance interactions varies interspecifically among species of Vespinae (Spradbery 1991). Whereas *V. vulgaris* and *V. germanica* show very little overt aggression, queens of *Dolichovespula maculata* seem to exert some degree of physical control over workers, in particular by mauling them if they attempt to oviposit. Nevertheless, workers become 'greatly excited' in the presence of the queen, suggesting a chemically-mediated effect on workers' behaviour, although there is as yet no demonstration of a queen pheromone.

Queen control in *Polistes* seems to be exercised via direct physical domination. Yet within the genus there appears to be a range in effectiveness of queen control, from the apparent absence of any direct dominance interactions in *P. bernardii richardsi*, *P. humilis* and others (Itô 1986*a*, *b*) to the very violent attacks by the despot queen of *P. canadensis* (West-Eberhard 1986*a*). Spradbery (1991) tabulates the nature of dominance interactions for *Polistes* and the other independent-founding genera. There is evidence of chemically-mediated queen recognition, however. Downing has shown that Dufour's gland (Fig. 8.1: 15) is the source of the cues used by dominant females of *P. fuscatus* to recognize eggs laid by subordinates (Downing 1991*a*).

As to the source of queen pheromones in the wasps, such evidence as we have points to the head as the source of queen recognition pheromones in the hornets and swarm-founding polistines (Ishay *et al.* 1965; West-Eberhard 1977). The source of the queen control pheromone in *Vespula germanica* is unknown, but it is clear from Spradbery's experiments that it is transferred onto the nest (Spradbery 1991) and, as he points out, this could implicate the sternal gland of the 6th metasomal segment.

Queens of parasitic species of *Dolichovespula* and *Polistes* typically perform excessive amounts of abdominal stroking of the nest, suggesting that they are applying a secretion to the surface (Jeanne 1977*b*; Greene *et al.* 1978; Cervo *et al.* 1990*a*). It is possible that Dufour's gland is the source of such a chemical; the fact that this gland is hypertrophied in vespine social parasites suggests that it plays some role in altering the behaviour of the host queen and workers (Jeanne 1977*b*; Greene *et al.* 1978). Interestingly, the Dufour's gland of *Polistes sulcifer*, a social parasite of *P. dominulus*, is not hypertrophied (R. Cervo, pers. comm.).

Further investigation into the chemical nature of queen control and of social parasite behaviour in the wasps would undoubtedly be amply repaid.

Trail pheromones and attractants

Swarm-founding polistines use chemical trails to guide swarms to new nest sites. The trail pheromone, produced by the gland at the base of the penultimate (5th) metasomal sternite, is wiped onto leaves and twigs by scout wasps (Naumann 1975; Jeanne 1981*b*). By hovering just downwind of such points or by landing on them and inspecting them, the wasps in the swarm determine whether the scent is present. If it is, they proceed farther from the cluster site in the same direction and repeat the search. This behaviour eventually brings them to the site selected by the scouts. The same scent is used to mark the site on which the swarm temporarily assembles prior to emigration and to mark the new nest site (Jeanne 1981*b*). Dragging behaviour has been described for several genera that apparently lack sternal glands (Naumann 1975; West-Eberhard 1982*b*; Jeanne *et al.* 1983; Francescato *et al.* 1993).

Is there any evidence of incipient trail function among the independent-founding polistines? There have been several reports for *Polistes* and *Mischocyttarus* of marked nestmates appearing together on newly founded colonies within a day or two of leaving their declining or destroyed natal colony (West-Eberhard 1969*b*; Jeanne 1972; Litte 1981). In these cases there are at least two possibilities for how the nestmates find the nest. First, they could be using random visual search followed by contact chemoreception (nestmate recognition cues) to distinguish former nestmates from non-nestmates. Alternatively, they could be using in addition a more specific, long-distance attractant odour. Evidence for the latter comes from two species of *Mischocyttarus*. Litte observed females of *M. labiatus* wiping their abdomens over leaves and twigs along the route between their natal nest and the new nest they had established some metres away (Litte 1981). Nestmates were observed landing on and antennating these spots, suggesting that they were responding to scent marks. Queens performed the same behaviour around the nest if nestmates were removed.

O'Donnell observed similar behaviour in *M. immarginatus* in Costa Rica, in which one or more of the co-founding females on a newly initiated nest rubbed the gaster on leaves within a metre of the nest (O'Donnell 1992). Within two days at least four additional females joined the colony, after which rubbing was no longer seen. Although linear scent trails were evidently not laid, O'Donnell suggested that the marking may act as a local attractant that helps potential co-foundresses orient to the nest vicinity.

This behaviour is remarkably similar to what is seen in *Polybia occidentalis*. When a nest is damaged or badly disturbed one or more females sometimes fly to nearby leaves and drag the gaster in a manner identical to trail and aggregation-site marking (Jeanne, pers. obs.). In other words, the pheromone appears to function as an attractant, to which adults respond when they are without a nest or when the colony is threatened. *Polybia* spp. have only the 5th sternal gland, so it is all but certain that the same pheromone is used in all these contexts.

If we assume that the 5th sternal glands (Fig. 8.1: 13) are homologous in independent- and swarm-founding polistines, it seems likely that it is the secretion of this gland that is being rubbed onto leaves by *Mischocyttarus* and probably onto the nest during abdominal stroking by both *Mischocyttarus* and *Polistes* and that it is serving as an attractant and possibly as a nestmate recognition odour as well. This hypothesis has yet to be tested.

West-Eberhard (1982*b*) posited a scenario by which swarming behaviour could have evolved from independent-founding ancestors. The steps she hypothesizes to have occurred involve increasingly sophisticated scent-marking, beginning with simple odours associated with the nest itself and progressing through assembly pheromones and orientation cues to full-fledged scent-trail marking. The few details we have learned in the past ten years suggest that this scenario may be correct, at least in broad outline.

The social wasps are the only one of the four groups of eusocial insects for which chemical recruitment of nestmates to a food source has not been demonstrated. Although there is evidence for low-level recruitment to food in one species of *Polybia* and two of Vespinae, there is no evidence that pheromones are involved (Lindauer 1961; Maschwitz *et al.* 1974). If social wasps indeed do lack the ability to communicate distance and/or direction of food, then the interesting question becomes why. On the surface of it, the ability would seem to be as advantageous to species such as *Brachygastra mellifica, B. lecheguana, Polybia scutellaris, Protonectarina sylveirae* and others that collect and store large amounts of honey as it is to *Apis* and the meliponine bees. Recruitment could conceivably also be advantageous to carrion feeders such as *Agelaia pallipes*. Lack of adequate genetic variability seems an unlikely reason, as there has been enough to have enabled the evolution of the trail pheromones described above, a mechanism similar to that used by some of the meliponine bees to recruit nestmates to a food source (Lindauer 1961).

Pupal thermoregulation

Late-stage pupae of *Vespa* and *Vespula* spp. elicit warming behaviour (abdominal pumping) in adults (Ishay 1972, 1973). The releaser of the behaviour is *cis*-9-pentacosene (Veith and Koeniger 1978). The source is unknown; it is possible it may simply be one of the many cuticular hydrocarbons. It is not known whether pupae of *Polistes* or any other independent-founding polistines produce a similar pheromone.

The logic of gland location

Is there any correlation between where exocrine glands occur in the body and the functions they perform? Exocrine glands in the wasps fall into four main locational groups, each of which appears to be associated with specific, limited functions. I describe these below and make some predictions about where particular gland functions may be located.

Glands associated with the mouthparts

Primitively, at least some of these glands probably had functions related to feeding, including the addition of digestive enzymes to food. In the social vespids, a gland in this group produces the nest construction secretion. Since this material must be mixed with the nestmaterial as it is manipulated with the mandibles, it is not surprising for such a function to have been evolved in an oral gland. The social sphecids are exceptions, using silk produced in the abdomen to bind the nestmaterial together.

Other possible functions of oral glands are the production of ant trap adhesive (*Leipomeles, Nectarinella*), antibiotics, and brood food. All these functions occur in the context of other behaviours involving use of the mouthparts (nest construction and feeding), so it would be most likely for these functions to be assumed by glands in the head.

Glands associated with the sting

The poison gland produces not only venom, but an alarm pheromone as well, a pattern not unusual in social insects (Hölldobler 1977). The odour of venom in the air, always associated with an attack on an intruder, would be a prime candidate to evolve alarm-signalling function. At least in *Polistes*, the venom also contains a sex pheromone (Post and Jeanne 1983*d*). Since only females produce venom, the odours associated with it would be reliable cues by which males could recognize females over short distances.

Oviduct-associated glands

Since it is the differential development of the reproductive system that distinguishes queens from workers, glands associated with the reproductive tract are logical candidates for queen-recognition pheromones. This is supported by Downing's recent demonstration in *Polistes fuscatus* that Dufour's gland secretion from a subordinate or non-nestmate female, when applied to eggs, significantly raised the probability that the eggs would be eaten by the dominant female (Downing 1991*a*).

Tegumental glands

Gland cells that open directly onto the outer surface of the integument may occur singly or be concentrated into sternal and tergal glands. These are more variable in occurrence and location throughout the social insects than are glands of the other three categories, suggesting that they are relatively labile in evolution. William Morton Wheeler recognized this 100 years ago when he stated that 'the entire ectoderm of Arthropods, excepting its nervous derivatives, is essentially à glandular layer' (Wheeler 1892, p. 116). In the wasps, glands in this category produce repellents, attractants (including trail), and, from evidently dispersed sources, nestmate recognition, colony odour, and thermoregulation pheromones. These are functions that do not involve sex or caste discrimination and are not related directly to behaviour involving the mouthparts (feeding, nest building) or use of the sting.

Evolutionary origins of chemical signal function

It has been tacitly assumed that the responses to chemical signals in insects do not have to be learned. Further, it is assumed that this has been true from the evolutionary inception of a chemical signalling system. These assumptions are probably based on the dogma that much of insect behaviour is innate and that the ability to learn is phylogenetically recent.

The opposite may be true. Tierney (1986, p. 339) hypothesizes that 'behavioural flexibility is phylogenetically primitive and that learned behavioural adaptations may commonly precede innate forms of the same behaviours.' She suggests that learned behavioural solutions to novel environmental challenges may be common evolutionary phenomena. Given genetic variability among individuals in the capacity to learn, those most able to adapt will leave the most offspring. Over many generations, providing that the best thing to learn remains constant, this can lead to genetic fixation, or canalization, of a character that previously had to be learned anew by each generation (Maynard Smith 1987). Hinton and Nowlan (1987) developed a simple model to show that learning can alter the fitness surface so as to lead to genetic assimilation of co-adapted traits much faster than in the absence of learning. The kinds of traits that could predispose the animal to perform the behaviour could include perceptual systems and feature detectors.

Certain chemical signalling systems in social insects may well have evolved via an initial learning stage. For this to happen, the odour that eventually evolves into a pheromone must regularly co-occur with the unconditioned stimulus that evokes the appropriate response, so that the association between it and the unconditioned stimulus can be learned. This condition appears to have been met in such systems as alarm and trail signalling in the social wasps. I use the example of the evolution of alarm signalling to illustrate how learning might have played a role.

Once the threshold of eusociality is crossed, the sting, ancestrally functioning to paralyse prey, evolves into an instrument of defence. At first, when colonies are small, individual colony members respond individually and directly to the stimulus of an intruder. With larger colony size, however, colonies that mount a more massive defensive response will be more likely to survive an attack. When a wasp stings an intruder the odour of venom is released. Colonies whose members learn to associate this odour with the presence of a predator and to respond with a greater probability of attack themselves, mount more massive attacks and enjoy greater fitness. Since the odour of venom always accompanies a defensive response to an attack by a predator, the association is ripe for genetic assimilation and the response can become genetically fixed in the population. Natural selection could subsequently modify both the chemical signal (greater volatility, volume) and the reception of it (more receptors; narrower tuning to the releasing molecule) so as to enhance the signal's effectiveness. A further evolutionary step may be the dissociation of the signal from the act of stinging. Thus venom may be released at the nest by the first worker to detect a predator, eliciting a massive defence by many workers against an approaching predator even before the first sting is administered.

Such a scenario for the role of learning in the evolution of chemical signals may appear to require that the wasps learn to associate a conditioned stimulus (odour of venom) with an unconditioned stimulus (cues issuing from the predator) without an immediate reinforcement. It could be argued that the reinforcement is the removal of the stimulus (predator) as a result of the wasps' attacks. In any event, immediate reinforcement is not necessary for learning to occur. An example is latent learning, as when a young honey bee forager learns the landmarks around its hive during its first outings (Winston 1987) or when a female bee wasp learns the location of its nest during a 6-second orientation flight (Tinbergen 1958).

A recent example of non-reinforced learning that is perhaps closer to the case in point in that it involves associative learning comes from a study by Mathis and Smith (1993). Naive fathead minnows (*Pimephales promelas*) exhibited a fright response (decreased activity level) when exposed to water from a tank of pike that had recently fed on conspecific minnows. Water from a tank of pike that had fed on heterospecific prey (swordtails) did not elicit fright, indicating that the fright response to pike odour is not innate. Mathis and Smith suggest that the minnows may have been detecting traces of their alarm substance, or 'Schreckstoff,' released from epidermal cells when they are damaged, as when a minnow is eaten. Minnows exposed to minnow-fed pike water subsequently exhibited a fright response to swordtail-fed pike water. In other words, the minnows appeared to learn, without any immediate reinforcement or punishment, the association between general pike odour and minnow-fed pike odour.

An analogous phenomenon that may help to make the point is nestmate recognition in wasps. During its first few hours a newly eclosed *Polistes* worker learns, without immediate reinforcement, the mix of hydrocarbons and other compounds peculiar to its own colony. Subsequently it recognizes non-nestmates by their different mix of cuticular odours and responds by attacking them. This differs from the scenario proposed for the evolution of alarm pheromones in that the recognition of nestmate odours has not become genetically assimilated. This is presumably because what is important is to discriminate your colony from others on the basis of the chemical cues that distinguish them and these cues vary from colony to colony and from generation to generation, precluding genetic fixation of any one combination of odours.

Concluding remarks

This brief overview of the biology of exocrine glands in the wasps perhaps does more to reveal what we don't know than what we do. It is clear that much remains to be done, even at the descriptive level, to deepen and taxonomically broaden our understanding of glandular functions and mechanisms.

Despite the extremely sketchy state of our knowledge of gland function and chemistry in this group, certain patterns are emerging and it will be exciting to see how they develop. For example, there are hints that the wasps are heterogeneous with regard to mechanisms of queen control. At a lower level, the independent founders as a group, and *Polistes* in particular, seem homogeneous with respect to possession of ant-repelling allomones, but heterogeneous with respect to alarm pheromones.

Is the absence of alarm pheromones in some species a function of phylogenetic inertia, lack of ecological pressure, small colony size, or something else? Any study of alarm/defence behaviour in the wasps should take into account not only whether pheromones are involved, but careful descriptions of the behavioural responses they elicit. Particularly among the swarm-founders there appears to be a tremendous diversity of ways in which colonies defend their nests against large predators. It will be extremely interesting to work out what roles such factors as predation, colony size, body size, social structure, and phylogeny have played in shaping this diversity.

The correlations between gland location and function suggests that natural selection has seized upon the most readily available, neutral chemical present in a given context to shape it into a structural product or chemical signal with a social function relevant to that context. With regard to pheromones, the wasps seem still to rely heavily on what may have been the primary odour cue evolved in each context, rather than having evolved secondary and tertiary pheromonal backup systems, as appears to have been the case in some of the higher bees and ants (Michener 1974; Hölldobler and Wilson 1990). For this reason, the social wasps are especially suitable for investigations into the evolution of the social functions of exocrine glands in general, and the role learning may have played in the evolution of chemical communication in particular.

These and other evolutionary, functional, and mechanistic questions beckon the next generation of insect sociobiologists.

Acknowledgements

My thinking on these issues has benefited from discussions with my past and present students, particularly Holly Downing and Karen London. Research supported by the College of Agricultural and Life Sciences, University of Wisconsin, and by the National Science Foundation, grants BNS-8517519 and IBN-9222108.

9

Kin recognition in social wasps

George J. Gamboa

During the past 15 years, a substantial literature has documented that a diversity of hymenopterans have the ability to recognize their conspecific nestmates (Fletcher and Michener 1987; Hepper 1991). Typically, colony members accept nestmates but exclude non-nestmates. Almost all studies of nestmate or kin recognition in social insects have concentrated on documenting recognition ability or investigating the mechanisms subserving recognition. As a result, there is a relatively good understanding of recognition ability and the proximate basis of recognition for several groups of social insects, including social wasps of the genus *Polistes*.

Although there is considerable knowledge about the proximate basis of kin recognition in social insects, little is known about the adaptiveness or sociobiology of recognition (Gamboa *et al.* 1991c). The vast majority of recognition studies have been laboratory studies. Recent evidence indicates that recognition is context dependent and that results from the laboratory do not necessarily reflect recognition in the field (Gamboa *et al.* 1991b). Consequently, there is a great need to investigate recognition in natural contexts. It almost certainly will be more difficult to gain an understanding of the sociobiology of recognition than its underlying labels and perceptual systems.

For the past 13 years, my students and I have been using *Polistes* as a model system for understanding kin recognition in social insects. We have investigated recognition ability, the mechanism of recognition, and the sociobiology of recognition in primarily one species, *Polistes fuscatus*. Although we and others have studied recognition in several other species of social wasps, *P. fuscatus* is by far the most intensively studied species. Based on studies of other *Polistes*, it appears that the recognition abilities and the mechanism of recognition in *P. fuscatus* can be generalized to other *Polistes*. The major features of recognition in *P. fuscatus* are probably the same as those of other genera of social wasps, and similar to those of ants and bees. I believe that our understanding of kin recognition in temperate *Polistes* is as complete as in any other group of animals.

The scope of my review is intentionally limited. There is a vast literature on kin recognition in ants and social bees, and a thorough integration of this literature with that of the social wasps would be more appropriate for a review of kin recognition in social insects. However, in some instances, and particularly in the last section of the review, I have identified several common features of recognition in social wasps and other taxa. These selective comparisons represent only a few of the important and interesting attributes shared by social wasps and other animals.

Certain aspects of kin recognition in social wasps are provided elsewhere. For example, the evolution of kin recognition and the adaptiveness of some features of its mechanism are considered by Gamboa *et al.* (1986*a*). Extensive discussions of recognition bioassays, experimental procedures, and the components of recognition are provided by Gamboa *et al.* (1986*a*; 1991*b*, *c*). Finally, Byers and Bekoff (1986), Waldman *et al.* (1988), and Gamboa *et al.* (1991*c*) discuss definitions of kin and nestmate recognition in considerable detail. In accordance with Gamboa *et al.* (1991*c*), in this review I consider nestmate recognition to be a subset of kin recognition.

I first review what is known about recognition ability in social wasps and then discuss its underlying mechanism, relying primarily on evidence from *P. fuscatus*. Later I discuss our present understanding of recognition in natural contexts, and, in the final section, identify common features of recognition in social wasps and other animals.

Recognition ability of social wasps

Adult–adult nestmate recognition

Nestmate recognition (the differential treatment of adult conspecifics on the basis of colony origin) has been documented in at least five of the approximately 200 species of *Polistes* (Reeve 1991; Table 9.1). It is likely that the ability to recognize nestmates is widespread, although not necessarily universal, in *Polistes*. Our knowledge of the recognition abilities of tropical *Polistes* is particularly deficient. Present evidence indicates that all castes or classes of *Polistes* adults can recognize their adult nestmates. Thus, foundresses, fall gynes, and workers can recognize their female nestmates (Noonan 1979; Post and Jeanne 1982*a*; Bornais *et al.* 1983; Pfennig *et al.* 1983*b*; Gamboa *et al.* 1986*a*; Pfennig 1990).

Males of *P. fuscatus* can recognize their male nestmates and probably their female nestmates also (Shellman-Reeve and Gamboa 1985; Ryan and Gamboa 1986). The possible adaptive significance of male–brother recognition is not obvious. Although there has been no rigorous study of females' abilities to recognize their male nestmates, it would be unexpected if they lacked this ability. Since females have well-developed recognition abilities and males obviously possess colony odours, it's highly likely that females can recognize their male nestmates. In all studies of *Polistes* that have documented recognition, nestmates have been more tolerant of each other than have non-nestmates (Gamboa *et al.* 1986*a*).

In addition to facilitating nepotism and preventing intraspecific parasitism, *Polistes* may use their recognition abilities to avoid mating with nestmates (Gamboa *et al.* 1986*a*). For example, Ryan and Gamboa (1986) found that males of *P. fuscatus* copulated significantly less often with nestmates than with non-nestmates in the laboratory. Although suggestive, these laboratory results do not establish that *P. fuscatus* normally outbreed in field conditions (Post and Jeanne 1983*b*; Ryan and Gamboa 1986). In contrast to *P. fuscatus*, virgin queens of *Vespula maculifrons* preferentially mated with male nestmates in the laboratory

Table 9.1 Classes of kin recognized, and recognition of related comb and brood, by social wasps. Unless otherwise noted, cited studies are laboratory studies.

Species	Recognition of nestmates	Recognition of non-nestmate kin	Recognition of comb or brood
P. fuscatus	Foundress–foundress (Noonan 1979: field; Post and Jeanne 1982a; Bornais *et al.* 1983) Gyne–gyne (Shellman and Gamboa 1982; Pfennig *et al.* 1983a; Gamboa *et al.* 1986b; Gamboa *et al.* 1987; Gamboa *et al.* 1991b: field) Worker–worker (Pfennig *et al.* 1983b) Male–male (Shellman-Reeve and Gamboa 1985) Male–gyne (Ryan and Gamboa 1986) Multiple classes of female nestmates (Gamboa 1988: field; Gamboa *et al.* 1991b: field; Fishwild and Gamboa 1992: field; Bura and Gamboa 1994: field)	First cousins (Gamboa 1988: field; Bura and Gamboa 1994: field) Aunts/Nieces (Gamboa *et al* 1987; Bura and Gamboa 1994: field) Sisters (Noonan 1979: field; Bura and Gamboa 1994: field)	Comb (Ferguson *et al.* 1987) Brood (Klahn and Gamboa 1983: field; Klahn 1988: field)
P. metricus	Foundress–foundress (Ross and Gamboa 1981) Worker–worker (Singer and Espelie 1992)		Comb (Espelie *et al.* 1990)
P. exclamans	Gyne–gyne (Allen *et al.* 1982) Worker–worker (Pfennig 1990)		Comb (Pfennig 1990)
P. carolina	Gyne–gyne (Pfennig *et al.* 1983a)		
P. annularis	Foundress–foundress (Strassmann 1983: field)		
P. gallicus			Brood (Cervo and Turillazzi 1989)
Ropalidia marginata	Female–female (Venkataraman *et al.* 1988)		
Belonogaster petiolata	Foundress–foundress (Keeping 1989, 1990a)		
Dolichovespula maculata	Gyne–gyne (Ryan *et al.* 1985)		Comb (Ferguson *et al.* 1987)
Vespula maculifrons	Male–gyne (Ross 1983)		

(Ross 1983). The significance of these presumably opposite mating preferences are unclear.

Wenzel (1987*a*) reported that kinship (nestmate vs. non-nestmate) did not affect male pursuit of females nor female acceptance of males in a laboratory study of *P. major*. The mating experiments of Wenzel (1987*a*) consisted of placing a male and female into a small (12 × 13 × 17 cm) mating cage and recording the behaviour of the wasps for 5 min after their first contact. It's not clear what effect, if any, the confining mating cage may have had on the wasps' mating behaviour. A more sensitive test of discrimination might be to provide males (or females) a binary choice of mating with a nestmate or a non-nestmate in a more natural context.

Nestmate recognition has been demonstrated in laboratory studies of social wasps other than *Polistes*: the tropical paper-wasp, *Ropalidia marginata* (Venkataraman *et al.* 1988), the South African social wasp, *Belonogaster petiolata* (Keeping 1989, 1990*a*), the temperate hornet, *Dolichovespula maculata* (Ryan *et al.* 1985), and the temperate yellow jacket, *Vespula maculifrons* (Ross 1983; Table 9.1).

There are at least two species of social wasps in which nestmate recognition has been investigated with negative results. Pratte (1982) examined nestmate recognition among spring foundresses of *Polistes dominulus* (cited as *P. gallicus*). He reported that foundresses did not preferentially associate with nestmates under laboratory conditions. Unfortunately, the study by Pratte (1982) had several flaws including the confusion of nestmates with non-nestmates that had overwintered together (Gamboa *et al.* 1986*a*). That the confusion of nestmates with non-nestmates could explain the negative results of Pratte (1982) is supported by the findings of Post and Jeanne (1982*a*). They reported that foundresses of *P. fuscatus* associated with nestmates, but not non-nestmate co-hibernators, in the laboratory. Because of the difficulties in the study by Pratte (1982), the status of nestmate recognition ability in *P. dominulus* remains unknown. It would be surprising, however, if *P. dominulus* lacked nestmate recognition ability since it appears to be both behaviourally and ecologically similar to those North American species that have well-developed nestmate recognition abilities.

Gastreich *et al.* (1990) investigated nestmate recognition ability in the tropical swarm-founding wasp, *Parachartergus colobopterus*. They found no evidence that females discriminated nestmates from non-nestmates. Although negative results can be due to factors other than a lack of recognition ability (Gamboa *et al.* 1991*c*), the findings by Gastreich *et al.* (1990) are compelling because their investigation consisted of both laboratory and field assays of recognition. Furthermore, their bioassays were very similar to those utilized for documenting recognition ability in other social wasps. Therefore, at the very least, it appears that *P. colobopterus* does not display recognition ability in the particular context in which it was studied. Gastreich *et al.* (1990) hypothesized that the differential in relatedness between nestmates and non-nestmates of *P. colobopterus* may be small, thereby decreasing the advantage of colony-level discrimination.

Brood and comb recognition

P. fuscatus and *P. gallicus* can distinguish between related and unrelated brood although it is not known whether the cues for this discrimination reside in the comb or brood. When confronted with unrelated brood as is experienced by conspecific usurpers, foundresses destroy those brood that are destined to become reproductives (Klahn and Gamboa 1983; Klahn 1988; Cervo and Turillazzi 1989). Foundresses of *P. fuscatus* may use their kin recognition abilities to avoid usurping colonies of close relatives (Klahn and Gamboa 1983) since field studies have documented that usurpers are almost never closely related to the colonies they usurp (Klahn 1988; Gamboa *et al.* 1992).

Pfennig (1990) reported that workers of *P. exclamans* can discriminate between a brood-filled comb fragment from their own nest and that from a foreign nest. Female *P. fuscatus*, *P. metricus,* and *Dolichovespula maculata* can discriminate their comb (or a fragment of their comb) from a foreign comb or a foreign comb fragment regardless of whether the comb contains brood (Ferguson *et al.* 1987; Espelie *et al.* 1990). These findings demonstrate that recognition cues must be present in the comb itself.

Recognition of non-nestmate kin

My students and I have examined the ability of female *P. fuscatus* to recognize non-nestmate kin, including first cousins, nieces, aunts, and sisters. In our only laboratory study of non-nestmate kin recognition, we found no evidence that gynes could distinguish between gynes that are non-nestmate aunts and nieces *and* gynes that are nestmate sisters. Thus, in a laboratory context non-nestmate aunts and nieces were treated like nestmate sisters, i.e. highly tolerantly (Gamboa *et al.* 1987; Gamboa 1988).

All of our subsequent investigations of non-nestmate kin recognition have been field studies of *P. fuscatus* in which we have tried to simulate a natural recognition context (Gamboa 1988; Gamboa *et al.* 1991*b*; Bura and Gamboa 1994). In these studies, we switched field colonies of known maternal lineages and recorded how returning females were treated by resident females of another colony. By controlling the 'relatedness' of switched colonies, we observed how resident females treated returning nestmates (controls) and non-nestmate sisters, aunts, nieces, first cousins, and unrelated wasps. Returning nestmates (sisters, mothers, and daughters) and returning unrelated non-nestmates were treated highly tolerantly and highly intolerantly, respectively. Most unrelated non-nestmates were evicted from the nest within 60 s of landing on the nest (Bura and Gamboa 1994).

Unlike their treatment of nestmates and unrelated non-nestmates, resident females are highly variable in their treatment of non-nestmate kin. Most cousins and nieces (50%–65%) are treated highly intolerantly, i.e. like unrelated non-nestmates. However, a minority of cousins and nieces are treated highly tolerantly, like nestmate sisters. Most aunts and non-nestmate sisters (80%) are treated highly tolerantly, i.e. like nestmate sisters. A minority, however, are treated very intolerantly, like unrelated non-nestmates (Bura and Gamboa 1994). Interestingly, despite the

symmetrical relatedness of aunts and nieces, they are asymmetrically tolerant of
each other. In particular, aunts are significantly less tolerant of nieces than nieces
are of aunts. In these field studies, aunts and nieces were foundresses and workers,
respectively, and their differential treatment may be a function of caste differences
in recognition (Bura and Gamboa 1994). The evidence for caste differences in recognition is discussed later.

Among the most intriguing studies of recognition in wasps are those of Pfennig
and Reeve (1989, 1993) on the solitary cicada killer wasp. In field and laboratory
studies, the authors documented that females discriminated between neighbours
and more distantly nesting conspecifics. In particular, females were significantly
more tolerant of neighbours than non-neighbours (Pfennig and Reeve 1989). Furthermore, DNA fingerprinting analysis revealed that females were significantly
more related to neighbours than non-neighbours (Pfennig and Reeve 1993). These
results suggest the fascinating possibility that kin recognition ability may have been
present in solitary ancestors of social wasps. It would be highly worthwhile to investigate the kin recognition abilities of other solitary wasps to determine if such
abilities are widespread. If so, it would suggest that all of the components of
nestmate recognition were present in the solitary ancestors of social wasps.

Intracolonial kin recognition

To the best of my knowledge, the ability of social wasps to discriminate among
contemporary nestmates on the basis of relatedness has not yet been investigated.
However, in a study closely related to the topic of intracolonial kin recognition,
Queller *et al.* (1990) reported that females of *P. annularis* do not preferentially
associate with more related former nestmates in founding spring nests. They concluded that either females lack the ability to make such discrimination or fail to use
it in a natural situation where it would be highly advantageous.

There are at least two aspects of the recognition mechanism in social wasps that
appear to preclude the possibility of adult–adult, intracolonial kin recognition.
First, *P. fuscatus*, *P. carolina*, *P. metricus* (Gamboa *et al.* 1986a; Singer and
Espelie 1992), and *Ropalidia marginata* (Venkataraman *et al.* 1988) do not learn
their own recognition odours, a requisite for intracolonial kin recognition subserved
by learning (Venkataraman *et al.* 1988; Getz 1991). Second, for *Polistes* there is no
evidence that females display a graded behavioural response that is proportional to
relatedness. Rather, females treat conspecifics either highly tolerantly or highly
intolerantly. It seems that such a binary behavioural response by females to vari-
ously related nestmates would be extremely disruptive to the colony unless intolerance of less related nestmates might somehow be moderated or modulated by
context.

Social wasps could discriminate between full- and half-sibs by employing a more
restrictive acceptance threshold and utilizing a self-referent template only in the
context of intracolonial kin discrimination (Gamboa *et al.* 1991c). Although such a
theoretical scenario is possible, I believe it is more likely that, if intracolonial kin
discrimination occurs, it would be manifested in adult-brood interactions. That this
possibility is at least feasible is suggested by a recent finding that queens of *P.*

fuscatus can recognize their eggs via a Dufour's gland secretion (Downing 1991*a*).

Queller (1994) recently developed a technique that utilizes statistical analyses of genetic markers for detecting intracolonial kin discrimination. This procedure may be especially efficacious because it appears to circumvent the problems associated with the use of artificially produced patrilines, a technique used to investigate intracolonial kin discrimination in honey bees. It would be especially interesting to use the techniques of Queller (1994) to examine the possibility of intracolonial kin discrimination in *P. fuscatus*.

The mechanism of recognition

Expression component of kin recognition

(This component involves the nature and production or acquisition of the cues (labels) used to identify kin.)

The physical nature of the recognition cue ✕

Since *Polistes* learn recognition cues from the natal comb and/or brood, recognition cues cannot be visual, tactile or auditory features of adult nestmates. Only chemical cues are likely to be shared by the nest (and/or brood) and adult nestmates. It has also been established that females can acquire recognition cues from the comb, a result that further points to chemicals as recognition cues (Pfennig *et al.* 1983*a*). Females that have been reared under conditions of homogeneous environmental odours and recently exposed to these odours fail to discriminate between nestmates and unrelated non-nestmates reared under the same conditions (Gamboa *et al.* 1986*a, b*).

Theresa Singer and Karl Espelie have shown that the removal of hydrocarbons from the natal nest disrupts the ontogeny of recognition ability in females of *P. metricus*. More specifically, wasps previously exposed to their unextracted control nest, but not wasps exposed to their hydrocarbon depleted nest, can recognize their nestmates (Singer and Espelie 1992). These results and others demonstrate that cuticular hydrocarbons are serving as recognition pheromones. Thus, all of the behavioural and chemical studies indicate that chemical cues, and only chemical cues, mediate kin recognition in *Polistes*.

Although Singer and Espelie (1992) demonstrated that the recognition pheromones of *P. metricus* are, indeed, cuticular hydrocarbons, it is not known which cuticular hydrocarbons are serving as recognition pheromones. In order to identify such compounds, we recently collaborated with Dr. Espelie at the University of Georgia to analyze the cuticular hydrocarbons of female *P. fuscatus* (Espelie *et al.* 1994). We identified 20 cuticular hydrocarbons, which represented about 94.4% of the hydrocarbons extracted from a female's cuticle.

Three of the 20 hydrocarbons appeared to be especially good candidates for recognition pheromones. Univariate and multivariate analyses of the 20 hydrocarbons indicated that the methylated hydrocarbons 13- and 15-MeC_{31}, 13-, 15- and 17-MeC_{33} and 11, 15 and 13, 17-diMeC_{31} are more diagnostic of colony origin than others. These compounds also appear to be genetically specified, a property that has been documented for recognition pheromones of *P. fuscatus* (Gamboa 1988; Bura and Gamboa 1994). The relative amounts of these pheromones did not differ between workers and foundresses of a colony, which is expected for pheromones that proclaim colony identity. Finally, these compounds have a distinctive stereochemistry, which might make them easier to distinguish than other cuticular hydrocarbons (Butts *et al.* 1993; Espelie *et al.* 1994).

Females of *P. fuscatus* may be using a relatively small number of cuticular hydrocarbons as kin recognition pheromones. For example, they may be using only the three methylated hydrocarbons identified above to recognize kin. If females can detect and respond to small differences in the concentrations of these three hydrocarbons, these compounds could generate a large number of chemical signatures. An alternative hypothesis is that female *P. fuscatus* use a much larger number of cuticular hydrocarbons to identify kin. According to this explanation, females would use information from many, if not most, of the identified cuticular hydrocarbons to identify kin. Even so, some of these cuticular hydrocarbons might be more important than others for recognizing relatives.

In order to identify the recognition pheromones of *P. fuscatus* and thus distinguish between the two hypotheses, we plan to combine data on hydrocarbon profiles of individual wasps with behavioural data. Specifically, we plan to compare the match of hydrocarbon profiles of wasps that accept each other with those of wasps that reject each other. It may be most profitable to conduct such a study on non-nestmate kin since they display high frequencies of both acceptance and rejection of related non-nestmates. The search for recognition odours is complicated by evidence that recognition odours in *P. fuscatus* also have an environmental component (Gamboa *et al.* 1986*b*).

The role of surface hydrocarbons in the social organization of paper-wasps and other social insects is extensively reviewed by Lorenzi *et al.* in this volume.

The proximate source of recognition odours

Kin recognition is mediated by both individually produced (endogenous) odours and acquired (exogenous) odours. Interestingly, either endogenous or exogenous odours by themselves can mediate recognition. For example, unrelated females of *P. fuscatus* exposed to different fragments of the same but unrelated comb treat each other like nestmate sisters (Pfennig *et al.* 1983*a*). These females could only have used exogenous odours to mediate recognition. Conversely, wasps isolated from their comb at emergence are recognized by their experienced nestmates. Since these isolates could not have acquired odours from the comb or nestmates as adults, they must have possessed endogenous colony odours (or possibly odours acquired as brood) in order to be recognized as nestmates (Gamboa *et al.* 1986*b*).

Venkataraman *et al.* (1988) investigated the mechanism of recognition in *Ropalidia marginata*. They concluded that females lack endogenous colony odours and must acquire such odours from the nest and/or nestmates after emergence in order to be recognized by their experienced nestmates. If this is true, *P. fuscatus* differs from *R. marginata* in this aspect of its recognition mechanism. Although female *P. fuscatus* can acquire colony odours from the nest, they clearly possess colony odours in the absence of prior exposure, as an adult, to their nest and nestmates (Gamboa *et al.* 1986*b*).

It may be that the disparate results of Venkataraman *et al.* (1988) and Gamboa *et al.* (1986*b*) are due to their different recognition assays rather than differences in the recognition mechanisms of the two species. Venkataraman *et al.* (1988) used a triplet design that consisted of a non-nestmate and two nestmates, only one of which had been isolated. The nestmates failed to show evidence of nestmate recognition ability. It may be that dyadic interactions between asymmetrically exposed nestmates do not reveal nestmate recognition when recognition ability is present in only one of the two interactants. This explanation is consistent with the results of Pfennig *et al.* (1983*a*) for *P. carolina*, which presumably has endogenous colony odours as does *P. fuscatus*. Like Venkataraman *et al.* (1988), Pfennig *et al.* (1983*a*) found no evidence of recognition when they used a triplet design in which only one of the two nestmates had been previously isolated. Thus, Pfennig *et al.* (1983*a*) failed to document recognition when both nestmates presumably had appropriate colony odours, but only one possessed recognition ability.

A potentially more sensitive assay for determining whether isolates can be recognized by their experienced nestmates (and thus whether isolates possess endogenous colony odours) would be to introduce nestmate and non-nestmate isolates to experienced females on their nest as was done for *P. fuscatus* (Gamboa *et al.* 1986*b*). This assay more closely approximates a natural recognition context and focuses more on the behavioural response of the experienced females than does the triplet assay. It would be informative to conduct such a bioassay for *R. marginata* to determine whether it conforms to the *Polistes* mechanism.

There is some evidence that the *Polistes* queen may be a primary proximate source of colony recognition odours. Dani *et al.* (1992*b*) reported that dominant foundresses of *P. dominulus* display high frequencies of stroking behaviour. During stroking, females press the posterior part of the gaster against the nest as they move over it. Dani *et al.* (1992*b*) hypothesized that stroking results in the application of a chemical substance to the nest, and that this substance may be involved in nestmate recognition. Consistent with this hypothesis is the fact that, in a simulation of usurpation, foundresses of *P. gallicus* and *P. biglumis bimaculatus* displayed high rates of stroking after their nest was experimentally switched with that of another conspecific (Cervo and Turillazzi 1989; Lorenzi and Cervo 1992). Presumably, usurpers attempt to replace the nest odour of the original queen with their own odour (Klahn 1988; Dani *et al.* 1992*b*). By doing so, subsequently emerging workers will learn the odour of the foreign queen and accept her.

The Dufour's gland, a large exocrine gland that opens into the dorsal vaginal wall near the base of the sting, produces a secretion that is involved in egg recognition (Downing 1991*a*). Downing (1991*a*) provided experimental evidence that secretions from the Dufour's gland are used by a dominant wasp to identify the eggs of subordinates or non-nestmates, which are eaten. It may be that the secretion also functions as a component of the colony odour, which might be applied by the queen to the comb via abdominal stroking.

The ultimate origin of recognition odours

The question of the ultimate origin (i.e. heritable or environmental) of recognition odours differs from the question of the proximate source of recognition odours. Although we know that both endogenous and exogenous odours can mediate recognition in *Polistes*, this doesn't tell us whether the odours are heritable (i.e. independent of environmental factors) or environmental, i.e. dependent on environmental factors such as food or nesting material (Gamboa *et al.* 1986*a*). The available evidence indicates that the recognition odours of *Polistes* have both heritable and environmental components (Gamboa *et al.* 1986*b*; Bura and Gamboa 1994).

Our evidence for heritable odours is actually more extensive than our evidence for environmental odours. It is my impression that researchers have mistakenly concluded that we have found environmental odours to be more important than heritable odours in *Polistes*. This may be due to our modelling of the potential information value of environmental recognition odours (Gamboa *et al.* 1986*a*). As we have stated in the literature (Gamboa *et al.* 1991*c*), it is extremely difficult to assess the relative importance of heritable and environmental recognition odours. In fact, their relative importance may differ in different habitats or in different species (Gamboa *et al.* 1986*a*).

In an attempt to determine the ultimate origin of recognition odours, *P. fuscatus* females were reared from egg to adult in an environmentally controlled laboratory (i.e. diet, nesting material, temperature, etc. were identical for all lab colonies). They treated all females highly tolerantly regardless of their colony of origin (Gamboa *et al.* 1986*b*). Thus, we found no evidence that these laboratory-reared females could discriminate between nestmates and non-nestmates when we controlled their environmental odours. We interpreted these results to indicate that the recognition odour of *P. fuscatus* has an environmental component because controlling environmental odours eliminated the manifestation of recognition.

However, if we removed the laboratory-reared females from their nestbox and isolated them for several days, they then did discriminate nestmates from non-nestmates (Gamboa *et al.* 1986*b*). Since the laboratory-reared females recognized their nestmates after isolation, the recognition odour must also have had a heritable component. Presumably, the environmental odours had decayed in the isolates thereby unmasking the females' heritable odours. These results suggest that frequent exposure to the source of the environmental odours, such as the comb, food or nestmates, may be necessary for environmental odours to serve as persistent recognition cues.

Although the previous experiment provided evidence of a heritable component to the recognition odour, we weren't able to eliminate the possibility that our results were due to maternal rather than genetic effects. We examined this possibility by investigating the ability of females to recognize non-nestmate kin that had different but related mothers, i.e. non-nestmate aunts, nieces, and first cousins. We found, in a series of studies, that relatedness had a significant impact on how non-nestmate females of *P. fuscatus* were treated, and thus we concluded that the recognition odour has a genetic component (Gamboa *et al.* 1987, 1991*b*; Gamboa 1988; Bura and Gamboa 1994).

Perception component of kin recognition

(This component includes the development of a kin template, the sensory processing of the recognition cues, and the algorithm used to match an encountered phenotype with the template.)

Evidence for learning

In a series of studies we found that manipulating the experiences of young female *P. fuscatus*, *P. carolina*, and *Dolichovespula maculata* immediately after emergence alters or disrupts nestmate discrimination. In the two species of *Polistes*, it was discovered that the exposure of a young female to its natal comb was both necessary and sufficient for the development of nestmate recognition ability (Shellman and Gamboa 1982; Pfennig *et al.* 1983*a*). These results indicated that the isolation of a wasp from its natal nest at emergence either disrupted the learning of recognition odours, the acquisition of recognition odours, or both.

In order to further investigate the role of learning, we exposed newly emerged, unrelated gynes of *P. fuscatus* to different fragments of the same but unrelated nest. We called these females pseudonestmates. Interestingly, pseudonestmates later recognized each other, i.e. they treated each other more tolerantly than non-pseudonestmates. These results unequivocally demonstrated that pseudonestmates must have learned common recognition cues in order later to recognize each other. The results also demonstrated that pseudonestmates must have acquired common odours from comb fragments in order to recognize each other (Pfennig *et al.* 1983*a*).

Venkataraman *et al.* (1988) provided evidence that *Ropalidia marginata* must learn recognition cues. Females previously exposed to their nest and nestmates, but not females isolated at emergence, were later able to discriminate nestmates from non-nestmates.

Timing and form of learning

Although we knew that the learning of recognition cues *could occur* during the adult stage, we didn't know if the learning of recognition cues was restricted to the adult stage. To investigate this, we examined *P. fuscatus* to determine whether isolates could be recognized by experienced females, i.e. females with extensive previous exposure to their nest and nestmates. Experienced females did recognize their isolate nestmates, and thus isolates clearly possessed appropriate colony recognition odours. Therefore, the failure of gynes to exhibit nestmate recognition after isolation from their natal nest at emergence must be due to the disruption of adult learning, not the adult acquisition, of recognition odours. These results also eliminated the possibility that the learning of recognition cues occurred in the larval or pupal stage. If such learning had occurred as brood, young gynes isolated at emergence should be able to recognize their nestmates. Thus, learning in *P. fuscatus* is restricted to the adult stage (Gamboa *et al.* 1986*b*).

Gamboa *et al.* (1986*b*) found that isolate females of *P. fuscatus* treated all females as nestmates regardless of relatedness. This result and the results of other experiments indicates that initially, newly emerged females are highly tolerant of all conspecifics. Over a period of time, however, they develop an intolerance of non-nestmates, i.e. conspecifics with unfamiliar odours (Gamboa *et al.* 1986*b*).

In a series of studies we found that female *P. fuscatus* learn recognition odours within a few hours after emergence. Gynes exposed to their natal nest for as little as one hour after emergence later recognized nestmates, although such gynes exhibited significantly weaker discrimination than gynes with longer exposure to their natal nest. However, increasing a gyne's exposure to her natal nest beyond four hours had no effect on her discrimination ability. These results indicate that the learning of recognition odours may be complete by four hours after emergence (Pfennig *et al.* 1983*a*).

The memory of the learned recognition odours in *Polistes* appears to be quite durable. Experienced gynes of *P. metricus* that have been isolated from their nest and nestmates for 99 days can discriminate nestmates from non-nestmates (Ross and Gamboa 1981). The learning of recognition odours by *Polistes* resembles imprinting since young adults learn recognition odours rapidly after emergence and retain the memory of these odours for long periods of time. Imprinting is also thought to be involved in the ontogeny of recognition ability in other social insects (Jaisson 1991).

We presently don't know if the learning of recognition cues is restricted to the four-hour period after emergence. We also don't know if *Polistes* wasps can update their learning of recognition odours over the colony cycle. It seems reasonable that wasps could update their learning of recognition odours. The odours of the comb, particularly its environmental odours, probably change over the colony cycle as the availability of various types of prey, nectar, and wood fibres change. However, it may be that the genetic component of the recognition odour remains relatively constant over the colony cycle. The genetically specified odour may be especially constant if one individual, say the queen, is a primary, proximate source of the recognition odour.

If the perception system of wasps is designed to detect and respond to familiar odours, wasps may be able to recognize a changing colony odour as long as they detect a familiar odour in the odour milieu. Our experimental results indicate that when encountering a wasp with both a foreign and familiar colony odour (pseudo-nestmates), wasps appear to ignore the foreign odour and, presumably in response to the familiar odour, respond tolerantly to the individual (Pfennig *et al.* 1983*a*). We are in the process of designing experiments to tease apart these and other aspects of the perception component of recognition.

Location from which the recognition cue is learned

The locus from which individuals learn recognition cues is poorly understood for many animals. In social insects, it is generally assumed that individuals learn recognition cues from nestmates although this assumption may not be justified. In the bald-faced hornet, *Dolichovespula maculata*, adult exposure to the nest affects the ontogeny of nestmate recognition ability. The exact role of the hornet nest in the development of recognition ability, however, is unknown (Ryan *et al.* 1985).

In *Polistes*, individuals learn recognition cues from their comb and not from themselves and not from nestmates. Gynes of *P. fuscatus* that have been isolated from their comb at emergence but provided extensive exposure to nestmates later fail to recognize nestmates (Shellman and Gamboa 1982). Gynes isolated from their nest and nestmates at emergence clearly possess recognition odours, and thus have the opportunity to learn their own odour. However, such isolates fail to recognize nestmates when tested later (Gamboa *et al.* 1986*b*). Gynes of *P. fuscatus* and *P. carolina* that have been exposed only to their natal nest are later able to recognize nestmates (Pfennig *et al.* 1983*a*). Thus, exposure to the natal nest is necessary and sufficient for the development of recognition ability.

Singer and Espelie (1992) have shown that female *P. metricus* also learn recognition odours from the nest because wasps exposed to a natal comb containing hydrocarbons later recognize nestmates. For *P. metricus* and *P. annularis*, the hydrocarbons present on the comb are the same as those on the wasp's cuticle (Espelie and Hermann 1990; Espelie *et al.* 1990).

Ropalidia marginata may also learn recognition cues from its nest. Females previously exposed to their nest and nestmates, but not isolates, later recognized their nestmates (Venkataraman *et al.* 1988). It would be interesting to expose newly emerged, female *R. marginata* to their nest only to determine if they can later recognize nestmates. If they could, this would establish that *R. marginata*, like *P. fuscatus*, *P. metricus*, and *P. carolina*, learn recognition cues from the nest. The nest (comb) is also intimately involved in the ontogeny of nestmate recognition ability in honey bees (Breed *et al.* 1988).

Cue similarity threshold property

All evidence indicates that females of *P. fuscatus* treat conspecifics tolerantly or intolerantly, and nothing in between. Presumably, a wasp matches the template of the odour learned previously with the odour of the encountered wasp; if the match between the learned template and perceived odour is greater than a minimum threshold similarity, the wasp treats the encountered wasp, in terms of mean tolerance, as a nestmate; otherwise, it treats the wasp as a non-nestmate (Gamboa *et al.* 1986*a*).

Our evidence that genetic and environmental, as well as endogenous and exogenous odours, are not additive in their effects on tolerance is consistent with the cue similarity threshold property. In addition, non-nestmate kin are treated either tolerantly, like nestmates, or intolerantly, like unrelated non-nestmates (Gamboa 1988; Gamboa *et al.* 1991*b*; Bura and Gamboa 1994). In other words, intermediately related conspecifics are not treated intermediately in tolerance. Thus, our evidence for *P. fuscatus*, as well as field evidence for *P. exclamans* (Pfennig 1990), suggest that tolerance of conspecifics does not increase continuously as a function of increasing similarity between the learned template and encountered odour. Unfortunately, we know virtually nothing about the matching process that wasps utilize to assess the similarity between their template and the encountered odour.

The adaptiveness of recognition

For the past several years we have concentrated on investigating the sociobiology of recognition in *P. fuscatus*. As I mentioned previously, this is the least understood aspect of recognition and it may be the most difficult area in which to secure definitive answers.

We were interested in knowing whether tolerance of conspecifics changes over the colony cycle, and if so, whether the changes in tolerance are consonant with the fitness interests of colony members. Since we have a reasonably good understanding of the field biology of *P. fuscatus*, we sought to know if changes in conspecific tolerance were related to certain ecological features of the colony cycle.

It was known from previous studies that conspecific pressures in *P. fuscatus* are intense, particularly during the latter part of the pre-worker stage of the colony cycle (Klahn 1988). For example, an average field colony of *P. fuscatus* experiences one usurpation attempt per day throughout the pre-worker phase of the colony cycle (Gamboa *et al.* 1992). Almost certainly, intraspecific usurpation pressure is one of the most significant ecological factors favouring nestmate recognition ability in *P. fuscatus*. We also thought that conspecific brood theft, which has been reported in other species of *Polistes* (Kasuya *et al.* 1980), would be more likely to occur during the pre-worker stage before the colony is defended by large numbers of workers (Gamboa *et al.* 1991*b*).

Based on our understanding of the timing of conspecific pressures, we predicted that females should be the least tolerant of non-nestmates early in the colony cycle when conspecific pressures are severe, and most tolerant of non-nestmates late in the colony cycle after the combs no longer contain brood. At this late time in the colony cycle colonies are no longer susceptible to usurpation or brood theft.

When we tested females in the field, we found that wasps did the opposite of what we predicted: females were generally the least tolerant of all classes of conspecifics, including nestmates, late in the colony cycle when combs no longer contained brood (Gamboa *et al.* 1991*b*). At first, our results were puzzling. After all, why defend an empty comb against conspecific intruders? We began to realize that, in our initial predictions, we had not considered the mechanism that might underlie changes in conspecific tolerance. I'll discuss this matter in the next paragraph. We had also failed to consider the fact that there is extensive food exchange among nestmate gynes late in the colony cycle, and that nectar consumed at this time may profoundly affect survivorship during winter diapause (Gamboa *et al.* 1991*b*). Thus, the consequences of admitting a non-nestmate into a colony late in the colony cycle could have substantial negative fitness consequences.

In our predictions, we had assumed that tolerance of non-nestmates was independent of the tolerance of nestmates. However, if tolerance of conspecifics is a function of the level of the acceptance threshold, then changes in tolerance would be effected by a change in the acceptance threshold. If this is the case, any change in the acceptance threshold would necessarily affect tolerance of both nestmates and non-nestmates (Reeve 1991). That is, if the acceptance threshold became more restrictive, the probability of rejecting both non-nestmates and nestmates would increase. Furthermore, such a change in the acceptance threshold would affect variances in tolerance in a predictable way. By increasing the probability of rejecting nestmates, a more restrictive acceptance threshold would result in an increase in the variance for tolerance of nestmates. In contrast, a more restrictive acceptance threshold would decrease the probability of accepting an unrelated non-nestmate thereby causing a decrease in the variance for tolerance of non-nestmates.

Our field results were consistent with the above explanation. All classes of conspecifics, including nestmates, were treated, in terms of mean tolerance, less tolerantly late in the colony cycle. The variance patterns were also consistent with our explanation: the variances in tolerance for nestmates increased significantly and the variances in tolerance for non-nestmates decreased significantly late in the colony cycle (Gamboa *et al.* 1991*b*).

Even though our empirical data provided a proximate explanation for decreasing tolerance late in the colony cycle, it wasn't clear why the observed tolerance patterns over the colony cycle might be adaptive. Why not have a highly restrictive acceptance threshold early in the colony cycle to minimize usurpation and brood theft? We began to realize that a highly restrictive acceptance threshold early in the colony cycle might have severe negative fitness consequences. Although a more restrictive acceptance threshold would decrease the probability of accepting non-nestmates, it would increase the probability of rejecting one's own nestmates. Since considerable evidence indicates that the presence of additional females on the nest greatly reduces the probability of a successful usurpation (Gamboa 1978; Klahn 1988), the loss of nestmates early in the colony cycle almost certainly reduces a colony's ability to repel usurpers. Thus, the level of the acceptance threshold may represent a compromise between the negative fitness consequences of accepting non-nestmates and rejecting nestmates (Gamboa *et al.* 1991*b*).

From our results in the field, it wasn't clear whether tolerance had decreased in individual wasps over the colony cycle. Most, if not all wasps present early in the colony cycle were no longer present near the end of the colony cycle. Since wasps present early and late in the post-worker phase of the colony cycle are mostly workers and gynes, respectively, the decrease in tolerance over the colony cycle may have been due to caste differences in tolerance (Gamboa *et al.* 1991*a*). Evidence from a subsequent study indicated that the late season decrease in tolerance reported by Gamboa *et al.* (1991*a*) was probably attributable to caste differences in tolerance as described below.

In a field study of *P. fuscatus*, we found that queens were significantly less tolerant of both nestmates and non-nestmates than were workers. In addition, the variance patterns for tolerance were consistent with queens having a more restrictive acceptance threshold than workers (Fishwild and Gamboa 1992). In other words, queens appear to require a better match between their template and encountered phenotype for acceptance than do workers.

The differential tolerance toward intruders by queens and workers appears to be consistent with each caste's fitness considerations. The fitness consequences of accepting an unrelated or distantly related intruder into the colony are almost certainly more negative for queens than for workers. Usurpers frequently evict or kill resident queens, and destroy their reproductive-destined brood (Klahn and Gamboa 1983; Klahn 1988). Resident workers, however, remain with nest after usurpation, and may have the opportunity to lay eggs. In fact, workers of usurped colonies have a higher probability of becoming a replacement queen than workers from a colony headed by the original queen (Klahn 1988).

In *P. fuscatus* the responsibilities for colony defence against foreign conspecifics reside largely with the queen. This is true for both pre-worker (Gamboa *et al.* 1992) and post-worker colonies (Fishwild and Gamboa 1992). For example, in pre-worker multiple-foundress associations, queens are significantly more likely to repel an intruder than a subordinate foundress when the queen and her subordinate(s) are together on the nest. Interestingly, queens and their subordinates rarely jointly defend their nest against conspecific intruders. This and other evidence suggests that the enhanced ability of multiple foundresses to defend successfully against usurpation compared with single foundresses is due to the decreased time that the multiple-foundress nest is left unattended (Gamboa *et al.* 1992).

Finally, we found in laboratory studies of *P. fuscatus* that individual wasps can change their tolerance of non-nestmate conspecifics rapidly in different contexts. For example, resident females were significantly more intolerant of unrelated non-nestmates when they were encountered on the nest than off the nest (Gamboa *et al.* 1991*b*). This modulation of tolerance appears to be consistent with the fitness interests of wasps since it's likely that the negative fitness consequences of encountering a foreign female are greater on the comb than off the comb. In fact, there may be little or no advantage for displaying differential tolerance toward nestmates and non-nestmates away from the comb as in the context of foraging (Gamboa *et al.* 1991*b*).

Common features of recognition in social wasps and other animals

As the kin recognition systems of more animals are investigated, it is becoming apparent that the *Polistes* recognition system shares fundamental features with the recognition systems of other animals, including vertebrates. For example, learning appears to be nearly universally involved in the ontogeny of recognition ability for both vertebrates and invertebrates. With a few possible exceptions, e.g. the sessile colonial ascidian (Grosberg and Quinn 1986), animals learn recognition cues in order to form a recognition template, which is then matched with the cues of an encountered conspecific. For *Polistes* and other social insects, this learning resembles imprinting since cues are learned rapidly by young adults and the memory of the learned cues are durable (Gamboa *et al.* 1986*a*; Jaisson 1991).

As in many other social insects, recognition odours in *Polistes* have both genetic and environmental components (Gamboa *et al.* 1986*a*). This is also true for the recognition odours of larval amphibians (Blaustein and O'Hara 1982; Cornell *et al.* 1989; Gamboa *et al.* 1991*a*). It may be that chemical recognition cues in vertebrates and invertebrates typically have both genetic and environmental components.

Finally, as in *Polistes*, such divergent animals as wood frog tadpoles (Cornell *et al.* 1989), naked mole rats (Reeve and Sherman 1991), and non-human primates

(Bernstein 1991) appear not to make fine grained estimates of relatedness and apportion tolerance/nepotism in a graded manner according to the coefficient of relatedness. These findings indicate that the cue similarity threshold property may be common in social animals (Gamboa 1992). Therefore, various social animals may make a binary decision whether to accept or reject (or to tolerate or be aggressive toward) a conspecific depending on whether the match between the template and encountered phenotype falls above or below a minimum acceptance threshold, respectively.

10

The role of cuticular hydrocarbons in social insects: is it the same in paper-wasps?

Maria Cristina Lorenzi, Anne-Geneviève Bagnères, and Jean-Luc Clément

On facing an individual entering its colony, a social insect must make a choice: residents are admitted and intruders rejected. This kind of nestmate recognition process has been favoured by selection presumably because it limits colony exploitation by parasites and conspecific individuals from other colonies while promoting colony integrity and success. Therefore, cues must exist which allow each colony member to discriminate between nestmates and strangers.

Within a colony, relationships between individuals follow precise rules so that each individual directs particular behaviours towards different colony members depending on their caste. Again, individuals must be able to detect specific cues in order to recognize the caste to which the interacting individual belongs. Extensive research following the formulation of the kin selection theory by Hamilton (1964a, b) was carried out to evaluate the ability of a social insect to recognize the degree of relatedness of an interacting individual; in keeping with the theory, the most co-operative behaviours should be directed towards kin. For such discrimination processes to operate, all individuals must be able to detect specific cues which carry information about the insect with which they are interacting. These cues have been named discriminators, recognition pheromones (Hölldobler and Michener 1980) and discrimination pheromones (for example Henderson *et al.* 1990; Downing 1991b), since, as suggested by Wilson (1971), most of them have proved to be chemical.

Among the wide variety of chemical cues, non-volatile compounds have been the focus of widespread research on semiochemicals used in recognition processes. Some volatile compounds used in these processes have been described by Jaffè (1983) in ants. Since the end of the 1970s, when Howard *et al.* (1978) and Blomquist *et al.* (1979) showed that non-volatile, cuticular lipids in termites discriminate species and caste, more and more clues have been collected about the possibility that cuticular compounds act as recognition factors in many species of insects. At the same time their composition and functions have also been studied.

The most crucial function of these compounds seems to be that of creating a layer over the cuticle to protect insects against desiccation, but they also prevent microorganisms and toxins from entering the body. It has also been demonstrated that these compounds are involved in chemical communication: in social insects they may be important cues in inter- and intraspecific relationships (Blomquist and Dillwith 1985).

Insect epicuticular compounds are usually complex mixtures of lipids containing large amounts of hydrocarbons. Their chemical composition has been described in detail by Lockey (1988). They consist mainly of *n*-alkanes, branched methyl-alkanes and alkenes (especially mono- and dienes) whose chain lengths range from 20 to 40 carbon atoms; esters, alcohols and fatty acids are usually minor components. The methylbranched components may be monomethyl-, dimethyl- or tri-methylalkanes. Cuticular hydrocarbons are usually very complex blends. They contain a wide variety of compounds, mixed in different qualitative and quantitative combinations and therefore can carry complex information.

Cuticular hydrocarbons have relatively high molecular weights, and are therefore non-volatile compounds. For this reason an insect may only perceive them through direct contact with the interacting insect, for example through antennation.

Because of their characteristics, epicuticular mixtures are potentially the semiochemicals used by insects to discriminate among individual nestmates, between nestmates and non-nestmates, and to detect whether a potential mate belongs to their own species. As variations in the composition and relative proportion of each compound can be detected by analytical instruments, hydrocarbon mixtures may also be used in research to discriminate between different groups. The complexity of hydrocarbon blends usually requires the application of multivariate statistics to compare different samples. Keegans *et al.* (1992*a*) state that the comparison of chemicals from exocrine gland secretions may lead to clearer biosystematic results. Owing to the greater variety of distinct products synthesized by these glands, their study does not require the complex statistical treatment needed for cuticular hydrocarbons. From the point of view of an ethologist, relevant compounds will be those that are relevant for the insect, i.e. those that the insect uses as recognition factors both in intra- and interspecific interactions. Up to now, behavioural tests, designed to assay whether cuticular hydrocarbons play a role in discrimination processes between insects, have shown that cuticular hydrocarbons do indeed function as recognition cues. However, behavioural tests on the role of cuticular hydrocarbons in discrimination processes between insects have been far more sporadic than chemical analysis of cuticular hydrocarbons, and consequently although research has often demonstrated that two species, population or colonies are chemically distinguished, we do not know whether the insects really use these cues or not when they interact.

Blends of cuticular hydrocarbons may differ both in quality (difference in chemical composition) and quantity of each compound (same chemical composition but each compound is present in a different quantity). Research has shown that each species is usually characterized by its own peculiar blend, so that qualitative variations may be found by comparing different species. Within a species, populations located in different geographical sites, different colonies, or sometimes even different castes may have cuticular blends of identical chemical composition. Gas chromatographic traces from two populations, colonies or castes, however, may differ because the relative proportions of each compound are different. Obviously, the extent of variation possible to measure between members of the same colony and caste or between individuals in the same physiological status is much lower than that between individuals from different colonies. However, there are exceptions to this rule. In the termite *Reticulitermes*, differences between castes are more important than those between colonies (Bagnères and Clément, in prep.). Multivariate analysis of the relative proportions of each product of the cuticular mixture is necessary to highlight the peculiarity of each group and to determine which of the compounds that vary in concentration are most characteristic of each group.

Ants, termites and bees

Chemical discrimination of species

Variations in chemical composition of the epicuticular hydrocarbon blends has allowed discrimination of species of termites in the genus *Reticulitermes* (Howard *et al.* 1978, 1982*b*; Bagnères *et al.* 1988, 1991*b*), *Zootermopsis* (Blomquist *et al.* 1979) and *Coptotermes* (Brown *et al.* 1990). Hydrocarbons are referred to as taxonomic characters that chemically distinguish species of *Nasutitermes* (Howard *et al.* 1988), although these species showed only quantitative differences in their hydrocarbon components. Hydrocarbon mixtures are distinct in ant species of the genus *Solenopsis* (Howard and Blomquist 1982; Vander Meer *et al.* 1985) and *Leptothorax* (Bagnerès *et al.* 1991*c*). Different proportions of hydrocarbons separate subspecies in honey bees (*Apis mellifera*) (Mar *et al.* 1987).

Chemical discrimination within a species

Within a species, the relative proportions of different hydrocarbons may vary from nest to nest in social ants, bees, termites and wasps. Quantitative variations may be linked to geographical variations (for example, Nowbahari *et al.* 1990) or to changes over time (in *Solenopsis invicta,* Vander Meer *et al.* 1989*a*; and in *Leptothorax lichtensteini,* Provost *et al.* 1993), but small quantitative variations were frequently found when comparing extracts of individuals from different colonies. In this case, each colony is said to possess a peculiar chemical signature. The chemical signature may be the nest odour responsible for nestmate recognition. The relative abundance of single hydrocarbon components varied from colony to colony in ants of the genera *Camponotus* (Bonavita-Cougourdan *et al.* 1987, Morel *et al.* 1988), *Solenopsis* (Vander Meer et al. 1989*a*), *Formica* (Henderson *et al.* 1990), and *Cataglyphis* (Nowbahari *et al.* 1990); however, experimental treatment of colonies can sometimes limit separation based on relative proportions of hydrocarbons (Obin 1986). A colonial signature was found in termites (Bagnères *et al.* 1990, 1991*b*; McDaniel 1990) and bees (Sasagawa and Kuwahara 1990). Page *et al.* (1991) showed that within a colony of *Apis mellifera* the quantitative differences of the hydrocarbon patterns between individuals are correlated to their degree of relatedness.

Chemical discrimination of caste

Within the same colony, the blend of cuticular compounds was found to contain caste-specific hydrocarbon proportions both in termites (*Reticulitermes*, Howard *et al.* 1982*b*, and Bagnères *et al.* 1990; *Coptotermes*, Haverty *et al.* 1990) and in ants

(Mintzer *et al.* 1987; Bonavita-Cougourdan *et al.* 1990), where even subcastes had peculiar hydrocarbon traces (Morel *et al.* 1988; Bonavita-Cougourdan *et al.* 1993).

Host and parasite relationship: chemical mimicry and camouflage

Although different species usually have different cuticular hydrocarbon blends, this general rule does not apply to parasites living in social insect colonies during some of their stages of development. The analyses of parasites, social parasites, and inquilines of social insects supply further clues that hydrocarbons may be relevant in interindividual communication: indeed, evidence exists that parasites and inquilines of social insects show cuticular blends similar to those of their hosts. In the papers dealing with this subject, both the term chemical mimicry and camouflage are used. Howard *et al.* (1990) observe that use of the term mimicry implies that a parasite biosynthesizes the cuticular hydrocarbons that it shares with its host; while, if a parasite acquires the hydrocarbons from its host, camouflage would be more correct.

In termites, the hydrocarbon profiles of four different species of termitophiles were found to be identical to that of their respective host species (Howard *et al.* 1980, 1982*a*) and cuticular compounds were similar in the ant *Hypoponera eduardi* and the *Reticulitermes* colonies on which it preys (Lemaire *et al.* 1986). Similarity between hosts and inquilines was thrown into evidence in ants by comparing the hydrocarbon pattern of a parasitoid (Vander Meer *et al.* 1989*b*) and a myrmecophile (Vander Meer and Wojcik 1982) of the fire ant. Identical hydrocarbon components and mimicry in percentage composition was found in a myrmecophile of a carpenter ant (*Camponotus modoc*, Howard *et al.* 1990). The hydrocarbon and fatty acid trace of the cuckoo ant *Leptothorax kutteri* proved to be exceptionally similar to that of its host *L. acervorum* by Franks *et al.* (1990). Chemical similarity was recently described in two other pairs of social parasite and host: *Formicoxenus nitidulus* and its host *Myrmica incompleta* and *Formicoxenus quebecensis* and its host *Myrmica alaskensis* (Lenoir *et al.* 1993) and also between the ant *Pseudomyrmex ferrugineus*, its host acacia plant and a cohabiting social wasp, *Parachartergus aztecus* (Espelie and Hermann 1988).

Chemical similarities in the cuticular hydrocarbon patterns were induced experimentally by creating artificially mixed colonies from different non-parasitic ant species (Bagnères *et al.* 1991*a*) and rearing a parasitic ant, *Formicoxenus provancheri*, with artificial hosts (Lenoir *et al.* 1993).

Analysis of another form of social parasitism, i.e. dulosis, showed differences between the cuticular hydrocarbon traces of *Formica* slave ants *and Polyergus* slave-making ants (Habersetzer and Bonavita-Cougourdan 1993).

Origin of parasite cuticular hydrocarbon resemblance with host

It seems that both a passive mechanism of camouflage acquisition through the transfer of chemicals (for example, Vander Meer and Wojcik 1982; Franks *et al.* 1990; Topoff and Zimmerli 1993) and active biosynthesis of the compounds involved in mimicry (for example, Howard *et al.* 1982*a*, 1990) both exist. When host similarity is achieved through the acquisition of compounds, specific behavioural and morphological patterns will assure contact between parasite and host. Such behavioural and morphological adaptations are not required if the parasite biosynthesizes the specific host compounds itself to obtain chemical resemblance (Dettner and Liepert 1994).

Evidence that cuticular compounds act as recognition cues

Only behavioural tests can in fact demonstrate the role hydrocarbons play in recognition processes. Being soluble in non-polar solvents, epicuticular hydrocarbons may easily be eliminated from a dead insect (acting as an inert lure) and transferred to another lure. Chemical analyses of epicuticular compounds were coupled with bioassays using lures treated in different ways with solvents and hydrocarbon extracts to test species recognition in termites (Howard *et al.* 1982*b*), and nestmate recognition in *Camponotus vagus* by Bonavita-Cougourdan *et al.* (1987). Similar experimental designs were used in the study of *C. floridanus* (Morel and Vander Meer 1987; Morel *et al.* 1988), *Cataglyphis cursor* (Nowbahari *et al.* 1990), *Formica montana* (Henderson *et al.* 1990) and in two species of *Reticulitermes* termites (Bagnères *et al.* 1991*b*). In all cases, behavioural tests showed that cuticular hydrocarbons are involved in the recognition system. Other papers have highlighted the role of cuticular compounds through behaviour experiments alone, as in paperwasps for example (Post and Jeanne 1984) and in the honey bee, where treatment of workers with synthetic hydrocarbons affected nestmate recognition by sisters (Breed and Julian 1992; Breed and Stiller 1992). Behavioural tests demonstrated the transfer of recognition labels during brief contacts between members of different ant colonies (Breed *et al.* 1992). Behavioural tests also showed that hydrocarbon extracts allow discrimination of the age of the host by a mite that parasitises honey bees (Phelan *et al.* 1991).

Social wasps

Research on ants and termites has thus largely documented that cuticular hydrocarbons do play an active role in recognition processes. In contrast to the abundance of papers dealing with cuticular hydrocarbons of these groups, the subject has only recently received attention from students of social wasps, so that evidence of their role is scarce.

Research has demonstrated that epicuticular hydrocarbon blends are species-specific in four species of Vespinae (*Vespa crabro, Dolichovespula maculata, Vespula squamosa* and *Vespula maculifrons*) (Butts *et al.* 1991). However, this last study showed a similarity in the hydrocarbon profiles of the workers of *V. maculifrons* and its social parasite *V. squamosa*. In each species, castes were chemically distinct, as worker and queen extracts had different proportions of cuticular hydrocarbons. Butts *et al.* (1993) showed that colonies of *Dolichovespula maculata* had quantitatively distinct hydrocarbon profiles, some quantitative differences between colonies (but not between castes) were also found in *Vespula germanica* (Brown *et al.* 1991).

Paper-wasps

Polistes wasps, like the other social insects, are highly efficient in that access to their nest is granted only to residents: they behave very aggressively towards any wasp-sized insects flying towards their own colony (West-Eberhard 1969*b*). The ability of wasps to distinguish nestmates and kin from stranger wasps has been widely documented in some species of the genus in extensive research by Gamboa and co-workers (reviewed by Gamboa *et al.* 1986*a* and Gamboa this volume). Odours, determined by both genetic and environmental components, are involved in the process (Gamboa 1988). In contrast to the abundance of experiments on nestmate and kin discrimination ability and of studies on the biology and behaviour of *Polistes* wasps, research on the nature of the labels used in discrimination processes has only recently begun.

The first clue that epicuticular compounds could be involved in species recognition in *Polistes* was found by Post and Jeanne (1984) who studied the interactions between males and females during mating. These authors referred to a 'surface pheromone' on the cuticle (at least on the thorax and gaster) of female wasps which was probably responsible for species discrimination by males. At that time, however, the chemical composition of the surface pheromone had not been studied and the relevance of epicuticular hydrocarbons in *Polistes* recognition processes was still unknown.

The first results of research on paper-wasp hydrocarbons were published in 1990.

Espelie and Hermann (1990) described the epicuticular blends of *P. annularis* females, males, and larvae, and found that their hydrocarbon compositions were similar. The hydrocarbon mixture covering the surface of the nest paper was also similar to the epicuticular mixture. The similarity between the hydrocarbon blend covering the cuticle of adult wasps and that covering the nest paper suggests that wasps could be responsible for nest odour: a female could apply her epicuticular compounds over the nest surface when she strokes her abdomen on the nest (stroking behaviour may serve to apply substances on the nest, Dani *et al.* 1992*b*). Nest paper hydrocarbons may play a recognition role (Espelie and Hermann 1990). Indeed, behavioural tests have demonstrated that *P. metricus* females recognized their nest by the hydrocarbon blend covering its surface; chemical analyses demonstrated that the proportions of the different hydrocarbons varied between nest papers as well as between adult wasps 'sometimes similarly according to colony identity' (Espelie *et al.* 1990). In fact, within a colony of this species, all wasps share a similar epicuticular hydrocarbon pattern (Layton *et al.* 1994). In accordance with this evidence, bioassays showed that the females of *P. metricus* needed to acquire experience with nest paper hydrocarbons soon after emergence to exhibit correct nestmate recognition responses (Singer and Espelie 1992). Multivariate statistical analysis of the variations in hydrocarbon mixtures of *P. exclamans* showed that adult wasps from different colonies had different epicuticular hydrocarbon patterns. Although hydrocarbon patterns were similar within a colony, surprisingly, the proportion of hydrocarbons in the mixture covering the comb surface differed from those of the wasps in the same nest (Singer *et al.* 1992*a*). These authors pointed out that possibly only some of the hydrocarbons in the blend played a recognition role (and in this case the variations of the other compounds are not relevant to the recognition process) and that it is also possible that in *P. exclamans* the nest is not involved directly in the learning of recognition cues by newly emerged wasps. In any case, it is unlikely that a female can apply a different blend over the nest surface from the one covering her cuticle, and thus the origin of the blend of hydrocarbons found on the nest paper is not known. These chemicals could be produced by adult wasp exocrine glands or could be acquired from the immature brood reared in the cells of the nest. Indeed, in a species of *Vespula*, after the brood had completed their development inside the nest cells, the nest carton had acquired a hydrocarbon profile that was typical of the pupae (and in turn different from adult wasps) (Brown *et al.* 1991).

Epicuticular hydrocarbons were also analyzed from monogynic colonies of *P. dominulus* by Bonavita-Cougourdan *et al.* (1991). They found a weak indication that hydrocarbons could serve in nestmate recognition because colonies were not always chemically similar (on the basis of multivariate statistical treatment of hydrocarbon relative proportions). On the contrary, these authors were able to demonstrate that, as in other social insects, dominant females, both foundresses and laying workers, carried a hydrocarbon pattern that was distinct from that of sterile workers, even if the differences were smaller than between ant castes. Females performing different behavioural tasks within the colony had no distinct hydrocarbon traces. Thus, apart from females with different ovarian conditions, it was impossible to discriminate subgroups within a colony on chemical differences. Different castes had different hydrocarbon profiles also in *P. metricus* (Layton *et al.* 1994). Colony specific patterns of hydrocarbons were recently demonstrated in *P. fuscatus*: three methylalkanes contributed to determine colony identity so that they could play a role as recognition pheromones (Espelie *et al.* 1994).

Colony odour varies within a colony both between members and between the different phases of the colony cycle also as a consequence of the presence of social parasites on the nest.

The analysis of epicuticular compounds extracted from females of *Polistes atrimandibularis,* the workerless social parasite of *P. biglumis bimaculatus*, has shown that the profile of this species changes with the season. In some phases of colonial cycle the parasite female mimicked the host species (Bagnères *et al.* 1994).

At the beginning of the season, before entering the host colony, the hydrocarbon trace of a parasite female is parasite-specific. The same profile was found in females that had just invaded the host nest. The peculiarity of this profile is the presence of alkenes (Bagnères *et al.* 1994), which are absent from the hydrocarbon pattern of the host species (Lorenzi 1992; Lorenzi *et al.* 1994*b*). However, the parasite female profile changed about one month after invasion of the host nest, i.e. one month after living on the host colony together with the host queen. After this period, the hydrocarbon trace of a parasite female became identical to that of the host species. In the post-emergence period, the profile of the offspring of the parasite female differs from that of the host species, and is similar to that of the parasite female on invading the host colony.

These results have raised two kinds of question regarding the relevance of hydro-carbons in colony recognition processes. First, similarity between host and parasite is verified at species level, i.e. the parasite female is camouflaged as a host female, not as the host colony. In this case, colony signature is not complied with. As a second point, the course of the hydrocarbon profile in *P. atrimandibularis* agrees with the observations on behavioural interactions between the invading queen and her host queen (Cervo *et al.* 1990*a*): the two females, after a short period of aggressive interactions, live peacefully on the nest together. However, the course of the hydrocarbon profile of the parasitic species does not explain the tolerance the host brood shows towards the parasitic offspring. It seems that *P. biglumis bimaculatus* offspring, which emerge before the parasitic brood, gain their first experience of the 'typical' parasite odour after living as adults for some days. Nevertheless, they accept the parasite offspring on the nest. This fact cannot be explained by the present hypotheses on the learning mechanisms of recognition cues in *Polistes* (Gamboa *et al.* 1986*a*). According to these hypotheses, a wasp learns the odour of her colony from the nest within a few hours of emergence and thereafter tolerates those wasps which bear a label similar to the one learned. Future research will investigate how *P. atrimandibularis* offspring override nestmate recognition cues without using a chemical camouflage.

It is possible that the recognition label a newly emerged wasp learns from her parasitized nest is that of the parasite species. Indeed a nest changes its odour once it is invaded by a social parasite. We analyzed the surface hydrocarbons of the paper of eight *P. biglumis bimaculatus* nests, four of which were parasitized by the social parasite *P. atrimandibularis*. The nests had been collected in the field at the end of the post-emergence stage and had no eggs, larvae or pupae. Extracts were obtained by directing a stream of approximately 5 ml of pentane onto the outer surface of each nest and collecting the solvent in a beaker. Each extract was concentrated to dryness under a stream of nitrogen and analyzed by gas-chromatography/mass-spectrometry. Preliminary analyses show that the hydrocarbon blends covering the surfaces of the parasitized and non-parasitized nests differ (Fig. 10.1). Differences in the hydrocarbon mixture are comparable to differences between host and parasite wasps. The cuticular hydrocarbon blend of a female of the host species is a complex mixture of *n*-alkanes and methylbranched alkanes (Lorenzi 1992; Lorenzi *et al.* 1994*b*). Analysis of extracts obtained from wasps washed in solvent (30 females and 10 males each dipped in 1 ml pentane for 75 s) pertaining to parasitized nests showed that the mixture of cuticular hydrocarbons from females of the parasite species also contains alkenes. Thus, by the end of summer, *P. atrimandibularis* females and *P. biglumis bimaculatus* females show qualitative differences in their hydrocarbon patterns. The hydrocarbon mixture covering the paper of parasitized nests, inhabited by both parasite and host wasps, is composed of alkanes and alkenes specific to the social parasite. Non-parasitized nest hydrocarbon patterns match those of *P. biglumis bimaculatus* wasps and both lack parasite-specific hy-drocarbons. These results indicate that this species of social parasite modifies the odour of the host nest.

The presence of alkenes on the nest paper could be due to contact by adult

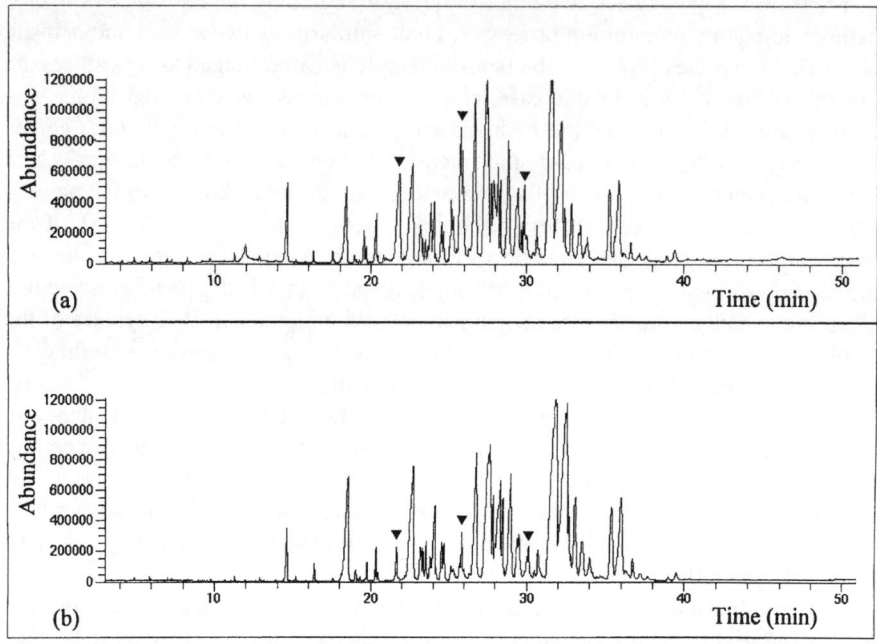

Fig. 10.1 Total Ion Chromatograms by gas chromatography/mass spectrometry of *Polistes biglumis bimaculatus* nests extracted with hexane: (a) a nest parasitized by the social parasite *P. atrimandibularis*, (b) a non-parasitized nest. Arrows show principal qualitative variations of compounds, unsaturated in a, saturated in b.

parasites with the nest. When stroking the nest with its abdomen (Cervo *et al.* 1990a), a parasite could transfer hydrocarbons to the nest from its cuticle or from its exocrine glands. Indeed Dufour's gland contains hydrocarbons (for example, Cane 1983; Dani, pers. comm., in *Polistes* wasps), but it should be documented that it contains the same hydrocarbons that are found on the cuticle and that the blend of epicuticular hydrocarbons and the blend of hydrocarbons from the Dufour's gland have the same relative proportions of compounds. Alternatively, the presence of parasite specific hydrocarbons on the nest paper could be due to the acquisition of compounds from the immature parasite brood developing in the nest. In *Vespula germanica,* Brown *et al.* (1991) observed that nest paper hydrocarbons change in composition over the annual colony cycle. At the beginning of the cycle, the carton is covered with a hydrocarbon blend similar to that of adult workers, but after some generations of brood have completed their development inside the cells, it acquires the characteristic profile of the pupae which is different from that of the adult wasps. It is also possible that a similar mechanism, working together with the transfer of compounds from the cuticle or glands of adult parasite wasps, helps modify nest paper odour in the nests of *P. biglumis bimaculatus* parasitized by *P. atrimandibularis.* Future research should however reveal whether in these two species of *Polistes* the epicuticular hydrocarbon traces of larvae or pupae matches that of the emerged individuals of each species.

Variations in nest paper hydrocarbon composition in parasitized colonies may influence the process of recognition cue learning in host workers, since it has been documented that paper-wasps learn colony odour from the nest (Gamboa *et al.* 1986*b*). It is therefore important to analyze how host workers deal with changes in colony odour during the colony cycle. However, we still need to discover at what point in the cycle of parasitized colonies does nest odour begin to acquire the parasite specific compounds and so begins to differ from that of non-parasitized colonies. If changes in nest odour occur before the host offspring emerge, they would learn the parasite specific label as a template. This would explain why host wasps are tolerant towards the parasite offspring. Otherwise, changes in nest odour require newly emerged host females to update the recognition system template. Vander Meer *et al.* (1989*a*) and Provost *et al.* (1993) stated that worker ants must 'update continuously the learned template' because the chemical signature of their colony changes. The way *Polistes* wasps learn recognition cues when they are young adults suggests imprinting, but Gamboa *et al.* (1986*a*) and Gamboa (this volume) point out that although learning of recognition cues may be completed shortly after emergence and is durable, there is no experimental evidence that the process is a form of imprinting. Gamboa (this volume) speculates that the environmental components of colony odour may change during the season and this may require updating the learning of recognition odours. This may apply in particular to a *P. biglumis bimaculatus* colony invaded by a *P. atrimandibularis* parasite, because the colony is subjected to dramatic changes in nest odour. In any case, at the end of the season, two odours coexist in parasitized nests (a parasite specific odour and a host specific odour) and host females accept individuals bearing one or the other label, and this is not easily explained by the present hypothesis on the recognition system of paper-wasps.

Cuticular hydrocarbons of *Polistes* wasps were shown to be species-specific mixtures of compounds, as in other species of social insects. In the species of paper-wasps examined, 18 to 61 compounds separated by gas-chromatography have been identified. The blend of each species is composed of hydrocarbons ranging from 23 to 35 carbon atoms: *n*-alkanes and methyl branched alkanes. In *P. dominulus* and *P. exclamans* extracts, small percentages of alkenes were also found. However, in all species the most abundant products are linear or branched odd alkanes from 27 to 31 carbon atoms. Analysis of the extracts from *P. atrimandibularis* wasps also showed that the cuticular blends of this species contain alkenes, and that blends differ between the sexes: alkenes are particularly abundant in males (Fig. 10.2). The same results were reported in *P. metricus* (Layton *et al.* 1994).

In the few species of *Polistes* wasps analyzed, colony identity was found when statistical analysis was applied to evaluate differences and similarities between relative proportion of hydrocarbons. In the paper on *P. dominulus* (Bonavita-Cougourdan *et al.* 1991) differences between colonies were very small. In this case, however, extracts of individual wasps were obtained from wasps born in the laboratory and from colonies founded in the laboratory; the differences between colonies

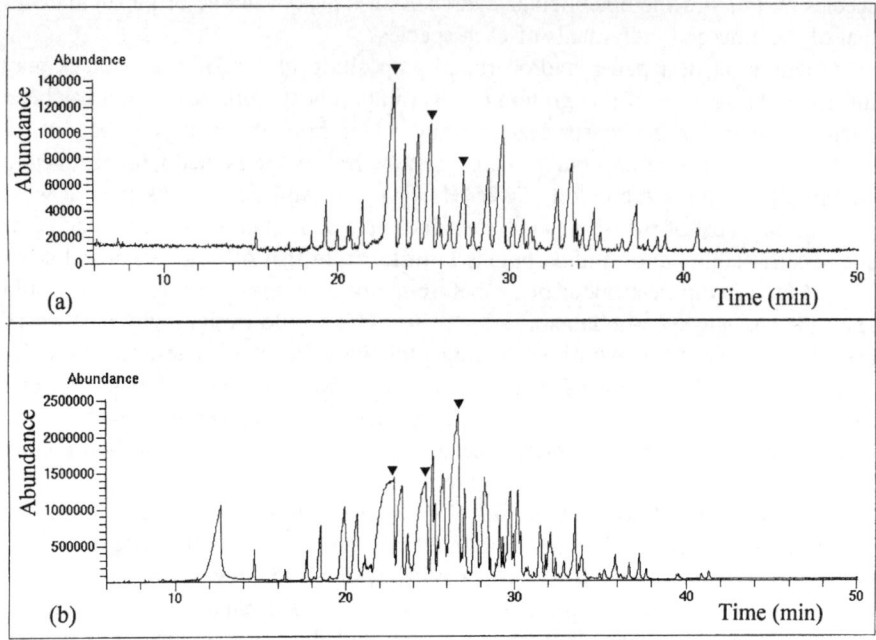

Fig. 10.2 Total Ion Chromatograms by gas chromatography/mass spectrometry of *Polistes atrimandibularis* wasps: (a) females, (b) males. Arrows show principal quantitative variations of unsaturated compounds between sexes.

were not large enough to allow chemical discrimination because they were in continuum. The homogeneous environment of the laboratory was probably responsible for the lack of colony odour identity. It has also been observed that non-nestmate wasps reared in the laboratory were more tolerant of each other than non-nestmates from field colonies (Gamboa *et al.* 1986*b*). Alternatively, other cues are responsible for nestmate recognition in *P. dominulus*.

Although only a few papers have been published on the role of cuticular hydrocarbons in wasps, it has been shown that these chemicals seem to play a fundamental role in their social life and the basis for the comprehension of the onthogenesis of nestmate recognition ability has been laid. Chemical analyses have demonstrated that a nest may carry the same odour of the wasps living on the comb, and behavioural tests have shown that the compounds covering the nest surface are really necessary for the correct learning of nestmate recognition cues.

The study of these compounds, including those which characterize the *Polistes* social parasites, will certainly contribute to completing the links between research on paper-wasp biology and recognition abilities.

Acknowledgements

We thank G. Dusticier for GC-MS injections. J. Wenzel, anonymous referee, M. J. West-Eberhard and S. Turillazzi provided many helpful comments on the manuscript.

11

Selective altruism towards closer over more distant relatives in colonies of the primitively eusocial wasp, *Polistes*

Joan E. Strassmann

One of the predictions of Hamilton's theory of inclusive fitness is that individuals should use relatedness information in behavioural decision making. In *Polistes* a number of circumstances give individuals the opportunity to favour certain individuals over others on the basis of genetic relatedness. These circumstances allow us to evaluate the ability of females to favour their genetic interests separately from the often overlapping common interests of the colony as a whole. Hamilton's Rule provides a structure for predicting the outcome of individual choices regarding where and how they should aid relatives (Hamilton 1964*a*, *b*). An individual should favour helping individual 1 instead of individual 2 if: $r_1 \Delta W_1 > r_2 \Delta W_2$ where r_1 and r_2 are the relatedness of the actor to individuals 1 and 2 and ΔW_1 and ΔW_2 are changes in fitness of 1 and 2 as the result of the actor's behaviour. If fitnesses in the two situations are equivalent, then Hamilton's Rule predicts that individuals should use relatedness alone to guide their choice of whom to aid.

Colonies in *Polistes* are typically begun by one to several females (e.g. Pardi 1940, this volume; West-Eberhard 1969*a*; Strassmann 1979). If the colony is formed by a group, one of the females dominates (Pardi 1942, this volume) and lays most of the eggs, which means subordinates act mainly as altruists that should favour dispensing their aid to the closest possible relatives. Hamilton's Rule has been used extensively to investigate this sort of joint nest initiation (e.g. Metcalf and Whitt 1977*a*; Noonan 1981; Queller and Strassmann 1988). After about six weeks worker females emerge (Pardi 1948*b*; Strassmann and Orgren 1983). These workers normally rear additional brood consisting of more worker generations and ultimately a generation of reproductive males and females (Pardi this volume). In some species a worker or the queen may leave the original nest and begin a satellite nest (Strassmann 1981*a*). This presents the other workers with a choice of caring for the brood in either of the two related nests (Strassmann 1981*b*). At the end of the season workers have a number of options, some of which have been investigated. One late season choice workers have involves the sex ratio of reproductives (Strassmann 1984; Mueller 1991). Workers on the nest of their mother will favour a 3:1 female-biased sex allocation while the queen will favour equal investment (Trivers and Hare 1976). Another choice of workers involves the possibility of becoming queen either in the current season or in the next (Strassmann 1989*b*). This possibility is available because caste is probably not fixed before adult eclosion in *Polistes* so individual females can become workers or gynes that overwinter (Solís and Strassmann 1990). In species where queens often die before the end of the season, workers can mate and become fully fecund queens in the same season even after working (Strassmann 1981*a*; Strassmann and Meyer 1983). However it appears that they cannot make the switch to overwintering gyne after working (Solís and Strassmann 1990). Females are predicted to be more likely to take either of these reproductive options if their relatedness to brood in the colony drops, as would happen after the death of their mother.

Other factors can cause reductions in relatedness that females might respond to. These include multiple simultaneous queens, though behavioural observations go against this, at least early in the season (Strassmann and Meyer 1983). Multiple mating would create multiple cohorts in the colony; detecting and favouring full sisters as honey bees may do (Page *et al.* 1989) would be advantageous. Workers themselves may lay eggs. However, only recently have molecular techniques sensitive enough to detect these factors been developed (Queller *et al.* 1993*a*), so this study will focus on changes in relatedness involving foundresses, responses to changes in queens, and sex ratios.

Terms for social individuals must be defined because roles in *Polistes* are so variable. I use **female** to refer to any female wasp, **queen** to refer to the principal egg layer in the colony, **gyne** to refer to a female that will become a foundress the following season, **foundress** to refer to a mated female capable of laying eggs in a spring nest (the same female just emerging in autumn would be called a gyne), and worker to refer to a female working (and not necessarily excluding the possibility of ultimately becoming queen), typically on her natal nest.

Genetic relatedness among nestmates in Polistes

Genetic relatedness is the probability that two individuals share genes that are identical by descent. It can be assessed in a variety of different ways. It can be determined from good pedigree information which can be based on observations of mating and egg-laying, or on allozymes. For example, Strassmann (1985*a*) and Lester and Selander (1981) came to very similar relatedness estimates for workers in colonies of *P. exclamans* using these two techniques. Allozyme frequencies can also be used to estimate relatedness without constructing pedigrees. Several methods to do this have been developed (Pamilo and Crozier 1982; Pamilo 1984; Queller and Goodnight 1989). The Queller and Goodnight (1989) estimate has the advantage of a sophisticated user-friendly program that uses a Macintosh computer for calculating relatedness estimates, a jackknife measure of the standard error, and F-statistics. The biggest problem with studies of relatedness in *Polistes* is the paucity of variable allozymes. The recent discovery of microsatellite DNA loci will revolutionize investigations into the role of relatedness in social interactions in social insect colonies (Litt and Luty 1989, Tautz 1989, Weber and May 1989; Evans 1993; Hughes and Queller 1993; Queller *et al.* 1993*a*; Choudhary *et al.* 1993).

Relatedness estimates are available for 13 species of *Polistes* (Metcalf and Whitt

1977*a*; Lester and Selander 1981; Queller *et al*. 1990; Strassmann *et al*. 1989; Table 11.1). All relatedness estimates are for nestmates. The largest sample available is for autumn gynes. These females are of similar ages, having emerged from the colony at the end of the season, usually over a period of about a month.

About half of these values do not include 0.75 in their confidence interval. Relatedness could be reduced because of multiple simultaneous queens, sequential queens or multiple mating by queens. Of these three, only sequential queens are well documented in *Polistes*. Recent advances that allow relatedness estimation for individuals will no doubt reveal whether or not multiple mating and simultaneous multiple queens also play a role in reducing relatedness among colony members. Multiple mating may occur in *P. metricus* where Metcalf and Whitt (1977*a*) found relatedness values that were consistent with mating twice, with one male fathering 90% of the progeny though their results could also be explained by multiple egg layers. In some species, levels of relatedness among females are high enough to preclude multiple paternity. Worker laying of male eggs has not been investigated but in any case it will not alter relatedness among females.

Inbreeding would elevate relatedness among nestmates, but appears to be unusual in *Polistes* (Strassmann and Queller, unpublished data). Only *Polistes gallicus*, the species with highest relatedness among autumn gynes, is inbred; females have been observed to mate on their natal nest with male nestmates (Beani this volume).

Genetic relatedness in foundress associations

Perhaps the most fundamental decision a foundress makes is whether or not to nest alone, or to join another female, and who that other female should be. Since these foundresses begin the spring inseminated and with developed ovaries, and subsequently pursue different options, this is a real choice. If relatives are available and can be recognized then foundresses should prefer them as nestmates over non-relatives unless a given foundress is assured of being queen, and can simply exploit non-relatives.

One of the best-supported generalizations in *Polistes* sociobiology is that gynes in spring voluntarily associate only with their natal nestmates that are relatives (e.g. West 1967; Metcalf and Whitt 1977*a*; Noonan 1978; Strassmann 1979; Klahn 1979). This polistine truism has been supported everywhere it has been investigated, though small numbers of females are occasionally reported on colonies of non-natal nestmates. Even when colonies are packed very closely together, as is the case for *P. annularis*, individuals discriminate natal nestmates from other nearby females, and preferentially begin new nests with them. It appears that nestmate recognition is based on the rapid learning of colony odours from the comb shortly after eclosion (Pfennig *et al*. 1983*a*; Gamboa *et al*. 1986*b*). However, individuals that are not from the same natal colony behave less aggressively towards genetic relatives than they behave towards non-relatives, even in the absence of any experience (Klahn and Gamboa 1983; Pfennig 1990).

When females join others who are not related, the behaviours look quite distinct from those of co-operative, related co-foundresses. This sort of joining usually

Table 11.1 Estimates of genetic relatedness among nestmates of *Polistes*.

Species (location)	Colonies (females)	Method	No. loci	Class	r	Reference
P. metricus (Illinois)	? (141)	pedigree based on allozymes	5	foundresses	0.63	Metcalf and Whitt 1977
P. metricus (Houston, TX)	19 (145)	Queller and Goodnight 1989	5	gynes	0.57±0.09	Strassmann *et al.* 1989
P. bellicosus (Houston, TX)	31 (289)	Queller and Goodnight 1989	1	gynes	0.34±0.15	Strassmann *et al.* 1989
P. bellicosus (South TX)	11 (200)	pedigree based on allozymes	1–4	workers and pupae	0.43 or higher	Lester and Selander 1981
P. carolinus (Houston, TX)	16 (59)	Queller and Goodnight 1989	6	gynes	0.63±0.06	Strassmann *et al.* 1989
P. dorsalis (Houston, TX)	21 (204)	Queller and Goodnight 1989	4	gynes	0.61±0.08	Strassmann *et al.* 1989
P. annularis (Houston, TX)	34 (156)	Queller and Goodnight 1989	2	gynes	0.31±0.07	Strassmann *et al.* 1989
P. annularis (Houston, TX)	29 (222)	Queller and Goodnight 1989	2	autumn gynes	0.45	Queller *et al.* 1990
P. annularis (Houston, TX)	29 (222)	Queller and Goodnight 1989	2	foundresses	0.47	Queller *et al.* 1990
P. exclamans (Houston, TX)	33 (277)	Queller and Goodnight 1989	3	gynes	0.69±0.10	Strassmann *et al.* 1989
P. exclamans (McAllen, TX)	46 (312)	Queller and Goodnight 1989	1	gynes	0.56±0.09	Strassmann *et al.* 1989
P. exclamans (Austin, TX)	16 (344)	pedigree based on allozymes	1–3	workers and pupae	0.39 or higher	Lester and Selander 1981
P. instabilis (Houston, TX)	42 (227)	Queller and Goodnight 1989	2	gynes	0.53±0.06	Strassmann *et al.* 1989
P. versicolor (Venezuela)	18 (107)	Queller and Goodnight 1989	4	females	0.37±0.08	Strassmann *et al.* 1989
P. canadensis (Venezuela)	20 (129)	Queller and Goodnight 1989	3	females	0.34±0.10	Strassmann *et al.* 1989
P. dominulus (Italy)	8 (57)	Queller and Goodnight 1989	4	gynes	0.65±0.09	Strassmann *et al.* 1989
P. gallicus (Italy)	12 (134)	Queller and Goodnight 1989	2	gynes	0.80±0.15	Strassmann *et al.* 1989
P. nimpha (Italy)	10 (103)	Queller and Goodnight 1989	3	gynes	0.54±0.16	Strassmann *et al.* 1989

happens later in the season, just before workers emerge when the joiners are likely to have lost their own nests to predators. The females join only as dominant queens, and we distinguish this joining and its marked lack of co-operation by calling it usurpation (e.g. Strassmann 1981*c*; MacCormack 1982; Klahn 1979; Cervo and Turillazzi 1989; Makino 1989*a*; see also Gervet *et al.* this volume).

Though foundresses choose former nestmates over non-nestmates as co-foundresses, elevated relatedness among gynes in the species is not sufficient to result in increased numbers of females nesting together (Hughes *et al.* 1993). Relatedness was estimated in autumn gynes for eight populations with mean foundress association sizes ranging from one to four females. A correlation between relatedness among autumn gynes and percent of those females behaving as subordinates on nests the following spring was not found, and if anything, the trend was in the opposite direction from that predicted (Table 11.2). This may be explained by costs and benefits outweighing relatedness in selecting for associative behaviour among foundresses. Another hypothesis that would explain the lack of correlation is that queens are ceding more reproduction in species where they are less closely related to the subordinates (Reeve 1991; Reeve and Ratnieks 1993; Hughes *et al.* 1993).

Genetic relatedness and preferential assortment among foundresses

An extension of the overall pattern that females nest with relatives is that they should choose closer over more distant relatives. This will usually involve within colony kin recognition and so may not occur because of the difficulties in detection, not lack of advantage. Studies of *P. fuscatus* to date indicate that there is a threshold response accepting individuals encountered shortly after eclosion on the natal nest. This acceptance extends to unrelated individuals transplanted into the natal nest as pupae. Such studies might indicate little scope for within-colony kin discrimination. However further investigation into the topic was indicated because the recognition assay in the laboratory studies might not have picked up all levels of discrimination, because sample sizes were small and because other studies such as Pfennig's (1990) study of *P. exclamans* indicated recognition in the absence of shared nest experience. Queller *et al.* (1990) investigated preference for closer over more distant relatives in a natural situation with both the opportunity to favour closer over more distant relatives, and a large fitness advantage as the result.

If gynes can distinguish degrees of relatedness among their nestmates, then they should preferentially found new nests with their closest relatives in the spring. *P. annularis* is a particularly good species for such a study because it has large colonies with many gynes in autumn, and those females usually begin new nests in groups (Strassmann 1979). Furthermore, relatedness among gynes is low enough to indicate that there are several different sibships probably resulting from different sequential queens. However, no such effect was found using an allozyme technique sensitive enough to pick up a difference in relatedness as small as 0.1 between natal nestmates founding a spring group together as compared with natal nestmates on different spring nests (Table 11.1 last two values for *P. annularis*, and Queller *et al.*

Table 11.2 Relationship between relatedness among autumn gynes and foundress association sizes the following spring (from Hughes *et al.* 1993).

Species	Location (Long. and Lat.)	No. colonies for r	No. individ. for r	No. colonies for group size	Relatedness	Foundresses per colony	% subordinates
P. metricus	95°4' W, 29°7' N	19	145	294	0.57±0.09	1.02±0.001	1.7
P. bellicosus	95°4' W, 29°7' N	31	289	124	0.34±0.15	2.09±0.08	52.1
P. carolinus	95°4' W, 29°7' N	16	59	254	0.63±0.06	2.10±0.09	52.3
P. dorsalis	95°4' W, 29°7' N	21	204	41	0.61±0.08	2.12±0.19	52.9
P. annularis	97°8' W, 30°3' N	34	156	637	0.31±0.07	4.46±0.12	77.6
P. exclamans	95°4' W, 29°7' N	33	277	135	0.69±0.10	1.07±0.01	6.2
P. exclamans	98°5' W, 26°3' N	46	312	66	0.56±0.09	2.02±0.02	50.4
P. instabilis	98°5' W, 26°3' N	42	227	23	0.53±0.06	2.52±0.57	60.3

1990). Even though it would apparently be advantageous for these foundresses to discriminate closer over more distant relatives from within colony members, they do not do so. The explanation for this that is most consistent with what we know about the mechanism for kin discrimination is that they lack the ability to discriminate within the colony.

Genetic relatedness and satellite nests

Other complexities of the colony cycle in subtropical polistines present females with circumstances ideal for investigating the role of relatedness in helping decisions because they approximate cases where the costs and benefits, or changes in fitness, are left constant and only relatedness varies. One such case is the occurrence of satellite nests in *Polistes exclamans* (Strassmann 1981*a*). In this species sometimes a queen or a worker leaves the nest when it is still flourishing and begins another nest nearby (Strassmann 1981*a*, *b*). Workers that become queens on either their original nest after the queen has died or on a satellite will first mate, and are therefore fully capable of producing female (and male of course) brood. Queens that were formerly workers are indistinguishable reproductively from queens that overwintered and never worked (Strassmann 1981*a*). The satellite may attract additional workers and grow. Initially there is an interchange of workers and the queen between the satellite and the original nest. If one or the other is destroyed, the surviving adults all join the surviving nest. Since queens often die before the end of the season in *P. exclamans*, there are colonies with different levels of relatedness between workers and queens in the population. If females are using relatedness as a motivating factor in either founding or joining satellite nests, then the frequency of these behaviours will vary according to whether the original queen is alive. Strassmann (1981*b*) found that relatedness played a role in both the likelihood of satellite initiation and joining. Workers are more likely to initiate satellites after the queen has died (Table 11.3). In this case there is a clear cue that relatedness has decreased since the workers are likely to be able to detect the loss of the queen rather directly.

An experiment was conducted to determine whether or not relatedness would guide worker joining of a satellite when the satellite was deprived of its initial workers (Table 11.4). In effect this makes the satellite especially needy of help, and increases the benefit a worker could contribute to the satellite. In the experiment were five satellites initiated by the queen, one satellite initiated by a worker in the absence of the queen, and two satellites initiated by a worker in the presence of the queen. In the first two cases relatedness of workers to brood was equal on the two nests and additional workers joined the satellite. In the third case workers were more closely related to brood on the original nest, and they did not join the satellite. Thus, in *P. exclamans,* workers use relatedness information to determine the frequency of both initating and joining satellite nests in ways that are predicted from Hamilton's rule.

Table 11.3 Worker satellite nest formation from nests with at least 20 workers compared for presence of original queen (from Strassmann 1981*b*).

Is satellite built?	Queen Alive	Queen Dead	Fisher Exact Test
Satellite built	1	7	P = 0.008
No Satellite built	11	4	

Table 11.4 Workers are more likely to leave the original nest and join satellites whose original workers were removed when relatedness of workers to brood is the same on the two nests (from Strassmann 1981*b*).

Behaviour of original nest workers when satellite workers are removed	Relatedness of workers to brood is		Fisher Exact Test
	equal on original and satellite nest	greater on original nest	
Satellite joined by workers	6	0	P = 0.036
Satellite not joined by workers	0	2	

Genetic relatedness and sex ratios

Allocation to male and female progeny is the arena where genetic conflicts of interest and their resolution have been best studied in social insects (Trivers and Hare 1976; Nonacs 1986). Workers may favour their interests by laying eggs themselves, or by biasing the sex ratio. Worker laying has not been investigated in detail in *Polistes*, so we will focus on sex ratio biasing. On single-queen colonies, workers in control of the sex ratio will favour three times as much investment in females as in males whereas the queen will favour balanced investment in males and females. In colonies where workers are rearing relatives that are more distant than full sibs, their preferences will converge with those of the queen. If colonies with full sibs occur in the same population as colonies with multiple queens, then full sib colonies may favour all female production while the other sort of colony will favour all male production. We can test the control of sex allocation by looking at the two types of colonies to see if relatedness predicts sex allocation.

In *Polistes exclamans* at the end of the season there exist two classes of colonies: those with the original queen (eusocial), and those where the original queen has died and has been replaced by a worker (parasocial). In the former case, workers are rearing sisters and brothers, and so they should prefer to rear more sisters than brothers. In the latter case, the workers that are daughters of the original queen are predicted to show no preference for nieces over nephews, since relatedness is equal in each case. In a population where both eusocial and parasocial colonies coexist a split in sex ratios is predicted to evolve, with workers on eusocial colonies preferring increasingly female-biased sex ratios and workers on parasocial colonies preferring increasingly male-biased sex ratios (Boomsma and Grafen 1991; Mueller 1991). Strassmann (1984) did not find significant differences between these two types of colonies. Ulrich Mueller (pers. comm.) reanalyzed these data excluding very small colonies (those producing only one or two sexuals) that have a high variance in sex ratios, and found a strong trend (that was not quite significant) in the direction predicted by Trivers and Hare (1976). In 1978 eusocial colonies produced 62% females and parasocial colonies produced 44% females, P = 0.064 (three colonies were omitted). In 1979 eusocial colonies produced 70% females and parasocial colonies produced 55% females, P = 0.187 (five colonies were omitted). These data may indicate that the predicted pattern is not occurring, or that it is somewhat swamped by other direct reproductive strategies of workers in this species, which include becoming queens on this or other nests, or in the following year. If females on parasocial colonies cease rearing workers early to themselves become the next year's gynes, then the very type of colonies for which we would predict a male bias in young reared would in fact exhibit a female bias but it would consist not of nieces of the workers, but the former potential workers themselves.

The lack of parasocial colonies in a northern population of *P. fuscatus* leads to a general prediction for female-bias among progeny (Trivers and Hare 1976). However Noonan (1978) did not find a preference for females over males in queenright colonies of *P. fuscatus*.

Genetic relatedness and queen succession

Another circumstance where workers and queens may have different interests involves the replacement of the original queen. In species with multiple foundresses, workers and foundresses will have different preferences regarding who becomes queen if the original queen dies. In *P. annularis* and other species we have looked at it is always a foundress who becomes queen (Hughes *et al.* 1987). This probably goes against worker interests since they are likely to share more genes with other workers than with other foundresses.

Another aspect of the response to a change in queenship has to do with the behaviour of workers that do not become queen. In central Texas, *Polistes annularis* stops rearing brood earlier than other conspecifics and well before the onset of cool weather (Strassmann 1989b). This may be attributable to the low relatedness among colony members in this species either because of increased conflict levels or because individuals 'decide' to be gynes rather than workers if there has been a turnover that results in decreased relatedness to brood. Colonies where relatedness is exceptionally low or suddenly reduced are predicted to be the first to stop rearing brood. I predicted that colonies would stop rearing brood earlier if the queen were usurped or if the queen were to die and be replaced by a worker. Usurped colonies stopped rearing brood on 16 August 1977 (±12 days, n=5 colonies) and on 10 August 1978 (±6 days, n=2 colonies), while non-usurped colonies stopped rearing brood on 31 August 1977 (±5 days, n=28 colonies) and on 14 September 1978 (±5 days, n=36 colonies), a difference in the right direction that was not significant.

The other sort of turnover involved the death of the original queen and her replacement by a worker queen. These workers mate and function as fully capable queens. Workers other than the one that becomes queen are predicted to respond to reduced relatedness to the brood by ceasing working or by not working in the first place and instead becoming the queens of the next year. Colonies that lost their queens earlier ceased rearing brood earlier in both years of the study (1977: b=0.22±0.13, d.f.=1 and 51, P<0.05; 1978: b=0.23±0.12, d.f.=1 and 44, P<0.03). These results support the prediction that colonies suffering a sudden reduction in relatedness will have females becoming gynes that might otherwise have worked to rear future individuals. The work of Solís and Strassmann (1990) indicates that the females most likely to have this flexibility are those that have recently eclosed and have not yet begun foraging.

Genetic relatedness and usurpation

A change in queens reduces the relatedness of current workers to new brood. If queenship is usurped by a female of another species, then workers are unrelated to the progeny of the new queen. If queenship is usurped by a member of the same species, relatedness will drop, often to zero. If the usurper is a natal nestmate of the original queen that simply did not initially join that nest, then relatedness may be the same as if a subordinate foundress superceded the queen, though the workers would not necessarily have any cues that would allow them to detect that this was the case.

Workers may respond to this reduction in relatedness in a number of ways.

1. They may leave the colony and begin their own nest.
2. They may begin laying their own eggs in the original nest.
3. They may feed only the older brood to whom they are closely related.
4. They may actively destroy unrelated brood.
5. They may refuse to work and instead become the gynes for the following year.

The loss of the queen provides a clear cue to the workers concerning their reduction in relatedness to the brood, making it more likely that an adaptive response will evolve than in cases where no such cues exist. Klahn and Gamboa (1983), Klahn (1988), Cervo and Turillazzi (1989), Makino (1989a, b) and Strassmann (1989b) have looked at the response of workers to conspecific usurpation.

P. fuscatus populations in Iowa have high rates of usurpation by conspecifics—as high as 20% in single foundress colonies (Klahn 1988). When an alien female usurps the colony she destroys eggs and young brood, destroying more on more advanced nests. The females that usurp are females that have lost their own nests. Usurping allows them to gain a head start over beginning a new nest and to use the nearly emerged workers. It appears that the workers can detect that they are on a usurped nest at least under some circumstances and respond in a number of ways. Usurping queens and workers were sometimes involved in intensive fights. On some colonies workers continued to work, on some they deserted and on some they took over (Table 11.5; Klahn 1988). Those that did work apparently did so at lesser rates, since usurped colonies had lower reproductive success than did colonies that were not usurped (Fig. 11.1). Workers became queens when the original queen or

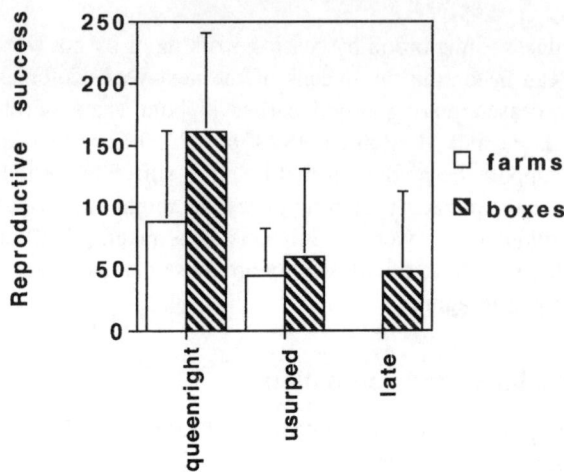

Fig. 11.1 Reproductive success in *P. fuscatus* following usurpation from 11 usurped and 31 queenright farm colonies, and 29 usurped, 59 queenright and 14 late-starting nestbox colonies, according to Klahn (1988).

Table 11.5 Response of workers to usurpation in *P. fuscatus* (from Klahn 1988).

Colony Type	Years	Worker takeover	Worker desertion	Successful	Chi-Square
Usurped	1977–1979	1	4	7	
Queenright	1977–1979	7	2	32	19.8
Usurped	1980–1984	16	4	33	(P < 0.001)
Queenright	1980–1984	15	0	129	

the usurping queen dies. Workers taking over from their mothers destroyed almost none of the eggs and young larvae, while workers taking over from usurpers destroyed nearly all the eggs and young larvae (Klahn 1981). Klahn and Gamboa (1983) also demonstrated that when nests were switched, females destroyed a smaller share of the brood when the nest they got was their sister's nest than when it was a non-relative's nest. Makino (1989*a*, *b*) found similarly uncooperative responses by workers to usurpation in *Polistes riparius*. In this species workers actively destroyed brood of the new queen.

Even species with low rates of natural usurpation respond to experimental nest switches. *P. gallicus* females destroyed eggs but not more developed brood when nests were switched in the field (Cervo and Turillazzi 1989). They also found that workers emerging from nests with unrelated queens had more developed ovaries than did workers emerging in colonies where their mother was queen (Cervo and Turillazzi 1989).

Taken together the studies discussed above make up persuasive evidence that where there are clear cues allowing individuals to assess relative degree of relatedness, females in *Polistes* favour their closest relatives. In contrast, in those cases where relatedness would have to be assessed within colonies in the absence of any obvious cues, individuals do not act on differences in relatedness. This is consistent with a kin recognition mechanism involving individuals contacted in the presence of the nest carton. With such a kin recognition system it is hard to imagine how within colony kin recognition might work. However it does not leave out the possibility for within-colony kin recognition for epiponine wasps whose nests are subdivided into multiple combs.

Acknowledgements

I thank Dave Queller, George Gamboa, Kern Reeve and Stefano Turillazzi for many helpful comments on the manuscript. This research was partially supported by NSF grants IBN-9210051 and BSR-9021514.

12

Behavioural screening and the evolution of polygyny in paper-wasps

Jacques Gervet, Marie-Charlotte Anstett, and Finn Kjellberg

The theoretical problem

How altruistic characters, i.e. characters costly to the individual and benefiting another could evolve was a neglected problem until thirty years ago when Hamilton (1964*a*, *b*) introduced the inclusive fitness idea. A gene, present in an individual, can increase in frequency in the next generation if it induces its bearer to help other individuals bearing the same gene. The transposition of this idea at the level of individuals (whose behaviour can be observed) led to the inclusive fitness concept for individuals: in the reproductive output of an individual should be included its own offspring but also the increase it may cause in production of offspring by other, related, individuals, thereby increasing transmission of genes identical to its own. The application of inclusive fitness to a number of situations has greatly enhanced our understanding of species' natural history.

As the notion of inclusive fitness is generally applied to individuals, the whole genome is involved, and thus measures of numbers of genes transmitted by related, interacting, individuals can be expressed only in terms of proportion of identical (by descent) genes in common, i.e. in term of relatedness. But if a 'gene for altruism' could recognize its presence in other individuals, helping could be restricted to individuals actually bearing the gene and would hence be much more efficient. This possibility was actually proposed by Hamilton (1964*b*): 'there still remains a discrimination which, if it could be made could greatly benefit the inclusive fitness. This is the discrimination of those individuals which do carry one or both of the behaviour-causing genes from those that do not.' Hamilton also stressed the conceptual difficulty of imagining such a gene that would be 'affecting (a) some perceptible feature of the organism, (b) the perception of that feature of the organism, and (c) the social response consequent upon what was perceived.' This idea has repeatedly been taken up again and then discarded due to the above objections (e.g. Maynard Smith 1976), notably as the 'green beard' hypothesis (Dawkins 1976).

In this contribution we want to show that, in the very particular case of the evolution of colony foundation by multiple queens in *Polistes* wasps, genes may exist such that 'structural assortment during association could substitute for Kin Selection to account for co-operation' (Eshel and Cavalli-Sforza 1982) and that the resulting phenotype would evolve so as to resist cheaters that only accept co-operation for their own benefit.

Reflection on the evolution and the fine tuning of eusociality has centred either on the concept of genetic similarity and altruism (Hamilton 1964*b*; West-Eberhard 1969*b*, 1975) or on the mutualistic hypothesis (Lin and Michener 1972; West-Eberhard 1978*a*; Itô 1993). While the models centred on helping relatives have been properly formalized and largely confirmed by biological observations, the mutualistic hypothesis has been much less formalized (but see Itô 1993). The mutualistic hypothesis is based on the assumption that a nest started by several foundresses fares better than one started by a single foundress: there is thus selection in favour of aggregation on a nest independently of relatedness. In a following step, the reproductively inferior individuals may turn into helpers: those are the mutualistic losers (West-Eberhard 1978*a*). In this chapter we will try to demonstrate formally that such selection for association may supplement kin selection as a driving force enabling the establishment of colonies by several gynes in *Polistes*.

Several terms may be used to define the number of queens present within a colony. We will use 'monogyny' and 'polygyny' to describe nests founded respectively by a single gyne (i.e. a mated female after over wintering) and by several gynes whether all lay eggs or not.

Our model does not try to describe the origin of eusociality through foundress association. The cladistic analysis of vespids (Carpenter 1991) shows that nest-sharing is not present in common ancestors of vespids and non-vespids, it is a convergent derivation. Eusociality is ancestral to both Polistinae and Vespinae. In Vespinae, matrifilial monogyny is ancestral while in Polistinae, short term, behaviourally enforced single egg layers among multiple foundresses is primitive. Available evidence does not reveal the type of eusociality present in the common ancestor of Vespinae and Polistinae. What we will be trying to describe in our model is thus limited, for the present purpose, to a potential route for the secondary reacquisition of polygyny in subgroups of *Polistes* that are presently exhibiting monogyny.

Potential polygynous foundation, especially under temperate conditions, probably provides one of the simplest systems that can be investigated because of the symmetry in the position of the interacting females. At nest foundation two or more queens may meet on the nest, the founder and intruders. This results in fights. In some species, here called *strictly monogynous*, the loser of the contest leaves the nest. In other species, here called *potentially polygynous*, the loser may stay on the nest and abandon reproduction, caring instead for the brood of the winner. Thus the ancestral condition, or a derived situation close to the ancestral, can be described and compared with the derived condition. *Polistes* is particularly useful because it contains some almost strictly monogynous phylogenetic groups. Thus through a careful investigation of the behaviours involved in strictly monogynous and potentially polygynous species (e.g. *P. jadwigae* and *P. dominulus*) it should be possible to propose a simple evolutionary scenario that can be tested through a simulation model.

Some behavioural aspects

The case of potentially polygynous species

We will describe the case of *P. dominulus*, the very species on which Pardi (1942, 1946, 1948*a*) first discovered the dominance relationships. We have often witnessed the establishment of a two-foundress society by introducing two formerly dominant females in an observation cage containing a comb with some brood (Gervet 1962, 1964).

Both females are strongly attracted by the comb. The encounter of the females on the comb leads to a fierce fight which can have the following results

1. A female is stung to death.
2. A female abandons the nest.
3. One of the females carries out a new display, the submissive attitude.

The other female reduces its attacks which turn into a dominance attitude. The dominance scale is established. This can again result in two outcomes. Either the subordinate leaves the nest (when the original founder leaves the nest this outcome is called an usurpation) or both females stay on the nest. In the latter case, the social situation is stabilized allowing the development of the two-foundress society. Dominance scenes become progressively less violent and less frequent. The social work of the two females becomes differentiated, the dominant laying the eggs and carrying out differential oophagy while the subordinate specializes on tasks less connected with the nest such as hunting. Differential use of space is expressed even during behavioural displays that are not directly linked with reproduction such as resting and self grooming, which are more frequently expressed outside the nest by the subordinate (Pratte 1990*a*, *b*).

Many other species, belonging to the American phylogenetic group (Carpenter this volume) show the same association of a high frequency of several foundresses initiating a nest and a division of labour within the society. These species include *P. fuscatus* (West-Eberhard 1969*b*), *P. erythrocephalus* (West-Eberhard 1969*b*), *P. apachus* (Gibo and Metcalf 1978), *P. metricus* (Gamboa 1978), *P. exclamans* (Strassmann 1981*c*), *P. annularis* (Strassmann 1983), *P. versicolor* (Itô 1985), *P. canadensis* (Itô 1985; West-Eberhard 1986*a*) and *P. lanio* (Giannotti 1992).

It is however known that the subordinate is not always excluded from reproduction, either because some of her eggs escape from the differential oophagy or because she may outlive the dominant and replace it as the queen.

The case of strictly monogynous species

In some species, e.g. *P. jadwigae* (Kasuya 1981*b*), only very rarely do several females found a nest together and then maybe only as a transient stage. In such species there is also a strong attraction of females by brood containing combs and encounter of two females on a comb leads to a fight. But in these species, the loser leaves the nest. The dominance ritual does not systematically differ from that of

potentially polygynous species. The fight can lead to the establishment of a dominance relationship, but afterwards the subordinate leaves the nest. Thus the main
difference between strictly monogynous and potentially polygynous species is not
the presence and form of dominance relationships but their consequences on social
integration.

A low level (below five per cent) of multiple foundress nests has been observed
in a number of strictly monogynous species such as *P. jadwigae* (Yoshikawa 1957;
Kasuya 1981*b*), *P. chinensis antennalis* (Yamane 1973), *P. riparius* (Makino and
Aoki 1982), *P. nimpha* (Cervo and Turillazzi 1985) and *P. biglumis bimaculatus*
(Lorenzi and Turillazzi 1986). Cohabitation among foundresses was even induced
experimentally in the strictly monogynous *P. gallicus* (= *P. foederatus*) (Perna *et al.*
1978). However, only scanty observations have been made and when available,
cohabitation does not seem to involve division of work as in truly polygynous species (Perna *et al.* 1978 on *P. gallicus*; Tsuchida and Itô 1991 on *P. jadwigae*). These
nests thus include several egg layers and are pleometrotic. However, the relative
contribution of each female to the brood is almost unknown. In this respect differential oophagy may play a central role, as in truly potentially polygynous species it is
the first expression of the differentiation of behaviours upon cohabitation on the
same nest (Pratte and Gervet, unpublished data).

Similarities and differences between the two kinds of species

Two kinds of species can be distinguished and their phylogenetic position are given
by Carpenter's cladograms (this volume). In strictly monogynous species, multiple
foundress nests are exceptional. They include all the species belonging to the
phyletic groups including subgenera *Megapolistes* (= *Gyrostoma*, see Carpenter this
volume) and *Polistella*, i.e. the Oriental and Japanese groups and all except one
species of the European subgenus *Polistes s.str.*. Potentially polygynous species
present a consequent proportion of multiple foundress foundation. They belong
mostly to the American Phylogenetic group, including subgenera *Aphanilopterus*
and *Fuscopolistes* (= *Aphanilopterus*, see Carpenter this volume). They also include one species belonging to subgenus *Polistes s. str.*

The main difference between the two kinds of species seems to be how a conflict
for the use of a comb is settled. The comb in itself is a strongly attractive object that
can trigger the take-over of a comb by a female that has lost its own nest (Kasuya
1982). In both kinds of species, the comb area triggers active defence as soon as two
reproductive females meet on it. Since usurpation can occur in both kinds of species, the conflict is biologically important. Fighting can be ritualized in the form of
dominance relationships as an outcome of aggressive displays. The scene of dominance is the typical contact ritual. Its intensity and frequency is lowered by habituation so that after some weeks it may become hardly recognizable. Dominance
relationships are linked with use of space. Priority of residence on a comb gives an
advantage in the establishment of dominance and dominance relationships entail
differential use of space according to rank (Pratte and Gervet 1992).

In strictly monogynous species, however, an aggressive component to the interaction persists. It is expressed by biting, chasing, intense licking (Makino and Aoki

1982) or even by the persistence of aggressive displays of the dominant whatever the attitude of the subordinate. The subordinate, furthermore, may display retaliatory biting. The dominant tends to stay on the nest and remains aggressive. The most frequent display of the subordinate is to leave the nest. The behaviour of strictly monogynous species is thus close to a territorial behaviour. The occasional persistence of cohabitation presumably would be due to a weak expression of territoriality rather than to a distinct regulatory mechanism.

In potentially polygynous species, the loser of the initial fight submits but usually stays on the nest. The winner becomes a dominant and ceases to aggress its subordinate. It performs antennal displays expressing dominance, but usually no more attacks. The new subordinate immediately abandons oophagy and slightly later on decreases its egg-laying. Scenes of dominance progressively become less frequent as cohabitation continues.

Thus the difference between strictly monogynous and potentially polygynous species involves the behaviour of both winner and loser of the fight. There is an evolutionary change from territoriality to social dominance during reproductive behaviour.

There is no relationship between the within-species frequency of multiple foundress nests and the violence of scenes of dominance. For instance *P. canadensis* shows very aggressive dominance (West-Eberhard 1986*a*) but always displays several female foundation whereas *P. jadwigae* (Kasuya 1981*b*), although much less overtly aggressive, almost always presents single foundress nests: there is a within-species adjustment of the sensitivity of subordinates and the violence of the dominant's dominance displays. Thus the main difference is not in the violence of the displays but, rather, in the transition from territoriality to social dominance. It can even be proposed that dominance relationships may become more violent in potentially polygynous species since prolonged maintenance of ranks is more important for the dominant in such species.

The difference in behaviour between dominant and subordinate foundresses in a potentially polygynous species parallels the relationships between queen and worker. The worker, especially in an orphaned colony can lay eggs, carry out territorial fight, dominate or submit and stay on the nest. Further, the difference between becoming a worker or an over wintering gyne (i.e. a future queen) can be a conditional response to post-emergence conditions in *Polistes*: it is not completely defined during larval development (Mead and Gabouriaut 1993). Thus in potentially polygynous species, the result of the interaction between two reproductives results in the dominant accepting the subordinate as it would accept a worker and the subordinate turning into a worker behaviour. The main difference between a potentially polygynous species and a strictly monogynous one would be related to the possibility or not of establishing queen worker type relationships between individuals that are both initially in the hormonal condition of an egg-laying foundress.

Quite interestingly, usurpation by the social parasite species *Polistes atrimandibularis* (formerly *Sulcopolistes*) of nests of the monogynous species *P. biglumis bimaculatus* and *P. gallicus* shows that the transition from egg layer to worker status is even possible in a strictly monogynous species (Cervo and Dani this volume). Parasitic females replace the host queen and induce the workers of the parasitized nests to take care of their brood. When a *P. atrimandibularis* female colonizes a *P. biglumis bimaculatus* nest, it progressively submits the original female which, nevertheless, remains on the nest (Cervo *et al.* 1990*a*).

Modelling the origin of potential polygyny

What is a likely behavioural mutation?

Following the preceding description of the differences between monogynous and potentially polygynous species, we may state that the difference involves the behaviour of the winner of the initial fight (it must accept a subordinate and not chase it away) as well as that of the loser (it must accept becoming a subordinate). The two changes must occur simultaneously as an altruistic loser could not one-sidedly decide to help the winner of the fight, for the winner would not accept the helper.

Considering the change as a transition from territoriality to dominance may offer a genetically simpler solution to the problem. If the behavioural shift only implies the shift from expressing a set associated with territoriality to a set associated with dominance, then the same behavioural shift would enable winner (accepting a subordinate) and loser (accepting to be a subordinate) to establish a two-foundress nest. Such a transition does not involve a complicated set of new behavioural characters: *instead it involves expression by sexually mature females of the dominance relationships that are already present in queen and worker.* Thus the transition from strict monogyny to potential polygyny would involve a single set of previously co-expressed traits: accepting a prolonged interaction with another sexually mature female. All the ensuing differentiation such as reduction of subordinate ovaries, change in the use of space, would be the expression on the one hand of the set of characters associated with the typical subordinate status on a nest, a worker status, and on the other hand, the set of characters associated with the status of reproductive female having workers. Alternatively, this change could be regarded as a reversion to the traits associated with the evolution of obligatory group life that may have characterized the origin of worker behaviour in primitively social wasps (as suggested in the 'epigenetic' hypothesis for this transition, West-Eberhard 1987*b*).

Would such a mutation in regulation of a behavioural trait set be sufficient to allow the establishment of potential polygyny within a species as has been previously suggested (Pratte and Gervet 1980; Gervet 1986; Gervet and Theraulaz 1991)? Would it resist invasion by a second, cheater, phenotype that would accept helpers but not itself contribute help? In order to examine the plausibility of this hypothesis we have developed a simulation model of the situation.

Models based on structural assortment

Structure of the models

A basic condition for the evolution of a beneficent trait is that help is directed towards individuals bearing the beneficent trait so that this trait will increase in frequency in the population. In the present case this is automatic for beneficent

Table 12.1 Outcome of the encounter between two females at nest foundation: pheno-type(s) of the foundress(es) remaining on the nest and productivity of nests.

		Wins the contest		
		SM	PP	Ch
Loses	SM	SM A	PP B	Ch B
the	PP	SM A	PP+PP C	Ch+PP C
contest	Ch	SM A	PP B	Ch B

Egg layer phenotype is underlined. SM= strictly monogynous phenotype; PP= potentially polygynous phenotype; Ch= cheater phenotype. A, B, C indicate nest productivity, with B<A<C.

mutual toleration because only reproductive adults accepting coexistence with other reproductives on a nest will give and benefit from help. This is a case of structural assortment during association as defined by Eshel and Cavalli-Sforza (1982). Thus if only a strictly monogynous and a potentially polygynous phenotype coexist, it will be relatively easy for the mutually tolerant phenotype to become fixed. A further, more realistic, complication is that if the tolerant behaviour reaches a sufficient frequency, a new deviant behaviour might be selected: electing to stay on the nest (and permitting the other to stay) only if dominant. Such individuals will be called cheaters in the following.

Thus we will use in the models three phenotypes: Strictly Monogynous (SM), Potentially Polygynous (PP) and Cheater (Ch). A preliminary analysis of an infinite population model showed that if no undue advantage was given to the Potentially Polygynous phenotype, it did not resist cheaters. Stochasticity due to finite population size and structuring into subpopulations had to be present. We therefore develop two models, one with a single limited population in order to visualize the intrapopulation dynamics, and one involving a metapopulation made up of a number of small populations.

In the single population model, a regular influx of immigrants was mimicked by a high mutation rate: the three phenotypes derive from each other with the mutation rate of 2%. An important element of the model is how the nests are established. We used a population of 1000 wasps of known phenotype. There was a first bout of encounters in which each wasp met another one, chosen at random, on a potential nest site. The winner of the fight was determined at random. The ensuing consequences were defined with phenotypes according to Table 12.1. If the SM phenotype won, it founded alone; if it lost it left. The PP phenotype tried to stay on the nest whatever its rank. It only left if it lost to a SM individual. It remained as a solitary foundress if it won the contest with a SM or a Ch phenotype. The Ch only stayed on the nest if it won. If the loser was a PP individual, a two-foundress nest was founded. Then a second bout of encounters was initiated: the wasps that had not settled on a nest either met with another free wasp or they tried to usurp a nest according to the relative frequencies of nests and free wasps. After this second bout, only the wasps that had managed to settle on a nest could reproduce, the other ones died without producing offspring. The egg-laying female of each nest was supposed to breed true so that the composition of the next generation of 1000 wasps was established according to the relative productions of the different nests. The variations in frequencies of the different phenotypes were followed for 1000 generations.

The relationships between the productivities of the different kinds of nests are given in Table 12.1. The following conditions were chosen.

1. SM individuals are (slightly) more efficient at founding single-foundress nests than those that potentially found in association: there is a cost to broadening the range of behaviours.
2. Two-foundress nests are equally productive whether they are produced by two PP individuals or by a Ch and a PP individual: no artificial advantage was given to the PP phenotype.
3. In order for polygyny to have a chance of developing, two-foundress nests are at least slightly more productive than single-foundress ones.

The second version of the model included structuring into a metapopulation of 100 populations containing 100 individuals each. The model was run as the preceding one for each population except that mutation was suppressed and replaced by migration. Migrants were drawn at random within the global production of all populations pooled, and migration rates was a parametre of the models. Further, a 'racoon' effect was added in the metapopulation. It was assumed that local populations sometimes became extinct, for instance because a racoon had specialized on feeding on *Polistes* nests. So, regularly, new populations were started by an initial small number of migrants. The frequency of population disappearance was another parametre of the models.

Results of the models

In the single population model, according to the different parameter values, more or less clear cycles in the frequencies of the different phenotypes were obtained. Figure 12.1 represents the case with A=1.5, B=1.3, and C=10 (respectively the productivity of a single-foundress nest of SM, that of single-foundress nests of PP or Ch, and finally, that of a two-foundress nest). Starting from an almost pure population of SM individuals, the PP phenotype begins to increase rapidly in number because of the high productivity of two-foundress nests. The population becomes almost totally composed of PP individuals. Then the Ch phenotype is highly favoured as there are many altruistic individuals with which to associate, eliminating in the process the PP phenotype. When the Ch phenotype is abundant, there are only single-foundress nests and hence the SM phenotype is favoured by its 1.5 productivity compared with the 1.3 of solitary Ch, replacing the Ch phenotype. So the system cycles, staying for a relatively long time close to one hundred per cent of SM individuals.

According to parameter values, the system may vary from very clear cycles close to the boundaries, at any time one phenotype being almost absent, to diffuse cycles that entail a large amount of stochasticity. In Figure 12.2, it is shown that there is no convergence towards a limit cycle, Figure 12.2A representing the first 500 generations of a simulation and Figure 12.2B the following 500 ones.

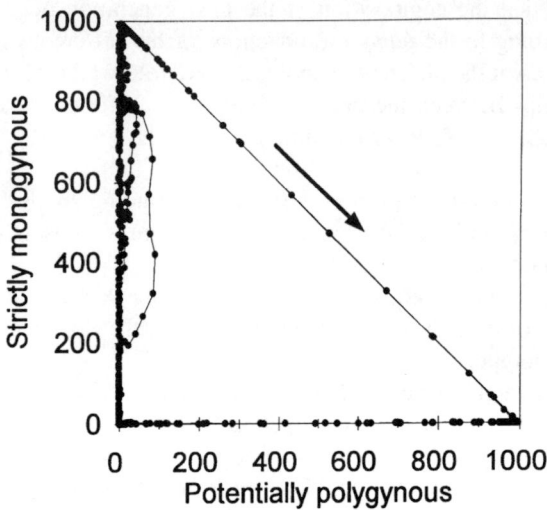

Fig. 12.1 Simulation results for a population of 1000 wasps with A=1.5, B=1.3 and C=10. The population cycles from almost only Strict Monogynes to almost only Potential Polygynes to almost only Cheaters and back to almost only Strict Monogynes.

Table 12.2 gives the general trends of the single population simulations. Except in one limit case, for B=1 or 1.3 and A=1.5, the PP phenotype will be increasingly frequent with decreasing values of C, i.e. with decreasing advantages of two-foundress nests compared with single foundress ones. This counter-intuitive situation results from the Ch phenotype becoming less frequent (see Table 12.2) and thus, as its frequency is the factor limiting the spread of the PP phenotype, this later one increases in mean frequency.

The PP phenotype will also increase in frequency with decreasing productivity of single-foundress foundation by Ch and PP individuals. As in the previous situation, this results from lowered frequencies of Ch that even repeatedly completely disappear at one point of the cycle with some parameter values. Quite intuitively however, the frequency of the SM phenotype increases.

Quite surprisingly, the PP phenotype can be maintained with a very low advantage in productivity of nests accruing to two-foundress nests. This is in part because there is some competition for space in the model: a PP phenotype is more likely to end up in a nest than a SM individual.

Table 12.3 gives the general trends of the 'metapopulation model' according to the amount of migration among populations and the frequency of local extinctions. In the absence of mutation, most often, a single phenotype becomes fixed within the metapopulation. Ch will be maintained for only a limited range of parametres as a border between pure SM metapopulations and pure metapopulations of PP. The factors favouring the fixation of the PP phenotype are increasing frequencies of local extinctions, decreasing migration rates between established populations and increasing productivity of two-foundress nests.

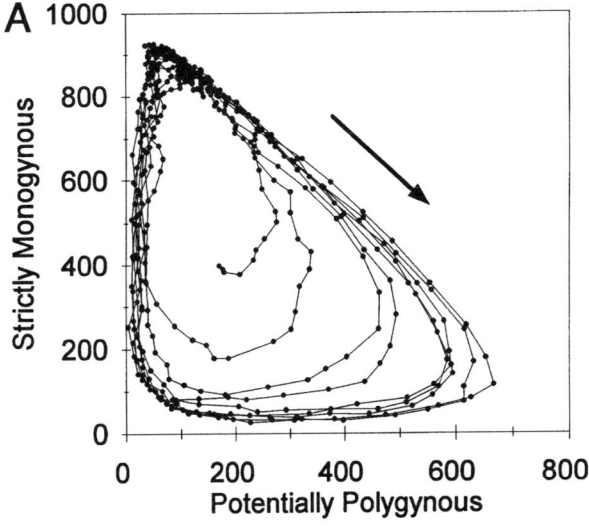

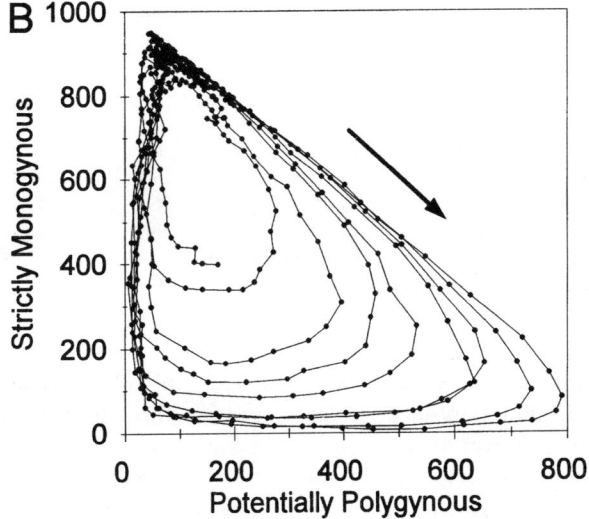

Fig. 12.2 The non convergence of the simulations towards a limit cycle. This figure represents the dynamics of the simulations for a population of 1000 wasps with A=1.5, B=1.0 and C=4. Figure 12.2A represents the first 500 generations and Figure 12.2B the following 500 ones.

Discussion of model results

The global result of the models is that a relatively simple shift in behaviour, the transition from a territorial to a hierarchical behaviour, is sufficient to allow polygyny to establish at non-negligible frequencies within a population. A slight difference in the productivity of single-foundress vs. two-foundress nests is sufficient

Table 12.2 Simulation results for populations of 1000 wasps according to the reproductive output of two-foundress nests (C), and reproductive output of single-foundress nests of PP and Ch (B).

	B=1.3				B=1		
	SM	PP	Ch		SM	PP	Ch
			C=3				
Min	458	40	82		852	45	0
Max	838	199	441		949	143	27
Mean	668	103	229		913	82	4.5
Std	55	29	57		18	17	3.8
			C=4				
Min	455	19	198		67	15	0
Max	738	145	488		926	552	840
Mean	576	67	357		675	151	174
Std	54	19	56		238	112	213
			C=5				
Min	310	9	277		15	8	5
Max	664	117	650		914	771	897
Mean	497	49	453		576	133	291
Std	59	17	61		273	124	258
			C=6				
Min	205	9	336		26	6	39
Max	618	112	758		895	514	903
Mean	444	40	516		560	102	338
Std	65	16	67		236	81	230

The reproductive output A of the single-foundress nests of SM was fixed at 1.5. Abbreviations as in Table 12.1.

and no stringent parameter set has to be envisioned. This is because the behavioural trait leads to structural assortment so that only individuals bearing the beneficent (tolerant) trait will remain together and thus benefit from the presence of others. The presence of polygynous foundations will however inevitably lead to the appearance of cheaters. In this 'Green beard' type model, the appearance of cheaters is not sufficient to suppress the beneficent behaviour.

In our models we have considered a primitive, unsophisticated cheater that does not accept being a subordinate but that accepts subordinates. Although such a behaviour leads to cycling of the different phenotype frequencies, it does not lead to the disappearance of the social trait. Thus there will remain room for the selection of a more subtle phenotype expressing kin preference. Such a phenotype, which accepts coexistence only with a relative, will resist the previous kind of cheaters and also eliminate the more broadly tolerant form. Hence the kin preferring pseudo altruistic phenotype, if it appears, will become the rule. It is an egoistic phenotype in the sense that it maximizes its own inclusive fitness (Lin and Michener 1972),

Table 12.3 Fixed phenotype in a metapopulation model. The model included 100 populations of 100 wasp. The productivity (B) of lone foundations of PP and Ch was fixed at 1.3 compared to 1.5 (A) for SM. The racoon effect is defined as the probability that any given population will go extinct. Usually a single phenotype got fixed within the metapopulation. In the table, SM represents a metapopulation of SM, Ch a pure metapopulation of Ch and PP a pure metapopulation of PP individuals. Ch/PP represents a mixed metapopulation and ? indicates that variable results were obtained due to drift.

Table 12.3(a)
C=4

Racoon Migration	10^{-3}	5.10^{-3}	8.10^{-3}	10^{-2}	5.10^{-2}	10^{-1}
10^{-4}	?	?	SM	SM	SM	PP
5.10^{-4}	SM	SM	SM	SM	PP	PP
10^{-3}	SM	SM	SM	SM	SM	SM
5.10^{-3}	SM	SM	SM	SM	SM	PP
$7.5.10^{-3}$	SM	SM	SM	SM	SM	SM
10^{-2}	SM	SM	SM	SM	SM	SM

Table 12.3(b)
C=6

Racoon Migration	10^{-3}	5.10^{-3}	8.10^{-3}	10^{-2}	5.10^{-2}	10^{-1}
10^{-4}	PP	PP	PP	PP	PP	PP
5.10^{-4}	SM	SM	PP	PP	PP	PP
10^{-3}	SM	SM	SM	SM	PP	PP
5.10^{-3}	SM	SM	SM	SM	Ch	Ch/PP
$7.5.10^{-3}$	SM	SM	SM	SM	Ch	Ch/PP
10^{-2}	SM	SM	SM	SM		

Table 12.3(c)
C=10

Racoon Migration	10^{-3}	5.10^{-3}	8.10^{-3}	10^{-2}	5.10^{-2}	10^{-1}
10^{-4}	PP	PP	PP	PP	PP	PP
5.10^{-4}	SM	PP	PP	PP	PP	PP
10^{-3}	SM	SM	Ch	Ch/PP	PP	PP
5.10^{-3}	SM	SM	SM	SM	Ch/PP	Ch/PP
$7.5.10^{-3}$	SM	SM	SM	SM	Ch	Ch/PP
10^{-2}	SM	SM	SM	SM	Ch	Ch/PP

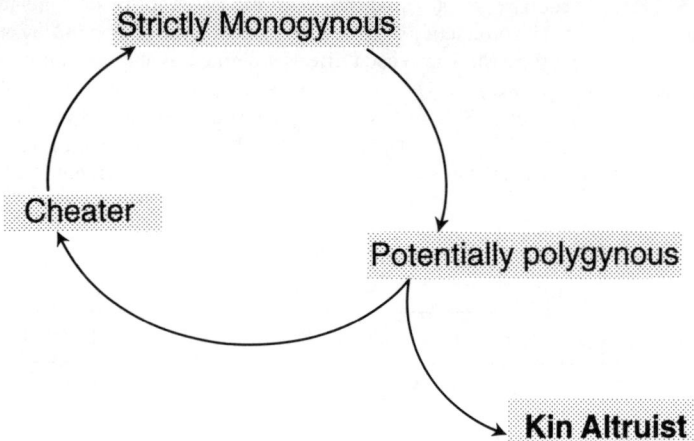

Fig. 12.3 The functioning of the models. The composition of the population cycles from mainly Strict Monogynes to mainly Potential Polygynes to mainly Cheaters and back to mainly Strict Monogynes. If a Kin Altruist phenotype appears, deriving from a Potentially Polygynous phenotype, it will become fixed and the cycling ends.

selection for the altruism is no longer a component of the system. We believe that behavioural screening enabling the establishment of this type of beneficent behaviour can be an important intermediate step in the evolution of kin directed altruisms that may evolve later on. The process is represented in Figure 12.3 with a cycle from Strict Monogynes to Potential Polygynes to Cheaters and back to Strict Monogynes, the kin 'altruist' being an absorbing side step in the cycle.

Discussion

Gene-inclusive fitness vs. individual inclusive fitness

In the present model, we have put the emphasis on structural assortment, that is the preferential association of beneficents with tolerant benefactors (including 'altruistic' defeated subordinates which remain in the group thereby benefiting others of their kind). A non-negligible level of kin interaction is however automatically included and is an obligate element of the model: there is a preferential association of individuals bearing the altruism gene within a limited population and hence there will be a non-negligible association between sisters, especially when the altruism gene is rare. Our model is however rather different from a classical inclusive fitness model as the automatic association of altruists suppresses the diluting effect of outbreeding on relatedness: altruists only help individuals bearing the altruism gene. In this respect it is more efficient in promoting association and therefore should necessitate less restrictive conditions to achieve it. However, the absence of kin preference within populations facilitates the establishment of cheaters.

A fair amount of stochasticity and/or spacial structuring is also necessary; it is because of a transient local rarity of cheaters relative to altruists that the altruists manage to increase in frequency. In an infinite population model without foundation effects in subpopulations, the egoistic phenotype inexorably eliminates the tolerant one, before being itself eliminated by the strictly monogynous phenotype. The effect of population structuring and local extinction can be seen in the context of interdemic selection much like the model by Wilson and Colwell (1981) on sex ratios: subpopulations containing mostly altruists will be more productive and hence contribute more to the foundation of new subpopulations than populations composed mostly of Strict Monogynes or Cheaters. The preferential association of altruists within the subpopulations allows, however, an initial increase in the frequency of the altruistic genes within the subpopulation, a feature that is absent in the Wilson and Colwell type of interdemic selection models. It is thus easier to achieve fixation of the gene for altruism than in a classical interdemic selection model.

An intrinsic biological limitation of the model is that it is based on the reversal of a gyne from a queen behaviour to a worker behaviour: it describes a shift in behaviour in a species that is already eusocial, it does not describe the origin of eusociality. Another limitation, that we have purposely neglected is the tendency of females to return to their original nest in spring. This biological feature, by favouring kin association, will facilitate the fixation of tolerant behaviour. Behavioural screening will still favour the establishment of altruism in this system of kin interaction. Quite interestingly screening of individuals bearing the altruism gene among individuals of equal relatedness (in this case full sisters), was the situation in which Hamilton (1964b) first suggested that co-carrier recognition could be important.

Territoriality/hierarchy in strict monogynes and the generality of a dominance-based model

The behavioural observations on strictly monogynous species underlying our model are based on a phylogenitically quite broad range of species belonging to the subgenera *Polistes s.str. Stenopolistes* (= *Polistella*, see Carpenter this volume) and *Megapolistes* (= *Gyrostoma*, see Carpenter this volume). However, long-term studies of colonies of potentially polygynous species are mostly restricted to subgenus *Polistes s.str.* and to the closely allied American subgenera. Hence our view of the behavioural aspects underlying the establishment of polygynous nests may be too restrictive.

One may speculate that in some instances, in the Australasian subgenera, a very different kind of foundation by several females has evolved in the context of habituation between queens, the behavioural shift being the loss of differential oophagy. Thus in *P. bernardii richardsi* in Australia (Itô 1986a), *P. stigma* in India (Suzuki and Ramesh 1992) and in *P. tepidus malayanus* in New Guinea (Yamane and Okazawa 1977) foundation by several females is frequent, involves large numbers of foundresses, dominance behaviour is rare, and this would be combined with an absence of specialization among foundresses for egg-laying, and nest construction. In *P. humilis synoecus*, the most common polistine wasp in Eastern Australia, mostly single female foundations are observed in spring. But, according to Itô (1986b), more often than in other species, females born on the same nest over winter together on the surface of the comb and reuse it in spring, founding together a new, multiple-foundress society. This lack of a dispersal phase seems to be associated with a lack of territoriality (there is neither oophagy nor ovarian regression). It should however be remembered that in the case of a swarm-founding American *Metapolybia* wasp, West-Eberhard (1978a) has shown that the occurrence of several egg layers is transitory: although there may be at times several egg layers, only one will remain long enough to contribute to the next production of queens.

Although prolonged observations and experiments are lacking, it may be suggested that within the phyletic group including *Megapolistes* (= *Gyrostoma*) and that including *Stenopolistes* (= *Polistella*) association among foundresses would have evolved through a loss of differential oophagy while in the phyletic group including European and American *Polistes*, association among foundresses would have evolved through queen reversal to a worker-like status. There would be two very different ways of being potentially polygynous and hence maybe two very different ways for a strictly monogynous species to become potentially polygynous.

The evolution of multiple-foundress association may seem easier to achieve through a mutualistic process in which several queens breed together as suggested above. Some experiments and observations show however the difficulty of such a scenario. Although strictly monogynous species have a territoriality-based behaviour on nests, it is possible under peculiar circumstances, to obtain stable hierarchical relationships between queens. Experimentally, females can be kept together during over wintering and until nest foundation. Then two-foundress nests can be observed. There is however no reversal of the subordinate female to a worker status: it keeps on laying eggs and perpetuates differential oophagy (Perna *et al.* 1978), so that globally such a nest has a very low productivity (Pratte and Gervet, unpublished results). Thus the establishment of a true multiple-foundress society necessitates more than just a forced mutual acceptance.

Perspectives

In the previous sections we have tried to reconstruct a scenario for the evolution of multiple-foundress foundation in the restrictive context of pre-existing eusociality with single foundresses. In order to do this we have followed two guidelines. On one hand we have tried to analyze which behavioural traits were really involved. On the other hand we have modelled the consequences of behavioural screening on the fixation of the behavioural trait. Further analysis, on other species, of the behavioural traits leading to association may enable us to propose different scenarios as suggested in the previous paragraph. More ambitious scenarios may also be proposed, for instance on a potential mutualistic origin of eusociality, through an understanding of the proximal causes enabling association. Then the behavioural analysis may enable proposal of a model based on the preferential interaction of potential altruists. This does not mean that such a model should be preferred to an inclusive fitness model, it just means that several selective forces, based on apparently different concepts, may interact to enable the establishment of a new trait as already suggested for eusociality by West-Eberhard (1978*a*). Ultimately only two conditions are required to explain the evolution of a new traits: the genes responsible for the trait must be capable of appearing and these genes must be maintained over time by selection. This emphasizes the role of history and the role of long-term variations in gene frequencies. In this perspective, inclusive fitness is a very elegant way to summarize long-term changes in gene frequencies since it can include all fitness effects, notably via aid to relatives, or mutualistic aid (including by sterile subordinates) of the kind discussed here.

13

The origin and maintenance of eusociality: the advantage of extended parental care

David C. Queller

In the eusocial insects, many individuals give up their own reproduction in order to help rear the offspring of relatives. Hamilton's (1964*a*, *b*, 1972) inclusive fitness theory has become the accepted framework for understanding the evolution of such reproductive altruism but there is no consensus on the nature of the selective forces involved. In this paper, I will argue that this may be because we have often been asking the wrong question, and when we do ask the right question, we have neglected what may be the most important kind of answer.

The question we should be asking concerns how behaviours are converted into fitness. How can the fitness benefit achieved through working be higher than the benefit of reproducing alone, even though workers and solitary reproducers perform the same kinds of tasks? The answers may lie less in the immediate effects of behaviour than in the larger framework of demography, a case I hope to make using the extensive demographic data on *Polistes* and related genera of primitively eusocial wasps.

A brief review of recent thinking on the evolution of eusociality will help put this question, and its answers, into perspective. Hamilton (1964*a*, *b*, 1972) showed that individuals are selected to rear a relative's offspring instead of their own if $Br > \frac{1}{2}C$, where C is altruist's loss of offspring, related by $\frac{1}{2}$, and B is the relative's gain of offspring, which are related to the altruist by r. That is, if an individual expects to transmit more copies of its genes indirectly through its relatives than it would expect to transmit by its own independent reproduction, helping is favoured by selection. Rearranging to an alternative form,

$$\frac{B}{C} > \frac{1}{2r} \tag{1}$$

highlights two classes of explanations of the evolution of eusociality that have emerged from Hamilton's rule. They are not mutually exclusive, but it is useful to consider them separately. One focuses on differences in the relatedness variables and the other on costs and benefits.

Besides devising the general inclusive fitness formulation, Hamilton (1964*a*, *b*, 1972) also developed a specific relatedness-centred hypothesis for the origin of eusociality. He noted that most eusocial insects are Hymenoptera which possess a haplodiploid genetic system (females diploid, males haploid). In haplodiploids, full sisters are related by $^3/_4$ instead of the usual $^1/_2$. Other things equal, it would there-fore be more profitable for females to stay and help their mothers to rear additional daughters than it would be for them to produce their own offspring. This hypothesis, known as the haplodiploid hypothesis, accounts for the fact that most eusocial in-sects are haplodiploid (because most are Hymenoptera) and that their workers are invariably female.

Though elegant, the haplodiploid hypothesis has encountered a series of objec-tions. First, it turns out that females have an unusually low relatedness to their brothers, only $^1/_4$ (Crozier 1970), so that a haplodiploid female rearing an equal mixture of sisters and brothers obtains an average r of $^1/_2$, exactly the same as to her own offspring, and exactly the same as in diploid systems. The haplodiploid advan-tage might be saved if female workers could avoid rearing brothers, either because of a female-biased sex ratio or because they lay the male-destined eggs themselves (Hamilton 1972, Trivers and Hare 1976). Worker laying of male-destined eggs oc-curs, but is far from universal (Trivers and Hare 1976; Nonacs 1986). Female-bi-ased sex ratios also occur, but their significance in selecting for eusociality is not entirely clear for two reasons. First, once workers succeed in biasing the sex ratio to their optimum (three sisters to one brother under the simplest conditions) the haplodiploid advantage disappears because each male becomes three times as valu-able as each female (Trivers and Hare 1976; Grafen 1986). So the maintenance of eusociality must be explained in some other way. The other problem is that the origin of eusociality requires two simultaneous changes, helping and female-biased sex ratios (Charnov 1978). Still, the haplodiploid hypothesis can work under certain conditions. There are some circumstances in which selection would favour female-biased sex ratios only in the broods of certain mothers, setting up the opportunity for the daughters in these families to rear mainly sisters (Seger 1983; Grafen 1986; Godfray and Grafen 1988).

While these adjustments make the haplodiploid hypothesis more logically sound, it is still questionable whether this advantage actually plays a critical role. The argument depends on there being a single once-mated queen and the objection has often been raised that multiple mating or multiple egg layers will lower relatedness among female colony-mates. In recent years, estimates of relatedness using protein electrophoresis have confirmed that it is usually lower than $^3/_4$ and often lower than $^1/_2$ (for polistine wasps see Metcalf and Whitt 1977a; Lester and Selander 1981; Queller et al. 1988; Strassmann et al. 1989; Ross and Carpenter 1991). The com-parative prediction can still be saved; it could still be argued that haplodiploids are more likely to evolve eusociality than diploids with an equivalent amount of multi-ple mating and egg-laying by multiple females. But what cannot be explained with-out recourse to costs and benefits is why individuals do not rear their own offspring instead of helping to rear other relatives related by less than $^1/_2$.

Perhaps because of the limitations of the haplodiploid hypothesis, the weight of current opinion seems to be in favour of explanations that stress costs and benefits (Lin and Michener 1972; Alexander 1974; Evans 1977; West-Eberhard 1975, 1978b; Eickwort 1981; Brockmann 1984; Andersson 1984; Stubblefield and Charnov 1986; Strassmann and Queller 1989; Alexander et al. 1991). For helping to be favoured when $r < \frac{1}{2}$ requires that $B/C > 1$. With the exception of egg-laying, workers perform the same kinds of tasks as solitary individuals. Therefore the question that needs to be answered is: How can caring for someone else's offspring yield higher returns than providing the same amount and kind of care for one's own offspring? Several kinds of answers have been suggested. Here I divide them into five categories of synergism:

(1) reproductive synergisms
(2) same-task synergisms
(3) coupled-task synergisms
(4) shared resource synergisms
(5) demographic synergisms

1. First, perhaps some females are poor at laying eggs and turn to helping as a last resort (West-Eberhard 1975; Craig 1983). This must be true for some of the highly eusocial insects with morphologically distinct castes; in the most extreme cases, workers lack ovaries entirely. It is less clear that this explanation works for primitively eusocial insects that lack morphological castes. It does not work if an ineffective (subfertile) egg-layer becomes an equally ineffective worker (Craig 1983). But if the subfertile female can be an effective worker, then a reproductive division of labour between an egg-layer and a worker can evolve. However, efforts at empirical demonstration in species without morphological castes have not generally succeeded (Strassmann 1979; Sullivan and Strassmann 1984; Queller and Strassmann 1989; review by Reeve 1991).

2. Second, some behaviours might have non-linear gains when performed by more than one individual. For example, a stinging attack by three wasps may be more than three times as effective as an attack by a single wasp. However, most studies of *Polistes* have failed to support this hypothesis; large colonies were not more successful at avoiding predation or parasitism than small colonies (Gibo 1978; Gamboa 1978; Strassmann 1981*a*, *c*; Strassmann *et al.* 1988).

3. Similarly, different tasks might be organized in ways that make them more efficient when performed by several individuals (Oster and Wilson 1978; Jeanne 1986*a*). For example, instead of a single individual having to perform each nest-building task in sequence, different individuals can specialize on separate parts of the tasks, and the numbers of individuals can be adjusted according to current needs. Such efficiencies might help explain the maintenance of eusociality, but they are less likely to explain eusocial origins. The ability to work together efficiently would usually not be an automatic consequence of grouping but instead would be a capability evolved after individuals have already been living together in colonies. An exception might be the way that foraging and nest protection interact. For a solitary female, foraging outside the nest means leaving the nest unprotected. If there are two or more adults, both tasks can be accomplished simultaneously (Michener 1974).

4. A fourth possibility involves resource limitation. Vertebrate helping systems are

frequently characterized by individuals that help only until they can gain access to some limited resource such as a mate or territory or nest site (Brown 1987). Mates are unlikely to be limiting for female social insects when the male provides no parental care. Some ants are territorial, but many other social insects are not (except for a very small area around the nest). A more likely possibility is that an expensive nest or nest site may sometimes be a limiting resource (Andersson 1984; Alexander *et al.* 1991). A newly emerged adult may be able to save the time, energy, and risk of acquiring this resource if it stays and helps at its natal nest. However, paper-wasps appear to have rather flimsy, inexpensive nests compared with some other social insects. The cost has never been adequately studied, but even if a new nest is costly, it still may not provide an advantage to helping because the costs of expanding a nest to accommodate a larger workforce may be just as large.

These hypotheses are reasonable and some undoubtedly play a role in the evolution and maintenance of eusociality. But each has limitations and no consensus has yet emerged. Perhaps this is because there are many causes, none of them approaching universality. But perhaps we have not yet exhausted the universe of possibilities. In this paper I want to concentrate on a fifth category of explanation that has been largely neglected: demographic synergisms.

5. Demographic synergism could take several forms. For example, Jeanne (1986*b*) suggested such a synergism exists in the common temporal caste pattern in which risky tasks are undertaken by older individuals. The colony becomes more productive when it shifts mortality risks to individuals who have shorter life expectancy. Another example concerns the age at which individuals are able to begin either direct or indirect reproduction. Workers may be able to reproduce (indirectly) at a younger age than solitary females, because they don't need to find a nest site and they don't need to mature their reproductive systems (termite juvenile helpers are an example; Alexander *et al.* 1991). Any mortality that occurs during the delay tends to select for helping, even if mortality rates are the same for workers and solitary females (Queller 1989; Strassmann and Queller 1989; Gadagkar 1990*b*, 1991*b*). But here I will focus on one particular demographic advantage that I believe to be particularly important in selecting for eusociality.

Perhaps the most striking feature of social insects, besides sociality itself, is the provision of extended care for the young. Of course parental care is very common if we include in this category any provision of resources for the young, either in the egg or through stored provisions to be consumed after hatching. But it is much rarer for insects to provide continued care after their young hatch. This could involve feeding, protection of the young, or both. Such care requires the long-term presence of a care-giver. It will therefore be ineffective if the parent has very low survivorship. But an alternative is for several adults to combine forces in the hope that at least one will survive long enough to provide extended care. In short, group nesting can be an effective means to provide extended care for the young in species where parents do not expect to survive very long.

As far as I know this idea has its genesis in studies of co-foundresses. It is a common finding that an advantage of joint nesting by co-foundresses is that it increases the chance that at least one of them survives long enough to rear the first brood of workers, at which point colony survivorship is much more assured (e.g. Metcalf and Whitt 1977b; Litte 1977; Gibo 1978; Noonan 1981). Later in this paper I will argue that this advantage, for co-foundresses, is less automatic than has been believed. But I also argue that it is much more general. Strassmann (1981c) showed that smaller colonies are also more likely to be orphaned during the worker stage. Recently it has been argued that this kind of advantage may explain why workers stay and help on their natal nest instead of leaving to nest on their own (Strassmann and Queller 1989; Queller 1989; Gadagkar 1990b). Why stay and work? Because if offspring benefit from extended parental care, but adult survivorship is low, a solitary female is unlikely to live long enough to be able to give her offspring these benefits. In the extreme case where offspring require extended parental care to survive, a solitary foundress that dies before her first offspring reaches independence suffers complete reproductive failure. Better to remain at the parental nest where effort is much less likely to be wasted in this way. There are two very closely related reasons why less effort is wasted. First, the worker can help to rear offspring that are already partially raised, so even if she dies young, she will have made an effective contribution (Strassmann and Queller 1989; Queller 1989). Second, if the worker dies young, her investments can be carried to completion by other workers (or by the queen) (Strassmann and Queller 1989; Gadagkar 1990b). In modelling these advantages, I will follow Gadagkar's (1990b) approach of treating them together. Either way, the key advantage is that offspring are able to receive the benefits of extended care, even if individual care-giving adults have short life expectancies.

I will treat the problem in three different contexts. First, I will address the question of selection on workers in currently eusocial species, focusing on polistine wasps. Then I see whether the ideas are applicable to the origin of eusociality. Finally, I will revisit the question of demographic advantages among foundresses.

Selection for working in currently eusocial species

When a solitary female dies, her partially raised young will receive no more parental care. In this paper I will model the case of complete offspring dependency, such that the partially raised young die. But it should be borne in mind that the arguments are more general. Even if the orphaned offspring do not always die, they will suffer in comparison to offspring that enjoy continued care and protection. The advantage of working in a colony is that death of the adult does not necessarily lead to abandonment of the young.

To model the advantage of working, I will assume that workers and solitary females are identical in every respect except whether they are members of a group. Specifically, I will assume identical mortality rates and identical patterns of investment. This will not necessarily be true in the real world, but it serves both to simplify the model, and to focus on the advantage being treated in isolation from other possible advantages.

First, I will assume that both workers and foundresses have constant instantaneous mortality rates, q. Letting time $t = 0$ be the time when an individual must decide

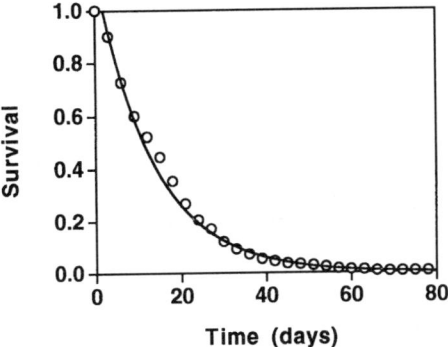

Fig. 13.1 Survivorship of 1230 marked workers in of *Polistes exclamans*. Data courtesy of Joan Strassmann. See Strassmann (1985*b*) for methods.

whether to stay and be a worker in her natal colony or leave to found her own colony, her probability of surviving to time t is e^{-qt} (and her probability of having died by that time is $1-e^{-qt}$). The assumption of constant mortality is made partly for simplicity and partly because the best survivorship data I know fit the assumption fairly well (Strassmann 1985b; see Fig. 13.1).

I will also make the simplest possible assumption about investment. At time $t = 0$, both workers and solitary foundresses begin investing at a constant rate. How this investment translates into productivity is more complicated, and is the crux of the model. Assume that, if none of the investment is wasted, it results in accruing independent offspring at constant rate m. The question is, how much investment is wasted, and how much is translated into actual productivity? Fig. 13.2 shows the patterns for three strategies. Part a if for a female in a strictly solitary species; she rears only reproductives. She invests a constant rate m (the slope of the line), but if she dies any time before the time required to raise an offspring to independence, t_0, she has no productivity (recall that I assume total dependence of offspring). If she survives past time t_0, she begins reaping the rewards of past investment, but productivity still lags behind investment. When she dies she still leaves her investments of the last t_0 time units unfinished and therefore wasted.

However, for our present purposes, we really want to compare the productivity of a worker with the productivity of a foundress which, if she survives long enough, becomes queen of her own colony. Let's exaggerate and assume that once workers appear, no investment is wasted. That is, in colonies, other individuals will always be present to carry on the investments made by our focal individual. This is a conservative assumption with respect to favouring worker behaviour because most workers eclose on large colonies where this assumption is closest to being met, while foundresses are generously granted the same advantage if they can manage to rear even one worker.

The resulting worker productivity schedule is shown in Fig. 13.2b. All investments are realized; whenever a worker dies, its investments are eventually carried

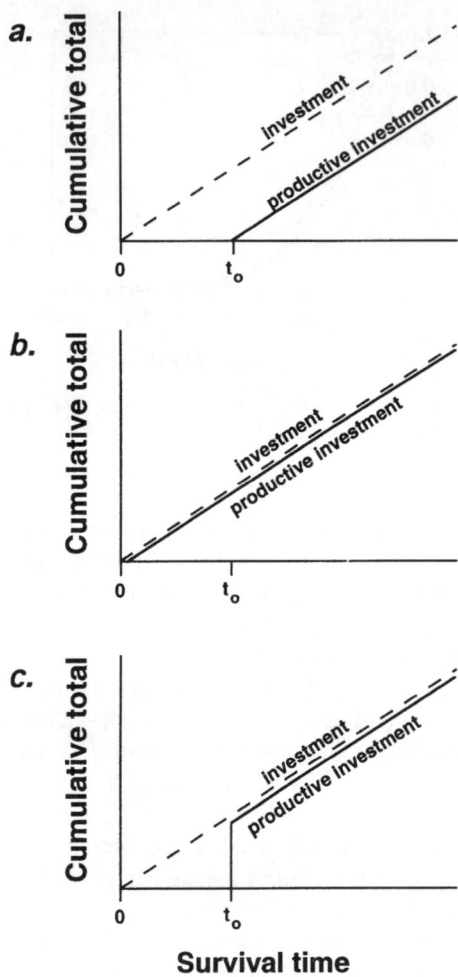

Survival time

Fig. 13.2 Investment/productivity model for three strategies (a) solitary female, (b) worker on established colony, (c) solitary foundress of a new colony. See text for details.

to completion by others. The investments of a potential queen (Fig. 13.2c), like those of a truly solitary female, are wasted if she dies before time t_0. But once she survives to produce workers, all of her investments will pay off including, as pointed out by Gadagkar (1990b), the investment made before time t_0.

Now that we have a mortality model and a productivity model, it remains only to put them together and compare the alternatives of staying and working versus leaving to nest alone. Summing up over all possible survival times, t, the productivity of a worker is simply

$$\int_0^\infty m e^{-qt} dt = \frac{m}{q}.$$

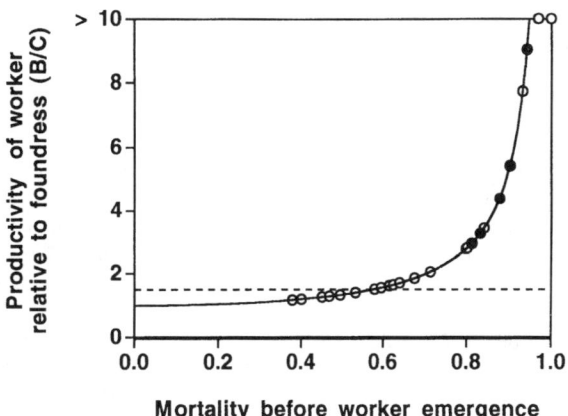

Fig. 13.3 The demographic advantage of helping described by equation (2), expressed as a function of adult mortality before worker emergence. The circles show estimates for polistine wasps based on two ways of estimating adult mortality. Solid symbols are estimates based on worker mortality rates and brood development times (see Queller 1989; Gadagkar 1990*b*). Open symbols are estimates based on success of single-foundress colonies (data in Table 13.1).

The productivity of a foundress is most easily expressed as this same total minus the productivity wasted when the foundress dies before time t_0:

$$\int_0^\infty me^{-qt}dt - (1 - e^{-qt_0})\int_0^{t_0} me^{-qt}dt = \frac{m}{q}e^{-qt_0}(2 - e^{-qt_0}).$$

Taking the ratio of worker to foundress productivity (B/C) we have:

$$\frac{B}{C} = \frac{1}{e^{-qt_0}(2 - e^{-qt_0})} \tag{2}$$

This ratio is plotted on Fig. 13.3, as a function of the amount of mortality before the time required to raise the first offspring to independence ($1-e^{-qt_0}$). Any mortality is enough to kick the B/C ratio above 1. When mortality is high, the advantage exceeds the maximum advantage possible through the haplodiploid relatedness advantage (dashed line at 1.5, the ratio of relatedness to a full sister over relatedness to own young).

How much mortality is there before time t_0? There are at least two approaches to estimating this; one can use either workers or foundresses. To use workers, one needs both a good worker survivorship curve and a good estimate of brood development times. The available data for five polistine wasps have been summarized previously by Queller (1989) and Gadagkar (1990*b*) and are plotted as closed circles on Fig. 13.3. Worker mortality rates are high. If they are accurate indicators of the probability of failure at independent nesting, then the B/C ratios are quite high.

Table 13.1 Failure rates of singly-founded and multiply-founded colonies of *Belonogaster*, *Mischocyttarus*, *Polistes*, and *Ropalidia*.

Species	% failure to reach worker stage *		% colonies multiply-founded	Reference
	Single	Multiple		
B. petiolata	100 (38)		53 (81)	Keeping and Crewe 1987
M. flavitarsus	45 (73)		2 (134)	Litte 1979
M. labiatus	80 (20)		22 (20)	Litte 1981, 1979
M. mexicanus	38 (76)	22 (41)	33 (171)	Litte 1977, 1979
P. annularis	80 (72)	20 (326)	86 (637)	Strassmann 1989a
P. biglumis bimaculatus	58 (24)		0 (24)	Lorenzi and Turillazzi 1986
P. canadensis	97 (143)	84 (393)	77 (532)	Pickering 1980, tabulated in Hughes 1987
P. chinensis	60 (421)		0 (421)	Matsuura 1977, cited in Miyano 1980
P. chinensis	50 (230)		0 (230)	Miyano 1980
P. dominulus	64 (125)		34 (185)	Nonacs and Reeve 1993 and pers. comm.
P. fuscatus §	93 (45)	83 (36)	44 (81)	Gibo 1978
P. fuscatus	40 (277)		37 (632)	Klahn 1981 and pers. comm. cited in Hughes 1987
P. fuscatus	47 (55)		52 (288)	Noonan 1979
P. jadwigae	53 (90)		0 (90)	Matsuura 1977, cited in Miyano 1980
P. nimpha	68 (37)		0 (37)	Turillazzi 1984
P. riparius (cited as *P. biglumis*)	71 (63)		0 (63)	Yamane and Kawamichi 1975
P. snelleni	84 (38)		1 (100)	Yamane 1969
R. fasciata	62 (109)	18 (129)	55 (251)	Itô 1993
R. marginata §	62 (24)	34 (41)	63 (65)	M. Shakarad and R. Gadgkar, pers. comm.

Values in parentheses are numbers of colonies studied.
* % failure data are reported only where the number of colonies followed exceeds 20.
Some survivorship values are approximations read off figures.
§ protected nests excluded from totals.

However, it is possible that workers work much harder, and take greater risks, than solitary foundresses. The very fact that workers have to worry less about colony failure may make them willing to take more risks if there is a higher payoff to be obtained by doing so. Though this would constitute an additional advantage of being in a colony, in the current context, it would mean that worker mortality rates might be misleadingly high. Therefore, I have also summarized data directly from single foundresses (Table 13.1). These data are not perfect either, as they are generally not true mortality rates. Instead, they are the probabilities that a colony initiated by a solitary foundress will fail before workers emerge. This is probably often a reasonable approximation for mortality, because in many species, the main cause of colony failure is death (or disappearance) of the foundress (e.g. Noonan 1979; Litte 1981). However, if a colony fails because of predation on the brood, the foundress may begin a new colony and eventually succeed at rearing workers. So the data in Table 13.1 will sometimes overestimate the probability that a foundress will fail. With that caveat in mind, the data are plotted on Fig. 13.3 (open circles). The B/C ratios average somewhat lower than those estimated using worker survivorship, but they are still substantial, with many still exceeding the maximum haplodiploid advantage. The lowest B/C ratio is 1.2, a 20% gain from working instead of nesting alone.

The B/C ratios plotted in Fig. 13.3 are inexact unless the study species meet the specific assumptions of the model. But the qualitative conclusion, that high adult mortality generates a significant advantage to grouping, seems quite robust.

The origin of eusociality

Even if the advantage of working is large for currently eusocial species, it does not necessarily follow that the advantage could have been important in the origin of eusociality. A worker that dies at an early age on a large colony has at least not wasted its investment. It may have already contributed to completing the prior investments of other workers and, if not, its own incomplete investments will be carried through to completion by the many surviving workers. However, at the origin of sociality, we must assume that there were few individuals, possibly only the foundress and one daughter. Can two individuals participate in the kind of synergistic interaction required?

It is useful to begin with a case where the advantage is not available. Consider the following scenario in which a foundress is joined by a single daughter. Prior to the joining, the foundress has made her reproductive decisions on the assumption that she would rear her offspring alone. She has not provided any extra work for a joiner. When the daughter joins, let's assume that the foundress, in addition to continuing her own investments as she would have, begins providing the daughter with a set of young to care for. The foundress lays extra eggs — the same number the daughter would have laid if she were nesting alone — and the daughter cares for them. Assume further that each individual cares exclusively for its distinct set of foundress young, even if the other individual dies (though the worker will begin laying her own eggs if the foundress dies). It should be evident that, in this somewhat artificial scenario, there is no demographic advantage to helping. The daughter worker simply replaces her own eggs with an equivalent set of her mother's eggs, and everything else proceeds as if the daughter were nesting alone. In particular, if the daughter dies before time t_0, she still wastes her entire investment. If this were the correct scenario, eusociality could not evolve under the kind of advantage described here.

There are two general ways around this conclusion. First, there might be ways in which a foundress sets the stage for the worker in a way that allows her to contribute before time t_0. Second, the death of one adult may open up ways for the other to reproduce more than it otherwise would have, if it has the option of caring for the dead adult's young. These kinds of advantages can come into play in at least six distinct ways. Mathematical models of these advantages are presented elsewhere (Queller 1994).

The first way is if brood-rearing conditions deteriorate so that the foundress can no longer raise the number of young she was originally anticipating (Queller 1989). She now has more young initiated than she can rear, and this opens up an opportunity for a helper to begin contributing immediately. Of course, this kind of helping would need to be facultative; it would not pay all potential workers to join, only those presented with indirect reproductive opportunities in the form of more eggs than the foundress can rear. Once a facultative worker response has evolved, foundresses may evolve to adjust their behaviour, laying more eggs than they can rear by themselves (see below). This would make the opportunity to help normal rather than occasional.

The second possible advantage is adoption (Nonacs 1991). If the foundress dies before the worker has reared young to adulthood, the worker can adopt the older orphaned young and rear them instead of her own younger set of foundress brood. In this way, she can rear some young to independence more rapidly and reduce her risk of failing completely. For example, suppose the foundress dies very soon after the daughter joins. In my original scenario, the daughter remains with her own set of young, which are still in the egg stage, and she must survive to time t_0 to rear any of them. But it is clear that it would be better to switch to the foundress' older orphaned young. More realistically, the two sets of young would probably be intermingled in the first place. The surviving adult, though unable to provide for the entire brood, would focus on rearing the oldest offspring, which have the greatest reproductive value.

The third kind of advantage is quite different. It applies if adult mortality is high because of risks taken. A joiner could take over the foundress' work, allowing the foundress to survive by being idle (except for laying eggs). The foundress would begin working again after the worker dies. Here, the colony does not invest at a rate greater than a single individual. Instead, the worker intercalates her lifetime of work into that of the foundress. This serial arrangement of lifetimes is another way to make it possible for two individuals to have only one event that leaves offspring orphaned (the second adult death) instead of the usual two. However, helping may not be the optimal solution for the potential joiner. Suppose she simply hangs around the parental nest, not working, but waiting for the queen to die. At that point she takes over rearing the queen's young and also begins laying her own eggs that will begin maturing t_0 time units later. The overall productivity of the nest is unchanged compared with the helping strategy. But by forcing the foundress to die first, the joiner will more often be able to rear her own young instead of the queen's young. This strategy should therefore be better as long as $r < \frac{1}{2}$. Therefore, the advantage of reducing the foundress's mortality may not be sufficient to select for helping, but it could be a factor that acts in concert with other advantages.

The remaining three advantages depend on some additional assumptions. I have implicitly assumed up until now that adults invest at a constant and presumably maximal rate. But in fact, the amount of investment required in an offspring may vary considerably with brood stage. In particular, it is likely that the provisioning stage is often the limiting stage, because foraging is particularly expensive in terms of time, energy, or risk of mortality. This expensive stage would come first for mass provisioners, but I will focus here on progressive provisioners like *Polistes* where the larval stage is likely to be limiting. The egg and pupal stages may be cheap but may nevertheless require some care, such as defence from ants or parasites. Under these circumstances, provisioning becomes a bottleneck that limits offspring production. A female wasp could lay more eggs, but she doesn't because she cannot feed all the larvae. She could also care for more pupae than she normally rears through the limiting larval stage. This opens up several opportunities for helping based on the inexpensive care needed for non-provisioning stages.

If the conditions outlined above hold, then once offspring reach the pupal stage, they may be protected against orphaning when one of the two adults dies. The remaining adult can rear the pupae at little or no additional cost beyond what it would do for its own set of pupae. Up until now, I have assumed that a surviving female cannot take over the investments of her deceased colony-mate without giving up investments of her own. Pupae may constitute an exception. This advantage may be larger for mass provisioners, because the inexpensive period following provisioning is longer than it is for progressive provisioners.

If very young brood (e.g. eggs) require very little investment a mother may lay extra eggs, either because of the chance that conditions may improve or to compensate for possible mortality of other young. Thus, if the parent often provides more larvae than she can rear, a worker could begin contributing to nest productivity sooner. This is similar to the first advantage discussed in that it depends on uncertainty about the offspring-rearing conditions, but it focuses on the special role of the inexpensive egg stage.

Finally, if the foundress does not provide extra eggs until after her daughter joins, then initially the number of larvae available for feeding could be handled by a single female. There would seem to be only enough work for one adult until the new eggs hatch. But even though the larvae initially available could be fed by one adult, they may still benefit from additional feedings. Such extra feeding could increase survival of the young (perhaps by speeding up growth) or it could provide them with extra resources with which to begin their reproductive lives.

One caveat should be added to the two advantages that depend on inexpensive eggs. I have assumed that adult mortality rates are fixed. But if they are flexible, with less mortality during the inexpensive egg period, then the cost of being a solitary foundress would be lowered during this period. This could act to counterbalance all or part of the advantages that workers can gain by helping during that period.

Evidently, there are a number of different ways for the extended-parental-care advantage to apply even at the origin of eusociality. Models for these ideas show that the advantages can be large, particularly when adult mortality is high (Queller

1994). If eusociality did evolve under any of these scenarios, then one interesting consequence is that the selective factor responsible for the evolution of eusociality may be the same as the reason for the ecological success of eusocial species (Queller 1994).

Foundress associations

As I noted earlier, it is in the context of foundress associations that demographic advantages have been most frequently mentioned. It is a common finding that associations of foundresses are more likely to survive to rear workers than are solitary foundresses. Some data culled from the literature are shown in Table 13.1.

However, it should be pointed out that enhanced colony survival does not automatically constitute a selective advantage for multiple-foundress colonies. With equal mortality rates, a colony of two is expected to survive longer than a colony of one, simply because it takes twice as many deaths to terminate the larger colony. But no advantage accrues if the amount of work that can be done by one individual is independent of its colony-mates. To see this, again imagine two individuals nesting together, but working entirely independently. Each tends its own set of brood and ignores the other set (which lays the eggs is immaterial for the argument). Such 'colonies' will last longer than single individuals, but the individuals involved cannot gain any advantage if they don't interact.

If we want to search for a demographic advantage from the increased survival of larger associations, we must address what happens when one of the colony members dies. If the dependent young it was caring for are orphaned and die, then there is no advantage. But if the surviving adult can somehow complete some of the unfinished investments of the dead one, without giving up equivalent investments of its own, then an advantage over single foundresses will have been gained. Some (but not all) of the kinds of mechanisms described in the previous section could come into play. The obvious candidate is the possibility that the survivor can rear extra pupae at little or no extra cost. It may also pay to adopt the older brood of the dead adult, even if this comes at the cost of some of the survivor's younger brood. Finally, an advantage might be gained if the subordinate's risky work permits the dominant to reduce its mortality rate.

Data from *Polistes fuscatus* suggests that at least one of these advantages does apply. Noonan (1979, p. 166) estimated the productivities of single-foundress and two-foundress colonies surviving to the worker stage. Colonies in which both foundresses survive to the worker stage are, not surprisingly, more productive than single-foundress colonies. But what of the two-foundress colonies in which one adult dies before workers emerge? These colonies have an intermediate productivity showing that the work of the dead foundress did not go for nothing. In one way or another, the surviving foundress is able to capitalize on some of the unfinished work of her deceased colony-mate.

If this kind of advantage is important in selecting for foundress associations, then it is possible that a single variable, adult mortality, might account for much of the considerable variation between species in association behaviour. Species with high mortality should show a higher percentage of multiple-foundress associations. The published data on actual mortality rates are too scant to allow much comparison, but it may be reasonable to substitute data on the percentage of single foundress colonies that fail to make it to the worker stage. If single-foundress failure is common, then a higher percentage of colonies should be founded by associations of several adults. Reeve (1991), using data from the literature on 12 species of *Polistes*, found a significantly positive correlation between these two variables.

This would appear to support the kind of demographic advantages I have been describing, but two cautions must be made. First, the result does not exclude certain other explanations. The correlation would also be expected if single-foundress colonies fail because of predation on the brood, and associations of adults help protect against predation. I will not attempt to sort out these two explanations here, other than to note again that it is often not true that multiple adults protect against predators and parasites (Gibo 1978; Gamboa 1978; Strassmann 1981a, c; Strassmann *et al.* 1988).

Instead, I will focus on another point; whether the correlation found by Reeve holds up with better data. Given the limited data available, Reeve was forced to make some compromises. However additional data are available in two forms. Some comes from unpublished theses on *Polistes*. I also broaden the sample base by including other polistine wasps, *Mischocyttarus*, *Ropalidia* and *Belonogaster*. By adding this data, I can be choosier about what data points to allow. My analysis makes the following restrictions. First, the two kinds of data point, survival to the worker stage and per cent of nests that are single, must be estimated from the same or very close populations. Both of these variables can vary considerably across populations, so it is important to make sure the two values reported for each population are really associated. Second, each data point (for both variables) is based on at least 20 colonies (which means 20 single-foundress colonies for estimating single-foundress success). This reduces the chance of erroneous conclusions based on sampling error. Third, I use only actual data on success from about the time of nest founding to very near the time of worker emergence. No extrapolations or interpolations are made from success data collected for shorter or longer periods of time.

The data are given in Table 13.1 and plotted in Fig. 13.4. The Pearson correlation coefficient of 0.40 just escapes significance ($P = 0.06$, one-tailed test). The ten *Polistes* species taken alone give a somewhat higher correlation, 0.52, but with the same P-value. The true significance level may be lower because I have treated each species as an independent point, when in fact some species may be similar for purely phylogenetic reasons. However, the trend is certainly consistent with the prediction.

Discussion

The models developed here, together with data from polistine wasps, suggest that the ability to provide extended parental care is a powerful force selecting for sociality in insects with high adult mortality. Only by grouping is it possible to ensure that such care is delivered. In fact, the model might seem to suggest that *any*

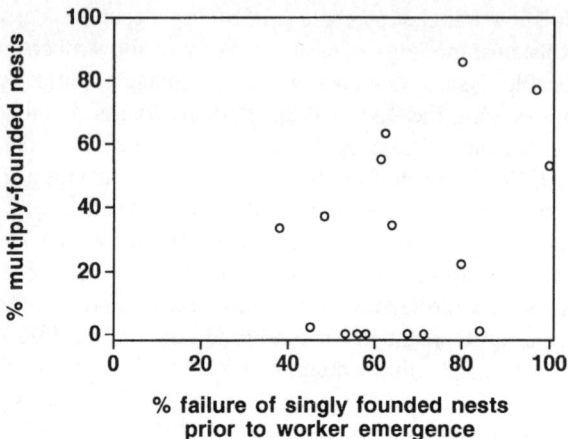

Fig. 13.4 The relationship between the probability of single-foundress failure and the percentage of multiply founded nests. Each point represents a species of *Belonogaster*, *Mischocyttarus*, *Polistes*, or *Ropalidia*. The criteria for including studies are described in the text. When several studies of the same species met the inclusion criteria, the data are combined (not doing so lowers the correlation somewhat).

adult mortality is sufficient to select for grouping (see Fig. 13.3). Since all species have some adult mortality, this would seem to raise the question of why isn't everything eusocial. But recall that the model *assumes* an advantage to extended care. In fact it assumes a very severe penalty for not providing extended care to time t_0: death of the offspring. Of course the advantage of providing extended care does not need to be this strong for the general idea to work, but it is important to remember that there does need to be an advantage. The optimal length of parental care is a complicated life history problem that I will not attempt to solve here. The model developed here simply shows grouping may make extended parental care feasible when it would be advantageous to have it.

Another reason why eusociality is not more common may be that other factors often counteract the advantage available through extended parental care. The argument assumed that other things are equal and this cannot usually be true. Relatedness between the potential worker and the progeny it would help is often below $1/2$, so selection for helping requires a B/C ratio correspondingly greater than 1 (see equation 2). And even if relatedness is high, it may often be true that other factors work keep the overall B/C ratio low, in spite of the potential demographic advantage. Group members may inadvertently interfere with each other's efforts, or they may compete with each other in ways that reduce efficiency.

An issue similar to the one raised above arises in the context of the life cycle of individual eusocial species. If being in a colony is so advantageous, why does anyone ever leave to nest alone? Again reduced relatedness and reduced efficiency on large nest may play a role. The advantage of dispersal (Hamilton and May 1977) must also be involved. But one other factor should be mentioned because is related to the demographic nature of the model. Temperate species, and also many tropical species, experience seasons that are unsuitable for rearing young. In temperate *Polistes* at least, independent founding comes at the end of this period. There is no intrinsic reason why this should always be so. In principle, new colonies could be founded at any time. And, in principle, colony identity could be retained over the winter. Colonies could stay together or re-aggregate in the spring and continue as if nothing had happened. Why is this time chosen for independent colony formation? I suggest that it is because individuals must start again at the egg stage. This eliminates part of the demographic advantage normally enjoyed by a worker: being able immediately to aid larvae and pupae begun by other individuals. As noted in the section on foundresses above, there are potential demographic advantages available when offspring production must be started anew, but they are more limited.

Nonacs (1991) has argued that the demographic advantages must be much smaller and can only account for the presence of one or a few helpers. This stems essentially from his assumption that, while helpers may insure against colony failure, they do not otherwise add to the productivity of the colony. For example, Nonacs would assume that surviving two-wasp colonies are no more productive than surviving one-wasp colonies. I assume in contrast that the two-wasp colonies can be twice as productive (though perhaps less if one wasp dies before time t_0). I make this assumption for reasons that are both empirical and analytical. Michener's (1964) often-cited conclusion that social insects show no *per capita* increase in productivity should not be taken to mean that there is no gain whatsoever. Many *Polistes* studies show clear evidence that more adults produce more brood (for example, see Noonan (1979) discussed above, and Strassmann (1989a)). But even if this were not so, our goal is to identify, and ideally quantify, the various factors selecting for and against sociality. It is a reasonable starting point to assume that two wasps *can* put in twice as much work as one wasp. If we find that they do not, or if they do work twice as much but still get less than the expected two-fold benefit, then this is a special finding requiring explanation. What are the competitive factors or the diseconomies of scale that cause this? These are separate factors that tend to select against eusociality, but their presence would not imply that the demographic advantages are absent. My goal in this paper has been to quantify the demographic advantage in the absence of other advantages and disadvantages.

The models require some additional development in several respects. First, some of the mathematical assumptions need generalizing to account for some important biological possibilities. One example has already been mentioned. The models assume complete dependence of offspring to time t_0. But modelling incomplete dependence is important both because it is likely to be more true in some cases, and because one implication of the general theory is that grouping may be what allows care to be extended for longer periods. One possibility is that total dependence sometimes evolves after, and because of, the protection afforded by eusocial systems.

Another useful modification would be to treat adult mortality as a strategic variable rather than a fixed constant. Often mortality will be a function of the risks taken, and individuals can to some degree choose their level of risk and with it, their level of reward. For example, increasing the time spent foraging may increase the risk of mortality and also increase the reproductive payoff, provided the individual survives. Some optimal level of effort should be selected, but the optimum may differ by group size. In particular, solitary individuals may need to be very conservative, because the costs of failure are so high. Individuals in groups may be able to afford greater risks because, if they die, not all of their previous work will be lost. The generally higher mortality of workers than solitary foundresses (see Fig. 13.3) suggests that workers do take greater risks. A model that assumes equal mortality rates may be correct for the origin of eusociality because conditional strategies of the appropriate type (work harder when you are a worker) would not yet have evolved. But the assumption may fail in currently eusocial species and the data should be interpreted with caution. Using worker mortality rates may overestimate the worker advantage if solitary rates are actually lower. However, using solitary foundress survival rates may underestimate the worker advantage when higher mortality is effectively traded for higher productivity.

Part of the appeal of Hamilton's haplodiploid theory is that it accounted for the multiple origins of eusociality in the Hymenoptera. Others have argued that the high frequency of subsocial behaviour in the Hymenoptera is the more important precondition (Andersson 1984; Alexander *et al.* 1991), an explanation that is consistent with demographic advantage proposed here. However, a major task awaiting the theory is to see if it can provide a more detailed account for taxonomic patterns of eusociality. It will be necessary to know much more about mortality rates in both eusocial and non-eusocial groups. One generalization may be that this advantage is not very strong in vertebrates. For example, in birds the probability of adult mortality during the time it takes to raise a clutch of eggs must be much smaller than it is in polistine wasps, and helping in birds probably has other causes. Perhaps this difference underlies the generally lower degree of social evolution in vertebrates compared with insects, which have a large and persistent demographic advantage.

Acknowledgements

I thank J. E. Strassmann, P. Nonacs, H. K. Reeve, M. Shakarad and R. Gadagkar for permission to use unpublished data. F. Kjellberg and P. Nonacs provided helpful comments on the manuscript. Supported by the John Simon Guggenheim Foundation and NSF grants BSR-8805915 and IBN-9210051.

14

Polistes in perspective: comparative social biology and evolution in *Belonogaster* and Stenogastrinae

Stefano Turillazzi

The previous chapters have considered and analyzed the decisive factors in the success of *Polistes* as a model organism for the study of social behaviour. But the biology of a group and its scientific importance can be understood completely only if it is studied in an evolutive context and its characteristics are analyzed through homologies and analogies with other groups. Independent-founding polistine species living in the tropical Old World, recently included by Carpenter (1993) in the tribe Ropalidiini (*Belonogaster*, independent-founding *Ropalidia*, *Parapolybia*), have—like *Polistes*—colonies of small size and nests without envelopes (Gadagkar 1991c). In the neotropics the genus *Mischocyttarus* is quite similar to *Polistes*, with which it competes in abundance and number of species (Gadagkar 1991c; Wenzel 1991). These wasps, however, together with species of tropical *Polistes*, are not 'sympatric' with most of the scientists interested in this kind of research. Until some decades ago this fact greatly limited studies on their biology and social behaviour. At present, only a few species of independent founding *Ropalidia* (see Gadagkar 1991c, this volume) are actively studied at the level at which *Polistes* are studied, whereas some species of *Mischocyttarus* have been the objects of thorough studies by North and South American researchers since the publication of Jeanne's monograph on the social biology of *M. drewseni* from Brazil (Jeanne 1972; Gadagkar 1991c).

Some of the features of *Ropalidia* are discussed by Gadagkar in this volume, and in this chapter I will examine a genus, *Belonogaster*, which has particular importance for the study of the evolution of social behaviour in wasps. Some traits of its biology and natural history make it very similar to *Polistes*, but other characteristics, especially an incipient morphological caste differentiation, are similar to those of more evolved social forms.

As Gadagkar (this volume) observes, the evolution of eusociality has two main phases: the origin of altruism and the origin of the traits that determine complete and irreversible reproductive differentiation of individuals. I maintain that the first problem cannot be fully comprehended through the study of genera, like *Polistes* or *Ropalidia*, which present species with very simple organization but with already fully eusocial characteristics. As we saw, *Polistes* represent good study subjects for determining the important features of primitively eusocial forms, for valuing the

costs and benefits of associative and solitary nest-foundations and those of subordinate sterile worker behaviour, and for understanding the importance of relatedness
in the evolution of sociality. But to understand the origin of altruism it would be
ideal to study an homogeneous group in which we can experimentally compare the
benefits and disadvantages of solitary and group life alternatively present in the
same population. For this reason, in the second part of this chapter I will examine a
very special group of wasps, those belonging to the subfamily Stenogastrinae
(hover-wasps). These wasps present colonies with some characteristics similar to
those of *Polistes* (small size colonies and nests generally without envelopes), but
have a simpler social organization.

   Belonogaster and hover-wasps are also particularly appropriate for this volume,
as they were the only social wasps, 'beyond the *Polistes*' to be studied by Pardi.

Belonogaster

The genus *Belonogaster* contains 69 species (according to a recent review by
Richards 1982), which are mainly distributed in Africa south of Sahara with two
species (*B. juncea* and *B. indica*) extending also to western India. In the following pages I will discuss some of the main characteristics of the social biology of
these wasps (see also Gadagkar 1991c), paying particular attention to caste
differentiation.

   Belonogaster were among the first social wasps whose behaviour was carefully
studied: Roubaud (1910, 1916), observing *B. juncea*, described for the first time
trophallaxis (ectotrophobiosis) in social insects—a phenomenon which, as Marino
Piccioli and Pardi (1970) observed, set off a series of important speculations and
studies (see Wheeler 1923, 1928). Research on species in this genus, though briefly
taken up by Bequaert (1918), was for all practical purposes interrupted until Iwata's
(1966) short note on a *B. grisea* nest from Abyssinia, and two papers in which
Richards furnished sparse but interesting notes on *B. juncea* from western Africa
(Richards and Richards 1951; Richards 1969).

   Pardi and Marino Piccioli's attention was at first focused on social behaviour,
and they demonstrated that the phenomenon of dominance hierachies (Pardi 1942),
already found in other species of *Polistes* (Yoshikawa 1956, in *P. jadwigae*;
Morimoto 1961, in *P. chinensis*), was also present in colonies of these beautiful
African wasps, although with a much lower frequency of interactions than observed
in *Polistes* (Marino Piccioli and Pardi 1970). Dominance hierarchies are established
in the associative foundations of *B. petiolata*—a South African species which
presents a marked seasonal cycle and a hibernation period for fertilized females
comparable in length (four and a half months) to that of temperate *Polistes*—in the
same way as in *Polistes* (Keeping 1989, 1992), and are also present in the
intertropical species *B. juncea* (Tindo 1991). Even though females have the capacity to distinguish kin from non-kin, foundress association between sisters is fairly
rare in *B. petiolata* (Keeping 1990a). In *B. juncea*, Tindo (1991) reports a case in
which two females, probably unrelated, associated to found a nest. In this last species we have recently demonstrated, in a population kept in captivity (Francescato

et al. in prep.), the existence of marked recognition among nestmates that associate for the foundation of new nests, with about 90% of the foundations (31 out of 34) involving females from the same colony (90 out of 104 associating females). In nest exchange experiments in the field, females were able to distinguish their own nest from an alien one (Francescato *et al.* 1994).

Pardi also saw in these wasps a perfect subject for studying the attributes which characterize egg-laying females compared with those of sterile females, and for examining whether pre-imaginal differences explained the roles played by the various females in the colony.

Pardi and Marino Piccioli's research brought about the rejection of the hypothesis advanced by Roubaud (1916) of societies without caste differentiation. This research confirmed Richards' (Richards and Richards 1951; Richards 1969) finding, in both *B. grisea* and *B. juncea*, of size differences between females with developed and undeveloped ovaries (Pardi and Marino Piccioli 1970). After that, the behavioural and size differences among fertilized and non-fertilized females were described and analyzed in detail. Pardi (1977) came to suggest that statistical differences in the measurements of some morphological traits were indicative of the presence of two incipient castes, and that the differences in fertility and behaviour could be a consequence of pre-imaginal inequality. He concluded, however, that the problem remained unresolved. Some years later, his last work on *Belonogaster* brought further contributions to the solution of the problem of caste differentiation in these wasps. In fact, Pardi and Marino Piccioli (1981) found that queen-like *B. grisea* females show statistically significant differences in thorax shape—broader with respect to the more elongated thorax of the worker-like females. This finding led the authors to conclude that *Belonogaster* cannot be considered representative of a very primitive form of social organization, nor as a transitional stage between solitary and social life. They decisively state that *B. grisea* 'has reached the formation of a sterile caste' and that since 'some of these [females] do associate in foundations this represents a transitional stage between independent foundations and true swarming.' However, the frequency of this last phenomenon needs to be examined in considerably more detail before any firm conclusion can be drawn.

In a recent and yet unpublished study on *B. juncea* from Cameroon, E. Francescato *et al.* (in prep.) showed significant differences in head width and ovarian development between fertilized (larger) and unfertilized females (smaller). A morphometric study of the same cage population of *B. juncea*, from five mother colonies reared for several months, has provided evidence of a significant difference of scutum shape between females with queen-like behaviour (QL) and females with worker-like behaviour (WL), despite the wide range of variation found in the two samples (Francescato *et al.* in prep.). Contrary to what Pardi and Marino Piccioli had observed in fertilized and non-fertilized females from a large *B. grisea* nest, in *B. juncea* it appears that the QL females are characterized by a more elongated scutum than that of the WL females (ratio Length/Width = 1.247 (n = 14) in the QL and 1.193 (n = 38) in the WL; $P<0.02$; t test). Because of the large variances in the samples, no significant dimensional differences were found between the two groups of females. These data seem to confirm Pardi's impression about an incipient pre-imaginal caste differentiation in species of this genus.

Keeping (1989, 1992) has demonstrated that colonies of *B. petiolata* show a very definite and permanent reproductive division of labour between foundress females and worker daughters, and that the latter emerge, at a well-defined time, before the reproductive individuals. He also found that the size of the emerged daughters increases towards the end of the colonial cycle, and, despite an overlap between workers and gynes, the latter are significantly larger and have more developed fat bodies. He concludes that in *B. petiolata*, castes appear clearly differentiated both at behavioural and reproductive-physiological levels, but their differentiation is mainly imaginal. Pre-imaginal factors are probably important because they influence the size and fat content of an adult through its larval nutrition (Keeping 1989). The problem of caste differentiation is thus of great importance in the study of the biology of *Belonogaster*, and its solution will require further research.

The temperate *B. petiolata* has been found to be a long-term monogynous species (Keeping 1989, 1992), as has the tropical *B. juncea* (Tindo 1991). In the latter species, Tindo observed an average colony life of 28.18 ± 3.61 weeks (7–8 months): colony duration is longer in colonies with larger foundress groups. The maximum colony size, however, does not exceed a total of 29 individuals.

Observations of the social behaviour of *B. grisea* also led to the discovery of two highly characteristic behavioural patterns. A certain rubbing of the abdomen that females perform on the nest peduncle, christened 'giravolta', awoke the interest of Marino Piccioli and Pardi (1970). In observations of various nests, they found that this behaviour was performed primarily by fertilized, egg-laying females belonging to the highest levels of the hierarchy. The behaviour was more frequent when the nest was in some way disturbed. The authors concluded that the manoeuvre, though not easy to interpret, might serve to distribute some kind of odorous substance on the surface of the nest (Marino Piccioli and Pardi 1970). At the same time, Bob Jeanne (1970) was publishing in *Science* a paper on a similar behaviour observed in *Mischocyttarus drewseni*, in which he demonstrated the ant-repellent function of a secretion from sternal glands applied on the nest peduncle. These findings opened a completely new line of research on wasps—the study of possible applications of pheromones and allomones to the surface of the nest. Keeping (1990*b*) has recently confirmed, in *B. petiolata*, the ant-repellent function of the secretions from the glands associated with the 6th gastral sternite. The rubbing manoeuvre has also been observed in *B. juncea* by Tindo (1991). This author reports that 43 out of 175 colonies of this species, followed in the field for 11 months, were destroyed by ants of the genus *Crematogaster*, which confirms the great predatory pressure exerted by ants.

Besides ants, enemies of these wasps include two species of parasitoids (tachinid flies) which have been studied by Keeping and Crewe (1983) in *B. petiolata*. In their interesting project the South African authors used these wasps to test the hypothesis that parasitism could represent a factor influencing colony size in social wasps, but find no consistent evidence to support this.

Another important behaviour observed in *Belonogaster* was that of the extraction of the larval peritrophic sac from the bottoms of the operculated cells, described by

Marino Piccioli (1968). This behaviour had already been described by van der Vecht (1962) in *Ropalidia* and is now also known to occur in *Parapolybia* (Vecht 1966) and *Polybioides* (Richards 1969); Evans and West-Eberhard (1970) and Jeanne (1980) indeed noted it as a distinctive characteristic of Old-World tropical Polistinae, as opposed to neotropical Polistinae (tribes Ropalidiini and Epiponini, respectively, (Carpenter 1993)).

The peculiar nests constructed by the species of this genus had already been briefly described (*cf.* du Buysson 1909, and Iwata 1966), but it is only in the work of Marino Piccioli and Pardi (1978) on *B. grisea* that we first find, after the important evolutionary review by Jeanne (1975), an attempt to codify the parametres of general measurements (such as divergence of the cell axes and displacement of the cells) of the nest architecture of the vespids. The Italian authors concluded that the *Belonogaster* nest undoubtedly represents an example of primitive architecture, and that its curved shape, almost unique among social wasps (but see the picture of the nest of *Polistes shirakii* reported by Starr 1992), could be attributed to an accentuated divergence of the cell axes (10%) and to a complete absence of cell displacement. This would lead to some corrective devices in the construction, such as not re-using and partially destroying the cells, to adapt the shape of the nest to the necessities of raising the immature offspring. Keeping (1991), concluding his description of the *B. petiolata* nest, also notes how the general constancy of the architectural design of the *Belonogaster* nest derives from the mechanical limits shown by Marino Piccioli and Pardi (1978). He points out that, according to du Buysson (1909), there exist only two exceptions to the *Belonogaster* architectural model: the *B. brevipetiolata* nest (very long and rectilinear), and that of *B. hildebranti* (almost sessile and with minimal cell-axis divergence). He observes as well that the measurement and comparison of parametres similar to those suggested by Marino Piccioli and Pardi can serve as a partial test of the importance of mechanical factors in the limitation of nest shape diversification.

A fundamental aspect of the natural history of a group is its mating system: if reproductive behaviour (as the scarce data produced by students of social insects suggest) is relatively tied to social behaviour, this is extremely important for understanding factors that influence group relatedness and the genetic structure of local populations (see Ross and Carpenter 1991). The mating system of these wasps is still largely unknown. While Keeping *et al.* (1986) find no special sexual behaviour in male *B. petiolata*, which usually mate on the nest, we have observed patrolling and gaster dragging in *B. grisea* kept in captivity (unpublished observation by Colombini and Turillazzi) and in male *B. juncea* both in captivity and in the field (D'Agostino and Francescato, in prep.). Males of the latter species follow aerial routes, which may include various trees or bushes, perching on defined leaves or other objects where they drag their gastral sternites. The inner surfaces of the 5th, 6th and 7th sternites are covered with a dense layer of unicellular glands. During their flights the males of this species sometimes make contact and try to copulate with dead females pinned on perches. This is very similar to what happens in most *Polistes* species studied at present (Beani this volume). Pardi (1977) also reports interesting observations of male behaviour on the nest in *B. grisea*. In particular he repeatedly observed in various colonies (both in the field and in captivity) an unusual contribution of males to the feeding of larvae.

While Keeping (1989, 1992) notes that at least those species of *Belonogaster* so far studied are eusocial and correspond to West-Eberhard's (1978*b*) highly social stage (IV), the characteristics of the genus' social biology and its phylogenetic proximity to swarm-founding *Polybioides* have recently induced Carpenter (1991) to declare that *Belonogaster*, being fully eusocial, may have species which represent the passage from short-term monogyny to swarm-founding polygyny, thus confirming the conclusions of Pardi and Marino Piccioli (1981).

Stenogastrinae

The stenogastrine wasps include about 50 described species divided into six genera distributed from South India to New Guinea. The reasons for which Pardi, in 1979, turned to the study of these wasps, involving me as well, were suggested in a famous article by Mary Jane West-Eberhard published the year before (West-Eberhard 1978*b*), in which the hover-wasps were described as 'a promising source of comparative data regarding the evolution of group living and castes in wasps'. As I observed in the introduction, the advantage that a group like this has over *Polistes* is that it furnishes more suitable material for studying the origin of sociality. At the same time it provides an evolutionary perspective on their social relatives (Polistinae + Vespinae) and on *Polistes* themselves.

In retrospect, our choice to begin to study Stenogastrinae in Java (Indonesia) was not particularly felicitous, since we found only two species of the genus *Parischnogaster* to study. One of these was *P. mellyi*, studied by Hansell just two years previously in Thailand, but on which only a brief note had been published (Hansell 1977). The other species, on which we concentrated our attention, was *P. nigricans serrei*, which constructs cells on threadlike substrates and defends them with formations of gelatinous secretions called ant-guards.

These slender, elegant little wasps, with colonies of such limited size, had something that might recall the societies of *Polistes*. Yoshikawa *et al.* (1969) had in fact described, in one species of *Parischnogaster* with long nests from Malaysia, the existence of a linear dominance hierarchy, correlated with the ovarian development of the various individuals, similar to that described in many *Polistes*. In the species that Pardi and I studied, interactions among the various females also led to the definition of a kind of linear hierarchy. The interactions between the females, however, were less marked and frequent than those found in *Polistes*: even the outcomes of encounters between individuals of the highest hierarchical levels were barely discernible. But the similarities to *Polistes* were evident. Though it was not possible to determine differences in egg-laying rate for the various females because of the reduced number of eggs produced, their ovarian development was correlated to hierarchical rank, and when an alpha female disappeared from a colony she was immediately replaced by the female that followed her in the hierarchy (Turillazzi and Pardi 1982).

The colonial dynamics of *P. nigricans serrei* became clearer some years later, when I had the opportunity to spend a longer time researching in the same area. The social organization (Turillazzi 1985*a*) is based essentially on long-term monogyny, but caste differentiation, exclusively behavioural, is regulated by the possibility that females have of adopting various strategies (Yamane *et al.* 1983*a*, report similar observations for *P. mellyi*). Callows waiting for full ovarian development can enhance their indirect fitness by helping their mother rear siblings. Then, a constant presence of helper females is assured as newly emerged females take over the role of foragers, while the older ones leave. This 'temporary helper' hypothesis could be tested by future researches against an alternative 'programmed delay' hypothesis, simply by measuring the callow period in species at a more or less high social level.

In *P. nigricans serrei* the sterile worker role is, in general, only temporary, but females with large ovaries mate sooner than females with smaller ovaries. This could be seen as a factor contributing to the evolution of a permanent sterile caste. A strong correlation exists between the number of females with developed ovaries and the number of females present in a colony (Turillazzi 1985*a*). In advanced colonies the potential egg-laying females reach a percentage of 43% of the total number of females, but in other species of Stenogastrinae this percentage is notably higher, up to a maximum of 51% in *P. alternata* (Turillazzi 1986, 1987) (see also Table 14.1).

Contemporaneously with our expedition to Indonesia, Mike Hansell and Charlotte Samuel were studying, in Malaysia, a species of another genus: *Liostenogaster flavolineata* (Hansell *et al.* 1982). In one of the most complete long-term studies ever carried out on social wasps, whose data unfortunately are available only in Charlotte Samuel's unpublished doctoral thesis (Samuel 1987), the principal characteristics of the colonies of this plainly eusocial stenogastrine wasp were photographed, showing:

(1) colonies of rather reduced dimensions;
(2) presence of pyramidal dominance hierarchies with the top occupied by an egg-laying female that is able to maintain leadership of the colony for a relatively short period (relatively also to the length of larval development which is unusually long) before being substituted by a daughter or nestmate;
(3) high frequency of usurpation of nests;
(4) presence of both haplometrotic and associative nest-foundation;
(5) absence of morphological differences between the various females;
(6) possibility of diverse behavioural options;
(7) presence of some females whose ovarian development is extremely slow and limited.

It is curious that *P. nigricans serrei* (and also *P. mellyi* according to a cage-study by Coster-Longman *et al.*, in prep.) presents long-term monogyny while *L. flavolineata* has, as a rule, short-term monogyny—this depends mainly on the duration of the immature life, which is around 45 days in the first two species (Hansell 1982; Turillazzi 1985*b*), and averages 107 days in the last (Samuel 1987).

L. flavolineata forms sizeable agglomerations of colonies under the arches of

Table 14.1 Characteristics of the colonies of some species of Stenogastrinae.

Species	Max. cells	Max. females	Max. males	Fem./ nest (n)	% PEL	Sex ratio	Ref.
P. nigricans	42	11	(26)	3 (10)	38	0.76	1
P. mellyi	35	5	4	1.28 (31)		0.66	2
P. alternata	47	13	16	3.27 (44)	51	0.33	3
P. striatula	20	6	–	2.7 (16)			4
P. jacobsoni	46	6	(18)	2.7 (20)	36	0.70	5
P. gracilipes	54	8	2	2.42 (10)			6
L. flavolineata	89	7	4	2.38 (49)		0.41	7
L. vechti	31	7	4	3.7 (34)	41.4	0.31	8
E. eximia	25	2	2	1.06 (39)		0.42	9
E. calyptodoma	13	4	1	1.8 (26)		0.49	10
E. fraterna	19	6	5	2.85 (20)	28	0.44	11
S. concinna	17	2	1	1.22 (9)		0.44	12
A. iridipennis	13	2	1	1.46 (13)			13
Anischnog. sp. A	20	2	–	1.2 (5)			14
A. laticeps	8	2	–	1.08 (22)			15
M. drewseni	12	2	2	1.5 (9)		0.42	16

References: 1) Turillazzi 1985a; 2) Hansell 1981; 3) Ohgushi et al. 1985, Turillazzi 1986; 4) Yoshikawa et al. 1969; 5) Turillazzi 1988; 6) Hansell 1986; 7) Hansell et al. 1982; 8) Turillazzi 1990a;
9) Krombein, pers. comm.; 10) Hansell 1987a; 11) Gerace and Turillazzi 1992; 12) Spradbery 1975;
13) Spradbery 1989; 14), 15) Turillazzi and Hansell 1991; 16) Turillazzi, pers. obs..

bridges, caves, or water conduits, as does *P. alternata*. In these two species associative nest foundation has also been found. In both species, however, the association has characteristics different from those shown in the pleometrotic associations of most *Polistes* and of other independent-founding Polistinae. First of all, the interactions between associated females are, as a rule, extremely mild: in 30 days of observation of seven associations of *P. alternata* I was able to show (Turillazzi 1985c) no net hierarchy among the associated females, a difference of no more than 24% in ovarian development, and an appreciable but not striking division of labour. Recently Coster-Longmann (1994), in a several-month-long study, has found dominance hierarchy in an associative nest founded in captivity. In *L. flavolineata*, Samuel (1987) finds that the greater part of the associated females do not come from the same nest and, most likely, have a very low level of kinship (this has been confirmed by a recent genetic study by Strassmann et al. 1994).

L. vechti is another species that forms dense aggregations of colonies (up to more than 250 per m^2) (Sakagami 1987; Turillazzi 1990a). Its nests are constructed with vegetable material and present a characteristic ring or horseshoe shape. In two populations of this species, situated in two different areas (one 610 m a.s.l. and the other about 1000 m a.s.l.), the mean number of females per nest was similar (means respectively of 3.86 (SD = 1.61) and 3.58 (SD = 1.31)), but while in the second population the majority of the colonies (n = 12) had only one female on each nest with ovaries of laying capacity, in the other population (n = 22) there were usually two or more (with significantly different regression coefficients) (Turillazzi 1990*a*). This leads one to think that colony leaders in some populations are better than those of others, at keeping the reproductive capacity of nestmates at a low level. Social organization in these wasps could differ in different populations. The factors which determine these differences should be clarified by future research.

Eustenogaster is the genus which presents species of the greatest size, which make distinctive nests with an envelope shaped like an upside-down flask. Notes on the biology of *E. eximia* from Sri Lanka have been furnished by Krombein (1976, 1991), who found that most of the colonies he collected contained only a single female. These data agreed with the hypothesis that the genus had some solitary species. Hansell, however, after a behavioural study of *E. calyptodoma*, concluded that this species presents eusocial characteristics despite the fact that the colonies rarely number more than two individuals (Hansell 1987*b*). Recently, studying *E. fraterna* from Malaysia, Gerace and I have shown that the number of individuals present in the colonies of this species can reach that of species of *Parischnogaster* and *Liostenogaster* (up to 11 individuals—six females and five males) (Gerace and Turillazzi 1992). The potential egg-laying females are characterized by elevated ovarian development and size larger than the mean, but, at present, only the fact that the majority of this species' colonies have only one egg-laying female is any evidence that this female exerts control over the younger females.

Stenogaster continues to be the least-studied genus of the hover-wasps, although one of its species, *S. concinna*, was the first to be a subject of fairly in-depth study. Spradbery (1975) notes that colonies of this wasp never number more than two individuals and, in particular, that they are made up of one foundress and one of her daughters, which remains with the mother for some time before abandoning the nest.

About *Metischnogaster drewseni*, formerly subject only of a brief note by Pagden (1962) (together with *M. cilipennis*), we now have somewhat more information. In this species, characterized by highly cryptic, thread-like nests, very limited in size precisely by their particular architecture, the colonies are also limited to a maximum of two females (sometimes with a male). Nevertheless, there is an evident division of labour and a generational overlap between the females (Turillazzi 1990*b*). *Metischnogaster*, as well as *Eustenogaster*, demonstrates how the level of sociality of this group can be limited by the particular cryptic architecture of the nest, which is a response to the selective pressure of predation and, as Hansell (1987*a*) suggests, by the poor quality of building materials.

Notes on *A. iridipennis* are reported by Spradbery (1989), who collected 16 nests (maximum 18 cells), only three of which had two females. The three species of *Anischnogaster* that Mike Hansell and I studied in Papua New Guinea (Turillazzi and Hansell 1991) present the same characteristics as the last genera discussed above. The normal colony size in the three species is one female per nest, and never goes beyond two females per nest. Social interactions were directly observed in only one species (*Anischnogaster* sp. A), and did not include any marked dominance behaviour, despite a clear differentiation of roles, with the older females staying on and defending the nest while the younger females spent a large amount of time away from the nest, foraging. The social situation of *A. laticeps* is still unclear, since of 27 nests collected 25 were inhabited by a single female and in the other two we found a fertilized female with developed ovaries and a recently emerged female. *A. laticeps* females also present a clear dimorphism in the length of the stinger, which suggests an attractive hypothesis (yet to be tested) about the specialization of a part of the population for the intraspecific usurpation of nests (Turillazzi and Hansell 1991). *Anischnogaster* also seems to make little use, in the raising of offspring, of the Dufour's gland secretion, an emulsion of hydrocarbons that is a unique and characteristic trait of Stenogastrinae. In genera other than *Anischnogaster* this secretion has various functions—from the support and protection of larvae to the principal constituent of nest-protection barriers (Turillazzi and Pardi 1981; Turillazzi 1985*d*, 1989, 1994; Keegans *et al*. 1992*a*, 1993). This fact corroborates my impression that this substance has been of importance in the evolution of sociality in the group, favouring group living (by allowing the possibility of storing, on eggs and larvae, a liquid food which can also be used by the adults, and transforming the nest in a nutritional source very attractive for adults) but also limiting it, owing to the great amount of substance necessary to rear a large number of individuals (Turillazzi 1989).

Table 14.1 reports some data on the species studied so far from all six genera of the subfamily, providing a summary of major points already covered. The average number of females per nest ranges from a minimum of 1.06 (*E. eximia*) and 1.08 (*A. laticeps*) to a maximum of 3.7 (*L. vechti*). Well above the average number of two females we find *E. fraterna* and all the species of *Parischnogaster* and *Liostenogaster* studied so far (except *P. mellyi*). Nest size (maximum cell number reported for the species) is roughly correlated with the number of females per nest. The sex ratio is female-biased in all the species, and eight out of 11 range from 0.31 to 0.49. Three species of *Parischnogaster* present the highest sex ratio while the species that nest in clusters (*L. vechti*, *L. flavolineata* and *P. alternata*) present the lowest value; for this reason this might be an interesting group to investigate the presence of a local mate competition effect (Hamilton 1967).

Male stenogastrine wasps take very little part in colony life but in *P. mellyi* (Hansell 1982; Turillazzi and Francescato 1989) and in *Parischnogaster nigricans serrei* (Turillazzi 1985*a*) they can influence the colony cycle, interfering with the rearing of the brood and eating eggs. In the second species, groups of males cruising from various nests can sometimes cause the collapse of the entire social organization of a colony, whereas colonies which are not 'invaded' by males persist longer (Turillazzi, unpublished observations). In some species (such as *P. mellyi* (Turillazzi and Pardi 1982), *P. gracilipes* (Hansell 1986), *Anischnogaster laticeps* (Turillazzi and Hansell 1991) and *A. iridipennis* (Spradbery 1989)), males abandon (or are chased from) the colonies and form unisexual clusters gathering on thread-like substrata. In *P. mellyi* these clusters seem to be used by the females as indicators of good substrata for nest foundation (Turillazzi and Francescato 1989). Males of many species perform patrolling flights which are part of their mating system. *Metischnogaster cilipennis*, *M. drewseni*, and various species of *Parischnogaster* (Pagden 1962; Turillazzi 1983a, b) fly in forest clearings at well-determined hours of the day, showing every now and then the white stripes on their gastral tergites. Mating occurs during these flights and the abdominal-stripe display, at least in *P. mellyi*, has the function of indicating individual strength to other competitors and females (Beani and Turillazzi 1994). In *Liostenogaster*, males stop on perches to drag their gastral tergites on the substratum (Turillazzi and Francescato 1990). Whereas in the Polistinae males have conspicuous clusters or layers of tegumental glands associated with gastral sternites (as we saw for *Polistes* (Turillazzi 1979, Jeanne this volume) or *Mischocyttarus* (Post and Jeanne 1982b)), in the hover-wasps this male apparatus occurs in association with gastral tergites, but the position and structure of the secretory organs can be highly variable even between species of the same genus (Turillazzi and Calloni 1983; Turillazzi and Francescato 1989, 1990)

Carpenter (1989a, 1991), using cladistic methods, recently analyzed and confirmed in its essential lines the evolutionary model of 'polygynous family' proposed by West-Eberhard (1978b) for explaining the origin and evolution of sociality in wasps. Carpenter observes that the Stenogastrinae present characteristics which place them in the III (primitive caste-containing) and IV (highly social) stages of the model. In fact Samuel (1987) finds that in *L. flavolineata* some uninseminated females 'experience permanently retarded ovarian development...' and 'function as workers on multi-females colonies.' This species is thus to be considered in the highly social (or eusocial) stage of West-Eberhard (1978b). Some other species could be probably placed at this stage but direct evidence is still missing owing to the paucity of long-term studies conducted to date. Even if permanently sterile females do not occur in *P. mellyi* (Hansell 1983) and in *P. nigricans serrei*, in the second species the reproductive potential varies among females of the same age (Turillazzi 1985a). According to the low percentage of potential egg-laying females in colonies of *L. vechti*, it is possible that in this species also there exists a permanently sterile caste (Turillazzi 1990a).

Carpenter (1991) maintains that the II stage of the West-Eberhard model is not represented in the Vespidae and that, on the contrary, nest sharing is contemporaneous to the onset of labour division modelled by intragroup competition, with the direct passage from the solitary stage (I) to the rudimentary caste-containing stage (III). A major problem is whether the hover-wasps present any solitary species. Evidence accumulated since Carpenter's (1988a, 1991) and Turillazzi's (1991) reviews, provides no solid grounds for disputing Carpenter's conclusion that stage III is the ancestral state for the subfamily, as all genera studied contain at least one species with nest sharing and division of labour. But if we look at some single species as *S. concinna* (Spradbery 1975), *A. iridipennis* (Spradbery 1989) and *A. laticeps* (Turillazzi and Hansell 1991), it is evident that solitary life is the rule for most females, the period of cohabitation on the same nest of two females of different generations is extremely reduced and, in this case, the presence of labour division is only hypothesized. It has been already demonstrated that plasticity of social behaviour is very common in the Vespidae. Cowan (1991) observes that intraspecific variations between mass and progressive provisioning are common in some eumenines on a geographic or seasonal basis, while *Zethus miniatus* displays a range of alternative behaviours from solitary foundation to nest usurpation, to adoption of orphaned larvae by temporarily non-laying females (West-Eberhard 1978*b*, 1987*a*; Cowan 1991). Various species of Stenogastrinae with more complex social habits (i.e. presence of dominance hierarchies) are known to present wide-ranging behavioural plasticity, with females following a large series of possible behavioural options ranging from solitary foundation and reproduction, to aggregation to parental or alien colonies. On this basis it seems reasonable to hypothesize, for some of the hover-wasps, a wide behavioural plasticity running from solitary to primitively social, influenced by ecological factors (such as the strong parasitoid pressure in the case of *A. laticeps* and *A. iridipennis*): could this represent the actual ancestral social trait in the group? Experiments in field and laboratory populations are needed, in the future, to test the limits of this plasticity, not only at the individidual but also at the populational level.

Thus the Stenogastrinae remain important for the study of the evolution of social wasps since they display a stage (III) which is phylogenetically intermediate between the solitary vespids and the eusocial Polistinae+Vespinae. However, I foresee that they will become increasingly important for studying the significance of ecological vs. genetic factors in social evolution. In this respect the biomolecular techniques which recently have begun to be used on this group (Strassmann *et al.* 1994), could shed light on differently structured kin groups and genetical similarities in populations. Behavioural and life-cycle studies, besides furnishing new information about different behavioural and life-cycle strategies, could also illuminate the general nature of certain phenomena such as variations on the theme of dominance-territoriality, tolerance vs. intolerance, different group sizes, and variants of mating systems and demographies. The importance of the hover-wasps from the perspective of phylogenetic studies remains high because of the distinctive evolutionary path that they have followed independently in the equatorial forest environment, and because of the constraints, such as the particular method of raising offspring and the use of low-quality building materials, which have channelled, limited, or even blocked their social evolution (Hansell 1987*a*; Turillazzi 1989).

Conclusions

The facility with which researchers today can move to tropical zones, and the birth of various groups which study the biology of vespids in various tropical countries have made it possible to use species of the genera *Mischocyttarus, Ropalidia,* and *Belonogaster* for the study of the evolution of social behaviour. These wasps, so similar to *Polistes*, have particular features that make them important for in-depth studies of some aspects of the evolutionary process. In the New World *Mischo-cyttarus* is especially important for the wide adaptive radiation it displays, while maintaining a very simple social organization and limited colonial dimensions (Gadagkar 1991c; Wenzel 1991). In the Old World tropics *Ropalidia* represents a key genus because it is the only one which includes both independent- and swarm-founding species, with a range of types of social organization that is unique in the Vespidae (from the almost solitary *R. formosa* (Wenzel 1987b) to the swarm-founding societies of *R. montana* (Yamane *et al.* 1983b)). By contrast, *Belonogaster*, for its phylogenetic proximity with *Polybioides* (a genus of Ropalidiini at present largely unknown but with very marked convergences with some neotropical Epiponini (Darchen 1976a; Francescato *et al.* 1993; Turillazzi *et al.* 1994)) and for some characteristics of its social biology—especially the possible incipient morphological caste differentiation—represents a very important genus in the passage from independent-founding to swarm-founding stages.

The Stenogastrinae represent a group whose social evolution has remained suspended at the first stages, limited by various factors intrinsic to their biology and by selective pressures which precluded an evolutionary process similar to the one followed by the Polistinae. However, the simplicity of their social organization is suitable for comparative work and, as Gadagkar (this volume) suggests, for studying the phenomenon of altruism free from the excessive schematism of definitions.

M. J. West-Eberhard ended her 1978b paper with a famous sentence (paraphrased from a remark C. D. Michener made in front of her country house in Colombia!): 'It is a good time to go look at the wasps across the road.' I think that after 15 years we have at last arrived on the opposite side, and knowing better who lives on this side of the road will help us trace, via comparative study, the evolutionary pathway back to its origin.

Acknowledgements

I wish to thank Elisabetta Francescato, Christina Coster-Longman, Patrizia D'Agostino and Maurice Tindo for permission to use some as yet unpublished results. Mike Hansell, Malcom Keeping, Jane Brockmann and Mary Jane West-Eberhard highly improved the manuscript with their criticisms and corrections. Researches on Stenogastrinae have been performed with funds from the Italian MURST (40% and 60%) and from CNR.

15

The evolution of eusociality, including a review of the social status of *Ropalidia marginata*

Raghavendra Gadagkar

What is eusociality?

Social insects, especially bees and wasps exhibit such a bewildering variety of social organizations that we would be quite lost without a sound classification and some technical terms with universally accepted definitions. A system of classification that is built along lines of progressively varying degrees of social organization and sophistication would be even more attractive. Michener (1969) has presented just such a system of classification that has been so popularized by Wilson (1971) that it has now the added virtue of being nearly universally acceptable. According to this system of classification, eusocial insects (the only truly social insects, by definition) are defined as those that possess all of the three fundamental traits of eusociality namely:

(1) cooperative brood care;
(2) differentiation of colony members into fertile reproductive castes (queens or kings as the case may be) and sterile non-reproductive castes (workers) (simply referred to hereafter as reproductive caste differentiation);
(3) an overlap of generations such that offspring assist their parents in brood care and other tasks involved in colony maintenance.

The system explicitly recognizes equally well-defined groups that are not eusocial. Omit the criterion of overlap of generations and we have the semisocial. Omit also the criterion of reproductive caste differentiation and we have the quasisocial. Omit all three criteria and we have the subsocial, if there are aggregates of parents and immature offspring (where parents care for their offspring and not vice versa) or communal, if there are aggregates of individuals of the same generation and solitary, if there are no aggregates at all.

It is customary to recognize two further subdivisions of the eusocial—the primitively eusocial and the highly eusocial—although there is relatively less clarity and agreement about the definitions here. The most widely accepted criterion for separating the primitively and highly eusocial stages is that of morphologically differentiated reproductive and non-reproductive castes in the highly eusocial species and their absence in the primitively eusocial ones. But the criterion of morphologically differentiated castes is not easy to define. Reproductive and worker castes

may exhibit morphological differentiation ranging from minor statistical differences in body size all the way up to morphs that have completely different morphological and anatomical structures, so much so that they may be classified as
different species by the unwary taxonomist (see Wheeler 1913, pp. 248–249).

The presence of morphologically differentiated castes in the highly eusocial species and their absence in the primitively eusocial species imply several other parallel phenomena. For example, if reproductive and worker castes are morphologically
differentiated, it suggests that caste determination must take place in the pre-
imaginal stages and be quite irreversible so that the adult insects should have little
or no flexibility in the social roles they can adopt. Conversely, if the reproductive
and worker castes are not morphologically differentiated, it is possible (though not
necessary) that caste determination takes place in the imaginal stage and that there
may therefore be considerable flexibility in the social roles that the adult insects
may adopt and indeed, caste determination could in principle be reversible—at
least there is no morphological barrier to reversibility. Although these parallel phenomena have been documented for several highly and primitively eusocial species
respectively, they are not so obvious as to be available for appropriate classification
of poorly studied species. It is therefore not surprising that there may be some am-
biguity in classifying, some species (bumblebees for example) as either primitively
or highly eusocial. Nevertheless, the distinction between the primitively and highly
eusocial is a useful, indeed an essential one (as we shall see below) if we are to
make any headway in understanding the evolution of eusociality.

The distribution of eusociality

When the definitions given above were being formulated, known examples of
eusociality were restricted to the class Insecta and even there to just two orders,
namely Isoptera (termites) and Hymenoptera (ants, bees and wasps). While all
known termites are eusocial, the distribution of eusociality in the Hymenoptera is
curious. The suborder Symphyta, consisting of several families of free-living
phytophagous species is devoid of eusociality. In the other suborder Apocrita, the
sub-group Terebrantia consisting of several families of parasitoid species is also
completely devoid of eusociality. It is only in the subgroup Aculeata that
eusociality is seen. But even here, while all ants are eusocial, most bees and wasps
are not eusocial. Nevertheless eusociality is believed to have originated at least
eleven (Wilson 1971) or twelve times (Alexander 1987) independently within the
Aculeata.

In recent times, eusociality has been demonstrated in three other orders of insects
namely Homoptera (in the aphids) (Aoki 1977; Itô 1989; Benton and Foster 1992),
Thysanoptera (thrips) (Crespi 1992) and Coleoptera (in an ambrosia beetle) (Kent
and Simpson 1992). There is also a claim (though not as well substantiated) of a
eusocial spider (Vollrath 1986) and a clear demonstration of eusociality in two
mammals, the naked mole rat (*Heterocephalus glaber*) (Jarvis 1981; Sherman *et al.*
1991) and the Damaraland mole-rat (*Cryptomys damarensis*) (Bennett *et al.* 1988;
Jarvis *et al.* 1994). The discovery of eusociality in any species of animal outside the

Isoptera and Hymenoptera has come to be regarded as sensational, usually warrant-
ing a report in *Nature* or *Science* (Matthews 1968; Jarvis 1981; Crespi 1992) but
claims and counter-claims about whether something should be classified as eusocial
continue (see, for example, Hölldobler and Wilson 1990, p.184; Benton and Foster
1992). All this has bestowed upon the definition of eusociality, a certain degree of
sanctity which is arguably unjustified (see below).

Problems with the definition of eusociality

In spite of the apparent soundness and wide acceptance of this system of classification, I think there are at least two quite serious problems with it. One has to do with
the rather unfortunate inclusion of the criterion of overlap of generations in the definition of eusociality because of its relatively trivial importance for an 'evolutionary'
classification of social life. For one thing, there is no less altruism in a semisocial
group of sisters, so long as they show co-operative brood care and reproductive caste
differentiation. The central problem of the evolution of eusociality is the evolution of
altruism, and by leaving out the semisocial species we are unnecessarily excluding
equally important and interesting model systems and especially model systems that
might represent the crucial transition from the selfish to the altruistic.

Another problem with the definition of eusociality has to do with its most important criterion namely reproductive caste differentiation. The problem is how to define a sterile worker caste? Should we demand that sterility and worker roles be
life-time properties or is it sufficient if they are only temporary? For instance Tsuji
(1990) who discovered age polyethism in reproductive division of labour in the
Japanese queenless ant *Pristomyrmex pungens* (where all individuals seem to begin
adult life as egg layers inside the nest and later forage outside the nest) and labelled
it as communal, has now been persuaded (by Furey 1992) to retract his classification and accept that *Pristomyrmex pungens* is not communal but has not been persuaded to call it eusocial; he insists that since there is no life-time sterility, the
species cannot be called eusocial (Tsuji 1992). Now all this has two unfortunate
consequences.

First, there are many species of 'primitively eusocial' wasps where sterility is not
necessarily for life for many workers. Second, the insistence on life-time sterility
has precluded the simultaneous consideration of the evolution of altruism in a
number of co-operatively breeding birds (Stacey and König 1990) and mammals
(Gittelman 1989) where individuals routinely stay on in the nests or territories of
their parents or those of others and help in rearing one or more broods of chicks or
pups. Only because many of these helpers will go on to raise their own brood (but so
do many primitively eusocial wasps), we exclude them from the eusocial fold.

The real problem is that primitively eusocial species not only have a great deal in
common with semisocial species, species that lack life-time sterility and co-
operatively breeding birds and mammals, but also that have rather little in common
with highly eusocial species. As we will see below, a *Ropalidia marginata* female is
really much more like the helpers in the Florida scrub jay for instance than she is
like a honey bee.

A possible solution

The evolution of altruism is a major unsolved problem in evolutionary biology and it is most likely to be understood if we have a way of classifying together all species that show a substantial amount of altruism—species where individuals give up at least some personal reproduction for aiding conspecifics. I therefore suggest that we expand the scope of eusociality to include semisocial species, primitively eusocial species, highly eusocial species as well as those co-operatively breeding birds and mammals in which individuals give up personal reproduction for aiding conspecifics. There should be no requirement of overlap of generations or of life-time sterility. Tsuji (1992) has worried that by not insisting on life-time sterility, we would be throwing open the flood-gates to include 'a great many species, including our own' and therefore that 'the sterility criterion should be interpreted as a whole-life phenomenon, and not as an ontogenetic one'. I think that will not be a problem as long as we use the criterion of giving up personal reproduction in order to aid conspecifics and not merely of sterility being restricted to an ontogenetic stage. I also suggest that the distinction between primitively eusocial and highly eusocial should continue based on the presence and absence of morphologically differenti-ated castes (Gadagkar 1994*a*). The primitively eusocial species may then be used to study the origin of eusociality and the highly eusocial species to study the mainte-nance of eusociality (see below).

I see no problems in further sub-dividing eusociality into semisocial, subsocial, communal, etc., but the critical entity in the study of the evolution of altruism, which is now eusociality, should include all the sub-groups and not include some (such as polistine wasps and honey bees) and exclude others (such as semisocial insects and co-operatively breeding birds). It may also happen that different nests or even different individuals of the same species represent different sub-divisions of eusociality; the same may be true of given nests and individuals at different times in their life. Indeed, such variation may even cross the boundaries of the liberal defini-tion of eusociality suggested here, to include completely selfish or solitary stages or individuals. All this variety could be usefully documented without forfeiting the advantage of having a single entity to include all groups showing altruism.

What do we mean by 'understanding the evolution of eusociality'?

Grafen (1991) has called the philosophy of behavioural ecology a phenotypic gam-bit where we assume quite brazenly (for we know it cannot be literally true) (1) that phenotypic characters of interest are determined by the simplest genetic system—perhaps even a haploid locus, at which each allele produces a distinct phenotypic character and (2) that enough mutations occur to produce the required variations. Natural selection is then expected to act on this variation to favour or disfavour specific variants as appropriate to the environmental conditions at hand.

The phenotypes of interest to us now are (1) a selfish individual that rears his or her own offspring and (2) an altruistic worker that cares for somebody else's off-spring rather than rear his or her own. The question we are interested in asking is, under what conditions the selfish and the altruist respectively would be favoured by natural selection. Our first and most important distinction is that between the origin of eusociality and its maintenance.

By the maintenance of eusociality we mean the conditions under which a selfish mutation cannot spread in a population of altruists so that eusociality is stably maintained. It seems reasonable to turn to the highly eusocial species when we are dealing with stably maintained populations of altruists. Thus the question is, why don't honey bee, ant and termite workers become selfish, stop working for their colonies and revert to solitary life? I suspect that this question is virtually impossible to answer. The problem is not that we have reduced the underlying genetics to an unrealistic level but that the second assumption of the phenotypic gambit that enough mutations occur to produce the required variation in characters is probably false. Because the reproductive and worker castes are morphologically differentiated and because caste determination is essentially irreversible in highly eusocial species, a worker cannot revolt against the queen and either drive her away or leave to found a new nest of her own. It is true that the cape honey bee (Anderson 1963) and ants such as *Pristomyrmex pungens* (Tsuji 1990) have evolved thelytoky and manage to get along without their queens, and that in many ponerine ants, workers can mate and manage to survive without their queens (Peeters and Crewe 1985, 1986; Peeters 1987). These examples suggest that there has been some reversal from the highly eusocial to the primitively eusocial state but there is no evidence that any eusocial species has actually ever reverted to the solitary state.

We cannot therefore rule out the trivial alternative hypothesis that workers in highly eusocial species remain altruistic because they have irretrievably lost their reproductive options and have no choice but to work for their colonies, even if the worker strategy is potentially invadable by a selfish alternative. Given this caveat, it is perhaps rather uninteresting to try and ask what maintains eusociality in the highly eusocial species today.

Let us turn to the question of its origin: under what conditions can altruists invade a population of selfish, solitary individuals? To answer this we would ideally like to be able to trace the evolutionary past of the eusocial species. But that is not really possible because there are no fossils and even if there were, they would tell us precious little about behaviour, altruism and social organization! Our next best bet then lies in turning to the primitively eusocial species to see if they are any good for our purpose. Fortunately we are on much better ground here. The primitively eusocial species are certainly in no evolutionary *cul-de-sac*. As we have seen, caste differentiation is not morphological and workers have clear reproductive options; they can replace their queens or leave to start their own solitary foundress nests. Why then do they not always do so? Now that is a much more promising question because we can rule out the trivial alternative hypothesis that they have no option but to continue to work for their colonies. In other words we have some hope of understanding why eusociality is maintained in the primitively eusocial species today. But is that equivalent to understanding the origin of eusociality? I think the answer is yes, but a guarded and qualified yes. It is perhaps reasonable to assume that highly eusocial species have passed through stages that might have resembled today's primitively eusocial species. If that is reasonable, we may explore the factors that maintain eusociality in the primitively eusocial species today and assume that similar factors may have been responsible for the origin of eusociality.

Inclusive fitness theory as a unifying theme

Our central question then is, why do workers in primitively eusocial species not all become selfish and start single foundress nests of their own rather than behave altruistically in somebody else's nest and indeed why do some do so? Hamilton (1964a, b) argued that the paradox of altruism was no paradox since natural selection is dependent on changes in relative frequencies of alleles without regard as to the pathway by which these changes are brought about. It follows then that producing offspring is only one way to increase the representation of one's genes in the population. Aiding genetic relatives that also carry copies of one's genes, identical by descent, is an alternate equally legitimate way of doing so. The attractiveness of this proposition comes from the fact that, since the probability of occurrence of copies of one's genes, identical by descent, in any class of genetic relatives, including offspring, can be relatively easily calculated, one can now compute a composite quantity called the inclusive fitness. Inclusive fitness then is the sum of all genetic relatives (including offspring) for whose survival a given individual is responsible, after appropriately devaluing the number of relatives in each class by the probability of sharing genes identical by descent, by that class of relatives with the individual in question. Thus inclusive fitness has two components, a direct or selfish component gained through production and care of offspring and an indirect or social component gained through care of genetic relatives. From this Hamilton deduced an elegant rule, which has come to be known as Hamilton's rule. Hamilton's rule thus specifies the conditions under which an 'altruist' allele would spread in a population relative to an alternate selfish allele. The condition is usually represented algebraically as:

$$B/C > 1/r \qquad (1)$$

Where B is the benefit to the recipient of the altruism, C is the cost to the altruist and r is the coefficient of genetic relatedness between altruist and recipient, the latter being the probability that genes present in the altruist are also present in the recipient, identical by descent. This expression can be conveniently rewritten as:

$$Br > 1 \cdot C \qquad (2)$$

meaning that the benefit to the recipient, devalued by the genetic relatedness between altruist and recipient should be greater than the cost to the altruist. Even more conveniently for our purpose,

$$B/C > r_0/r_i \quad or \quad Br_i > Cr_0 \qquad (3)$$

Where r_i is the genetic relatedness of the altruist to the recipient's offspring and r_0 is the genetic relatedness of the altruist to his/her own offspring. Here B can be thought of as the additional numbers of offspring produced by the recipient as a result of the help given by the altruist and C can be thought of as number of his own offspring given up by the altruist in order to help the recipient.

Hamilton's inclusive fitness theory offers a unifying theme to evaluate a multitude of factors that might be responsible for this. Inclusive fitness theory predicts that workers do not become selfish and start their own solitary foundress nests if,

$$\Omega > W \tag{4}$$

where Ω and W are the inclusive fitnesses of a worker and a solitary foundress respectively. The question then is, under what circumstances would the inclusive fitness of workers be greater than that of solitary foundresses? To understand this let us break up inclusive fitness into its constituent components and rewrite inequality (4). The inclusive fitness of workers and solitary foundress can be broken up into at least three components (Queller 1989; Gadagkar 1990b, 1991b) so that we have the prediction that

$$\beta \rho \sigma > b r s \tag{5}$$

where β is the intrinsic productivity of a worker, defined as the number of individuals she can rear to adulthood provided she survives for their entire developmental period, ρ is the coefficient of genetic relatedness of a worker to the brood she rears and σ is the demographic correction factor for a worker, defined as that factor by which a worker's intrinsic productivity should be devalued because of the probability of her dying before the brood under her care complete development. b, r and s are the corresponding parametres for a solitary foundress. Clearly at least three classes of factors can contribute to inequality (5).

Predispositions to eusociality

Ecological or physiological predisposition: $\beta > b$

One reason why the inclusive fitness of workers may be greater than that of solitary foundresses is that workers may be able to rear more brood per capita than solitary foundresses. This may happen because of better protection from parasites, predators and conspecific usurpers in a group nesting situation compared with a solitary nesting situation (see for example, Lin and Michener 1972; Gamboa 1978; West-Eberhard 1978b; Litte 1977, 1979, 1981; Suzuki and Murai 1980; Gadagkar 1985, 1991c; Itô 1986c; Strassmann $et\ al.$ 1988) or because individuals that opt for worker roles may be subfertile whose b would be relatively small if they became solitary foundresses but whose β as workers would be relatively high (see for example, West-Eberhard 1975; Craig 1983; Gadagkar $et\ al.$ 1988, 1990).

Genetic predisposition: $\rho > r$

Another reason why the inclusive fitness of workers may be greater than that of solitary foundresses is that workers may have access to brood that are more closely related to themselves than a solitary foundress is to her offspring. The haplodiploid genetic system in the Hymenoptera, coupled with an ability to bias investment in favour of female brood can make this possible (see for example, Hamilton 1964*a*, *b*; Trivers and Hare 1976; Gadagkar 1990*c*, 1991*a*).

Demographic predisposition: σ > s

Yet another reason why the inclusive fitness of workers may be greater than that of solitary foundress is that workers may have lower mortality rates compared with solitary foundress but more importantly, the consequence of similar mortality rates may be quite different for workers who function in groups compared with a solitary foundress that works alone (see for example, Queller 1989, this volume; Strassmann and Queller 1989; Gadagkar 1990*b*, 1991*b*; Nonacs 1991).

The use of the word predisposition perhaps needs some justification. To predispose means to render susceptible or liable, beforehand. This connotation is perfectly valid in the case of genetic predisposition caused due to the genetic asymmetries created by haplodiploidy. Since nearly all known hymenopterans including the solitary groups most closely related to eusocial species are haplodiploid, haplodiploidy can be inferred to have preceded eusociality. Hymenopterans may thus be legitimately said to be potentially genetically predisposed to the evolution of eusociality.

The situation with ecological, physiological or demographic predisposition is somewhat different however. The kinds of factors mentioned above, as examples of these forms of predisposition to eusociality (such as variations in fertility, life spans, etc.) cannot really be said to be more primitive characters of the Hymenoptera compared with eusociality. We know almost nothing about the distribution of these factors among different groups in the Hymenoptera. Nevertheless I think we can justifiably use the term predisposition even for ecological, physiological and demographic factors as long as we are dealing with primitively eusocial species. Recall our formulation of what is meant by understanding the evolution of eusociality. Since we cannot really hope to retrace the actual events in the evolutionary progression towards eusociality, we have decided to look at primitively eusocial species, where workers behave altruistically in spite of having the option of direct reproduction, and equate the factors that maintain eusociality in primitively eusocial species today with those that might have been responsible for the origin of eusociality. Given this caveat, any ecological, physiological or demographic asymmetries that we may discover in today's primitively eusocial species may analogously be considered to constitute predispositions to the evolution of eusociality. As a model system for studying predispositions to eusociality, *Ropalidia marginata* has provided a rich source of new insights.

Ropalidia marginata as a model system

Ropalidia marginata is an Old World, tropical, primitively eusocial polistine wasp abundantly distributed in peninsular India. The life cycle and many aspects of the biology of *Ropalidia marginata* are quite similar to those of *Polistes*, especially the tropical species (see Reeve 1991 for a review). New colonies may be founded throughout the year by one or a group of females wasps. In single foundress colonies, the lone female builds a nest, lays eggs, cares for her larvae by foraging

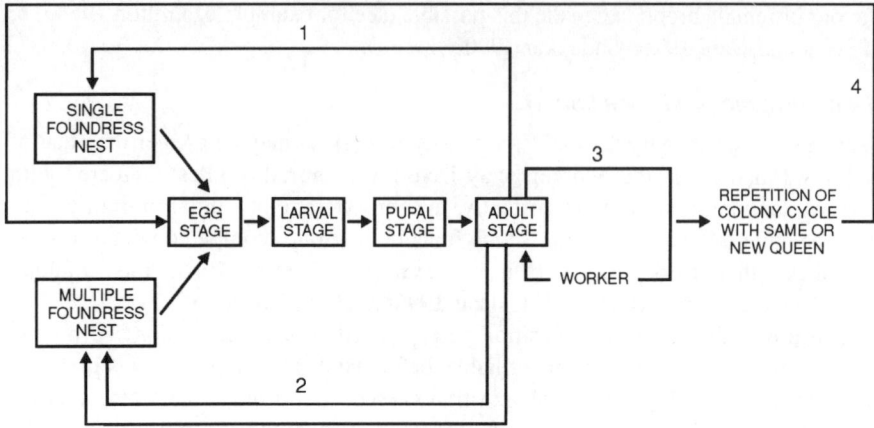

Fig. 15.1 The perennial, indeterminate nesting cycle of *Ropalidia marginata*. Female wasps have at least four different options shown by numbers 1—4 namely, leave their natal nests to initiate single foundress nests, leave in a group to initiate multiple foundress nests where they may become queens or workers, stay back in their natal nests as workers and stay back in their natal nests to eventually take over as new queens. For schematic convenience, the egg, larval, and pupal stages are shown as being distinct. In reality, there is considerable overlap between them. Similarly, change of queens can take place at any time of the colony cycle. Note also that new colonies may be initiated at any time of the year and may also be abandoned at any time of the year and at any stage in the colony cycle.

for them as well as guarding the nest and brings them to adulthood, unaided by conspecifics, much like a solitary wasp. In a multiple foundress nest however, there is a dominance hierarchy, leading to division of labour such that only one individual functions as the queen and lays all eggs while the rest perform all the tasks involved in colony maintenance and brood care. Male offspring stay on their natal nests for about a week before leaving the nest. They lead a nomadic life near places likely to be visited by foraging female wasps and mate with them.

   Daughters eclosing from single- as well as multiple-foundress colonies have a number of options open to them. They may either leave soon after eclosion to found their own single- or multiple-foundress colonies or join other newly initiated colonies, where they may act as queens or workers. They may stay on their natal nests and spend their entire life performing the role of a sterile worker. They may stay and work for some time and then leave to found new colonies or join other colonies or after some period of work, they may drive away the queen of their natal colony and take over the role of the queen. Colonies are initiated throughout the year, and the nesting cycle is indeterminate: there is no consistent natural end to the colonies, which may persist for years, with a gradual turnover of the individuals present. This is made possible by a number of factors. The tropical climate in peninsular India makes colony life and brood rearing possible through out the year, the cells in the nest are reused many times over and workers can replace old and weak queens from time to time (Fig. 15.1).

All of these features make *R. marginata* a particularly useful species to investigate the factors that favour the spread of eusociality. Its primitively eusocial status is only one of the reasons for this. Many well-studied species of *Polistes*, for example, occur in the temperate latitudes but where colony life is restricted to the few favourable months of the year (see Yamane this volume). All queens and workers die at the onset of winter and only females eclosing late in the season and which have not acted as workers usually mate and hibernate. These overwintered female wasps then initiate new nests more or less synchronously the following spring. Early initiation of nests can be so crucial to success in eventually producing reproductive offspring that subordinate wasps, which lose out in fights for the dominant, egg-laying status, seem not to have the option of then leaving and founding their own nests (West-Eberhard 1969*b*; Gadagkar 1991*c*; Reeve 1991). However, this option is common in *R. marginata* just as in the case of *P. erythrocephalus* (West-Eberhard 1969*b*). Foundresses move from one nest to the other and seem to choose the appropriate set of co-foundresses to nest with and they have the freedom to do this as time is not that critical; colony success does not depend critically on the time of initiation. This difference in the importance of time of nest initiation automatically reduces the options open to different female wasps in temperate latitudes. Besides, females eclosing early in the year in temperate species, usually have no males to mate with and thus cannot produce female progeny. Even in the well-studied tropical polistine wasps such as *Mischocyttarus drewseni* the nesting cycles are determinate (Jeanne 1972), thus preventing these species from having as rich a milieu for co-operation and conflict among individuals as *R. marginata* does.

It is thus the combination of the primitively eusocial status, the tropical climate that guarantees nearly equal opportunities to wasps eclosing at all times of the year and the highly indeterminate and prolonged nesting cycle, that makes *R. marginata* especially suited for the study of the evolution of eusociality.

The evolution of eusociality is a complex problem and has almost certainly involved a multitude of factors. It is not surprising that the insights gained from studying different species are likely to be different and even contradictory, at least in the beginning. Presenting a unified picture therefore can be hazardous and there is merit in making explicit the picture that emerges from detailed studies of individual species before attempting a unifying picture. Here I will attempt to illustrate the evolution of eusociality by exploring potential ecological, physiological, genetic and demographic asymmetries between workers and solitary foundresses in *R. marginata*.

Ecological predisposition in *R. marginata*

Do ecological asymmetries contribute to the evolution of eusociality in *R. marginata*? Is $\beta > b$ for ecological reasons? This is not easy to test empirically but at least one can ask if workers in multiple-foundress nests raise more brood than solitary foundresses manage to do. In a recent study, we surveyed two localities where *R. marginata* is known to nest regularly and discovered 145 pre-emergence nests in their early egg stage and monitored them until the eclosion of the first adult

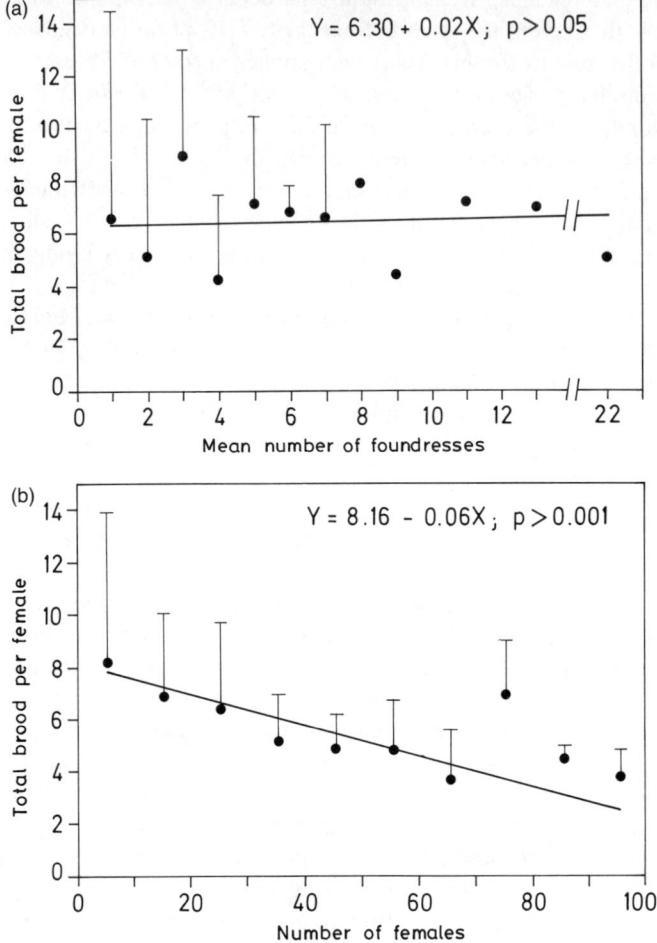

Fig. 15.2 (a) Per capita productivity (measured as the numbers of eggs+larvae+pupae per female, on the day of the eclosion of the first adult) does not increase as a function of number of foundresses in pre-emergence nests (n = 145 nests) (after Shakarad and Gadagkar 1993). (b) For a random sample of pre- and post-emergence nests (n = 244 nests) per capita productivity (measured as the numbers of eggs + larvae + pupae per female, on the day of collection of the nest) decreases with number of females. Thus larger nests do not produce more brood than smaller ones, and if they do it is only in proportion to the number of females present on the nest.

(Shakarad and Gadagkar 1993). As mentioned earlier, these nests are perennial with an indeterminate nesting cycle so that one cannot hope to monitor nests for their entire duration. We therefore chose the time of eclosion of the first adult offspring as a reasonable time to terminate monitoring because the number of adult wasps on the nest changes at that time. The number of foundresses per nest varied from one to 22 with a mean of 2.84. About 35% of newly initiated nests were single foundress nests. Given the distribution of group sizes in multiple foundress nests, it turns out that about 35% of the founding population become egg layers and the remaining 65% become workers. There was a statistically significant increase in productivity (measured as eggs+larvae+pupae) with increase in group size but the per capita productivity did not vary with group size (Fig. 15.2; upper panel). This suggests that there is no particular advantage in being in a large group as the productivity of larger groups increases only in proportion to the number of individuals contributing to labour in these colonies. The probability of survival until the production of the first adult offspring was higher for larger groups but this was taken into consideration in computing the mean productivity for each group size.

Larger group sizes might conceivably have the advantage of making it possible for the brood to complete development earlier. We tested this hypothesis but it was not supported; the time taken from the hatching of the first larva to the eclosion of the first adult offspring did not vary as a function of group size. Another possibility is that while there may not be any effect of group size at the time of the eclosion of the first adult offspring but subsequently, there may be an increase in per capita productivity, with group size. To test this we used 244 nests that had been collected over the years for various experiments in our laboratory, and found that there was a significant negative correlation between per capita brood content and number of adult female wasps present on the nest at the time of collection (Fig. 15.2(b)). The nests used here represent the range of natural colony sizes in different stages of development. We realize that all the females present on a nest at all times of the nesting cycle may not be actually working for the colony but, at least, we can say that so far we have been unable to discover any per capita increase in productivity as a function of group size in *R. marginata*.

In these arguments we are making the assumption that all wasps are identical and that the workers would have been as productive as the observed solitary foundresses had they decided to found nests solitarily. This assumption may not be valid and there may be inherent differences between individuals that opt for solitary foundress roles and those that opt for worker roles, as I discuss below. Another potential problem with the foregoing analysis has to do with the possibility of direct pay off for successor queens. Because total productivity (as opposed to per capita productivity), increases as a function of group size, queens in multiple foundress colonies are better off than solitary foundresses. It follows then that if workers have some chance of becoming queens in their life time, they may gain more fitness than solitary foundresses. Workers in *R. marginata* are sometimes successful in driving away the queens in their colonies and taking their place. The hope of becoming a queen may thus be what selects for worker behaviour in *R. marginata* and those that realize this hope must compensate for those that die as workers (West-Eberhard 1978*b*; Gadagkar 1990*d*).

Cautioning against an exclusive concentration on the role of genetic relatedness in driving social evolution, Lin and Michener (1972) drew attention to a large number of insect species where sterility is absent or incomplete and suggested the possibility that individuals in such groups may be selected to come together for mutual benefit. This so called theory of Mutualism has sometimes been dismissed as incapable, by definition, of explaining the evolution of a sterile worker caste because the term mutualism suggests that both or all participants benefit but the sterile worker caste and the fertile queens are not usually thought of as benefiting equally from the associations (e.g. Itô 1989; Crozier 1977). This argument however deserves a second examination. Consider a situation where an individual that nests in a group may obtain more fitness, on the average, than it would as a solitary individual. If we replace the concept of alleles programming individuals into workers with alleles programming individuals to take the risk of being part of the group, then, under certain ecological conditions, the 'Gamblers' will be fitter than the risk-averse solitary individuals (West-Eberhard 1978b; Gadagkar 1991a). The losers in the 'Gamble' will leave behind no offspring and we will see them as 'sterile workers'. Nevertheless, it is a form of mutualism that has given rise to this situation. One advantage of such a model is that it requires no assumption of increased genetic relatedness or parental manipulation.

An important assumption of this so called 'Gambling Hypothesis' (also called the 'mutualistic loser' hypothesis in West-Eberhard 1978b) is that the productivity in the group mode is higher (or more reliable; see Wenzel and Pickering 1991) than in the solitary mode. In *R. marginata* a solitary foundress produces on the average no more than one or two offspring (Shakarad and Gadagkar 1993) whereas a queen of a multi-female colony produces on the average, 76 offspring (Gadagkar 1990a). I argue therefore that, although there appears to be no per capita increase in productivity as a function of group size, the opportunities for queen turn overs provide some ecological predisposition to the evolution of eusociality in *R. marginata*. We are not yet in a position to assess the magnitude of such asymmetry as we do not have good estimates of the probability with which workers become queens in their life time.

Physiological predisposition in *R. marginata*

Do physiological asymmetries contribute to the eusociality of *R. marginata*? To test this one must know if there are physiological differences between individuals choosing solitary foundress roles and those choosing worker roles. Again it is not easy to test this directly but at least one can ask if all eclosing female wasps are capable of developing their ovaries and initiating single-foundress nests. We now have evidence that mating is not only unnecessary for the development of a female's ovaries but is also unnecessary for an individual to assume the role of the queen of a colony, prevent all other individuals from laying eggs and maintain normal social organization (Chandrashekara and Gadagkar 1991).

We therefore tested a large number of freshly eclosed virgin females for their ability to develop their ovaries and initiate single-foundress nests in laboratory cages with an *ad libitum* food supply. To our surprise we found that only 97 out of 197 wasps tested initiated nests and laid eggs whereas the remaining 100 wasps died without doing so, in spite of living, on average, longer than the time taken by the egg layers to lay their first eggs (Gadagkar *et al.* 1988). We repeated the experiment with an independent sample of 102 wasps and obtained essentially the same results;

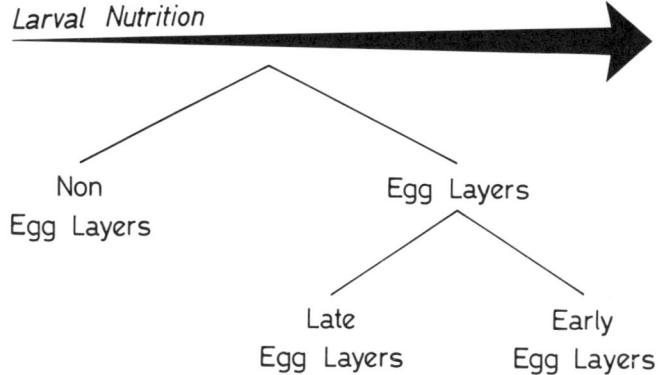

Fig. 15.3 When virgin, freshly eclosed female *Ropalidia marginata* are isolated into laboratory cages, about 50% of them initiate single foundress nests and lay eggs while the remaining die without doing so. Among those that build nests and lay eggs, the time taken to do so varied from 14 to 191 days after eclosion. Larval nutrition influences these capabilities of the adults such that poorly nourished larvae develop into non-egg-layers, somewhat better nourished larvae develop into late reproducers while even better nourished individuals develop into early egg layers (Gadagkar *et al.* 1988, 1990, 1991*a*).

53 wasps initiated nests and laid eggs and the remaining 49 died without doing so (Gadagkar *et al.* 1990). With these results we postulated that there is at least some pre-imaginal biasing of caste leading to the production of potential egg layers and non-egg-layers.

In a subsequent study we were able to determine the mechanism of such pre-imaginal biasing of caste. Using six naturally occurring nests we measured the rates at which larvae were fed in them and then we tested individuals eclosing from these nests for their ability to lay eggs under laboratory conditions, as above. Once again only about 50% (47 out of 87) of the wasps initiated nests and laid eggs. But here we were able to show that the probability of becoming an egg layer was influenced by the rates at which larvae were fed in the nest from which these individuals eclosed. Thus larval nutrition seems to play a role in pre-imaginal biasing of caste such that better nourished larvae become potential egg layers whereas relatively poorly nourished larvae become potential non-egg-layers (Gadagkar *et al.* 1991*a*) (Fig. 15.3).

These experimental results suggest that indeed there are inherent differences between individuals and these differences might well influence their choice of roles. But how would the results of these laboratory experiments translate to natural conditions? Although, I do not believe that our results would apply quantitatively under more natural conditions, I think they would qualitatively. I tend to think of the egg layers in our experiment as individuals that would have a higher chance of capitalizing on an egg-laying opportunity under natural conditions and the non-egg-layers in our experiment as those that might have a smaller chance of doing so.

These observations suggest a substantial physiological predisposition to the evolution of eusociality in *R. marginata*. Individuals with a high probability of

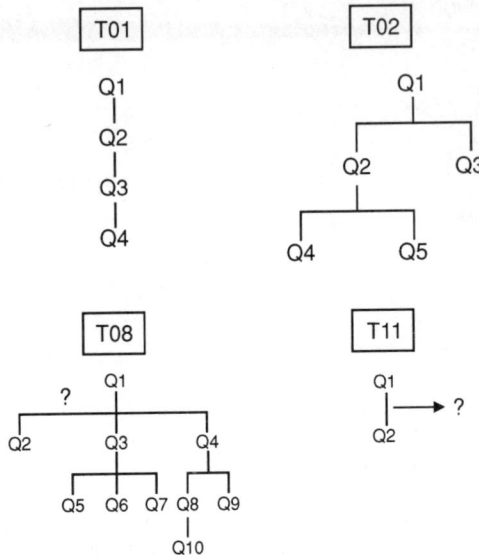

Fig. 15.4 Pedigrees of queens in four colonies of *Ropalidia marginata*.The relationship between queen 1 (Q1) and queen 2 (Q2) was unknown for colonies T08 and T11 because both queens 1 and 2 were among the animals present on the nests at the time of their collection and transplantation. Note the variety in the kind of pedigrees seen in different colonies. In colony T01 for example, new queens were always daughters of their predecessor queens. In colony T08 on the other hand, new queens were daughters, sisters, nieces or cousins of their immediate predecessor queens (Gadagkar *et al.* 1993).

capitalizing on egg-laying opportunities would have high intrinsic productivities as solitary foundresses and would thus be selected to choose that pathway; those with lower probabilities of seizing egg-laying opportunities would have relatively higher intrinsic productivities as workers than they would as solitary foundresses and would be selected to opt for worker roles. The possibility of pre-imaginal caste bias has not been explicitly investigated in *Polistes* and other polistine wasps. However, circumstantial evidence suggests that a similar phenomenon as that seen in *R. marginata* may well be occurring in other species (see Gadagkar *et al.* 1988 and Gadagkar 1991c for reviews).

Genetic predisposition in *R. marginata*

Is *R. marginata* genetically predisposed to the evolution of eusociality? Do workers have opportunities to rear brood more closely related to them than a solitary foundress would? In principle this is possible if, in addition to genetic asymmetries created by haplodiploidy, workers have an ability to skew investment in favour of female brood. But much would depend on whether queens are singly mated and whether colonies are monogynous with daughters working to rear the queen's

Table 15.1 Genetic relationships between successive queens and between workers and brood observed in the four colonies (Gadagkar *et al.* 1993).

Observed genetic relationships	
Relationship between queens and their immediate predecessors	Relationship between workers and brood*
a. Daughters	1. Sisters (0.75 or 0.53)
b. Sisters	2. Brothers (0.25)
c. Nieces	3. Nieces and Nephews (0.375 or 0.265)
d. Cousins	4. Cousins (0.1875 or 0.1325)
	5. Cousins' offspring (0.0938 or 0.0663)
	6. Mother's cousins (0.0938 or 0.0663)
	7. Mother's cousins' offspring (0.0469 or 0.0331)
	8. Mother's cousins' grand-offspring (0.0234 or 0.0165)

* Brood were sisters, brothers etc. of the workers. Values of relatedness given average r for sisters = 0.53, based on electrophoretic data from Muralidharan *et al.* (1986) and Gadagkar (1990*d*).

brood. This is of course not possible in pre-emergence colonies as these are likely to be initiated by groups of sisters where the workers rear their nieces and nephews rather than their siblings. More importantly, there is considerable drifting of females during colony foundation so that cofoundresses are probably not always sisters (Shakarad and Gadagkar 1993). To test if queens mate singly, we analyzed isozyme polymorphism at the *esterase* locus of the egg layers and their daughters from four colonies. Our results showed that queens mate with at least 1–3 males and use sperm simultaneously from different males to produce mixtures of full and half sisters. Our data yielded an average coefficient of genetic relatedness among sisters of 0.52, a value not very different from a solitary foundress's relatedness to her offspring (Muralidharan *et al.* 1986; Gadagkar 1990*c*).

Even more important, there is frequent queen turnover leading to a system of serial polygyny. Long-term studies of another sample of four colonies have shown that new queens may be daughters, sisters or cousins of their immediate predecessor queens and there is sufficient overlap between offspring of different queens so that workers care for a complex mixture of their full and half siblings, nieces and nephews, cousins, cousin's offspring, mother's cousins, mother's cousin's offspring and even mother's cousin's grand-offspring (Fig. 15.4; Table 15.1). Our data yield values of genetic relatedness among female wasps in a colony ranging from 0.22 to 0.46 as a combined effect of polyandry and serial polygyny (Table 15.2). A computer simulation model of serial polygyny suggests that in *R. marginata* eusociality is not associated with asymmetries in genetic relatedness (Gadagkar *et al.* 1991*b*, 1993) (Fig. 15.5).

In spite of such low values of genetic relatedness brought about by polyandry and serial polygyny, the genetic asymmetries created by haplodiploidy may be effectively restored if the workers have an efficient mechanism of intra-colony kin recognition. We investigated this possibility by an indirect method and came to the

Table 15.2 Effects of serial polygyny in *R. marginata* on worker-brood genetic relatedness (Gadagkar *et al.* 1993).

Colony	Number of Queens	Relationship between successive queens	Single mating Grand mean genetic relatedness of workers to: Female Brood	Male Brood	Weighted mean relatedness of workers to brood (b)	Multiple mating (Relatedness between sisters = 0.53) (a) Grand mean genetic relatedness of workers to: Female Brood	Male Brood	Weighted mean relatedness of workers to brood (b)
T01	4	Known	0.65	0.28	0.54	0.46	0.25	0.38
T02	5	Known	0.53	0.28	0.44	0.38	0.24	0.32
T08	10	All but one known; one unknown relationship, assumed daughters	0.35	0.28	0.32	0.25	0.20	0.23
		All but one known; one unknown relationship, assumed sisters	0.32	0.24	0.29	0.22	0.18	0.20
T11	2	Assumed daughters	0.63	0.29	0.52	0.45	0.26	0.38
		Assumed sisters	0.57	0.23	0.47	0.40	0.21	0.34

(a) Data from Muralidharan *et al.* (1986); Gadagkar (1990*d*).
(b) Weighted mean relatedness to brood obtained if worker skew investment in sisters and brothers in the ratio of their relatedness to them (Gadagkar 1990*c*, 1991*a*).

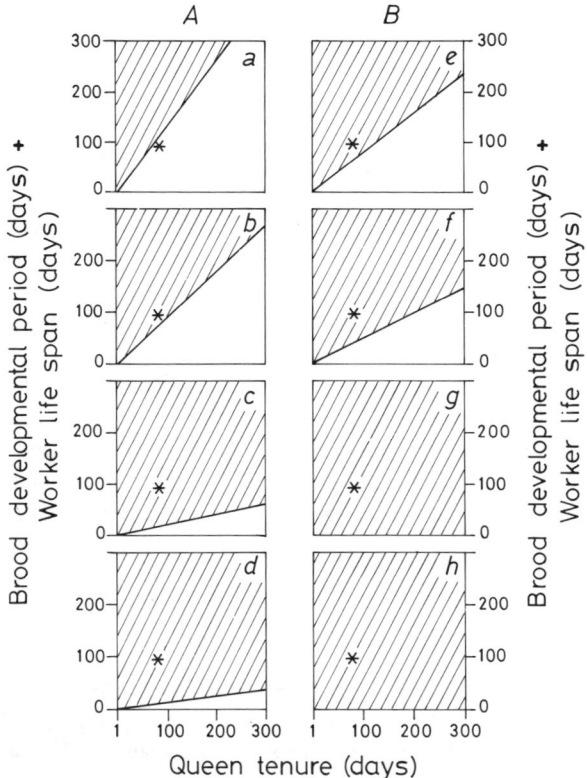

Fig. 15.5 A computer simulation model explores regions in the queen tenure and brood developmental period plus worker life span parameter space where haplodiploidy cannot (hatched region) and can (unhatched region) lead to genetic predisposition for the evolution of worker behaviour. In panel A, the condition for genetic predisposition is that the relatedness of workers to female brood is greater than 0.5. In panel B, the corresponding condition is that workers should obtain a weighted mean relatedness to brood of the two sexes of at least 0.5 after they bias investment in female and male brood in the ratio of their relatedness to them. (a) and (e): queens mate singly and new queens are daughters of their predecessors. (b) and (f): queens mate singly and new queens are sisters of their predecessors. (c) and (g): queens mate multiply and new queens are daughters of their predecessors. (d) and (h): queens mate multiply and new queens are sisters of their predecessors. The lines separating the hatched and unhatched regions are given by the equations: (a), $Q \times 1.33 = B + W$; (b), $Q \times 0.9 = B + W$; (c), $Q \times 0.2 = B + W$; (d), $Q \times 0.13 = B + W$; (e), $Q \times 0.8 = B + W$; and (f), $Q \times 0.5 = B + W$ where, Q is the queen tenure, B is the brood development period and W is the worker life span. The position of *R. marginata* is shown in each part by an asterisk to illustrate that haplodiploidy does not lead to genetic predisposition for the evolution of worker behaviour in this species except in case (a) (Gadagkar *et al.* 1993).

conclusion that the labels and templates used by *R. marginata* workers in nestmate recognition are acquired by all individuals from a common external source such as the nest. Thus the possibility of wasps discriminating between different classes of relatives within a colony seems remote (Venkataraman *et al.* 1988). This conclusion is supported by more recent experiments showing that unrelated conspecifics, if introduced early in life, are not only accepted by alien colonies but become well integrated into their foster colonies and even become replacement queens (Arathi and Gadagkar 1993). In addition, there is a wealth of evidence from other species of primitively eusocial wasps suggesting that intra-colony variations in genetic relatedness are unlikely to be detected by the workers (Gamboa *et al.* 1986*a* and this volume; Queller *et al.* 1990). Nonetheless, all experiments so far have been indirect; no one has actually measured whether workers in primitively eusocial species feed larvae differentially, based on their relatedness to them. Besides, we have recently obtained the tantalizing result that queens are more productive when the worker-brood genetic relatedness during their tenure is relatively high (Gadagkar *et al.* 1993). Nevertheless, our tentative conclusion is that the genetic asymmetries potentially created by haplodiploidy are so broken down in *R. marginata* that there is not much scope for the existence of a strong genetic predisposition to the evolution of eusociality.

Demographic predisposition in *R. marginata*

Is there an asymmetry in *R. marginata* between workers and solitary nest foundresses in the values of the demographic correction factor in inequality (5)? I have developed a hierarchy of models to test this possibility (Gadagkar 1991*b*).

Delayed reproductive maturation

The first model shows how any delay in the time taken to attain reproductive maturity will affect solitary foundresses more than it will affect workers because workers are provided eggs by the queen and do not have to wait for their own reproductive maturity. Under laboratory conditions, the time required to attain reproductive maturity by isolated *R. marginata* females varies from 14 to 191 days after eclosion, with a mean and standard deviation of 48 ± 31 days. Using the mean delay of 48 days, and a value of 62 days for the brood developmental period (Gadagkar 1990*b*) and survivorship data from natural colonies, I have computed the demographic correction factors for workers and solitary foundresses. The demographic correction factor s for a solitary foundress is the probability of survival for 48 days (time taken to attain reproductive maturity) + 62 days (brood developmental period) = 110 days, which is 0.015 (Gadagkar 1991*b*). The demographic correction factor σ for workers is simply the probability of survival for 62 days (as workers do not have to wait for their own reproductive maturity), which is 0.12. These different values for the demographic correction factors obtained for a solitary nesting female on the one hand, and a worker on the other, illustrate the disadvantage of delayed reproductive maturation for a solitary foundress compared with a worker. Assuming $b = \beta$, and because r is expected to be 0.5, the threshold ρ value required for satisfying inequality (5) is given by the equation:

$$\text{threshold } \rho = s/2 \qquad \sigma = 0.06 \qquad (6)$$

Alternatively, assuming that $\rho = r = 0.5$, the threshold b/β value required for satisfying inequality (5) is given by the equation:

$$\text{threshold } b/\beta = \sigma/s = 8.0 \qquad (7)$$

This means that other things being equal, workers would break even with solitary foundresses in spite of rearing brood related to them by a mere 0.06, or if $\rho = r$, then they would do so in spite of solitary foundresses being capable of performing eight times more work per unit time. Compared with the maximum threshold b/β of 1.5 obtained under haplodiploidy (when the brood consists entirely of full sisters), delayed reproductive maturation is thus 5.3 times more effective than haplodiploidy in promoting the evolution of a worker caste.

Variation in age at reproductive maturity

The delayed reproductive maturation model suggests a substantial advantage to workers over their solitary nesting counterparts but it does not explain why a mixture of single-foundress and multiple-foundress nests is common in a variety of primitively eusocial species (Michener 1974; Gadagkar 1991b; Reeve 1991). However, the variation in age at reproductive maturity may provide just such an expectation. With increasing delay in reproductive maturation, the disadvantage for solitary foundresses increases, thus favouring early reproducers to take up the solitary founding strategy and late reproducers to take up the worker strategy. In effect, a polymorphism with single-foundress and multiple-foundress associations will be favoured by natural selection.

Mixed reproductive strategies

The models and data considered above also suggest the possibility of selection for an individual to adopt a mixture of queen and worker strategies. If there is likely to be a delay in attaining reproductive maturity, then such an individual would maximize her inclusive fitness by first being a worker and then, approximately at the time of attaining reproductive maturity, changing over to the role of foundress or queen. Our data show that about 28% of the wasps could complete rearing one entire brood (by working for 62 days) before they become reproductively mature. Indeed, we have often seen *R. marginata* females adopt such mixed strategies, i.e. first be a worker and later take-over the colony as the next queen or leave to found or join other colonies.

Assured fitness returns

The asymmetry in the demographic correction factors for solitary foundresses and workers seen so far was due to the requirement that solitary foundresses have to survive longer than workers to rear the same number of brood. I have also developed a model that shows how there will be a difference in the demographic correction factor even if solitary foundresses and workers survive for the same period of time. While solitary foundresses have necessarily to survive until the end of the developmental period of their brood, failing which they will lose all their investment in it, workers have a special advantage. If a worker cares for some brood for a part of its developmental period and dies before bringing the brood to independence, there is a good chance that another worker from that colony will continue to care for the same brood. Thus workers in multi-female nests are assured of some fitness returns for their labour even if they work only for a fraction of the brood developmental period. The demographic correction factor for solitary foundresses, s, once again is the probability of survival up to end of the brood developmental period of 62 days, which is 0.12. To allow for the fact that workers can get fitness returns in proportion to the fraction of the brood developmental period for which they survive, I have computed the demographic correction factor for a worker σ as,

$$\sigma = \sum_{i=1}^{n-1} \rho_i(i/n) + \sum_{i=n}^{\infty} \rho_i = 0.43 \qquad (8)$$

where p_i is the proportion of workers that have a life span of i days. Using equations (6) and (7), the threshold ρ and b/β required for satisfying inequality (5) are 0.14 and 3.6 respectively (Gadagkar 1990c). Thus assured fitness returns can by itself make the worker strategy more advantageous than a solitary founding strategy. Its relative strength compared with haplodiploidy is 2.4 (Gadagkar 1991b).

In concert with the first three models considered above, Assured Fitness Returns provides an even more powerful force in selecting for worker behaviour. For example, when Assured Fitness Returns and Delayed Reproductive Maturation act in concert, the demographic correction factor s for a solitary foundress is 0.015 (because a solitary foundress has no Assured Fitness Returns) but the corresponding value of σ for a worker is 0.43. Now equations (6) and (7) yield values of 0.017 and 28.7 respectively showing that in concert, these two factors have a relative strength compared with haplodiploidy of 19.1. Just as Assured Fitness Returns and Delayed Reproductive Maturation can act in concert, Assured Fitness Returns and Variation in Age at Reproductive Maturity can also act in concert and provide, for any given age, a more powerful selective advantage for the worker strategy. In the absence of Assured Fitness Returns, the advantage of Mixed Reproductive Strategies can be exploited only by individuals that have a delay of 62 or more days in attaining reproductive maturity. When Assured Fitness Returns and Mixed Reproductive Strategies act in concert however, the advantage of mixed reproductive strategies become available to individuals with a variety of values of age at reproductive maturity. Their labour is not wasted even if they leave their natal nests to become solitary foundresses or queens, because Assured Fitness Returns will give them fitness returns in proportion to the fraction of the brood developmental period that they have worked.

We have good reason to believe therefore that *R. marginata* is strongly demographically predisposed to the evolution of eusociality.

The role of developmental plasticity

I have argued above (see also Gadagkar 1991*b*) that demographic factors such as initial variations in time taken to attain reproductive maturity will make it worthwhile for some individuals to adopt worker-like roles and for others to assume queen-like roles, leading to the establishment of eusociality. I have also suggested (Gadagkar 1991*b*) that such demographic factors will co-evolve with eusociality and become more pronounced because late reproducers will, for example, have a smaller selective disadvantage in a eusocial species compared with a solitary species because of the possibility of gaining indirect fitness in the former.

I argue that we can also envisage the evolution of highly eusocial species starting from completely solitary ancestors through selection for developmental plasticity. In solitary species, any character such as time taken to attain reproductive maturity, ovary size, mandible size, etc., would have limited developmental plasticity on account of stabilizing selection, because, both queen and worker functions would have to be optimized under a single developmental program. But even here, under the right ecological, genetic or demographic conditions (Gadagkar 1991*b*), individuals at the extremes of the distribution of values for these characters would be selected to take up predominantly or exclusively queen-like or worker-like roles. For example, individuals that have to wait only a short time before attaining reproductive maturity or have larger than average ovaries would be fitter as queens compared with individuals at the opposite ends of the distributions. Conversely, individuals with delayed attainment of reproductive maturity, smaller than average ovaries or larger than average mandibles (useful say, in transporting food back to the nest) would be fitter as workers than individuals at the opposite ends of the distributions. This stage may thus be thought of as the origin of eusociality. As the worker-like individuals begin to rely increasingly on the social component of inclusive fitness and queen-like individuals continue to depend on individual component, there would begin a quite different regime of selection.

First there would be a relaxation of stabilizing selection on genes that regulate the making, in workers, of structures needed for mating and reproduction and in queens, of structures needed for foraging, nest building and brood care. Such relaxed stabilizing selection would lead to divergence of queens and workers. In the next step such relaxed stabilizing selection would make previously impossible levels of directional selection on genes that regulate the making of structures needed for mating and reproduction in queens and structures required for foraging, nest building and brood care in workers. This is because the two kinds of structures, those needed for mating and reproduction on the one hand and those needed for foraging, nest building and brood care on the other hand, would no longer need to be optimized in the same individual. In a process analogous to evolution by gene duplication (Gadagkar 1994*b*), genes involved in making caste-specific structures (or behaviours) can evolve to new and extreme levels. As West-Eberhard (1979) has argued, this process can go far enough to make intermediate individuals to be good neither at being queens nor at being workers, thus leading to disruptive selection to reinforce the process of morphological caste differentiation (Fig. 15.6).

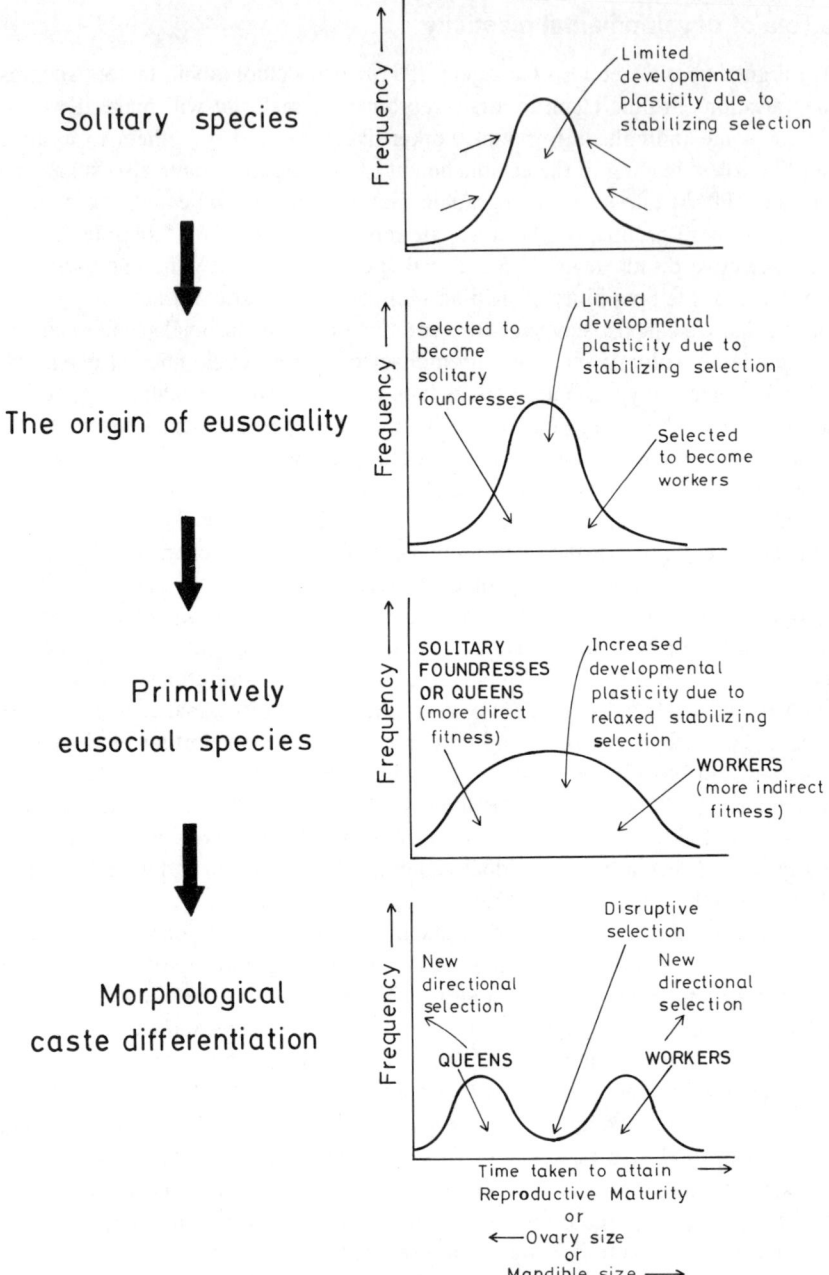

Fig. 15.6 The origin and evolution of eusociality through selection for developmental plasticity (see text for details).

The ideas in this section are not all new. They are intimately related to West-Eberhard's discussion (1979, 1987b, 1988b, 1989, 1992) of the general role of 'alternate' phenotypes in evolution. I do believe however that the 'gene duplication' model suggested here, which is linked to the fact that some individuals in a species begin to rely on the social component of inclusive fitness while others, in the same species, continue to rely on the individual component, is different from (and, of course, in my opinion, simpler and superior!) to West-Eberhard's 'epigenetic' model.

Conclusion

In conclusion, our present state of knowledge, suggests that *R. marginata* is not genetically predisposed to the evolution of eusociality but potential ecological, physiological and demographic asymmetries between solitary foundresses and workers can provide a strong selective advantage for the worker strategy while at the same time permitting the coexistence of single-foundress and multiple-foundress colonies. All of this should be viewed as no more than a first approximation of the shape of one piece in the jig-saw puzzle of the evolution of eusociality. Needless to say, much more detailed investigations of many more species are needed.

Acknowledgements

Some paragraphs in this review have been taken with slight modifications from Gadagkar (1994a,b). My research reported here was largely supported by the Department of Science and Technology, and Ministry of Environment and Forests, Government of India and the Homi Bhabha Fellowship. This paper was largely written while I was visiting the University of Würzburg. I thank Bert Hölldobler for the invitation and financial support through the Leibnitz Award of the German Research Council DFG. I thank C. D. Michener, D. C. Queller, M. J. West-Eberhard and S. Turillazzi for many helpful comments on a previous version of this chapter. I also thank Barbara König and Wolfgang Kirchner for many stimulating discussion of the ideas discussed here.

16

Wasps make nests: nests make conditions

Michael H. Hansell

I am always struck by the unselfconscious way in which the words *natural history* are used by researchers of social insects. This approach to biology has come to be regarded by some as unfashionable and rather equated with stamp collecting. This view probably gained currency with the great expansion of molecular biology, a discipline in which generalizable concepts have developed from the discovery of similarities between organisms rather than attempting to see order in their differences.

The attitude is of course mistaken. Natural history at its best never was about cataloguing differences for their own sake. The English naturalist John Ray published in 1691 his book *The Wisdom of God Manifested in the Works of Creation*. Now we look to the evolutionary process to explain the variety of living things; hence it is entirely right that the first half of the title of this International Workshop should be *Natural History and Evolution of an Animal Society* and it is this which has encouraged me to concentrate on aspects of nest variation in this chapter.

The second half of the Workshop title stresses the organisms which were the focus of the distinguished research career of Leo Pardi, *The Paper-Wasp Case*. As Pardi himself recognized, these are an important model system for evolutionary biologists because of the very great variability they exhibit in their social systems.

From my point of view there is a particular significance attached to the fact that they are identified as *paper*-wasps. The nest is a key part of their biology. It cannot be divorced from their social life and, since the nest is the focus of social life, then no social system can be fully understood in isolation from it.

If this is accepted then variation in nest material and architecture between and within species—and there is a great deal of variation in the Vespidae in both—becomes an important aspect of natural history and deserving of careful examination.

Topics being presented in the later chapters stress the role of paper-wasps as a model system through comparison with other groups. I have taken this as an opportunity to widen the scope of my own presentation to reiterate a theme which has been central to my research interest: Animal building behaviour and the resulting artifacts have a distinct and important biology. Nests, in fact, have a significance in behavioural ecology which has in certain respects been neglected. I therefore wish to place wasp nest building in the context of building behaviour as a whole. In this way I wish to identify generalizations emerging from the study of wasp nests which are supported by evidence coming from other animal groups, and to show that further work on wasp nests may enhance not only our understanding of social wasp biology but also the biology of nest building.

Nest building has consequences

I have chosen my title to illustrate two things. The first is that by making a nest the builder immediately gains some measure of control over the environment. This allows individuals to become more adapted to the special conditions which, through their behaviour, they have created. This may lead to an evolutionary progression in which the artifact becomes more sophisticated in its ability to isolate the builder from the unpredictable and dangerous world outside. The builder, as a consequence, adjusts other aspects of its biology to the environment provided by the nest.

The second aspect of the *conditions* in my title refers to the constraints which are then placed upon the builder as a consequence of its growing dependence upon the nest. This may impose on nest builders limitations such as confinement to a particular habitat or restriction in possible social interactions. It may also impose an evolutionary constraint on the builder. Evolutionary change in social life or habitat may require accompanying innovations in nest construction to be achieved. There is of course an obverse side to this; innovations in nest construction may facilitate other evolutionary changes in biology.

I therefore want to consider the consequences of nest building under four subheadings:

1. Nests reinforce social life.
2. Nests facilitate changes in habitat range.
3. Nest material specializations have ecological consequences.
4. Nests, as valuable resources, attract usurpers and squatters.

Nests reinforce social life

The attraction of wasps rather than say ants or termites in the study of the evolution of eusociality has been the occurrence in living species of examples which represent all stages of social complexity from solitary to advanced eusocial. This gradation in social complexity is, not surprisingly, positively correlated with an increased complexity in nest architecture. This has led to speculation on the possible relationship between the two.

I have suggested (Hansell 1987*a*, 1993) that the creation of a nest has consequences which act as a positive reinforcement on the evolution of social complexity. I pointed out that the transition from mass to progressive provisioning was accompanied by a slower rate of reproduction and longer adult life (Malyshev 1968; Evans and West-Eberhard 1970), both features of a K-selected biology. The model I proposed argues that nest building, by creating a more constant environment, generates a population nearer to the carrying capacity of the environment. This makes colony founding more difficult and inhibits the leaving of the home nest resulting in a more complex colony life and further elaboration of the nest.

The association between nest preparation and features of a K-selected biology has been noted by other authors. Halffter and Edmonds (1982) describe a negative correlation in dung beetles (Scarabaeinae) between nest burrow complexity and fecundity, with ovarian reduction reaching the extreme of a single ovary with a single ovariole in one species. Kent and Simpson (1992) discovered that in the ambrosia beetle *Australoplatypus incompertus* a foundress female of low fecundity is later supported by a small retinue of workers living in galleries excavated deep in the heartwood of living trees. These colonies it seems may stay active for at least 35 years. Nevo (1979) includes K-selected biology as one of the consequences of competition in populations of subterranean mammals. The extreme example of this is the naked molerat *Heterocephalus glaber* which has a termite-like colony with one royal pair living permanently in a highly elaborate environment of their own creation (Jarvis 1981).

It is too early to say whether this *habitat saturation* model has any validity but it is not the only theory to propose that nest construction may have contributed to the evolution of greater social complexity. There are certainly two alternative theories although they are not entirely distinct from one another. These can perhaps be called the *cost of foundation* and *wealth of inheritance* hypotheses.

I attribute the first to Mary Jane West-Eberhard who was the first person to explain it to me but it is perhaps most widely regarded as an intuitively acceptable theory. It identifies the costs of nest foundation firstly as a risk of predation in an initially poorly defended situation and secondly the energetic cost of creating a secure nest environment.

The wealth of inheritance theory, expressed by Myles (1988) in the context of termite social evolution, differs chiefly in the emphasis it gives to the benefit of staying. This consists largely of the great prize that can potentially be won by inheriting the home nest compared with costs of emigration.

These three theories, clearly, are not mutually exclusive but they certainly do produce different predictions and are therefore testable. For example it would be possible to use a stenogastrine species like *Liostenogaster flavolineata* to test between the cost of foundation and habitat saturation hypotheses. Nests of this species contain a handful of females with flexible life history options and occur often in large aggregations under rock overhangs (Samuel 1987; Turillazzi this volume).

Destruction of a proportion of *L. flavolineata* colonies should lead to subordinate females on surviving nests leaving to become foundresses but only if the habitat saturation theory were true, since the cost of nest foundation measured either as predation risk or construction costs would be unaltered. The introduction of unoccupied nests from another site would lower costs of foundation, so encouraging females to leave their home colonies provided that other resources, most notably food, were not limiting. An illustration of this approach is the study of the effect of artificial watering of the soil in which females of the sphecid wasp *Cerceris antipodes* make their burrow nests. This encouraged new burrow construction (Fig. 16.1) and was therefore recognized as a possible variable influencing burrow sharing (McCorquodale 1989).

So the role of the nest in promoting social evolution is amenable to investigation but does need a more ecological approach. That means more detailed knowledge of

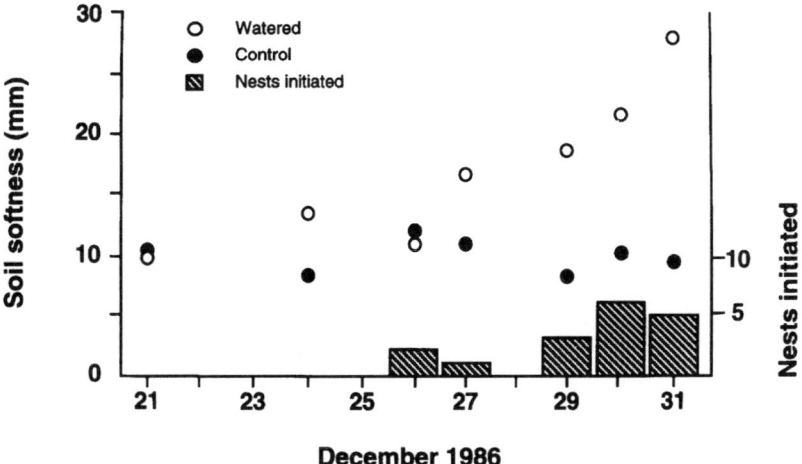

Fig. 16.1 The effect of experimental watering of the ground on the number of new burrow nests initiated by female *Cerceris antipodes* (Sphecidae). Soil softness increased in the experimental plots leading to a number of nests being initiated. In control plots the soil remained hard and no nests were initiated. (Adapted from McCorquodale 1989.)

the food supply and the costs of nest foundation, but it also offers scope for experimental manipulation most obviously of population levels and nest availability.

Nests facilitate changes in habitat range

It has been pointed out by Starr (1991) that, whereas wasp nests may initially simply have provided brood cells, natural selection favoured the modification of architecture to incorporate secondary features such as protective envelope and storage areas. Such changes may not only have had the result of making nest builders more secure in their existing environment but of allowing them to survive in areas which were previously inhospitable (Hansell 1989). Such an example is the use of food stores of honey, in the case of honey bees, allowing the colony to survive the winter in temperate regions, and of seeds in the desert ant species *Messor* (*Veromessor*) *pergandei* allowing colonies to survive a recorded 12 years of drought (Tevis 1958).

More important than changes in architecture in terms of consequent shifts of ecology have been evolutionary changes in building materials. The most important of these for wasps has clearly been the innovation of a paper nest technology.

I have proposed elsewhere (Hansell 1987*a*) that small colony sizes in Stenogastrinae may be due to a constraint imposed by the inferior nest building materials used by them, compared with the Polistinae and Vespinae; that is rotted vegetation or mud rather than unrotted fibrous material. I argued that this constraint might exist because a change in mandible design which allowed tougher vegetation to be chewed would also alter the nature of the preparation of larval food or, one could argue, disrupt specialized prey capture techniques. This constraint argument has been criticized by Wenzel (1991) who points out that other factors, such as

predation pressure, might equally be the limiting factor. It is also true that certain
architectural devices could have evolved to overcome the technological limitation
such as the multiple small combs seen in *Ropalidia revolutionalis* (Itô 1993).

There are problems in demonstrating constraint (Maynard Smith *et al.* 1985),
however, in the case of the stenogastrine nest material, a structural innovation
which strengthened the material without affecting mandibular design would at least
support the argument. Such evidence is provided by *Anischnogaster laticeps* in
which the nest material composed of fine rotted vegetation fragments supports a
rich growth of fungal hyphae which penetrate the whole fabric apparently as an
integral component of the nest structure (Hansell and Turillazzi 1995). It seems that
the fungus is being tolerated for the extra strength it provides. A similar compound
material possibly with the same function is reported in *Eustenogaster eximia*
(Krombein 1991).

The selective advantage of a tough nest material for the Polistinae may initially
have been the ability to construct a nest petiole creating secure nest sites in a much
wider range of locations than can be used by Stenogastrinae, which remain dependent on natural nest suspensions. This tough paper also offered scope for a wide
range of nest architecture. This, as I will point out again later, gave ecological
opportunities to Polistinae unavailable to Stenogastrinae.

Hanging paper (carton) architecture has also evolved in a number of ant taxa (e.g.
Azteca, *Tapinoma* and *Crematogaster*) allowing nests to be placed high in trees and
relatively inaccessible sites such as on single leaves. These sites became available
to species of the ant genera *Dendromyrmex*, *Camponotus*, *Polyrachis* and
Oecophylla by nest construction using larval silk (Hölldobler and Wilson 1983).

Nest material specializations have ecological consequences

I have already pointed out the need to know more about nest building costs. Nest
materials have two obvious costs: travel costs and preparation costs. One way to
reduce these is to reuse what has been collected already. Colonies in a growth phase
may find this difficult to do; however, as the colony approaches more nearly to its
maximum size, we would expect to see evidence of reuse of existing structures or
recycling of nest material.

There is widespread evidence of both of these in wasp nests. The recycling of
nest material is well known in Stenogastrinae (Hansell 1981) and in Vespinae both
paper and pupal cocoon silk may be reused to strengthen suspension (Matsuura and
Yamane 1990). Keeping (1991) records similar recycling of materials in the
polistine *Belonogaster petiolata*.

A similar phenomenon occurs in certain stingless bees (Meliponini) where brood
cells are separate globular structures which, after they have been lined with silk,
have all the wax removed from them by worker bees to make new brood cells before
the original bees have hatched (Hodl, pers. comm.). In *Trigona terminata* where
brood cells are arranged as a spiral comb, material from used cells is removed to
build cells elsewhere (Khoo and Yong 1987).

The widespread occurrence of material recycling suggests that nest architecture
may be designed to make a little go a long way. This is typified in Stenogastrinae

where a cell is frequently only as big as appears necessary to contain its particular stage of brood. The adaptive significance of the lack of cell wall sharing seen for example in *Parischnogaster jacobsoni* or *P. nigricans serrei* (Turillazzi 1991) may in part be due to the flexibility such designs give to the recycling of nest material. This interpretation contrasts with that of Jeanne (1975) who argued that the more economic nest design would maximize cell wall sharing, however, these stenogastrine examples show that with active nest material recycling and suitable nest architecture, a different economic solution may prevail.

Travel costs will be affected by the availability of the nest material; so, more accurate identification of collected materials is needed in order to determine its distribution in the habitat. To these travel costs must be added costs of preparation which for most wasp nest materials includes the addition of a salivary secretion. However, as Jeanne points out (this volume), there is currently conflicting evidence on the consistency of this matrix. The evidence for the Polistinae preponderantly suggests that protein is the major constituent, nevertheless analysis by Schremmer *et al.* (1985) on the nest envelope *Pseudochartergus chartergoides* indicates significant amounts of chitin.

I want to suggest that we can use the assumption that natural selection will tend to favour a reduction in nest material costs to predict the likely composition of the nest material matrix. This can be illustrated first by considering the evolution of mechanical defences by certain plant-sucking bugs. For example, the woolly alder aphid, *Prociphilus tessalatus* protects itself from predators by the secretion over its body surface of long plumes of waxy material (Eisner *et al.* 1978), identified as a long-chain keto-ester (Meinwald *et al.* 1975). Similar protective secretion is produced by a variety of other aphid species (Pope 1983). These specialized secretions must have arisen from a combination of two circumstances. Firstly, the presence of secretory glands on the surface of ancestral insects upon which natural selection could operate; secondly, I would suggest the ability of these glands to produce large quantities of secretion at relatively low cost. Cuticular waxes have been recorded in a variety of insects, produced by sub-cuticular oenocytes (Lockey 1988) and forming part of a protective barrier against water loss. These cells seem in some species to have been the origin of more specialized wax-secreting glands; for example, in the woolly aphid *Erisoma lanigerum* developing wax glands have a basal layer of oenocytes (Waku and Foldi 1984) and in worker honey bees oenocytes are at their peak of activity when wax secretion is at its peak (Kramer and Wigglesworth 1950).

Plant-sucking bugs are generally sedentary but highly fecund. Their dietary needs are therefore more for amino acid than carbohydrate, so they must ingest large quantities of sap to filter out the small proportion of amino acid, giving them a large surplus carbohydrate. Some aphids secrete this as 'honeydew', which is largely a mixture of sugars (Hölldobler and Wilson 1990) and secures for them the protection of ants against other insect predators. In the case of woolly aphids this carbohydrate surplus is converted to wax to provide mechanical protection.

The argument that aphids have a cheap carbohydrate material that can readily be adapted for defence, could be applied to social bees since surplus nectar might be metabolized to wax to create a cheap nest building material. But we would predict

that for carnivorous species like social wasps the economics should be quite different and protein might well be the cheapest secreted material for nest construction.

The presence of chitin in the envelope of *Pseudochartergus chartergoides*, as
claimed by Schremmer *et al.* (1985) would then not be expected and therefore deserves more detailed examination of its benefits and costs. It may be that the identification of chitin is mistaken or, more interestingly, that the unusual transparent
envelope of this species has required a rather more expensive material to effect,
although, as Jeanne (this volume) points out, the same type of envelope has been
achieved by *Ropalidia opifex* by means of a secretion largely composed of protein
and without chitin (Maschwitz *et al.* 1990).

In addition to low cost, a nest material should also be selected for its mechanical
effectiveness. The result should in general be that wasp nest technology is consonant with the level of social organization and hence the size and architectural complexity of the nest of the species. This is well illustrated by the stenograstrine wasp
Anischnogaster laticeps whose nest of rotted vegetation fragments permitted by
fungal hyphae has already been described. Colonies in this species rarely exceed a
single adult female (Turillazzi and Hansell 1991) and the nests typically have only
three or four cells. The wasp, it appears, is making use of the fungus which gives
additional strength to the nest but ultimately consumes it; a technology adapted to a
colony life of small size and limited duration.

At the other social extreme, the nest of *Polybia rejecta* can contain over 10 000
adult wasps (Itô 1993) and may last for many years, so very different nest technology must be required for such a nest compared with that seen in nests of *A. laticeps*.
Between these extremes of nest size there may indeed by intermediate technologies
either in the collected or secreted component of the material.

What might be revealed by an investigation of this problem is exemplified by
unpublished results of my own on the use of arthropod silk as a structural component of bird's nests in relation to nest size. My analysis using the expanded family
Corvidae (Sibley and Monroe 1990) shows not simply that silk is a material suited
to small rather than large nests but, more interestingly, that there is quite sharp cut-
off between the size of nest where silk is a valuable nest material and that where it
is unused (Fig. 16.2). This boundary may be set by the mechanical properties of the
silk or the costs of gathering sufficient amounts of it, in either case it is a material
with clear technological limitations.

If the usefulness of silk for breeding birds is set by its availability then suitable
habitats must be ones that include sufficient silk as well as other resources such as
food. So, for central place foragers, as nest builders inevitably are, nest materials
are likely to be more and more ecologically important the more specialized the
material and the more scarce its occurrence. Consequently, in the case of social
wasps, it is really necessary to determine as precisely as possible the identity of the
collected component of nest paper.

We know already that nest papers vary. Many include woody plant stem fibres
but a number of species make use of plant hairs (Wenzel 1991). However, the commodity *plant hairs* is itself very heterogeneous, it may apply either to pappus hairs
which aid seed dispersal or to hairs covering leaf and stem surfaces and even leaf

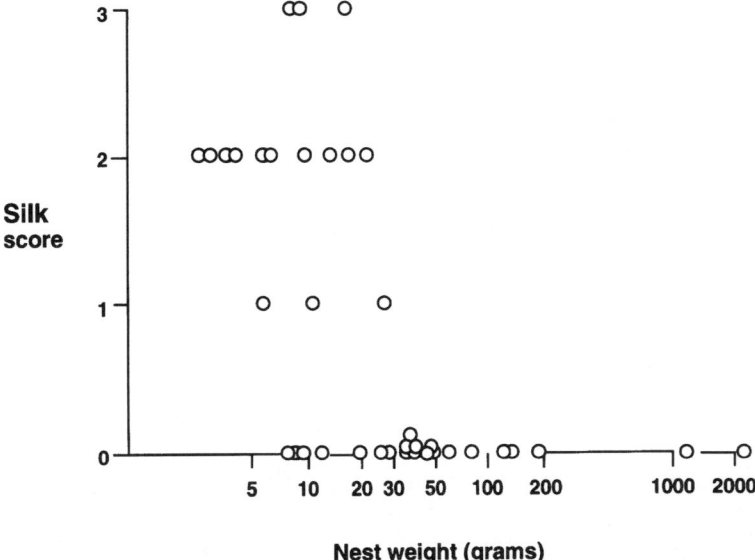

Fig. 16.2 Nest weight and silk use by birds of the family Corvidae (Sibley and Monroe 1990). Silk score is a scale in which: 3 = abundant use of silk which is essential to the nest structure and 0 = no silk present in the nest. Silk is shown to be an important material for nests of 30 grams or less but unused in nests of a larger size. (Each point represents the nest of a different species; n = 38.) (Hansell, unpublished.)

hairs have widely differing morphology between plant species. I found a sample of the nest material of *Protopolybia sedula* to be composed entirely of boat-shaped leaf hairs from a plant of the family Malpighiaceae (Fig. 16.3a) (Hansell, unpublished), whereas nest material of *Ropalidia guttatipennis* contained significant amounts of stellate hairs. Similar hairs were found among the more usual decayed plant fragments of the material of *Anischnogaster* sp. B (Fig. 16.3b).

As I discovered, stellate plant hairs are the material used to make the nest cup of *Eugenes fulgens* (the Magnificent Hummingbird) (Fig. 16.3c). This I believe is not just a curious coincidence. The nests of a number of hummingbird species are composed of plant hairs, either of the pappus or leaf type and, with a diameter of 3–4 cm and a weight sometimes of less than 1.0 g, they are built on a similar scale to the nests of many wasps. This suggests to me again not only that certain technologies may be appropriate to a particular scale of engineering but also that generalizations concerning nest building may usefully be made across phyla.

The Black-chinned Hummingbird (*Archilocus colubris*) also makes its nest cup of plant leaf hairs but of a complex branching kind obtained from the under-surface of the leaves of the California Sycamore (*Platanus racemosa*) (Kiff, pers. comm.). Such an ecological constraint is also recorded for social wasps not of the Vespidae but of the Sphecidae. *Microstigmus comes* is a eusocial sphecid in which small groups of wasps construct a nest by gathering the waxy material from the undersurface of a particular palm (*Chrysophila gargara*) and moulding it *in situ*

Fig. 16.3 The use of leaf hairs in the nests of wasps and birds: (a) Elongate hairs (probably family Malpighiaceae) in the nest material of *Protopolybia sedula* (Polistinae) (Hansell, unpublished). (b) Intact stellate hair among shattered plant fragments in the nest material of *Anischnogaster* sp. B (Stenogastrinae) (Hansell and Turillazzi 1995). (c) Stellate leaf hairs which comprise the nest cup of the Magnificent Hummingbird (*Eugenes fulgens*) (Hansell, unpublished). In all cases the distance between white bars = 10 microns).

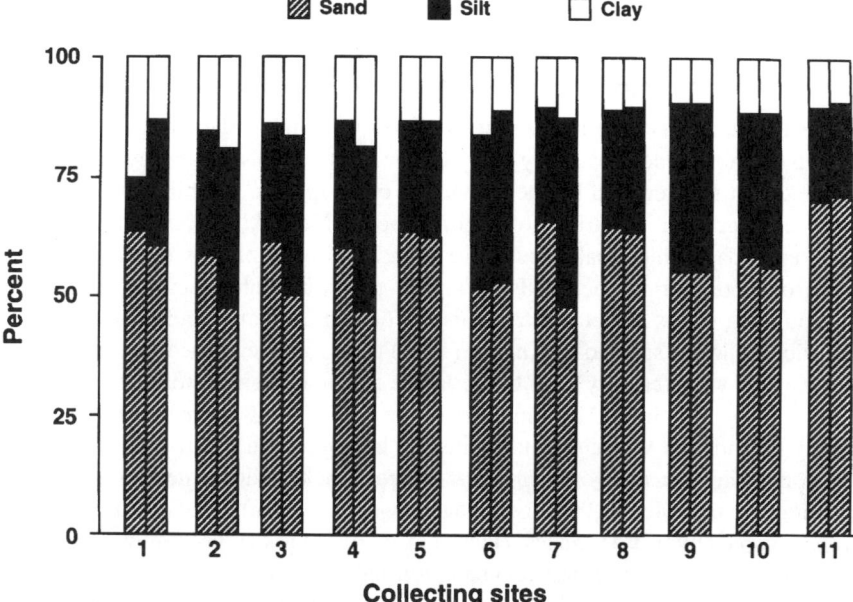

Fig. 16.4 Differences in the composition of the mud of nests of the Cliff Swallow (left half) and Barn Swallow (right half) at 11 different collecting sites. Values are mean percentages for each component. Cliff Swallows are seen to use more sand and less silt than Barn Swallows. (Adapted from Kilgore and Knudsen 1977.)

with a silken secretion of their own. Matthews and Starr (1984) found that if they destroyed these nests, 75% of the dispossessed wasps returned to the same palm tree from which their nests had been removed. A further example of dependency on a single source of nest material is described by Prance (1992). An ant of the genus *Hypoclinea* builds nests in trees of the species *Parinaria excelsa* composed of hairs harvested from the undersurface of the tree's leaves. Prance (1992) claims that this is a bilateral dependency in which the function of the hairs on the leaves is to provide nest material for ant colonies which will protect the tree from folivores.

These are extreme examples of nest material specialization but any specialization may limit habitat choice. Mud for example may seem like a fairly widespread commodity, yet Tomialojc (1992) suggests that the European Song Thrush (*Turdus philomelos*) may be prevented from extending its breeding range into drier more Mediterranean habitats by the shortage of mud to line the nest cup. Mud may even, it seems, show structurally important varieties. For example, Kilgore and Knudsen (1977) studying Cliff Swallows (*Petrochelidon pyrrhonota*) and Barn Swallows (*Hirundo rustica*) found nesting together at 11 different sites, showed that in building their retort shaped, rock attached nests, the former species used mud with more sand and less silt than the latter (Fig. 16.4). The authors speculate that this allows Cliff Swallows an improved ability to mould a complex shape, albeit at some expense to nest strength. Such a study emphasizes again the need to examine nest materials in greater detail.

Valuable nests attract usurpers and squatters

As nest building has costs then a nest represents wealth. It is therefore not surprising to find that abandoned nests may be reoccupied or that individuals may make use of suitable nest sites created by the activity of other species. Roubik (1989) summarizes evidence for a variety of solitary bees (Megachilidae, Euglossinae and Anthophoridae) which make use of the abandoned nests of bees or wasps and the holes of wood-boring beetles. Field (1992) reviewing intraspecific parasitism in nest building wasps and bees cites examples of nest usurpation in five ground-nesting subfamilies of sphecidae. Knisley (1985) even found evidence that a eumenine wasp would kill predatory tiger beetle larvae to obtain their burrows.

Usurpation may indeed be a more effective option in situations where the nest is poorly defended or where defenders are few in number. In the tropical harvestman (opilionid) *Zygopachylus albomarginis* where a male builds a nest to attract ovipositing females, Mora (1990) found that between one quarter and one third of nests were occupied by usurper males. Hansell (1987*b*) found that usurpation was a typical if minority method of nest foundation in the stenogastrine wasp *Eustenogaster calyptodoma* and it seems likely that this pattern is widespread in the subfamily.

Usurpation is also likely to be selected for where the value of the nest is especially high. This may be the case in the carpenter bee *Xylocopa pubescens* which excavates nest burrows in wood. Van der Blom and Velthuis (1988) found 50% usurpation of burrows that were occupied by a lone female.

The accumulation of valuable resources in nests has also led both in wasps and other social insects to the evolution of specialist social parasites, frequently species closely related to those that they parasitize. Cervo and Dani in this volume provide a detailed account of the evolution and biology of such species in the genus *Polistes*.

Nests which are occupied by the same colony over long periods of time stand the risk of attracting a wide variety of sometimes highly specialized lodgers and parasites. Hölldobler and Wilson (1990) list arthropod symbionts of ant nests extending across six classes and 10 insect orders which includes 35 families of beetles. Nests of wasps have attracted similar kinds of cohabitants (Wilson 1971) but generally rather little seems to be known about the extent of the damage, or indeed benefit, which they confer. Among the wasps, the nests of genus *Polistes* are also shown by Yamane in this volume to attract a variety of parasitoids, scavengers and predators and the range of species responsible for attacks on wasp nests in general was reviewed by Jeanne (1975).

Wasps have responded in this evolutionary arms race with defences which include aspects of nest architecture but, if these are to be fully understood, a more detailed study of the nature of predatory attacks now seems necessary. Three different types of investigation are possible: first, direct observation of predatory attacks; second, careful examination of nest remains after attacks by identified predators; and third, studies on nest survival in wasp species which exhibit variation in nest site or nest architecture.

Accumulating this information is time consuming, direct observations of predatory attacks being particularly difficult to obtain. Nevertheless, Jeanne (1975) was able to compile a list of social wasp brood predators including ant, wasp, reptile, bird and mammal species. Some of these species are also known to be to a greater or lesser extent social wasp specialists. *Vespa tropica* is a predator of social wasps (particularly Polistinae) in the Asian tropics (Matsuura and Yamane 1990), the Old World army ant *Aenictus laevis* (Dorylinae) is a predator of social wasps and termites and the New World army ant *Eciton hamatum* is yet another specialist on social wasps (Hölldobler and Wilson 1990). Among the birds we know that the Honey Buzzard (*Pernis apivorus*) depredates temperate Vespinae (Cramp and Simmons 1980), and the Red-throated Caracara (*Daptrius americanus*) specializes in South American Polistinae and Meliponini (Thiollay 1991). To understand how nests provide defence against these attackers now requires more concentration on the predator. This might best be achieved through collaboration between wasp biologists and, for example, ornithologists.

The insights that might come from a more detailed study of the evidence remaining after a predator attack is exemplified by observations on damaged nests of *Parischnogaster nigricans serrei* after attacks by *Vespa tropica*. This shows that although the majority of pupae and older larvae are depredated, eggs and young larvae are largely untouched. Turillazzi (1985d) speculates that while younger brood may be ignored because of their small size, they could be gaining protection from the white sticky Dufour gland secretion applied to these stages in most species of Stenogastrinae. Experimental removal and transfer of the secretion could be used to test between these alternatives.

The third approach, that of studying nest survival in species where there is variation within the population in nest site or architecture is discussed in more detail below.

Nest variation in space and time

In the Vespidae there is a great variety of nest designs (Wenzel 1991). Much of this between-species variation may be a result of differences in colony sizes necessitating differences in architecture. However, evidence from the Stenogastrinae, for example, suggests that this is only a partial explanation. This subfamily is notable for the variety of nest designs found among its species (Turillazzi 1991) yet colony sizes are uniformly small (i.e. less than 10 females). Great diversity of nest design is also shown within the polistine genus *Mischocyttarus* (Wenzel 1991) where again colony sizes are relatively small and uniform. Our inability to explain these striking architectural differences exposes the weakness of our current data and the methods I have described above would certainly help to remedy the situation.

It may of course be that between-species differences in the nest architecture of Stenogastrinae are of no adaptive significance, a by product of divergence between species in other characters. But such a negative hypothesis is essentially untestable without detailed phylogenetic and ecological evidence, and I am reluctant to resort to such an explanation until adaptive hypotheses have been tested. Furthermore, we

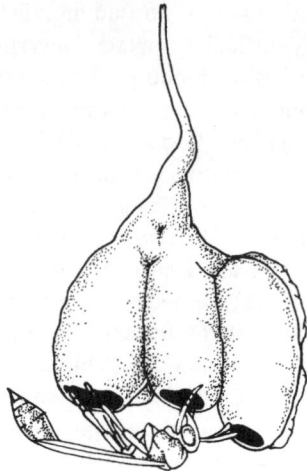

Fig. 16.5 The nest of *Anischnogaster laticeps* (Stenogastrinae) is a small cluster of cells with entrances all at the same level. These can be protected from parasitoids by the single female adopting her typical resting position across the bottom of the comb. (Redrawn from Turillazzi and Hansell 1991.)

do know enough already to produce a general explanation for this variation which generates some testable predictions. If we make an assumption that nest architecture is influenced by multiple selection pressures, which seems probable, then between-species differences may result from differences in the importance of the various selection pressures between them.

The linear arrangement of brood cells along the length of the nest suspension by *Parischnogaster nigricans serrei* and *Parischnogater jacobsoni* may give a stick-like appearance to the nest which reduces detection by visual predators such as *Vespa tropica* (Turillazzi 1991). The same stick-like profile is achieved by *Metischnogaster drewseni* and *Parischnogaster timida*, probably for the same reason, by attaching only the first cell to the suspension and each subsequent cell directly below the previous one. This design therefore places increasing stress on the walls of the upper nest cells which may explain why the nests of both species have one or more conical hats placed on the suspension (tentatively described by Williams 1919 for *P. timida* as 'umbrellas') which apparently deflect rain water from the fragile cells (Turillazzi 1990*b*).

The nest design of *Anischnogaster laticeps* can be explained as protection of brood against parasitoids. It is a cluster typically of only three or four cells. The exits of the cells are all at the same level allowing the single resident female resting on the comb to protect larvae from tachinid flies which seem to be an important source of brood loss (Fig. 16.5) (Turillazzi and Hansell 1991). The single comb nest of *Eustenogaster calyptodoma* surrounded by an envelope which tapers to a narrow spout may, however, reflect the importance of nest usurpation as a method of colony foundation in this species. The most reproductively mature female in the colony rests at the top of the spout, repelling frequent attempts by strange females to gain access (Hansell 1987*b*).

Some within-species variation may also be explained as differing colony responses to conflicting or locally variable selection pressures. This is also illustrated by a stenogastrine example. The nest design of *P. mellyi* varies between a single comb attached to the end of a suspension and a row of cells extended along it. The preferred design (i.e. the design adopted if a long suspension is used) is the elongate nest form but the comb arrangement is resorted to on short suspensions (Hansell 1981). It appears that while the elongate design may render nests less detectable to visual predators, the flexible architecture allows a nest to be constructed on a suboptimal nest suspension in a habitat where better suspensions are in short supply. This should be testable by the manipulation of suspension availability in laboratory or field experiments.

The dependence on natural suspensions for nest attachment is a general one for Stenogastrinae which, unlike the Polistinae are unable to construct a suspension of their own. This may provide an explanation for the apparent within-species variation in *Parischnogaster alternata* (Coster-Longman and Turillazzi 1995). The nest takes two distinct forms, the one attached to a fine downward-projecting suspension, the other attached to a cave roof.

In the former the nest narrows to a point around the fine suspension whereas in the latter it spreads out to form a flat top as it meets the cave roof. There are also other differences; the material of the flat topped form contains more mineral material than the pointed form and its brood cells are slightly larger. But, as these two designs may occur together on the roof of the same cave or storm drain, are they the work of two separate species, examples of within-species polymorphism or of behavioural flexibility contingent upon nest site availability? In the absence of further evidence, contingent response to extend the range of suitable nest sites seems the most parsimonious explanation, indeed *Liostenogaster flavolineata* appears to produce two different attachment designs in its nests to achieve the same options (Samuel 1987).

The problem of nest attachment to variations in nest site topography is of course a general one. Consequently we would predict more variation in that component of nest architecture concerned with attachment than in other aspects of nest design. This is confirmed for a variety of bird species which attach their nest to tree branches (Nickell 1958).

Within-species variation in a widely distributed species could come about through different designs being adapted to slightly different climatic conditions resulting in architecturally distinct regional morphologies. Jacklyn (1991) has shown that the mounds of the termites *Amitermes meridionalis* and *A. laurensis* show slight regional differences in the overall north–south alignment apparently to improve internal nest temperatures in response to local wind conditions. The mounds of *Cornitermes bequaerti* in South America appear to be composed of two completely distinct designs. In one there are many ventilation openings in a low mound, in the other there is a single tall chimney through which the air is drawn (Fig. 16.6a and b). The two forms are found in different areas, in different soil types

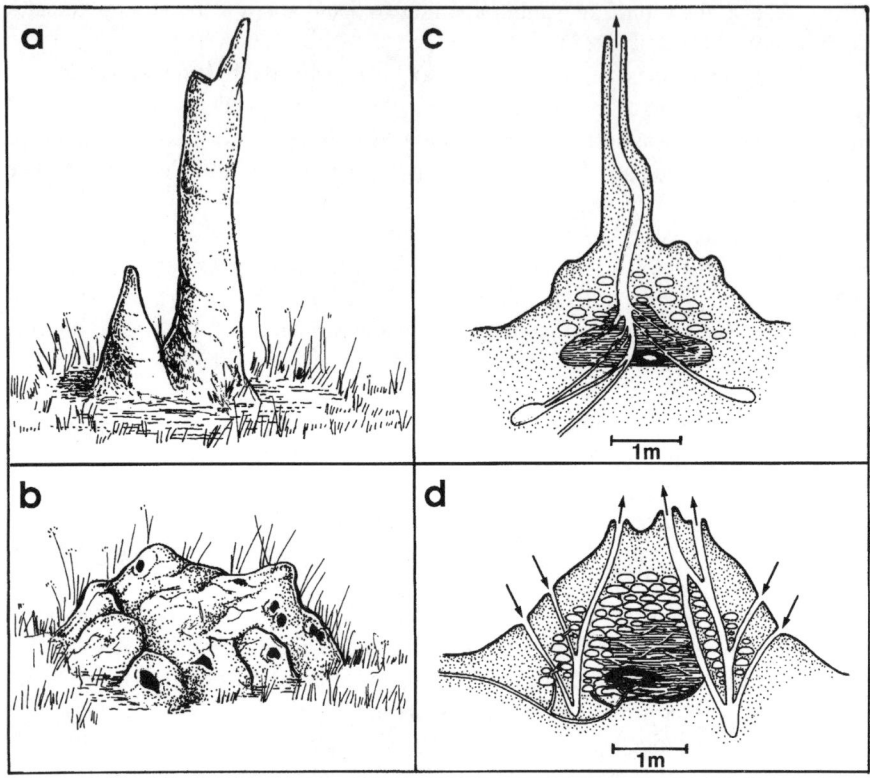

Fig. 16.6 Apparently convergent polymorphic architecture in the mounds of two termite species, *Cornitermes bequaerti* (a and b) and *Macrotermes subhyalinus* (c and d). (a) and (c) show ventilation through a single tall chimney out of which air is drawn whereas (b) and (d) show ventilation through a number of apertures scattered over the surface of a lower mound.

(Cancello 1991). These geographic variants are closely paralleled by ones found among the more varied architecture of *Macrotermes subhyalinus* (Fig. 16.6c and d) (Darlington 1984).

The presence of distinctive nest variants within one population raises the question of whether nest building is important in facilitating the creation of new species. West-Eberhard (1986*b*) proposes that the presence of alternative adaptive phenotypes in a population could provide a mechanism for rapid speciation. In the above termite examples the alternative phenotypes are architectural. Geographic separation of the two types could lead to conditions in both sub-population in which either only one phenotype prospers or in which only one can be expressed, leading to phenotypic fixation. Fixation could then accelerate evolutionary change by allowing innovations previously selected against, because they were incompatible with other architectural specializations in the population as a whole. If sympatry were now restored, speciation might result because worker offspring of matings between two distinct nest building traditions would produce disorganized architecture.

Nests result from behaviour which is an important agent of evolutionary change (Wcislo 1989). In particular, animal mobility can greatly extend habitat choices, exposing animals to new selective environments. These could reveal new phenotypic variability among existing genotypes (Bateson 1988). These need not, of course, be new architectural or even behavioural phenotypes; however, a shift in habitat range might result in the simultaneous generation of a new architectural variant, leading to speciation in the manner proposed by West-Eberhard (1986*b*).

Bateson (1988) also identifies behaviour changes which alter physical or social conditions as having substantial evolutionary consequences. Nest building is a very important aspect of such behaviour since, as described under the heading *Nests reinforce social life*, the construction of a nest alters both the physical and social environment with important evolutionary consequences, and so provides opportunities for speciation.

One other behavioural mechanism could have had consequences for speciation through divergence in nesting behaviour. The mechanism is imprinting and the way it might achieve its effect is through nest site preference. Ten Cate and Bateson (1988) point out how the behavioural mechanism of imprinting could, in a single generation, lead to differences in mate choice which could restrict the free flow of genes within the population. Such a mechanism, say in choice of nest site, could result in the rapid development of distinct within-species nest site traditions. Wenzel (this volume) found evidence of this in *Polistes annularis* in an experiment where some females were obliged, in one year, to make use of artificial nest sites which became the chosen nest site of their daughters the following year.

West-Eberhard (this volume) also draws attention to the association between within-species and between-species variation. The implication being, in the context of nest architecture, that for nest architecture to have been important in wasp speciation there should be ample evidence of within-species variation. As has been shown above, in spite of a lack of studies concentrating on this issue, such variation does clearly exist. If this within-species variation in nest building behaviour has indeed been influential in the speciation of wasps, then we would expect that the phylogeny of a group could be substantially deduced by nest architecture alone. The phylogeny of 28 genera of Polistinae, based on evidence obtained from both the morphology of the insects and the architecture of the nests (Fig. 16.7) (Wenzel 1993), does at least show a general correspondence between the two. So, while it can be argued that diversification in nest design has been a consequence rather than a cause of speciation, various other kinds of evidence does suggest that within-species variation in nest architecture could have been a significant factor in the speciation of social wasps and other nest building groups.

One final aspect of nest variation merits consideration: the ontogeny of a nest from foundation to termination. During ontogeny, changes may occur in the size or spatial relationships between different nest parts. The allocation of materials to these various nest components requires decision rules which may alter during nest development. Simple nest expansion mechanisms have been adopted by some species, for example the modular system shown by *Polybia* (Wenzel 1991). This design may be very effective where the selection pressures acting upon younger nests

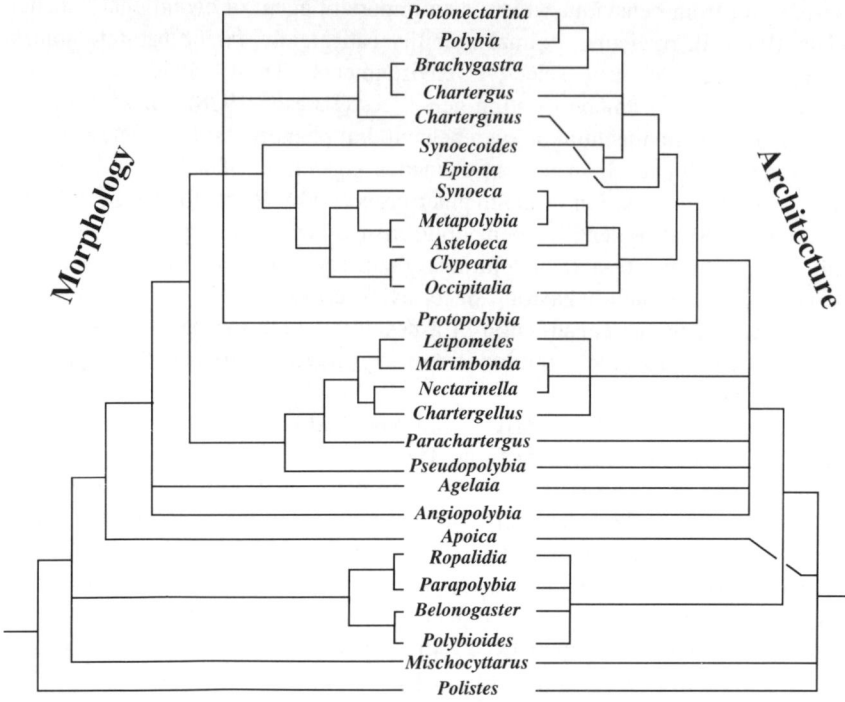

Fig. 16.7 Consensus trees for wasp morphological traits (left side, Carpenter 1991) and nest architectural traits (right side, Wenzel 1993) in the 28 genera of Polistinae.

and older ones are similar, but in some species this may not be the case. For instance, temperate Vespinae nests at the end of the season are very large compared with their initial size. The climate may change substantially through the season and the priorities of the colony alter from one of raising workers to the production of reproductives. So, is the proportion of comb to envelope, to take only one nest growth relationship, optimal throughout these changes? I suspect it may well not be.

Instead, allocation of material to comb or envelope might be determined by some relatively unsophisticated rule of thumb; for example, some simple allometric growth rule with perhaps limited scope for modification through the response of workers to changing environmental cues. The rule might represent a compromise between different optima of comb in relation to envelope at different stages of nest ontogeny, or ensure that an optimal proportion was created at some critical developmental stage. Nest growth rules could be deduced from the collection of nests from different sites at different stages of the season, as well as from the growth of nests built by captive colonies in controlled environments. Measures of colony fitness are less easy to determine but could be indirect, such as evidence that colony growth was unaffected by experimental reduction of the amount of envelope at a certain stage of nest growth.

Conclusions

1. The evolution of social wasps appears to have been influenced in a number of important ways. In particular it is suggested that nest construction may have encouraged the evolution of social living, brought about substantial changes in habitat range and been an agent of speciation.

2. The study of colony biology should consequently include an examination of the relationship between wasp colonies and their nests. This could reveal, for example, the degree of specialization of nestmaterial which could affect the mechanical properties of the nest, the cost of nestmaterial collection and even the suitability of a habitat for colonization. Understanding the defensive architecture of the nest could help establish the importance of specialist parasites and predators to colony biology, while insights into changing priorities of the colony during growth may result from a study of the ontogeny of nest architecture.

3. Marked differences in nest architecture between species of similar colony size could reflect differences in dominant selection pressure but this at present is largely a matter of speculation.

4. Greater attention needs to be given to the activities of wasps away from the nest to determine, for example, if new colony foundation is limited by competition for food or the costs of nest foundation (such as predation risk or nest material collection).

5. More information is needed on the extent and nature of within-species nest variation, if its ecological and evolutionary significance is to be established.

6. The relationship between nest builder and nest has generalizable biological features. Consequently insights into this relationship in wasps may come from knowledge of work in other fields (e.g. ornithology). Equally, greater understanding of wasp nest building behaviour may have implications beyond the study of social Hymenoptera.

Acknowledgements

I am grateful to the following referees for their helpful comments: James Carpenter, Mary Jane West-Eberhard, Howard Evans and Stefano Turillazzi. I would also like to give special thanks to John Wenzel for his constructive encouragement.

17

Wasp societies as microcosms for the study of development and evolution

Mary Jane West-Eberhard

Introduction: Darwin's dilemma

Many discussions of insect social evolution begin with 'Darwin's Dilemma'—the question of sterile castes raised by Charles Darwin (1859, pp. 203–4) in his chapter on 'problems' for the theory of natural selection. Darwin's Dilemma is by now familiar to students of the social insects:

... one special difficulty ... at first appeared to me insuperable, and actually fatal to my whole theory. I allude to the neuters of sterile females in insect-communities: for these neuters often differ widely in instinct and in structure from both the males and fertile females, and yet, from being sterile, they cannot propagate their kind.

For Darwin the 'Dilemma' was twofold. First, it concerned how selection could favour traits associated with sterility in the performer (altruism). That he solved by visualizing selection on families, so that traits costly to one would benefit others of the same line or group of relatives (a 'kin selection' explanation).

The other aspect of the dilemma is usually overlooked. It has to do with the fact that individuals of similar hereditary makeup—members of the same family—represent phenotypes evolving in different ways. Not only are the neuters different from the queens, but the neuters themselves can show different adaptively specialized morphologies. So multiple simultaneous directions of evolution occur within a single lineage.

Darwin focuses on this second aspect of the social insect problem after having resolved the problem of relatedness (1859, p. 238) [Sixth edition, 1872, p. 205]:

But we have not as yet touched on the climax [acme] of the difficulty; namely, the fact that the neuters of several ants differ, not only from the fertile females and males, but from each other, sometimes to an almost incredible degree ... It will indeed be thought that I have an overweening confidence in the principle of natural selection, when I do not admit that such wonderful and well-established facts at once annihilate my theory.

Modern discussions in terms of fitness and relatedness do not deal explicitly with this difficulty, but they imply a solution: a certain developmental pattern must evolve, such that social (altruistic) traits are carried by queens but expressed in workers. Furthermore, the cost-benefit side of kin selection (Hamilton 1964*a*, *b*) implies a regulatory mechanism with an adaptively adjusted threshold such that conditional altruism is expressed in circumstances where its benefits are likely to exceed its costs. That is, a particular kind of condition-sensitive developmental mechanism is required for kin selection to work. In the social insects the divergent characteristics of workers and queens are 'caste limited' and adaptively regulated so as to be expressed only when likely to be advantageous. And the result of disruptive selection against intermediates is developmental: discrete phenotypes are the result, not of bimodal distributions of genotypes, but of bifurcate developmental pathways (disruptive development) (West-Eberhard 1989 and 1992).

This and other 'developmental' aspects of social evolution are the subject of this chapter. I will argue that all major areas of social wasp evolution—those concerned with natural selection, adaptation, speciation and macroevolution—are clarified if development (patterned, conditional gene expression) is considered alongside genetics (gene transmission) as an aspect of inheritance. My aim is to use wasps to show how genetics, development, behaviour and selection are related during evolutionary change.

A developmental view of the wasp

'Development' is the process by which phenotypes are produced. The phenotype includes not only morphology but also behaviour and physiology. Thus, I will consider the neuromuscular co-ordination of movement an aspect of the development of a behaviour pattern.

Development is condition-sensitive. So relating development to evolution leads to a more general problem—that of incorporating environmental influence into one's view of phenotypic form, selection (differential reproductive success of different phenotypes) and genetical evolution (evolution involving gene-frequency change). One way to do this is to visualize the phenotype as composed of many hierarchically arranged 'subunits' of gene expression or gene-product use, each co-ordinated by a regulatory mechanism that is to some degree condition-sensitive.

Consider a wasp—it begins life as an egg. Even at that stage phenotype development could not be accurately said to be 'dictated' by the 'genetic program' of the zygotic genes. If the egg were just a sack of genes, phenotype development would not occur. Rather, the initiation of insect development depends on intricate relations between the genes of the new individual and specific elements of its environment, especially, proteins supplied to the egg by the maternal nurse cells prior to oviposition (e.g. see Davidson 1986). From the beginning, the environment is an active director of development—not just a source of materials or of favourable conditions.

As a larva the wasp is already specialized to the peculiar social niche of a dependent offspring: it stretches forward in its cell to receive food and to transfer liquid droplets to adults. In a still later stage, it stops feeding and spins its own cap to the cell. Then, as an adult, the morphology and behaviour of the same wasp is completely different—it walks and flies and occupies a distinctive ecological and social niche. A single female can have a variety of distinctive behavioural phenotypes: at different times it is a predator, a nest builder, and an aggressive defender of the colony. And there is a striking phenotypic divergence between worker and queen.

Given the multiple phenotypes of a wasp, we can restate the second part of

Darwin's Dilemma in more general terms (R. Burian (this volume) calls this problem 'Lillie's paradox' with reference to Lillie (1927)): genetical evolution involves a change in gene frequencies; yet a given genotype can produce all of these distinctive forms, each apparently evolving in its own way. How can there be adaptive change in all of these phenotypic subunits at once? Given the complexity of gene interaction, what keeps one improvement from disrupting or diluting the others?

Indeed, this question has reinforced the idea that major change is difficult to achieve; that evolution must proceed by small steps; that the entry into new niches may require a period of genetic instability or lowered fitness (the maladaptive 'valleys' between adaptive peaks); and that major divergence requires genetic re-organization only possible in genetic isolates—that is, associated with speciation (e.g. see West-Eberhard 1992, p. 58 for a brief review). The punctuated equilibria hypothesis is an extreme version of the last idea.

A vision of populations and organisms as intricately cohesive and resistant to change would seem to complicate, not solve, Darwin's dilemma of diversification within social insect species. This is where development enters the story of wasp evolution. To visualize how 'intragenomic' divergence can occur, it is necessary to have a developmental view of the wasp, alongside a genetic one.

The subunit organization of phenotypes

To facilitate a developmental analysis, picture the individual phenotype as composed of subunits, each controlled by a regulatory mechanism or switch. This is an over-simplification in various ways: it does not acknowledge overlap (or genetic correlation) between traits, but I will return to that later; and it may obscure the fact that these switches can occur at all levels of organization; larval traits, for example, although a co-evolved set at one level of analysis, at lower levels may be seen as a mosaic, with a nested series of switches controlling ever smaller subunits of the larval phenotype. Furthermore, the development of a subunit 'trait' may be initiated by one regulatory event (or switch) and terminated by another. The important aspect of 'subunits' for this discussion is their co-ordinated expression or use by whatever means and at whatever level of organization. Co-ordination by a switch mechanism makes them dissociable and somewhat independently subject to selection and evolution.

The origins of novelty: the principle of dissociability and phenotypic recombination

A developmental or behavioural decision point is a potential uncoupling point during evolution: a switch can turn a character off or change the timing of its expression; it can cause an established trait to be lost or a lost trait to be revived. Owing to this dissociability at switching points, novel traits can originate via re-organization and recombination of pre-existing subunits. New characters can be made from old pieces via a reordering of development. By this means, as stated by Hansell in a discussion of the origin of vespid secretory organs (this volume), evolution 'recruits' from what is already there.

Kinds of dissociation and phenotypic recombination, include duplication, deletion, atavism, shifts in timing of expression (heterochrony), shifts between the sexes
(sex transmutation) and novel co-ordinated reductions due to pre-existing flexibility of response (see West-Eberhard 1989). The example I will discuss here—the
origin of reproductive 'castes' (queens and workers) in wasps—might be classified
as a 'reciprocal deletion,' for it probably involved the decoupling of ancestral reproductive cycle into two parts with one expressed in workers and the other in
queens (West-Eberhard 1987*a*). Each caste lacks the 'deleted' set of traits expressed in the other, and the two alternatives are mutually 'dependent' or complementary morphs (see West-Eberhard 1979 and Gadagkar this volume): they stay
together and co-operate (or parasitize each other!) in the same colony, compensat-
ing each other's deficiencies.

It might seem that a deletion is an odd form of progress. It creates a novelty by
subtraction of formerly essential traits. But there is an advantage (also mentioned
by Gadagkar) of compartmentalization that has interesting general consequences:
origin by deletion means that the new form, as in the case of 'phenotype fixation' of
a former alternative (West-Eberhard 1986*b*, 1989 and below), is 'released' to specialize in a narrowed set of functions, without conflicting selection on the omitted
traits. To some degree this is true whenever a new developmental or behavioural
bifurcation occurs: it creates a new semi-independent subunit of selection. But it is
especially clear in the case of origin by deletion, for genetic correlations are broken
without new ones being formed as may occur in some other kinds of dissociation
and re-organization.

One challenge of a developmental evolutionary biology is to demand more preci-
sion in pinpointing the actual effects on phenotype ontogeny of the genes that influence a particular phenotypic change. This requires attention to the mechanisms of
regulation. As a first step, one can ask 'What is the least possible genetic or environmental change that might account for this transition?' Is it correct to suppose that a
particular innovation and its fitness effects are due to genetic variation? Or might
they be due to environmental factors such as nutrition and social milieu?

The ovarian groundplan and the origin of workers and queens

To show how a novelty can originate via an environmentally influenced re-ordering
of old pieces—and to show how the old pieces can remain, even though dissociated,
to influence many aspects of the evolution of a radiating group—consider the evolution of the reproductive division of labour in the wasps. This is a story that can be
told in terms of ovaries. As Pardi and others have shown, a wasp ovary is a kind of
biography. This is because its visible degree of development or degeneration is
connected via hormonal regulatory links and complex feedback relations to behavioural traits such as reproductive behaviour and aggressiveness (e.g. see Strambi
1985). This set of associations—among ovarian development, hormone activity,
and behaviour—I will refer to as the 'ovarian groundplan.'

Research on wasp hormones (e.g. see Röseler *et al.* 1985) suggests that the correlation between ovarian state and behaviour is mediated at least in part by juvenile
hormone and perhaps other physiological factors. The emphasis on ovarian state in

the present discussion, then, is not intended to imply direct ovarian control of behaviour and social organization. It simply refers to a common set of correlations effected via neurohormonal co-ordination.

The evolutionary establishment of a novel trait is always a two-step affair: first there occurs (or 'originates') a new phenotypic variant, then the variant increases in frequency (e.g. under selection or drift) until it is as a regularly occurring or (in some cases) fixed feature of the population.

So the first task of an evolutionary explanation for a novel trait like the worker phenotype is to describe the trait's origin—its initial occurrence as a variant, prior to or independent of its spread. Furthermore, an 'origin' always involves developmental change—a new phenotype presumes a new developmental pathway. Phenotype 'origins' are quintessentially developmental events.

The worker phenotype probably had its origins in the ovarian groundplan of the solitary wasps. Given that all social wasps are progressive provisioners, consider the ovary of a progressively provisioning solitary female. Such an ovary produces one egg at a time, with each oviposition being separated by several days of provisioning and caring for the growing larva. For example, in a progressively provisioning eumenine wasp, *Zethus miniatus*, I observed that the average time between ovipositions was 11.5 days (range 9–15 days; n=19). Although this species was not abundant enough to permit dissections, a figure by Iwata (1955 in Evans 1966a) showing the ovary of another progressive provisioner (*Bembecinus* species) indicates an ovary dominated by a single mature egg, with only a few very reduced immature oocytes and the accompanying nurse cell clusters. This implies a pronounced ovarian cycle of development followed by relatively less development (an absence of mature eggs following oviposition). This cycle is accompanied by a behavioural cycle of cell building just before oviposition (as the egg approaches maturity), then brood guarding and provisioning (during the ovary-reduced phase following oviposition) (West-Eberhard 1987a).

Another characteristic of some solitary and primitively social wasps is that they show territorial aggressiveness at the nest and even fairly stereotyped fighting if approached by a conspecific invader (e.g. see Brockmann and Dawkins 1979 on the sphecid wasp, *Sphecius ichneumonius*; Elliott and Elliott 1987 on *Cerceris cribrosa*; Evans 1973 on *Philanthus gibbosus*; Field 1989 on *Ammophila sabulosa*; Kurcziewski and Miller 1983 on *Philanthus sanbornii*; Wcislo *et al.* 1988 on the primitively social pompilid wasp, *Auplopus semialatus*). In *A. semialatus* aggressive usurpation of prey from conspecifics was most frequent by the females who laid the most eggs (Wcislo *et al.* 1988, table I). So, in at least some noneusocial nest building wasps, aggressiveness is another characteristic of egg-laying (reproductive) females and can be considered part of the ovarian groundplan. The aggressive dominance of *Polistes* and other social wasps—a cornerstone of their social organization and divisions of labour well known to be associated with developed ovaries—is probably an ancient trait.

Another important quality of the ovaries of female Hymenoptera has been demonstrated in parasitic wasps as well as social ones: if oviposition is prevented, developing ovarian eggs may be resorbed (Flanders 1942 and reviewed in

Bell and Bohm 1975). Factors known to cause oocyte resorption are starvation and lack of an oviposition site (e.g. lack of a suitable host) (Richards and King 1967; Bell and Bohm 1975).

Given these qualities of solitary wasps, I have argued elsewhere (West-Eberhard 1987a) that the reproductive division of labour in the eusocial wasps originated when social dominance led to a decoupling of the solitary ovarian cycle into a queen (developed) and a worker (undeveloped) phase, with the corresponding division of previously cyclic behavioural roles (see also Flanders 1969, p. 50). This could serve as an example of a major new phenotype derived from an old one via developmental re-organization that is, making something new (worker and queen phenotypes) out of old pieces—in this case, the different phases of the solitary ovarian cycle and its behavioural correlates.

If social wasps are just rearrangements of the solitary wasp phenotype, what, if anything, is really new? What in genetic terms has evolved?

A tentative answer is suggested by comparative study of primitively social wasps like *Auplopus* and *Zethus* and eusocial wasps like *Polistes*. As already shown, fighting occurs in primitively social and even solitary wasps, but workers do not result. Losers in *Zethus* miniatus whose cells are usurped do not become subordinates. Instead, they leave the nest. In effect, they revert to solitary life, so the solitary ovarian cycle is maintained. In worker-containing species the losers remain on the nest, and as a result have their reproduction (but not their brood-care activities) curtailed. The solitary ovarian groundplan is de-coupled into two extreme states, those of worker (ovary undeveloped) and queen (ovary developed), with the worker performing the brood-care activities ancestrally associated with cyclic change.

The genetic difference between the behaviour of a wasp like *Zethus* and that of one like *Polistes* may be small, effecting a relatively simple change in the threshold for leaving the nest (e.g. in a population having poor success of solitary nesting) or environmental circumstances (e.g. frequency of independent nesting opportunities) may change, so that females only leave their natal nests if they are destroyed. By either means, reproduction in groups can become 'fixed' rather than optional (West-Eberhard 1987a). As in *Polistes*, nests may still be founded by lone females, but they are joined by others (co-foundresses or worker offspring) prior to production of sexuals.

This is far from saying exactly which genes—or precisely which characters or how many of them—have actually changed to bring this innovation about. But if this is a reasonable interpretation of the observations at hand, then the worker phenotype may have originated as an indirect effect of selection for group life, and part of the mechanism producing it was an evolved change in the social environment.

The ovarian groundplan and colony organization: a critical period of adult development

An echo of the solitary ovarian cycle reverberates through the behaviour of all of the major groups of social wasps as well as the ants and bees. It takes the form of a widespread association of relatively developed ovaries with aggressiveness and intranidal duties, especially (in wasps) building, and of relatively undeveloped ovaries with extranidal foraging and provisioning. For example:

1. In a stenogastrine wasp (*Parischnogaster* sp.) ovarian development is associated with building and with dominance, and ovary reduction is associated with a greater proportion of time spent foraging and subordinance (Yoshikawa *et al.* 1969, fig. 2).
2. In *Belonogaster juncea*, foraging workers do not participate in the construction of new cells, an activity restricted to mature egg-laying females (Roubaud 1916, p. 120).
3. In *Polistes*, queen-like behaviour includes not only dominance and oviposition, but also building behaviour, especially the initiation of new cells. Often building activity by the queen immediately precedes oviposition (West-Eberhard 1969b). The same association between building and ovary developments occurs in queenless Vespa orientalis workers (Motro *et al.* 1979). In *Polistes* ovariectomized females do not build new cells (Deleurance 1955 and Röseler *et al.* 1985).

The same association appears to underlie the 'worker age polyethism' (Jeanne *et al.* 1988 and Jeanne 1991a)—sometimes called the 'temporal division of labour' or 'temporal castes'—a statistical shift in the activities of workers over time.

Worker age polyethism seems to take the same general form in all groups of social Hymenoptera where it has been observed.

In *Polybia occidentalis* (Jeanne *et al.* 1988) young workers perform relatively more building and brood care, and older workers spend relatively more time in foraging away from the nest. In *Vespula vulgaris* (Spradbery 1973, p. 152 after Potter 1965) young foragers specialize in pulp and water collection, whereas older workers gather more food. Age polyethism in the honey bee, *Apis mellifera*, and the ant, *Pheidole dentata*, follows a pattern similar to that of wasps, with younger workers performing relatively queen-like activities (handling eggs and staying near the brood) and older workers more often found in the nest periphery or foraging outside the nest (see Hölldobler and Wilson 1990, p. 316). Hölldobler and Wilson (1990, p. 320) call this the 'typical age polyethism' of ants, and provide an extensive review. In stingless bees (*Meliponinae*) it is the younger workers (10–25 days old) that lay 'trophic eggs' (Sakagami 1982).

This common pattern of temporal change in worker behaviour appears to be associated in some group with a slight but definite ovary-development cycle that occurs even if the workers do not lay eggs. Slight ovarian development in young workers has been documented in *Polistes* (Pardi 1946), in five genera of tropical swarming species (West-Eberhard 1978a), and in numerous ants (Billen 1982 and Hölldobler and Wilson 1990, p. 317). In the ants several authors have noted a close correspondence between this 'critical phase' of ovarian development and the position (tasks) of individuals within the nest. Ant workers 'typically' undergo a marked development of the ovary (Hölldobler and Wilson 1990, p. 317). In *Formica sanguinea* yolk development begins on about the tenth day of adult life and is at a maximum at 26–35 days of age (Billen 1982).

In *Ropalidia marginata* there is also an association between well-developed ovaries and aggressiveness, intranidal tasks such as building and brood care, and idleness; poorly developed ovaries are associated with foraging (Gadagkar and Joshi 1983 and Chandrashekara and Gadagkar 1990).

Hölldobler and Wilson attributed the similarity in age polyethism between ants and bees to 'convergence', implying that it is independently evolved in the two lines. But these examples showing a similar pattern of age-polyethism in a wide variety of wasps, ants and bees including both primitively and highly eusocial species, indicate that it may be a kind of symplesiomorphy, the result of a common ancestral ovarian groundplan.

The effect of the ancient ovarian groundplan probably extends to many other aspects of vespid social organization. For example, it may be responsible for the pattern of queen succession in the tropical epiponines, *Metapolybia aztecoides* and *Synoeca surinama* (West-Eberhard 1978*a*, 1981). In these species (and probably other tropical multi-queen societies, see West-Eberhard 1978*a* and Strassmann *et al.* 1991 and 1992), there is an alternation between a large number of queens and a small number or only one. Then, when the single highly developed queen dies or disappears, she is suddenly replaced by a large number of young queens. The underlying role of the ovarian groundplan is revealed by the fact that all of the replacement queens are recently emerged females—females marked within 10 days of the disappearance of the dominant queen (West-Eberhard 1978*a*). At that age dissected females of this species show a slight but measurable degree of ovarian development. If a queen is present, they behave as predicted by the ovarian groundplan: they either build or show idle behaviour at the nest, only later, as their ovaries decline, leaving the nest to forage. But if the colony is queenless young females in this ovary-developed phase mate and become queens—their unsuppressed ovaries continue to develop in the absence of dominant females.

In other species of wasps (e.g. *Polistes pacificus*—see Raveret Richter *et al.* 1987) and also in honey bees (Saleh-Mghir 1989) young worker females are the subject of strong aggression by older females: nestmates seem to recognize this 'critical period' for reproductive development (which occurs in young queenless workers of *Apis* as in wasps—see Hepburn *et al.* (1991)). Similarly, in *Belonogaster petiolata* young workers show a peak frequency of dominance interactions at 8 days of age, and after 17 days of age do not perform dominance acts (Keeping 1992).

The most thorough examination of the adult female critical period was published by Pardi (1946, fig. 9) who noted the effects of both age and order of emergence. Pardi found that not only is there slight ovarian development in young workers at 9–25 days of age, peaking at about 16 days, but also the earliest to emerge dominated their younger sisters and developed larger ovaries. Just as in *Metapolybia*, these older, more highly ovary-developed workers—those in the ovary-developed phase during the 'power vacuum' left by a missing queen—were the ones that became replacement reproductives, effectively suppressing younger nestmates. Thus in this respect there may be a fundamental similarity between queen determination in tropical *Metapolybia* wasps and the behaviour of *Polistes*. The same mechanism could underlie the 'gerontocracy' of *Polistes exclamans* (Strassmann and Meyer

1983) and *P. instabilis*. In those species the workers that became successor queens, while usually the oldest present, were also usually (11 of 13 cases in *P. exclamans*, all in *P. instabilis*) within the 9–25-day age category when the ovary development described by Pardi occurs. Their average age (17 days in *P. exclamans*) corresponds almost exactly to that (16 days) of the ovarian peak for non-laying young workers of *P. dominulus* (Pardi 1946, fig. 9). In *P. exclamans* and *P. instabilis* successors foraged prior to becoming queens, but in *P. instabilis* even queens of worker-containing colonies regularly leave the nest (1.7 minutes per hour, SD=2.2 min; Strassmann and Hughes 1988, p. 7). In contrast, *Metapolybia aztecoides* queens never leave the nest except (possibly) to mate and with swarms.

There are notable and interesting departures from the usual patterns described here. For example, deviations from age-typical task performance can occur in response to colony conditions (see Jeanne 1991*a*), and young females may take up foraging and subsequently lay eggs, for example in queenless colonies lacking workers (Litte 1979). Some exceptional patterns seem to represent adaptive, evolved tactics. For example in some species (*Mischocyttarus flavitarsis*, *Polistes exclamans*, *Synoeca surinama*) prolonged idleness with neither foraging nor oviposition persists in some females long past the usual critical period for adult caste determination. In all of these species queen mortality and/or satellite nest foundation frequently allows such 'waiting' females eventually to become queens (Litte 1979; West-Eberhard 1981 and Strassmann *et al.* 1984). So idleness may be part of alternative reproductive tactic in these colonies.

The 'groundplan,' then, is not a rigidly conserved set of correlations. Rather, it is a presumed ancestral state whose elements, in more or less modified form, are the basis for diverse aspects of vespid sociality. A common pattern is a three-phased ovarian cycle during the 'critical period' of young adulthood, when ovaries are first undeveloped, ascend to a peak, then (in subordinates) decline, with a correlated behavioural cycle of idleness, then 'queen-like' nest duties including building, then the laying of eggs (in unsuppressed females) or foraging (in suppressed ones). (For a more thorough review of the age-correlated behavioural phases, and some possible exceptions to this pattern in wasps, see Jeanne 1991*a*.)

The three phases of adult ontogeny, then, appear to underlie both the age polyethism and the reproductive division of labour, a developmental bifurcation that is socially mediated during the 'critical period' of young adulthood.

The ovarian groundplan and the evolution of regulation

Although much can be deduced from dissections and comparative studies of natural history, a developmental approach to the evolution of sociality invites a combination of such research with laboratory and experimental studies of regulatory mechanisms, especially hormones.

The 'ovarian groundplan' is a mosaic of correlated responses including both ovarian development and broodcare behaviour, organized into phases as just described. In the worker and queen phenotypes of wasps the groundplan correlations are dissociated so that ovarian development is maintained in queens, and broodcare behaviour without ovarian development occurs in workers. How, in terms of regulatory evolution, is this dissociation achieved? And how is its mechanism related to that of the age polyethism of workers?

The evolution of juvenile hormone function and the division of labour

In the social Hymenoptera juvenile hormone (JH) figures prominently in the regulation of social traits, including the reproductive division of labour between workers and queens (reviewed by Röseler 1991) and the division of tasks among workers (Robinson 1992). There are apparently no available studies of JH function in solitary wasps or in nest-sharing progressive provisioners like *Zethus miniatus*.

Robinson *et al.* (1992) note that the stimulation by JH of ovarian development is 'traditional' (widespread and likely ancestral) in the primitively eusocial Hymenoptera and other insects; and JH is also well known to influence behaviour in many adult insects (e.g. see Truman and Riddiford 1974; Nijhout 1994), including aggressiveness and dominance in *Polistes* foundresses (e.g. see Turillazzi *et al.* 1982; Strambi 1985; Röseler 1991), oviposition behaviour in numerous social Hymenoptera (references in Robinson *et al.* 1992, p. 476), and the speed of progression through the stages of age polyethism in *Apis* (see Robinson 1992) and *Polybia occidentalis* (O'Donnell and Jeanne 1993). It seems reasonable to hypothesize, therefore, that JH plays some role in the co-ordination of the behaviours and physiological changes associated with reproduction in pre-eusocial wasps, for example, their aggressive defence of cells against conspecific competitors; and reproductive 'maturation'—the age-related changes in behaviour and ovarian development that occur in nesting Hymenoptera females as they pass from callow idleness to reproductive maturity (nest building and the laying of their first egg).

Other broodcare activities—especially, provisioning of larvae with food—occur later in progressively provisioning solitary wasps when the female is already reproductively (and hormonally) mature; and they appear to be more strongly context dependent than those, like building and aggressiveness, associated with presence of a maturing ovarian egg. For example, in the solitary nest building wasp *Ammophila pubescens* (formerly *campestris*, see Baerends 1976) mature females provision several larvae at a time, indicating a lack of synchrony with cyclic egg-maturation and oviposition. Baerends (see Evans 1966*b*) saw the nidification cycle as under the influence of an endogenous 'brood-rearing drive' (a term that, like 'positive egg pressure' (Tsunecki 1957) is suggestive of hormonal mediation); he showed experimentally that foraging and provisioning behaviour, however, depend importantly on the status of larvae and provisions in each cell (Baerends 1941; summarized in Baerends 1959 and 1976).

During the reproductive maturation of a newly emerged solitary adult female, the onset of provisioning could, even if context dependent, be 'age dependent' as well since the context stimulating it (the presence of brood) comes as a sequel to ovarian maturation, nidification and oviposition. Thus, the ancestral ovarian groundplan could have involved a kind of 'age polyethism' associated with hormonally mediated reproductive maturation—increasing ovarian development, aggressive defence of cells, and nidification behaviour accompanied (and co-ordinated) by rising JH and followed by context-dependent foraging in response to the condition of the brood. Even if context dependent, provisioning would be performed only by reproductively mature females with relatively high JH compared with younger females that have not yet oviposited.

Beginning with the hypothesis that the ancestral ovarian groundplan involved a JH-mediated maturational age polyethism during early adulthood, then how, in terms of JH function, could this have given rise to a reproductive division of labour? What physiological mechanism might have decoupled the reproductive and brood-care activities of workers and queens?

Röseler et al. (1985; see also Deleurance 1955) found, by ovariectomizing Polistes foundress females, that building behaviour (an early phase of the matu-rational age polyethism) and JH production continues in the absence of ovarian development and suggested (p. 12) that 'this uncoupling of cell initiation and ovary development is a first physiological step in the evolution of cell initiation by sterile workers common in highly eusocial species'. It would be important to know if the two functions could be uncoupled by experimental ovariectomy even in some solitary or Zethus-like species. If so, it would suggest that this susceptibility to de-coupling preceded the evolution of sterile workers, facilitating the origin of workers via 'social ovariectomy'—the ovarian regression that occurs in group-living females that lose out in food-transfer interactions and have their oviposition attempts blocked when territorial females defend cells and fill them with their own eggs.

In primitively eusocial wasps social ovariectomy occurs during the critical period of caste indeterminacy of young adulthood. It is accomplished by several means (see Pardi 1946; Hunt 1994): 'trophallaxis' transfers food from subordinate to dominant females; energetically costly egg production and associated maternal behaviours are allowed (and then cancelled) by oophagic dominants which thereby glean a triple (nutritional, laboural, and suppressive) profit (see Velthuis 1990; Röseler 1991); and territorial dominants eventually occupy all cells, causing egg resorption in subordinates unable to lay eggs. Castration in the adult stage also may be facilitated in some species by low larval nutrition (see Gadagkar et al. 1991a on Ropalidia; West-Eberhard 1969b on Polistes). Once these measures have success-fully ovariectomized or castrated a female, her 'critical period' is over and she is less often attacked, as already discussed (above). Then JH level, which can be de-pressed as a result of subordination (see Turillazzi et al. 1982; Röseler 1985; Strambi 1985), might be relatively free to rise, unchecked by the negative effects of either dominance or ovarian feedback (see below).

'Social ovariectomy' may achieve worker sterility while leaving intact the ances-tral hormonal maturation phase and the potential for age polyethism. The effect of JH on ovarian development in many insects is via two pathways: JH (and ovarian ecdysteroids, see Wheeler 1994) can stimulate vitellogenesis (yolk protein produc-tion) by the fat bodies, as well as stimulating uptake of vitellogenin by the ovary (e.g. see Bell and Barth 1971; Robinson et al. 1992). The process of vitellogenesis is essential to egg production, and it requires adequate protein nutrition either from fat body reserves or dietary intake; starvation or fat body reduction can cause steril-ity (e.g. see Richards and King 1967). Although protein storage by adult fat bodies has been demonstrated in only a few insects (D. Wheeler, pers. comm.) ants

(*Lasius*), wasps (*Polistes*) and bees (*Apis*) are among them (Pardi 1938, 1939; Snodgrass 1956). Pratte *et al.* (1982) noted a correlation between haemolymph protein concentration and the size of fat-body adipocytes in *Polistes*; and West-Eberhard (1969*b*, p. 23) and Pratte *et al.* (1982) discuss observations indicating that the enormous fat bodies of overwintered females are used not during hibernation but during maturation of the first eggs. In some primitively eusocial bees (*Lasioglossum zephyrum*) 'environmental, hormonal and nutritional factors . . . synergistically stimulate yolk deposition' with either JH application or a protein diet alone initiating ovary development (Bell 1973, for other evidence of JH-nutrition synergism see Wheeler 1994, p. 254). This may explain the findings of Strambi and Girardie (1973) that experimental JH stimulation of overwintering wasps can induce ovarian development even in some parasitized females with reduced fat bodies, where (as expected) the response of parasitized wasps is less frequent and less marked than that of unparasitized females with larger fat bodies. Additionally, in some parasitoid wasps prolonged blocking of oviposition by lack of an oviposition site (host) leads to irreversible oocyte resorption (Bell and Bohm 1975, p. 378), a process that, if it occurs in social Hymenoptera, could contribute to the refractoriness of worker ovaries to high JH.

These facts suggest that a decoupling of the ovarian and behavioural influences of JH in workers and queens could initially have been achieved by incidental and/or socially imposed nutritional differences between the two castes that differentiated their responses to JH, producing two classes of females: protein-nourished ones (queens) able to respond to rising or high JH with vitellogenesis and ovary development; and poorly nourished ones (workers) either unable, or slow, to show a full ovarian response but still showing hormonal maturation accompanied by temporal changes in behaviour, sometimes but not necessarily accompanied by slight ovarian development (as already documented, above).

This 'split-function' or 'maturational' hypothesis for the role of JH in the evolution of castes differs from another interpretation, which I will call the 'novel-function' hypothesis. It proposes that during the evolution of eusociality and especially with the advent of marked age polyethism, the function of JH changed from being primarily reproductive to being primarily important in governing a new trait—the age polyethism of workers (Robinson *et al.* 1992; O'Donnell and Jeanne 1993). By this view *Polybia occidentalis* and other swarming species with a marked age-dependent polyethism have a new role for juvenile hormone (JH) in place of the 'traditional' one seen in wasps like *Polistes*. It thus considers that 'JH does not play a major role in both processes in the same organism' (Robinson *et al.* 1992, p. 478) and (see also Wilson 1971; O'Donnell and Jeanne 1993); that (1) 'Age polyethism is considered to be a derived trait that occurs in the largest and most complex insect societies' due to evolutionary convergence; and that (2) the regulation of worker activities is 'incompatible' with a traditional, reproductive role for JH in highly eusocial species. Although both hypotheses visualize a divergence of JH function in workers and queens, with JH effects becoming less ovarian and more behavioural, they differ in the degree to which this involved a break with the past and in certain predictions for future research (see below).

The novel-function hypothesis is based on comparisons of highly eusocial wasps (e.g. *Polybia*) and bees (e.g. *Apis*) with primitively eusocial genera (e.g. *Polistes* and *Bombus* species).

In highly eusocial species high JH activity has been implicated in the regulation of the worker polyethism, with JH (or methoprene) application causing an acceleration of the inception of foraging in both *Polybia* wasps (O'Donnell and Jeanne 1993) and honey bees (Fluri *et al.* 1982; Robinson *et al.* 1987; Huang *et al.* 1991).

In primitively eusocial wasps the maturational age polyethism is a relatively weak quantitative shift in predominant activities (e.g. see Post *et al.* 1988; Chandrashekara and Gadagkar 1990; Tsuchida 1991). It is less distinct than in highly eusocial species like *Polybia occidentalis* (see Jeanne 1991*a*). In some primitively social bees (*Bombus* species) worker behaviour has proven insensitive to experimental application of JH (Cameron and Robinson 1990). This has led to the impression that the maturational age polyethism is absent and not an ancestral effect of JH. But JH titre is only one of many task-determining factors that can vary under different conditions, as well as owing to selection. Others are the number and activities of worker nestmates, resource supply, and the size and demands of the brood (Jeanne 1986*b*, 1991*a*; Reeve and Gamboa 1987). In the small colonies of primitively eusocial species, where all tasks are performed by a relatively few versatile workers, these factors would put an especially high premium on task flexibility and individual responsiveness to changing conditions. In large colonies, with more workers available for each task, however, there may be both less context-dependent 'noise' in the expression of a maturational age polyethism and less selection for task-performance flexibility. So it is not surprising that it is in the large-colony species that one observes not only morphologically distinct workers and queens (less flexibility in adult caste determination, allowing greater caste specialization) but also a more consistent expression of the maturational age polyethism. Nonetheless, the relatively marked age polyethism of honey bees can be experimentally weakened by artificially consituting small colonies with a narrow age-range of workers, in effect mimicking some of the qualities of primitively eusocial colonies including their weak age polyethism (see Michener 1974, p. 126; Gary 1975, pp. 193–4). This reinforces the suggestion that parallel responses are involved in both primitively and highly eusocial species.

The occurrence of the same general pattern of age-correlated (maturational) change in behaviour in all major groups of social Hymenoptera (wasps, ants and bees) and its presence in primitively eusocial genera, plus the common correlation of certain behaviours with ovarian state even in solitary and primitively social wasps (discussed above) argues against the idea that the mechanism of the age polyethism is independently derived in different highly eusocial species. Rather, it appears to be related to a common, ancestral 'groundplan' as already discussed. Michener (1974, p. 130) reached a similar conclusion following a review of task ontogenies of solitary, primitively eusocial and highly eusocial bees. He states categorically that primitively eusocial species lacking such a division of tasks among workers 'probably never existed', and that such divisions of labour simply 'become more prominent with higher levels of social development'. Thus, the developmental mechanisms of the age polymorphism may be to a degree homologous in the three families, involving a maturational age polyethism and other 'groundplan' traits present in their common solitary ancestors. Rather than being convergent, the evolution of a marked age polyethism appears to be an example of parallel evolution, with large colony size and the associated exaggeration of the age polyethism occurring similarly in each of the three groups.

The 'traditional' function of JH refers, in the novel-function hypothesis, to its stimulatory effect on ovary development in primitively eusocial species like *Polistes* and other insects; and it implies the absence of an effect on age-dependent tasks in those species. To my knowledge the influence (if any) of JH on worker age polyethism in primitively eusocial wasps like *Polistes* has not in fact been investigated. The studies of *Polistes* cited by Robinson *et al.* (1992) and O'Donnell and Jeanne (1993) as indicating a strictly reproductive role for JH did not evaluate possible JH effects on a maturational age polyethism of workers under worker-like nutritional conditions. All of them utilized females that probably had fat bodies and/or protein nutrition supportive of vitellogenesis: some (Röseler *et al.* 1984) were temperate-zone *Polistes* foundresses early in the nesting season, which typically have large fat bodies (Eickwort 1969), and some were both young and fed high-protein diets: Bohm (1972) provided a diet of calf serum, and Barth *et al.* (1975) used early-summer workers fed *ad libitum* proteins. The finding in these studies that higher JH correlated with greater ovarian development (the 'traditional' role for JH) accords with the interpretation given here that when nutrition is adequate, JH stimulates ovarian development in maturing females, and graded responses to different JH titres may be expressed.

Social ovariectomy can explain the failure of hormonally mature workers to reproduce, but it cannot explain the failure of mature queens to work. Behavioural maturation—the age polyethism—is associated with rising JH. If there were to be negative feedback between ovarian development and JH titre following a JH-stimulated oviposition-initiation phase or 'critical period' in the adult stage, then well-developed established queens would be expected to have relatively low JH titres (compared with foragers) while still producing eggs (as in honey bees, Robinson *et al.* 1992). Röseler *et al.* (1980) found that in *Polistes* foundresses the corpora allata decline in size following a peak at the initiation of oviposition, which (if c.a. volume reflects JH activity—cf. Röseler *et al.* 1985; Huang *et al.* 1991) suggests that negative feedback is possible in wasps (see Nijhout 1994 for example in other insects). Old (foraging) workers, lacking feedback from developed ovaries, might experience a JH 'overshoot,' yielding high JH titres like those of honey bee foragers (see Fluri *et al.* 1982). But, due to their low nutritional status, the old (post-critical-period) foragers would not respond to high JH with ovarian development.

Hildebrandt and Kaatz (1990) found that during the first eight days of adult life queenless honey bee workers experience a higher rate of JH synthesis than dominated (queenright) workers, and Robinson *et al.* (1992) hypothesized that this may represent a 'commitment peak' of JH activity that initiates worker reproductive development in queenless workers. If so, this would suggest a hormonal contribution to the 'critical period' for ovarian development, which is similarly timed in young queenless wasps (e.g. in *Metapolybia* and *Polistes*, see above); and it would support the hypothesis that regulation of adult reproduction may have a similar hormonal basis in all these taxa.

The possibility of feedback between ovary and corpora allata is a reminder that hormones and neurosecretory products other than JH may interact with it and provide important clues to understanding the origin and evolution of castes. For example, Formigoni (1956) long ago showed that worker honey bees undergo neurosecretory changes that correlate with age and task performance: secretions produced in the perikaryons of the pars intercerebralis are low at the time of imaginal ecdysis, high in nurses and young foragers, intermediate in ventilator bees and wax workers, and absent in old inactive workers and at the onset of hibernation. The same organs affect the ovaries in *Polistes*: Strambi and Strambi (1973*b*) studied two neurosecretory products of the perikaryons whose appearance correlates with ovarian activity in *Polistes*. These neurosecretory cells are connected to the corpora allata by nerve fibres (the nervi corporis allati), and early ligation experiments showed them to have an inhibitory influence on corpora allata activity; the nervi corporis allati were also shown to carry neurosecretory material in many insects including *Apis* (Formigoni 1956) and the eumenine wasps *Synagris* and *Eumenes* (Thomsen 1954 after Highnam 1964). In *Polistes*, castration, whether by ovariectomy or strepsiteran parasitism, leads to a decrease in the secretion of certain pars intercerebralis cells (Strambi 1967; Strambi and Strambi 1973*a*). Another candidate for research on co-ordination of ovarian groundplan traits is ecdysone, a hormone produced during oogenesis by the ovary and, along with JH, stimulating dominance in *Polistes* (Röseler *et al.* 1985).

It would be of interest to know whether the foraging activity of specialized old *Apis* and *Polybia* workers is exaggerated by their high JH activity. Do allectomized workers continue to forage? If JH affects foraging rate, high JH titres of honey bee foragers (and unusually low levels in queens) may have been favoured under selection for exaggerated worker efficiency. Alternatively, adjusting task ratios may have involved lengthening the pre-foraging phase by raising the JH threshold of the switch to foraging. If JH does not affect foraging rate, then sustained high JH levels in honey bee foragers may simply reflect lack of corpus allatum inhibition, for example due to the undeveloped ovary or the reduced fat body (Haydak 1957), a source of JH-degrading esterases in many insects (Kort and Granger 1981) and may have no (or some unknown) adaptive significance.

The 'split-function' hypothesis predicts that JH may be shown to accelerate the behavioural maturation and the onset of foraging of offspring workers in some primitively eusocial genera such as *Polistes*. In the normally solitary bee *Ceratina japonica* (Anthophoridae) JH-treated diapausing females whose oviposition is socially suppressed show worker behaviour (Sakagami *et al.* 1993). However, the novel-function hypothesis predicts that JH should have no behavioural effect in maturing offspring workers on natural (worker) nutritional regimes in primitively eusocial species. There may, however, be complicating factors, like responsiveness to brood condition which might alter the response to JH in some circumstances. And, as already discussed, in some (especially, small-colony) species selection may virtually obliterate age-correlated JH effects that would interfere with flexibility in responding to the numerous factors known to influence worker task performance (see Jeanne 1986b and 1991a). In species apparently lacking an age-polyethism, aggressiveness and ovarian development during the young-adult critical period may nevertheless occur as an indication of hormonal 'maturation' (see especially Keeping 1992). The general similarity of the age polyethism in *Polistes* and the tropical polistines, best studied in *P. occidentalis* (Jeanne 1991a) but also evident in *Metapolybia*, *Synoeca* (West-Eberhard 1981) and other genera (Jeanne 1991a), as well as its correlation with ovarian state (West-Eberhard 1978a and 1981) and responsiveness to JH when that has been studied (O'Donnell and Jeanne 1993) suggests that a similar dual-function role for JH will be found in swarming polistines.

Nutrition has long been regarded as a crucial element in the social organization of the wasps (reviewed in Hunt 1994). Earlier observations could profitably be extended to the level of physiological mechanism and possible interaction with hormones (e.g. see Wheeler 1994) to better define the evolutionary role of nutrition in adult female castes. Especially needed are hormone studies of solitary wasps. As already mentioned, early writings on the ovarian cycles of solitary wasps referred to an endogenous 'brood-rearing drive' (Baerends 1941) or 'positive egg pressure' (Tsuneki 1957) (see Evans 1966b, p. 143), suggesting in almost mystical terms an underlying motivational and co-ordinating (physiological) mechanism. A primitively social sphecid wasp (*Cerceris antipodes*) begins laying eggs at about 10 days of age (McCorquodale 1990), a timing of reproductive maturity tantalizingly similar to that of unsuppressed (queenless) social wasps. More data of this sort would better illuminate how the critical period for caste determination in eusocial species may have evolved from solitary-ancestral maturational patterns and enable more definitive tests of the hypotheses discussed here.

Whatever the outcome of these tests, the social insects offer under-exploited opportunities to see how non-genetic factors like nutrition and social rank can interact with genotypically influenced processes (like vitellogenesis, reproductive maturation and cyclic broodcare) to induce phenotypic novelties, even without (or prior to) associated gene-frequency change. Whatever the mechanism behind the transition to eusociality, it can show how environmental factors (such as those influencing caste), as well as genes, can be originators of evolutionary change.

The evolution of regulation in eusocial species

Once a decoupling of JH-influenced reproduction and broodcare has occurred, these two functions, expressed separately in post-caste-determination workers and queens, may diverge and take different forms in different groups of social insects (see Wheeler 1986 and 1991). This follows from the rule of independent selection of independently expressed traits (below). Thus, the present hypothesis does not predict absence of change in the role of JH in workers vs. queens of highly eusocial species. Rather it predicts that the two ancestral functions (reproductive and behavioural) will evolve semi-independently following the origin of workers and queens.

It is not surprising that in highly eusocial species the differences between the castes are exaggerated. Morphological differences between castes are achieved by moving the critical period for caste determination and JH action earlier in development (see Wheeler 1994, p. 252 ff.), and this alone may render the patterns of JH activity in adults relatively free to evolve in other contexts. Selection prior to the evolution of larval caste determination may, however, have acted to reduce JH levels or JH sensitivity in mature queens as a means of limiting the expression of maternal behaviours other than oviposition. The extreme of pre-adult social ovariectomy occurs in some ants, where adult workers lack even the rudiment of an ovary (Hölldobler and Wilson 1990).

Some reasons that selection might favour an increased expression of the age polyethism in workers of large, highly eusocial colonies have already been discussed (above). But how this has been achieved in terms of the evolution of JH is not yet clear. Are JH titres higher in *Polybia* foragers than they are in *Polistes* foragers? Or does the exaggerated age polyethism of *Polybia occidentalis* indicate that workers are more sensitive to JH, due to selection for increased individual task specialization and reduced attention to contextual 'noise' (colony needs)? If selection acts to partition tasks throughout the average lifespan of workers, as suggested by Jeanne (1986*b*), then by what mechanism can this occur given populations with different demographies (life expectancies) and task demands? One can imagine drawing out the phases of the maturation polyethism to cover the entire worker life span, by raising the JH threshold for switching between tasks; and this might make JH titres of foragers relatively high (as in honey bees). Is it possible, given the fact that honey bee age polyethism can be weakened by experimentally altering colony conditions (above), that little genetical evolution of JH functions and responses has occurred? Data comparing the hormonal profiles of adult castes in primitively and highly eusocial wasps and bees are still insufficient to answer these questions. In any case, the characteristics of workers and queens differ greatly in different groups of social insects, and this may render generalization about the evolution of regulation difficult. Some highly eusocial workers (e.g. the trophic-egg-laying workers of stingless bees) show well-developed 'reproductive' traits; and some highly eusocial queens (e.g. spring foundresses in the Vespinae) work, albeit relatively inefficiently (see Greene 1991).

In a broader perspective, juvenile hormone has multiple functions in adult wasps, not just one or two. What happens to the 'traditional' function of JH in the regulation of social dominance (Röseler 1985) in wasp species if, as in honey bees, the reproductively subordinate (worker) caste has a high JH titre relative to queens? In honey bees the mandibular gland 'queen substance' used to maintain dominance reflects ovarian development (see Velthuis *et al.* 1990) not, as in the dominance behaviour of *Polistes* foundresses, JH titre. So the evolution of this JH-independent indicator of rank may have been an important step in the evolution of extreme castes. Such a signal reflects reproductive quality (egg-laying capacity) and as such may influence adaptive worker decisions that affect the timing of selfish egg-laying and queen succession (see West-Eberhard (1983) on 'worker choice' of queens; Keller and Reeve (1994) on the possible uses by workers of queen dominance signals).

Three principles of development and selection

The most commonly discussed means of phenotypic divergence under natural selection involves speciation, or the reproductive isolation of the contrasting forms. There is a very large number of models to explain the *maintenance* of alternative forms (polymorphisms), by balancing selection, frequency dependent selection, and disruptive selection against intermediates (see Levins 1968 and Mather 1973 for a partial review). But these models do not consider increasing specialization or modification. They do not explain how natural selection can cause the kind of divergence, referred to by Darwin (above), between forms producible by the same genotype.

A solution is to be found in the relationship between co-ordinated gene expression (or co-ordinated gene-product use) and natural selection. Consider the pattern of gene expression that obtains for a pair of alternative phenotypes like worker and queen. The regulatory mechanism involved is evidently genetically complex since it can involve things like juvenile hormone and the capacity to assess and respond to differences in nutrition and dominance among nestmates. Such a genetically complex mechanism is probably subject to genetic variation and selection, favouring advantageous thresholds for switching between the two alternatives.

A switch mechanism determines which of a pair of developmental or behavioural pathways is activated. Each entails the co-ordinated use of a somewhat different set of genes or gene products. So what a switch does is to control gene expression or gene-product use. Hormones (like JH), involved in caste determination and aggressive dominance are famous for playing a role in gene regulation (e.g. see Nijhout and Wheeler 1982), and even when the nervous system orchestrates the performance of different behaviours (such as building vs. foraging), it mandates the co-ordinated use of somewhat different structures (gene products) and can stimulate the expression of genes (e.g. see Christensen *et al.* 1991 and Steward *et al.* 1991). Thus, different alternative phenotypes—like worker and queen or builder and forager—are in this special sense genetically distinctive; they represent the co-ordinated use of somewhat different sets of expressed genes.

The genetic-regulatory architecture of development has profound consequences for evolution because it connects developmental pattern with evolutionary pattern by organizing the co-expression of genes and thereby directing the action of selection. And, since switches can be condition-sensitive, traits expressed together may be associated consistently with certain conditions. These points can be summarized as three simple 'rules' or principles which, along with the principle of dissociability and phenotype recombination (above), describe the relation between development and evolution.

The co-evolution of co-expressed traits

Traits that are expressed (or used) together are selected together and therefore evolve together. This may seem a trivial or obvious idea, but it explains why developmentally and behaviourally co-ordinated traits are also 'functional' units—a relationship that has to be explained, even though it is so common that it is usually taken for granted. In terms of evolutionary cause and effect it acknowledges the fact that co-ordinated use (a developmental association of traits) precedes selection, which in turn produces functional adaptation.

The principle of co-evolution of co-expressed traits helps to explain the perfection of 'design' of such complex organs as the vertebrate eye, and, in fact, of all phenotype 'subunits' as they have been defined above.

The independent selection of independently expressed traits

The 'Rule of Independent Selection' (West-Eberhard 1992) follows from the co-evolution of co-expressed traits, and solves Darwin's Dilemma. It states that to the extent that traits are independently expressed, they are independently subject to selection. Therefore evolutionary divergence can occur between alternative forms, like workers and queens, males and females, or the different life-stages of a species. Similarly, the same genotype can produce contrasting 'antagonist' behavioural phenotypes, analogous to those occurring in different species in parasite-host 'co-evolution'. In the same *Polistes* species there can be one specialized set of socially parasitic traits shown by usurpers and another specialized set for defence against them (reviewed by Turillazzi 1992).

Perhaps the most general significance of development for evolution is that it is an environmentally sensitive phenomenon that organizes or compartmentalizes the phenotype and thereby determines in what combinations and in what circumstances genes are exposed to selection. In this way, development directs evolution.

You might ask: if worker and queen diverge as if they were independently evolving, why aren't they more divergent in the social wasps? The most striking discontinuous caste differences known in the vespids are the discrete size differences found in the vespines (e.g. see Spradbery 1973) and the allometrically distinct gaster differences found in *Agelaia* (Ihering 1904 and Jeanne and Fagen 1974). But these pinnacles of wasp specialization do not approach the extremes seen in certain ants and bees (see Wilson 1971).

Independent expression is necessary for independent selection and evolutionary divergence, but it is not sufficient. In the case of the social wasps divergent morphological specialization of worker and queen phenotypes appears to be limited by counter selection to maintain adult behavioural flexibility in the face of predation and parasitism by a wide variety of natural enemies (see Chadab 1979 and Jeanne 1979*b*). In keeping with this, the most extreme caste dimorphisms (above) are found in species with relatively large well-defended colonies. [A possible exception occurs in the small-colony species *Ropalidia ignobilis* (Wenzel 1992*b*), if the apparently seasonal extreme change in size and proportions proves to represent discrete size morphs and/or caste differences in allometry (as in Jeanne and Fagen (1974); Eberhard and Gutierrez (1991)).]

Self-accelerated increase in frequency and modification of form

As in the evolution of morphological castes, the evolution of obligate social parasitism in *Polistes* is associated with (1) a loss of behavioural flexibility in the adults and (2) increased morphological specialization. It is not surprising that an organism specialized to a particular behaviour may evolve increased morphological adaptations associated with that behaviour. What may not be so obvious is that a self-accelerating evolutionary process can be involved as an alternative becomes increasingly free of the genetic correlations represented by the 'nonspecific' shared modifiers that constrain its divergence from other alternatives. This can occur in two different ways: due to increasingly earlier determination in the ontogeny of the alternative developmental pathways, as in the case of pre-adult caste determination in highly eusocial species (see Gadagkar this volume and above) and due to increasing frequency or fixation of a single alternative in a population (see West-Eberhard 1986b, 1989 and below) as when social parasitism becomes obligatory.

The basis for self-acceleration is in the shared, nonspecific traits whose expression is not controlled by the switch between 'morphs' or alternative behaviours (West-Eberhard 1989, 1992). In a population of developmentally flexible individuals those traits may be under conflicting selection, or, to put it another way, they are selected to be compatible with different alternatives expressed in different selective contexts. But the greater the effect of a particular alternative on individual lifetime fitness, the more strongly are the nonspecific modifiers selected to accord with the dominant alternative. This means for example, that the more frequent and important the fitness effects of usurpation, the more strongly are morphology and behaviour for specialization to parasitism favoured. The trend can be 'self-accelerating' because there is an expected correlation between increased trait efficacy due to increased modification (specialization) and selection on regulation of expression to increase the frequency of its occurrence. So there can be positive feedback between improvement and frequency of expression: the more improved a trait, the commoner it is expected to become (barring negative frequency-dependent effects), and the more common it becomes, the more it continues to be improved by selection (see West-Eberhard 1986b, p. 263, for examples).

This principle was formulated as a quantitative model by Clarke (1966) in a discussion of the evolution of genetic morph ratio clines, with special reference to Batesian mimicry in butterflies. And it was applied by Endler (1977) to explain population divergence (speciation) in clines. I believe it to be of much more general importance since it should apply to the evolution of all adaptive traits that evolve toward fixation during the alternative-phenotype phase of their evolution including all cases of 'genetic assimilation' (Waddington 1961). The principle of self-accelerated frequency and modification is best understood by considering what happens if an alternative (for example, usurpation behaviour vs. independent nesting) becomes fixed in a population.

Phenotype fixation and punctuated change
When frequency of expression reaches 1.0 ('phenotype fixation'), overlapping morphology and behaviours of alternatives are more free to specialize to the fixed form, being developmentally 'released' from the conflicting selection on (genetically correlated) alternatives that formerly retarded their evolution. In a genus like *Polistes* adult females are notable for their behavioural flexibility in switching among alternative reproductive tactics, all accommodated by a single morphology. If a single

alternative behaviour such as usurpation is fixed in a population, morphology may then specialize to the fixed form, and this may produce a burst of morphological change by the self-accelerating process as the behaviour approaches fixation.

This process could lead to a 'punctuated' pattern of morphological evolution associated with loss of behavioural flexibility. But it would reverse the causal relationship with speciation assumed by proponents of the 'punctuated equilibrium' hypothesis (e.g. see Gould and Eldredge 1977, p. 137) who see speciation as the cause of a 'genetic revolution' and phenotypic change. Instead, the phenotype-fixation hypothesis suggests that gradual specialization, especially in behaviour, and gradual increase in frequency of expression occurs during an alternative-phenotype phase, with punctuation accompanying phenotype fixation with or without speciation. The character release and 'punctuated' burst of change may often be associated with speciation because the self-accelerated process, by speeding divergence between populations, could accelerate the evolution of reproductive incompatibility (West-Eberhard 1986b and 1989). Thus, punctuated change may be a cause, not an effect, of speciation.

This does not, of course, deny that reproductive isolation is an important cause of divergence. But it does predict that punctuated change will especially be associated with loss of developmental flexibility in behaviour and morphology (for possible examples, see Moran and Whitham 1988, West-Eberhard 1986b, and the discussion of social parasites below). The punctuated equilibria hypothesis, however, does not predict that any particular class of characters will show punctuated change.

Some authors (e.g. see Matsuda 1982 and 1987) would call phenotype fixation of a conditionally expressed trait a kind of 'genetic assimilation' (Waddington 1961), implying that increased frequency involves a genetical change in threshold of expression. The 'self-accelerating process' discussed above would involve such genetic change. But it is important to realize that fixation of a facultatively expressed tactic can occur without genetic determination of the trait, due to increased frequency of the environmental condition(s) inducing it. Thus, increase in predation frequency and nest loss would increase usurpation frequency due to both selection (a lowered threshold for parasitism) and increased frequency of induction. And either process by itself could lead to phenotype fixation. The final transition to obligatory parasitism, for example, could conceivably occur without genetic change, with the use of non-conspecific hosts an 'opportunistic' (not necessarily genetically distinctive) consequence of rarity of conspecific colonies. Subsequent evolution of specialized behaviour and morphology would, however, make interspecific parasitism increasingly genetically specialized and increasingly 'obligatory'—the option of independent nesting as a profitable alternative in some circumstances would be lost.

Alternative phenotypes as a phase of evolution: the social parasites

Occurrence as an alternative phenotype (sensu West-Eberhard 1986b, 1989) is an inevitable stage in the evolution of a novel trait. As soon as a novelty originates in a population, at any level of organization, it exists as an alternative to one or more other, pre-existing forms; for even if strongly advantageous or induced by a wide-spread environmental factor, a novelty could not usually be immediately ubiquitous. So it is not surprising that all of the major events in the social evolution of the vespid wasps have begun as intraspecific alternatives, including life in groups, sterile workers, and social parasitism.

Calling an alternative a 'phase' in evolution does not mean that a novel alternative is destined to evolve to fixation. It may spread, remain at some intermediate frequency, or disappear. But as long as it is expressed and has genetically variable fitness effects, it is subject to selection and modification—it can evolve.

Expression as an option in a variable environment can have a protective or 'buffering' effect permitting change in a new direction without dropping older established adaptations (West-Eberhard 1986b, 1989 and references therein). This buffering effect applies especially to conditionally expressed alternatives which may evolve so as to be expressed primarily when advantageous relative to other options. As a result of these properties, alternative phenotypes may achieve a high level of specialization, whether or not they eventually become fixed or 'obligatory' in a population.

The evolution of social parasitism in *Polistes* illustrates well how a major novel adaptation may have been elaborated as an alternative phenotype, then become fixed as a species characteristic. Cervo and Dani (this volume) present a synthetic review on the evolution of social parasitism in *Polistes*, which I shall socially parasitize here.

The three social parasite species of *Polistes* (*atrimandibularis*, *sulcifer* and *semenowi*) are 'obligatory' workerless social parasites or 'inquilines.' They do not found their own nests, instead usurping young colonies of other species and utilizing the nest and brood-care services of the host workers. As Cervo and Dani show (see also Turillazzi 1992), most of the parasitic behaviours of the three socially parasitic *Polistes* species can be found in non-parasitic congeners where intraspecific parasitism (nest usurpation) is a common alternative reproductive tactic. The parasite-like behavioural characteristics of intraspecific usurpation include two different invasion patterns (queen-expelled, queen-tolerated as a subordinate); timing of usurpation, usually just before or just after host worker emergence; unusually violent aggressiveness compared with that among nestmates, with the host queen often killed (in the case of queen-expelled takeovers); discriminatory destruction of young host brood (eggs, young larvae) with the older brood spared; intense gastral rubbing of the nest surface; and effective dominant suppression of worker reproduction. A similar situation occurs in the vespine wasps, where intraspecific usurpation is exceedingly common and shares many characteristics with that of obligatorily parasitic species (see Greene 1991).

As pointed out by Turillazzi (1992, p. 263), 'The way in which the social parasites control the host colonies seems to be similar, but is more efficient, than that used intra-specifically by *Polistes* wasps'. Intraspecific usurpers are like 'super queens', and specialized obligatory social parasites are, in turn, super usurpers. So the specialized parasitic behavioural phenotype may have been largely consolidated during a phase of evolution as an alternative within species.

Then a hypothesized transition between primarily independent nesting to obligatory parasitism involved gradually increased frequency of usurpation at the expense of other alternative nesting tactics until finally parasitism is a complex reproductive tactic and the only alternative expressed (Taylor 1939, Cervo and Dani this volume).

In relatively mild climates intraspecific usurpation is one of several alternatives following nest loss in *Polistes*. Others include re-nesting, working as a subordinate on another nest, and activation of abandoned nests (Dani and Cervo 1992; Klahn 1988 on *P. fuscatus*; Makino 1989*a* and *b*; Makino and Sayama 1991 on *P. riparius* and *P. snelleni*; Yoshikawa 1955 and 1962*b* on *P. jadwigae*; Gamboa 1978 on *P. metricus*; and Strassmann 1981*a* on *P. exclamans*).

In regions of short nesting season, and in the absence of associations among relatives, usurpation may become the primary alternative following nest loss due to lack of time for brood rearing by any other tactic. Finally, in the obligate interspecific social parasites, usurpation is expressed in all females (phenotype fixation).

Predictions

The expected positive feedback between phenotype frequency and rate of modification, and the morphological consequences of phenotype fixation suggest some predictions regarding the evolution of social parasitism that can be tested with comparative study.

1. If obligate inquilinism (social parasitism) evolved from facultative usurpation, traits used in facultative usurpation by social species should be those exaggerated in obligate parasite species where usurpation is fixed. This appears to hold in wasps where some traits exaggerated in obligate social parasites (abdominal stroking of the host nest, and enlargement of the Dufour's gland (in vespines, Jeanne 1977*b*)) are also associated with facultative intraspecific usurpation or hyper-aggressiveness in non-parasitic species (see Cervo and Dani (this volume) and Turillazzi (1992) on abdominal stroking in independent and parasitic *Polistes* species, and Downing and Jeanne (1983) on enlargement of Dufour's glands in aggressive *P. fuscatus* females).

2. Some traits exaggerated in social parasites should show evidence of conflicting selection when expressed facultatively as part of intraspecific alternative. That is, there should be some reason to believe that the exaggeration seen in parasitic species is selected against when additional alternatives occur. This would support the hypothesis that developmental 'release' from genetic correlations among alternatives contributes to their exaggeration in obligate parasites.

As just mentioned, possible examples of selection on nonspecific, shared traits 'released' following phenotype fixation are the enlarged Dufour's gland of the Vespinae and the abundant nest-rubbing behaviour of *Polistes* species. Glandular secretions can be signals of dominance (e.g. a 'queen recognition' pheromone), territorial markers, and/or nestmate recognition and intruder 'camouflage' cues (see Lorenzi *et al.* this volume). Over-development of the glands that produce them may be selected against in workers, not only because of the cost of producing inappropriate secretory tissue and secretions but also because an overactive gland might subject them to aggressive attack by nestmates; whereas in a usurper greater development of the glands may be favoured as an aggressive weapon. When usurpation becomes the only alternative in a workerless parasite, such secretions are free to evolve toward levels not attained in worker-containing species.

It would be of interest in this context to know if the broad-grooved mandibular shape characteristic of workerless parasitic *Polistes* species is somewhat disadvantageous for nest building or some other worker activity. If so, this and other morphological specializations to parasitism, once fixed in a species, could reduce the likelihood of a reversal to independent nesting.

The bright white, un-camouflaged pupal caps of *Polistes atrimandibularis*, considered by Cervo *et al.* (1990c) to function possibly as a kind of 'warning' signal to other potential usurpers, might similarly be selected against in 'normal' wasps attempting to camouflage the nest, then released to evolve toward conspicuousness in parasites.

3. Traits characteristic of social parasites, like mandibular form, clypeal apex flattening, sharply pointed clypeal margins, and Dufour's gland enlargement, may show geographic variation within species showing geographic variation in frequency or commonness facultative intraspecific usurpation. The same may apply to behaviours (nest stroking, efficacy of domination, ability to inflict injury) employed by usurpers. This might be revealed by comparing populations, especially somewhat isolated ones, at the latitudinal or altitudinal extremes in species showing north–south and/or altitudinal variation in frequency of nest usurpation, providing the variation can be considered of sufficient duration to involve an evolutionary/genetic response. Examples of such species are *P. fuscatus* in the US and *P. riparius* in Japan. Alternatively, very closely related species showing consistently high vs. low frequencies of usurpation could be compared.

Field (1992, p. 115), proposes this same means of testing the frequency of expression/phenotype fixation hypothesis. He cites studies indicating that some vespine wasps with a high frequency of temporary social parasitism 'appear to have some of the morphological specializations of obligate parasites.' However, he cites as a 'flaw' in this hypothesis the absence of recorded cases of obligate *temporary* social parasites.

In *Polistes* temporary obligate social parasitism, in which the usurper produces some worker offspring of her own, is expected to be rare or absent if, as just discussed, obligate (interspecific) parasitism originates in extreme climates with short nesting seasons, and invasion occurs at about the time of first worker emergence (see Turillazzi 1992 and Field 1992). At that time temperate-zone *Polistes* females are already beginning to lay reproductive-producing eggs (e.g. see West-Eberhard (1969b) on *P. fuscatus*). Usurpers in such a species would soon produce reproductive offspring, not workers, and 'temporary' parasitism may seldom occur. This is in contrast to the situation in vespine wasps, where most (500–800) worker-producing eggs are laid after first worker emergence and the pre-emergence queen produces only about 40 workers prior to that date (data from Greene *et al.* (1976) on *Dolichovespula arenaria*). In any case, temporary parasitism is not a necessary step in the transition from facultative to obligatory parasitism. Nor is facultative interspecific parasitism expected to be common (for a review of the few known cases in *Polistes*, see Cervo and Dani (this volume) and Carpenter *et al.* (1993)): it may occur only when usurpers are running out of conspecific host colonies in a population where usurpation is the most frequently productive option—a transient situation expected to change rapidly toward obligate parasitism in the presence of suitable hosts.

The significance of annual migrations

Based on work still in progress, Cervo and Dani (this volume) note that:

useful information on social parasitism in Polistinae [*Polistes*] could be deduced from their migratory habits...Many females and males of the three species of parasites—as well as wasps of some of the host species—were observed, at the end of summer, at the tops of various Italian mountains, which are non-nesting zones for *Polistes* (except for the mountainous species, *P. biglumis bimaculatus*). These wasps undertake altitudinal migrations from the place of [nesting and] emergence, usually at the base of the same mountains [where they must return in search of host nests in the spring], to the tops of the mountains . . . [references omitted].

Increased length of migrations could be associated with increased dispersal in search of a host nest as part of an intraspecific usurpation tactic. The ability to migrate between late summer mating/feeding/hibernation sites and spring nesting sites may have been a crucial adaptation for the transition to obligatory parasitism: as usurpation increases in frequency and importance, it may decrease the size of an already rare population due to the setbacks occasioned by brood destruction and (in some species) plundering of other nests. This would increase rarity and further reinforce selection for searching dispersal and effective migration, which would thus be an integral part of the parasitic tactic.

It is characteristic of long-distance seasonal migrations in animals that they connect distant sites of two different sets of critical resources (e.g. see Brown 1975). It is not always obvious, however, how selection could have gradually increased the distance travelled and endowed a species with the ability to return or 'home' following a long absence. Independent nesting *Polistes* are perhaps pre-adapted for the kinds of migration described by Cervo and Dani by their well-developed homing ability (see Ugolini and Cannicci this volume), and their ability to home from overwintering sites following months of hibernation (Noonan 1981 and West-Eberhard 1969*b*). If this pre-existing ability is used in the parasitic migrations, one might predict that individual parasite females return to areas near their natal nest sites at a greater than random frequency.

Since several closely related *Polistes* species, both parasitic and non-parasitic, engage in these annual migrations in the same geographical region, multiple occurrence of this migratory habit, favourable to the evolution of obligate parasitism, in the history of these wasps is not unthinkable. This is a point in favour of the possibility of multiple origins of obligate parasitism in European *Polistes*, a possibility discussed in the next section.

Obligate interspecific parasitism: one episode of phenotype fixation or more?

There is a general correlation, in a wide variety of plants and animals, between phenotypic flexibility and evolutionary lability. Alternative phenotypes (like usurpation versus independent nesting) may show adaptive temporal and geographic variation in phenotype ratios (West-Eberhard 1986*b* and 1989), and when an alternative is common in numerous species of a taxon, as in the case for usurpation in *Polistes*, recurrent phenotype fixation may occur, producing recurrent parallel forms (e.g. the 'iterations' of Rensch (1960) and see also Yablokov (1986) on Vavilov's (1922) Law of Homologous Series).

It is therefore not surprising that obligate, interspecific social parasitism has evolved independently at least three times in the social wasps, a group in which facultative intraspecific parasitism is quite common (Greene 1991, Cervo and Dani this volume). Obligate inquilines evolved at least once in *Polistes* and at least twice in the Vespinae (in the genera *Vespula* and *Dolichovespula*) (Carpenter 1987 and Carpenter *et al.* 1993). Such lability gives reason to consider seriously the possibility that social parasitism originated more than once in the three known closely related social parasite lineages of *Polistes*.

This is not to challenge the monophyly of the socially parasitic species (Carpenter *et al.* 1993; Choudhary *et al.* 1994; Carpenter this volume) but only to point out that traits held in common by closely related species can evolve at different rates in different ones. Given the commonness of usurpation behaviour in the subgenus *Polistes*, to which all three parasite species belong (Carpenter this volume), it is likely that the common ancestor of the three was either a facultative or obligatory social parasite. But it is impossible to tell for certain which was the case. Since facultative parasites can become highly specialized and therefore poised for a self-accelerating rapid transition to obligatory parasitism, three closely related species disposed to this transition by their common ancestry and common location in a climatic zone favourable to high frequencies of intraspecific usurpation could have made the phenotypically small step from facultative to obligatory social parasitism at somewhat different times with parallel morphological consequences.

The law of parsimony requires attributing the common features of a monophyletic group to their common ancestry only in the absence of biological evidence to the contrary. The evolutionary lability of facultative traits, and their propensity to reversals within a lineage as well as to rapid and parallel fixation in related species give reason at least to consider the possibility of multiple fixations of obligate parasitism in *Polistes*. Evolutionary lability of conditionally expressed traits means that it may sometimes be impossible confidently to track character 'fixation' points using cladistic methods, because the character can come and go (change polarity) and rapidly become fixed rather than optional *between* the branching points of a cladogram (see also Frumhoff and Reeve 1994).

The distinctive behaviour and hosts of *P. atrimandibularis* compared with those of the other two socially parasitic *Polistes* species (which have shared hosts) (see Cervo and Dani this volume) could be evidence of two independent origins of parasitic behaviour. Alternatively, the distinctive behaviours of *P. atrimandibularis* could have evolved following a common origin of obligatory parasitism.

The fact that all three species bear similar morphological accoutrements of parasitism (enlarged mandibles and associated head modifications, thickened cuticle) should not, of course, be taken as evidence that obligatory parasitism occurred prior to their speciation since these characteristics can be convergent (or parallel) developments (see Carpenter 1987, p. 424, on similar kinds of morphological convergence in the socially parasitic Vespinae).

Whether it occurred once or several times, the fixation of social parasitism could have accelerated speciation of the parasitic species, as mentioned in the section on phenotype fixation (above and see West-Eberhard 1986*b* and 1989).

Conclusion

The social wasps provide a microcosm for understanding general principles of development and evolution: the origin of new traits via the re-organization of old ones, the separate divergence of independently expressed complex characters, the origin of major novelties as intraspecific alternatives, and the accelerating evolutionary effect of phenotype fixation on the morphology of formerly optional traits.

Darwin's treatment of the social insects in *The Origin of Species* (1859, first edition, p. 242) concluded that the divergent evolution of worker and queen depended upon the sterility of the workers:

As ants work by inherited instincts and by inherited tools or weapons, and not by acquired knowledge and manufactured instruments, a perfect division of labour could be effected with them only by the workers being sterile; for had they been fertile, they would have intercrossed, and their instincts and structure would have become blended.

Had he left the matter at that, we would have to see this as one of Darwin's rare mistakes. Divergent alternative phenotypes within populations are not lost owing to blending inheritance. If this were so, male and female differences could not persist nor would the differences between large- and small-horned beetles and other polyphenisms.

The error was deleted, however, in later editions of the *Origin* (e.g. see the sixth edition, 1872, p. 207) where Darwin added examples to several previous ones showing the analogy between social insect castes and other developmentally similar phenomena, such as alternative phenotypes in plants having sterile-double vs. fertile-single flowers in artificially selected stocks, 'Malayan butterflies regularly appearing under two or even three distinct female forms' and 'Brazilian crustaceans likewise appearing under two widely distinct male forms' (p. 207).

During the years between the first and last editions of the *Origin* Darwin wrote a book (1868) on *Variation in Animals and Plants under Domestication*. In Volume II of that book he developed a concept of inheritance where he related developmental pattern to selection and evolution much as I have tried to do here. In his much maligned theory of 'pangenesis'—really a sophisticated theory of development and evolution—he insisted that inheritance includes both transmission (genetics) and development and that transmission and development are 'distinct powers'. Although he did not return to the social insects, he clearly arrived at a developmental solution to the former dilemma of their maintenance when (p. 485) he discussed other cases of intraspecific dimorphism:

With those animals and plants which habitually produce several forms, as with certain butterflies described by Mr. Wallace, in which three female forms and one male form coexist, or, as with the trimorphic forms of *Lythreum* and *Oxales*, gemmules [genes] producing these different forms must be latent [unexpressed] in each individual.

Thus, alternative phenotypes influenced by the environment (like those of social insects) were important in the formulation of Darwin's theory of inheritance, just as discrete alternative phenotypes influenced by genes of large effect were important in Mendel's. Darwin's theory of inheritance was more ambitious in that it attempted to encompass both. Having adopted the more limited (but perhaps temporarily less confusing) Mendelian approach, modern evolutionary biology is only now getting back to where Darwin left off.

So by following a trail of wasp ovaries and social parasites marked out by Pardi and his students through the complexities of dominance relations and modified mandibles to current problems in evolutionary biology, we arrive at a solution only to find (as in the case of kin selection) that Darwin was already there.

18

Some epistemological reflections on *Polistes* as a model organism

Richard M. Burian

It is a rare privilege for me, as a philosopher of biology, to be invited to contribute to a volume such as this, on a topic that I know mainly from a brief acquaintance with some recent literature. I thank the organizers for their invitation to reflect on some of the virtues and problems of recent comparative work employing *Polistes* and related organisms. In doing so, I will be extending some work I have previously done on the use of model organisms, mainly in laboratory contexts (Burian 1992, 1993). I have learned a great deal from the contributors to this volume; I can only hope that this chapter provides some small return for the education they have given me. As a usurper at the biologists' nest, I offer this chapter in hope of warding off agonistic interactions; after all, it is in my best interest to avoid a grappling fight and expulsion from the nest—or worse.

The cultures of biology and of philosophy are very different (at least in the United States). All too often, philosophers, even philosophers of biology, fight over *words* rather than trying to understand *things* or *processes*. When we make public presentations, the program is usually organized so that we are followed by a hostile commentator who tries to find flaws in our arguments. So we are very careful about our words—and very argumentative in public. (This is why we tend to read carefully prepared texts rather than to talk from slides—after all, when programs are set up this way, every word counts. One fears that, in some cases, words are *all* that counts.) None of this is typical for biologists. Biologists engage less often in structured public argument and focus far less than philosophers do on *words* (though, of course, there are exceptions!), for their primary concerns are with the organisms with which they deal or the underlying mechanisms and processes that govern the events and interactions they study. On this occasion, I shall try to leave the worst of my professional culture behind, engaging as constructively as I can with some substantive questions about the methods I have observed in use by *Polistes* workers (*pro tempore* my model organisms).

I especially hope to overcome the barriers of language and terminology that separate us. Much of the terminology in this volume is new to me, and much of mine is foreign to the primary audience of this book. The differences in our cultures—national and professional—make communication difficult. But we can overcome those barriers with some patience and learn to ask useful questions of each other, questions that will help each of us to reflect on our own work and to get a bit clearer about what we are up to. To this end, I will set forth a few things from recent philosophy of biology that may be of use to comparative biologists in their work on *Polistes* and other organisms. In so doing, I draw on work that has focused mainly on experimental biology rather than comparative biology, field studies, or natural history. It will be up to the reader to ask whether the suggestions I make can legitimately be exported from contexts of the first sort to those of the other sorts.

A problem regarding model organisms

Evolution is a branching process. At each branch point there are important contingencies that leave their mark on the organisms in at least one of the resulting branches. Indeed, this fact is a key to the feasibility of systematics. At least sometimes, some contingencies remain fixed as suboptimal solutions to functional problems, as 'frozen accidents', or as 'vestigial' characters, traces of which remain in subsequent branches. Controversially, such features can be found in most lineages. In fact, the extent to which marks of history remain in various clades is an empirical question about rates—one that cannot be easily resolved. But since Darwin published the *Origin*, it has been clear that there are enough such features to reveal a great deal about the historical pathways followed by particular lineages. Virtually any organism we choose to study is a mosaic of features, some proportion of which are widely shared with other organisms (though perhaps somewhat transformed), others of which are peculiar to the lineage of the organism, or a small group of organisms. Functionally necessary features (such as streamlining for large-bodied rapid swimmers) are at one extreme, while vestigial characters are at the other. As Darwin taught us, it is the peculiarities (e.g. the femur in whales and the intermaxillary bone in humans), not the optimally functional features, that serve as the best indicators of the historical pathway followed by the lineage of an organism.

There is no *a priori* method for determining the extent or distribution of such features. This is especially clear since, as R. C. Lewontin (pers. comm., cited in Maynard Smith *et al.* 1985, p. 282) has argued, biological evolution can be seen as a history of organisms finding devious ways of getting around constraints. Thus, finding a way to evade a constraint can trigger a radiation, the size of which may not depend at all on the specific character of the 'device' by which the constraint in question was surmounted. Nor need that particular device always be preserved in the descendant organisms of the lineage that evaded the constraint at issue. In this regard, the problem of evolutionary contingency is that of determining which traits (and, indeed, which evolutionary regularities) are, in this sense, primarily 'historical' in character.[1] All of which raises an interesting general problem, namely, how to evaluate the generality of what we learn when we study a particular organism or group of organisms in order to gain an understanding of an evolutionary issue such as the origin and maintenance of eusociality.[2]

The point may be connected to the topic of model organisms by drawing on a formulation due to the biochemist Hans Krebs (see Holmes 1993). Krebs employed frog muscle tissue in his work on the physiology of respiration. He extols this choice retrospectively in an article entitled 'The August Krogh Principle: "For

Many Problems There is an Animal on Which It Can Be Most Conveniently Studied"' (Krebs 1975). Krebs quotes Krogh (1929) as follows:

[W]e used to say as a laboratory joke that this animal [a certain kind of tortoise] had been created expressly for the study of respiration physiology. I have no doubt that there is quite a number of animals which are similarly "created" for special physiology purposes, but I am afraid that most of them are unknown to the men for whom they are "created" and we must apply to the zoologists to find them and lay our hands on them (Krebs 1975, p. 221).

Krebs later adds:

It is an important point that a relatively minor modification of a standard situation [e.g. based on the anatomical size of an organism or part of an organism, the magnitude of a particular process, or a convenient anatomical arrangement] may present great advantages in studying a phenomenon without affecting basic principles. . . . A general lesson . . . is the importance of looking out for a good experimental material when trying to tackle a specific biological problem. . . . [W]hile extensive knowledge of the living world no doubt helps, a deliberate search for a suitable species is [often] needed (Krebs 1975, pp. 225–6; see Holmes 1993).

These comments are easily applied to comparative biology,[3] particularly when one is dealing with well-defined issues such as the origin(s) and maintenance of eusociality, but also, for example, the roles in vespid (and other) social structures of various features such as nest architecture, stereotyped dominance behaviour, etc. The problem of the choice of model organism or group of organisms is clearly a deep one of great importance. Thus, in dealing with the origin (though perhaps not the maintenance?) of eusociality, *Polistes* seems a much wiser choice than any social insects with permanently differentiated social castes. For one thing, the social traits of the latter are more highly derived and may mask the traits relevant to the onset of highly social behaviours. Of course, it does not follow from these considerations alone, or even from the dozens of others in the literature, that *Polistes* will turn out to be a *good* model for studying the origin of eusociality, just that it is better and more convenient than some of the more obvious and widely studied alternatives. Unknown contingencies that entered into the origin(s) of eusociality in Hymenoptera may or may not be well reflected in the biology of polistine wasps.

Krebs also employs the obverse of Krogh's principle, in the form of a warning. Paraphrased, the point of **Krebs' Warning** is this: the very features that make an organism into a useful tool for dealing with a particular problem, i.e. that give it 'special advantages', may or may not be widely distributed. In other words, when looking for *basic principles* or for *historical pathways* pertaining to a wide class of organisms, we must be very careful about generalizing from special cases, lest we make the classic mistake of searching for lost keys where there is sufficient light rather than where they were lost.

To be sure, one could overstate Krebs warning. Obviously there is great deal that is worth knowing about *any* organism. And obviously (though controversially in this heyday of molecular biology), it is imperative to know a great deal of natural history to pose realistic questions and arrive at sound knowledge of pathways, or of justified regularities, principles, or theories in biology. But when we seek 'basic principles' and historical pathways, we *do* run into Krebs's warning. This chapter consists mainly of some reflections on the issues raised by that warning.

Let me put a little biological flesh on these abstract bones with an illustration. Krebs' mentor, Otto Warburg, began to study photosynthesis around 1919, using the unicellular alga, *Chlorella* (see Warburg 1919; Krebs, 1975, p. 221; Zallen 1993). Warburg argued along the following lines for the suitability, indeed necessity, of using something like *Chlorella* to study processes that take place in the leaves of higher plants: Given the primary biochemical technique available for this purpose (at the time manometry), one must have reliable diffusion of gases in the material employed. The cells of leaves, even after being ground into a brei, have complex, ill-understood membranes that delay the diffusion of the gases produced in photosynthesis—the very gases whose production we wish to study. In addition, they block out an unknown amount of light, adding considerably to experimental uncertainty. In contrast to a leaf brei, *Chlorella* has the requisite 'special advantages', namely, it is easy to keep, easy to grow, and, above all, it responds to light, which it absorbs far more readily than leaves, by diffusing gases instantaneously and with maximum efficiency. Until one can isolate naked chloroplasts and maintain their functions [something it took 30 more years to do], unicells like *Chlorella* are the best choice available.

Chlorella (and later *Chlamydomonas*, which had some further advantages, see Zallen 1993) thus made it possible to study photosynthesis by biochemical means. But once those studies had progressed far enough and new techniques had been developed, it was necessary to ask how well *Chlorella* represents what happens in higher plants. The answer proves generally positive; although the latter's efficiency in the capture of photons in leaves is lower, and the products made are sometimes different, the basic mechanism is essentially identical, as it is not, for example, in photosynthetic bacteria (see Zallen 1993, p. 278). Thus the entire cycle (including features bearing on the number of quanta required for photosynthesis in leaves) could be studied in *Chlorella* and *Chlamydomonas*, but not in photosynthetic bacteria. This is the kind of limitation about which Krebs (1975) was warning.

Polistes as a model organism

One of the reasons for which Leo Pardi was awarded the Balzan Prize was that his work, plus the work he inspired, turned *Polistes* into a model organism for the study of eusociality and, indeed, for the study of sociality in general, (E. Mayr, pers. comm.). It seems worth reflecting a little on what it means to turn an organism into a 'model organism' and, given this use of *Polistes*, to ask whether *Polistes* workers should pay particular attention to Krebs' Warning.

I will approach this from two sides, one biological, one philosophical, starting with the biological. Here is a list of some of the advantages of *Polistes* that I have seen in the literature.

1. *Polistes* are large, easily marked, and easily identifiable.
2. Their defences do not interfere seriously with field or laboratory studies.

3. They nest in the open.
4. The combs of the nest are exposed.
5. The nests contain a manageable number of individuals.
6. Their life history is (relatively) simple.
7. Their life span is short compared with social mammals and birds.

These advantages can be summed up simply by stating that one can observe all individuals, all behaviours, and the entire life cycle with relative ease. Together, the items on the above list ensure the convenience of working with *Polistes*. Since, in addition, *Polistes* are interesting in their own right and strikingly beautiful, it is clear that they are very attractive animals on which to work. But does all this show them to be a good model organism? Thus stated, this question cannot be answered, for one must first answer, 'A model for what?' Note that if the topic is the origin of eusociality, the advantages listed above seem, arguably, irrelevant, though if the topic concerns the current range of social behaviours among vespids, some of the points above are clearly relevant (as they may also be if the issue is how sociality is maintained). Putting the matter in a slightly more philosophical mode, the above list of features makes *Polistes* into a lovely object of study for natural history (one that deserves a community of workers comparable to ciliatologists or mammalogists), but it does not show *Polistes* to be a useful model organism until one knows what problem is at stake and the relevance of the features listed to that problem.[4] Before the advantages on the list can be made to provide strong reasons for supposing that *Polistes* can be turned into a vehicle especially well suited to trying to answer questions about the origins or maintenance of eusociality, they must be joined to others specifically appropriate to the problems in question.

Since the focus here is on the use of *Polistes* in studies of the origins of eusociality, I shall deliberately abstract from many interesting features of *Polistes* (e.g. certain physiological ones) that do not seem relevant to this particular set of problems.[5] Since the discussion is partly philosophical (for it concerns what goes into the use of organisms as 'model organisms' as much as it concerns *Polistes*), I have the luxury of greater freedom of abstraction in this investigation than a biologist working in the field on a particular organism might have. In order to disentangle what makes an organism into a good model organism for a particular purpose, it is necessary to simplify the account by setting aside for the purpose of discussion properties of the organisms in question that a field biologist could not afford to ignore.

Given this, I turn to three further advantages of *Polistes* that are more directly relevant to the question whether *Polistes* is a wise choice of model for pursuing the problem of the origins of eusociality. These are:

1. Castes are not anatomically differentiated.
2. (Nearly?) all females can found (or co-found) nests.
3. There are about 200 species of *Polistes*, distributed on all continents (except Antarctica if that is counted) and in nearly all climates.

This list is surely incomplete, but it clarifies the point at stake. Since caste formation almost surely preceded anatomical differentiation of castes in evolutionary time, non-differentiated castes are almost surely the primitive condition. Many workers hold that Hymenoptera (and other social insects) with anatomically differentiated castes are so highly derived that their properties are not likely to shed light on the origin of eusociality. (Many would add, I believe, that studies of the maintenance of eusociality are subject to a similar problem; it is very difficult to determine which of the relevant traits of these organisms are primitive, which derived.) Another important argument for the relevance of studies of *Polistes* to questions concerning the origin and maintenance of eusociality runs as follows: Because (nearly?) all *Polistes* females are potential queens, the mechanisms by means of which they are prevented from abandoning the group nest in favour of becoming solitary, from becoming dominant egg-layers, and, in many species, from laying eggs at all are likely to be pertinent to the primitive condition in which eusociality arose. Finally, as the contributions to this volume demonstrate, the wide distribution of *Polistes*, the varied ecologies in which they are found, and the variety of specific controls on female–female interactions allow very fruitful comparative work aimed at the range of questions on which I am focusing. The 'special advantages' on my second list thus support the claim that *Polistes* is a wise choice as a model organism with respect to questions about the origin and maintenance of eusociality.

It is important to stress that the relevance of such considerations depends on the state of our knowledge (including our theories). Attempts to demonstrate relevance should be connected to the best available evolutionary theorizing and best available knowledge of evolutionary scenarios. Even then, the same organism may be used in connection with *many* different theories, problems, and questions. The arguments just provided depend on the soundness of the relevant background assumptions. If successful, they delimit a certain range of purposes for which *Polistes* might count as a wise choice of model organism, but they do not make further claims to generality. In this respect, claims about the suitability of an organism or model system are specific to the subject matter under investigation.

First epistemological interlude: the character of research problems

Let me digress from *Polistes* for a while to talk about the nature of basic research. Scientists who ask a fundamental question and turn to natural history or experiment seeking an answer to that question face a very delicate situation. They are searching for *what is not yet known*. The organisms or experimental system they employ must be able to yield enormous surprises. After all, they are seeking some unknown feature(s) of the systems with which they work, features that lead them to an object or process that can guide them to an answer to the question with which they started or force them to revise the question because it has a false presupposition (Jerne 1966, p. 301; Rheinberger 1995). There is no guarantee of success in this job, for one does not know exactly what one seeks and the system that one is working with may prove recalcitrant or unsuited to resolve the problem. But there are some *minimum conditions* that should be met if one is to solve the problem at hand:

1. The system (or model organism) should behave consistently and yield a stable or regularly patterned response.
2. The behaviour of the system should resemble or mesh with that of other systems appropriate to the problem in hand [otherwise one has only a special case].
3. One should be able to develop a model, explanation, or theory that ties together all the diverse results bearing on the (revised) problem.

Typically, when we start to work on a problem, we start with *some* ideas about how an explanatory model, theory, or explanation might go. We had *better* have some idea, or we won't know what to look for. Thinking of *Polistes*, from about 1964 on, much work on the origin and maintenance of eusociality was guided by or interpreted in terms of inclusive fitness theory—or better, in terms of *models* of inclusive fitness and kin selection (Hamilton 1964*a*, *b*).

We start with expectations and, where convenient and feasible, devise null hypotheses to test these expectations. But when we turn to nature—to field studies, model organisms, or artificial experimental systems—we usually find surprises.

Gadagkar's chapter in this volume is set up very nicely along just these lines. He begins with a question: Given the fact that primitively eusocial wasps such as *Polistes* and *Ropalidia* have the option of direct reproduction, does kin selection theory enable us to understand how they (and primitively eusocial wasps in general) maintain their eusociality? Then, working with *Ropalidia marginata* as a model organism, he obtains two surprising results—first, that the coefficients of relatedness are low enough to raise questions about the effectiveness of kin selection and, second, that there is strong evidence against assortative mating or sorting by degree of kinship in colony formation.[6] If these results stand up well, something that should be tested by techniques such as those presented by Strassmann (this volume), they would force us to reassess the applicability of inclusive fitness theory to explain the maintenance of eusociality at least in this system. Given general background knowledge about *Polistes*, stenogastrine wasps, etc., if this reassessment revealed significant difficulties in applying inclusive fitness models to *Ropalidia*, it would cast a long shadow on the belief that inclusive fitness models provide a general framework for handling the origin and maintenance of eusociality.

Surprises and anomalies

Now surprises like this may be crudely divided into three kinds: interesting, but not relevant to the problem at hand or related problems; artifacts; and anomalies. When they are first encountered, it is usually very hard to tell what a surprise will amount to. In the case of Gadagkar's results, it is likely that we can rule out one of these possibilities: if they hold up, the findings *will* be relevant to the issue at hand. Low degrees of relatedness within colonies, especially when there is little assortative mating, would pose serious problems for the application of kin selection and inclusive fitness theory. There remain the possibilities of artifact and anomaly. The latter term is common in the philosophy of science. It is used for a surprise like this (especially one with theoretical import) that withstands scrutiny in spite of serious efforts to dislodge it, a surprise which, when probed, won't go away and threatens the standard explanations derived from, or concordant with, pre-existing theoretical conceptions. This use of the term *anomaly* was made prominent by Kuhn (1962). Put somewhat more formally, an **anomaly** is an unexpected result, perhaps conflicting with a prior prediction or with expectations based on current theory, which, if it withstands scrutiny, *should* force one to reconsider one's expectations and the arguments and conceptions supporting those expectations.

Handling anomalies

Kuhn (1962) treats anomalies as if they (or their accumulation) typically bring about a **crisis** in the science in question, a crisis that has to be resolved in one of three ways:

1. The anomaly is shown to be an artifact or is reconciled with the pre-existing theory or background expectations. (Two ways in which the latter can be done are by demonstrating that the antecedent or boundary conditions believed to obtain were incorrect or by showing that some specific interfering factors altered the otherwise expectable outcome.)

2. After much effort, the anomaly is left unresolved, but set aside to be returned to in the indefinite future.

3. Work on the anomaly yields a radically new model, theory, or world view which becomes the basis for the complete overthrow of the preceding theory. This is a **scientific revolution**. The result is a new 'gestalt' in the relevant scientific community's conceptions and perhaps perceptions, typically associated with a generational change since those scientists entrenched in the old view often cannot make a conversion to the new one. The new world view or large-scale theory yields reinterpretation of facts (and loss of some 'facts'), changes of methods and standards, and change in the training of new scientists so that, after some time, the displaced theory and world view are at best difficult and at worst impossible to understand or recover.

In fact, however, Kuhn gives an overly dramatic account of scientific revolutions, particularly of the way in which the old theory is *totally* overthrown by an 'incommensurable' successor. The resultant account of the gulf between the predecessor and successor theories is so extreme that it provides no footing for the use of anomalies to localize the difficulty and thus *improve* or *transform* theories by changing a *part* or, piecemeal, *a series of parts* of a theory. In a more biological terminology, Kuhn's portrait of scientific theories does not allow mosaic evolution of theories to yield drastically altered progeny; rather, it requires that all major change-of-theory be revolutionary substitution of a radically new theory for an abandoned old one.[7] Many philosophers of science have been sceptical of this non-locality of criticism, of the resultant claim that we are unable to make piecemeal theoretical progress in science. (Incidentally, similar problems infect Sir Karl Popper's falsificationist philosophy of science. A falsifying result should yield the total overthrow of the falsified theory but, in its own right, gives no guidance as to what should replace that theory other than it must be something that can account for the falsifier as well as all of the true claims derivable from the predecessor theory.)

Kinds of anomalies

A number of philosophers of science disagree with this epistemological pessimism about piecemeal change. For recent expositions of alternative views, see Darden 1991, Bechtel and Richardson 1992, Schaffner 1993. On this occasion, I can illustrate only a small fragment of a large corpus of work. For ease of exposition I will concentrate on the work of Darden (1991, 1992), as recently amplified (Darden 1995). One virtue of her scheme is that it does not depend crucially on whether the work is highly experimental or is based on natural history and field studies. The point at stake is to show that anomalies can be put to work in the effort to *localize* the difficulties that they reveal. Darden sets forth a number of ways to use unexpected observations or experimental findings, locally if possible, to correct the account of particular cases, to improve a subsidiary part of a model or theory, or to replace a central part of that model or theory. She suggests that such piecemeal change is characteristic of much forefront science. I suggest that it is more characteristic of such science than the dramatic overthrow of theories on which Kuhn focused.

In her forthcoming work, Darden classifies anomalies in terms of their effects. I add one class, the Kuhnian anomaly, not explicitly on her list, and present her classification in Table 18.1.

I cannot here go into a detailed analysis of the classes of anomalies in this table. It will be useful, however, to make a few brief comments and to illustrate each class briefly. First, note that this classification of anomalies is unusual in grouping them by their effects, not by their intrinsic characteristics. I consider this a virtue of the classification, for when scientists encounter a potential anomaly there is no way to tell how far it will reach. Thus, in the context of ongoing research, the import of a (potential) anomaly remains to be discovered; on obtaining a surprising finding, one is obliged (indeed, the relevant communities are obliged) to determine whether

Table 18.1 Kinds of anomaly, after Darden 1995.

Kind of anomaly (in terms of effect)	Cases eventually affected by the anomaly	Consequence
Kuhnian anomaly	Paradigmatic cases under existing theory	Complete overthrow of a major theory
Monster anomalies	'Malfunction' cases	No theory change required; peculiarities leading to 'malfunction' classified
Model anomalies	'Normal' cases	Theory change required; new models of 'normal' cases
Special case anomalies	'Occasional' or 'special' cases	Minor theory change required; new models of 'special cases'

it is an artifact or a genuine anomaly and, if the latter, whether it is due to some interfering circumstance (in which case it may be a monster anomaly), a rare peculiarity (yielding a special case anomaly), some significant problem in the original theory (and thus a model anomaly), or the fundamental inapplicability or malconstruction of the whole theoretical framework. In this way, newly found (potential) anomalies—which count as such only against a very wide background of factual and theoretical knowledge (see Schaffner 1993, chap. 2)—open up a domain of problems within which to work, problems whose specific character is crucially delimited by background knowledge, but to which a variety of approaches may be taken and about which a variety of intuitions and conflicting stances should be expected.

It should be clear that there can be no sharp boundary between the various classes of anomalies, especially when they are first explored, for their import cannot be known in advance. The most that is available in advance is a set of heuristics that provide guidance for scientists attempting to diagnose the problem in hand or to localize the source of difficulty. But scientists with different backgrounds will (and should!) employ different heuristics. Accordingly, in confronting an anomaly they will generate a variety of approaches to the problem(s) at hand. The different heuristics applicable to a given case are based on alternative procedures and intuitions and often vary greatly in the degree of 'depth' they assign to a potential anomaly. This leads scientists with different backgrounds or employing different heuristics to put forward quite different 'solutions' for a given problem. These various solutions need to be examined and, if possible, tested. Later, I will reinforce the point that the resultant pluralism is healthy and desirable in comparative biology (as it is elsewhere as well).

Let me offer a brief illustration of each class of anomalies so that this classification obtains some biological content. Since I am sceptical about the existence of **Kuhnian anomalies** (at least in biology), the first example is provided for the sake of the argument, based on assumptions that I reject. In *On the Origin of Species*, Darwin (1859) offered several examples of results that would 'overthrow his theory entirely'. Among these was a change (not imposed by, e.g. domestication or domination) in the characteristics of organisms of one species brought about solely for the benefit of another species. Thus, once a Darwinian theory of evolution by means of natural selection became established as the orthodox theory of evolution,[8] genuine findings fitting this description might (according to Darwin) undermine the entire foundation of that theory. Such a result would then count as a Kuhnian anomaly, producing the complete overthrow of a fundamental theory.[9]

Recent work on mass extinctions (summarized interestingly in Raup (1991), helpfully discussed by Gayon (1995)), provides an interesting example of a **monster anomaly** for the theory of natural selection. Assume that there has been at least one major cometary impact. The resultant environmental excursion is entirely outside the range of ordinary environmental change (including a wide range of terrestrial environmental catastrophes). The theory of natural selection does not allow organisms to have developed specific protective mechanisms against such a catastrophe, nor does it provide any means for them to anticipate such events.

Accordingly, the fact that certain organisms and lineages survive according to the accidents of circumstance or in virtue of fortuitous preadaptations, but not in virtue of selection for the features necessary for survival in the face of such a drastic environmental change, does not call for any change of theory. Nor (though there are semantic issues here) should it count as a special case. Rather, it is something outside the domain of environmental change entering into accounts of the make-up of biological populations based on a selectionist theory of evolution. Such a 'resetting' of the biotic base from which evolution proceeds turns, contingently, on an environmental anomaly outside the ken of natural selection or the theory of natural selection.[10] Raup's treatment of the matter reinforces this perspective. He supports a strong selectionist position, indeed, a fairly orthodox Darwinism, while regarding the paleontological record as containing major excursions, 'resettings' of the fauna and flora due to catastrophes—resettings that cannot be accounted for within the theory of natural selection. What we have here, at least on Raup's account, is a classic monster anomaly.

Let us turn to **model anomalies**. According to Darden, these 'turn out to be examples of a typical, normal pattern that had not been included in the previous stage of theory development' (Darden 1995). Her example of such an anomaly is the discovery of linkage between genes on a chromosome. Departures from the expected 9:3:3:1 ratios in dihybrid crosses proved to be sufficiently common that linkage clearly should count as a 'normal case'—indeed, they provide the conceptual basis for genetic mapping of chromosomes. Examples of linked 'characters' (really, genes) came to be central to the theoretical elaboration, and the textbook accounts, of the chromosome theory of heredity. Over the period of roughly a decade (starting from Bateson et al.'s (1905) report of aberrant ratios in the sweet pea, *Lathyrus odoratum*), it became clear that purity of the gametes had to be separated from independent assortment. These labels indicate two presuppositions built into Mendelian doctrine (they formed the basis of 'Mendel's two laws'), and were not easily isolated from each other in the thinking of early Mendelians. Linkage quite upset the 'law' of independent assortment. This example nicely illustrates what is meant by a model anomaly; it adds a 'normal' process, linkage, to the models of Mendelian inheritance and does so with considerable effect on theory.

An example of a model anomaly especially appropriate to the readers of this volume is the existence of sterile ('neuter') castes in social insects, famously first treated as an anomaly (though not with the use of that term) for the theory of natural selection by Darwin (1859, pp. 236–42). Darwin himself, by treating selection in these instances as operating on *families* rather than *individuals* took a long step toward turning the case of neuter castes into what, in the present classification, should be classed as a model anomaly. Selection between families provided a new mechanism, potentially applicable quite generally. Indeed, in the 1960s (though with precursors in the 1930s), the development of models of kin selection and the concept of inclusive fitness, which further expanded the class of relatives taken into account in calculating selection coefficients, completed the job of turning this anomaly into a model anomaly. Inclusive fitness theory expanded the class of models for treating 'normal' calculations of individual (inclusive!) fitness.

Finally, there are **special case anomalies**, or those that call for a new model or mechanism, but one that applies only to a small range of special cases, not to normal cases.[11] A classic example concerns the anomalies in the distribution of the t-allele in the house mouse, *Mus musculus*, given extended treatment in a well-known paper by Lewontin and Dunn (1960). This allele, extensively studied by cytologists, developmental biologists and geneticists, is a homozygous lethal with important developmental effects, and is prevalent in startlingly high frequencies in natural populations. The high prevalence can be partly explained by an unusual distribution of gametes, systematically biased toward over-representation of the t-allele in contrast to the wild type ('meiotic drive').[12] Meiotic drive is a special case, known to be rare for well-studied genes and to involve unusual interactions between alleles at the same locus. Thus, the anomalously high frequency of the allele came to be explained by a special mechanism, meiotic drive, that applies in a small range of cases, but not in normal (Mendelian) ones.

Anomaly driven theory redesign

To illustrate the fuzzy border among these classes of anomalies, consider one example, chosen at random from those discussed in this volume. It may be the case that obligate interspecific parasitism in *Polistes*, such as is discussed by Cervo and Dani (this volume), falls in the last category, i.e. that it depends on mechanisms that are not normally present (or utilized in the same manner) in non-parasitic species. Thus, to the extent that the behaviour of the relevant species is anomalous, it may well prove that we have a special case anomaly. But it may also turn out that work with the parasitic species will turn up mechanisms that they exploit that are present in (nearly) all polistines but not yet widely appreciated. Which will prove to be the case is an open question, calling for an open-ended investigation. Darden's work seeks to provide a methodology that will help to structure such investigations.

I can only say a few words about this methodology, which she calls 'anomaly driven theory redesign' (Darden 1995). The methodology proceeds along the following lines:

1. **Anomaly detection**. One begins with a (potential) anomaly, examining whether it is relevant to an issue of current interest and, if so, whether it should be taken seriously or can be set aside as either (a) an artifact or (b) a 'monster'. It is worth emphasizing that a long and difficult process must intervene between finding a potential anomaly and determining that it cannot be accommodated by existing theory—i.e. that the anomaly is an appropriate vehicle for producing at least a certain amount of change in the available theory and that it is a useful tool for guiding theory redesign.[13]

2. **Anomaly localization**. Once one is convinced that the anomaly is one that raises serious questions calling for redesign of available theories or models, it is crucial to determine, if possible, whether it calls for adding or modifying a basic pattern or mechanism or whether it can be isolated to a small range of special cases. One of the key steps here, and one of the most delicate, is determining whether the problem lies within or outside the theory—and, if it should be the former, attempting to

locate a *part* or component of the theory that bears responsibility (Darden 1991, pp. 269-75; Darden 1992, pp. 258-61, 266-9).

3. **Redesign of models or theories.**[14] In particular, at this stage, one attempts to provide the revised exemplars, models, or (portions of) theories called for in accordance with the attempts to localize the problem in (2). The 'reach' of the anomaly, that is whether it calls for a new or revised model of normal cases or only a model for an occasional situation determines how fundamental the theory change will prove to be. Thus, a central problem for the investigator is to estimate the centrality of the anomalies faced by the community.

4. **Comparative assessment of the new and the old models or theories.** To evaluate the proposed solutions to whichever anomalies, it is necessary to compare them with the pre-existing models and theories. To achieve full success in handling a model anomaly, one would have to account not only for the anomalous findings, but also for all (or the vast majority[15]) of the findings that were successfully handled by the pre-existing theories or models.

One more point completes this exposition of Darden's account of anomaly driven theory design. It is important to recognize that the process is an iterative one, with feedback loops from each step to the preceding ones. Thus, for example, an attempt to redesign a model may lead to a reassessment of the centrality of the anomaly that instigated that redesign—in an extreme case either reducing it to the status of an artifact or heightening it from a minor 'cloud on the horizon' to a major model anomaly. The continued traffic among these various phases of theory redesign is a central feature of the activity of a scientific community successfully engaged in the attempt to fathom the unknown. It is in this continuing traffic that the community evaluates the centrality of such mechanisms as gene linkage, meiotic drive, or kin selection for the understanding of particular problems.

Second epistemological interlude: model organisms and experiments

Robert Brandon (1994) recently made some important points about the place of experiment and of reasoning from experiment in biology. Two of his major points are particularly germane here. One is that the concept of experiment is more complex than is generally recognized. In particular, he identifies two components to experimental work that are often run together, much as the early Mendelians ran together purity of gametes and independent assortment. The other is that descriptive work of the sorts common in natural history and comparative biology play a much more important experimental role in many biological domains than many advocates of experimental methods recognize. I shall discuss these points briefly.

According to Brandon, one can recognize two orthogonal axes in the usual treatments of experiment—**manipulation of putative independent variables** and **hypothesis testing**. These vary independently of one another. It is important to gain a clear understanding of what 'manipulation' means in this context, for a great deal of laboratory manipulation may be required for descriptive work that is not experimental in either of the above senses. For example, in preparing soft-bodied fossils for interpretation, a great deal of laboratory-intensive manipulation may be required, but the purpose of this manipulation is, in the first instance, to achieve a description of the fossils about which interpretative hypotheses are subsequently to be constructed.[16] Again, consider the immense industry of gel electrophoresis that invaded population genetics beginning in the mid-1960s. In this context, electrophoresis was used for about twenty years primarily as a tool for determining the amount of variation in the structure of soluble proteins within and between natural populations. The work is highly manipulative; one grinds up or extracts material from large numbers of organisms, submits the resulting material to powerful biochemical manipulations, and follows fairly rigid laboratory protocols in running and analyzing gels. But in spite of all this manipulation, the purpose of this work is to obtain a *description* of the extent of (soluble protein) variation (and an estimate of genetic variation) within natural populations. This is shown in part by the fact that alteration of the proteins from their native state in ways that would affect their positions on the gels must be avoided *on grounds that it would interfere with the intended findings*. The manipulation in question is not manipulation of a putative independent variable, but a means of ascertaining the actual values of a parameter of concern in other contexts. Thus, in spite of the large amount of manipulation, so far forth this counts as descriptive work.

At least sometimes, in contrast, one can manipulate an independent variable with very little, if any, laboratory intervention—or even find it manipulated in a 'natural experiment' or by comparison of different populations or species. In general, such manipulation of an independent variable is typically intended to test an hypothesis about the effect of changes in the value of the putative independent variable on various dependent variables. Classical hypothesis testing along these lines is at the heart of one strand of traditional experimental method. But (as Brandon 1994, argues in detail), there is no obvious correlation between whether or not an experimental design serves to test an hypothesis and whether it is strongly manipulative in Brandon's technical sense. And with a shift in background knowledge, the very same experimental protocol can at one time serve as a test of an hypothesis (e.g. to the effect that natural selection is effective in altering certain features of natural populations) and at a later time (when the hypothesis is generally accepted) as a means of fixing a parameter (e.g. the magnitude of the selection coefficient). Thus, *the state of the relevant community's knowledge* is relevant to the interpretation of a given protocol as an experimental test of an hypothesis or as a technically sophisticated means of fixing a parameter.

To recapitulate: the following three items are conceptually and factually independent: the amount of laboratory manipulation; the extent of manipulation of putative independent variables, and the use of such procedures to test hypotheses. Given this, one must complicate one's understanding of the nature of experiments in biology and of the variety of ways in which to test hypotheses. In particular, much descriptive field work in natural history and comparative biology is appropriately designed for the testing of hypotheses and, in this respect, is as legitimately experimental as much of the work that takes place wholly within the confines of a laboratory. And there are important cases in which the manipulations that take place within a laboratory do not serve as part of an experiment in the sense elaborated by Brandon, either because they are not manipulations of independent variables to determine what happens to dependent variables or because they do not serve to test hypotheses. In short, it is helpful to represent putative experiments on a continuum along two axes according to the degree to which they manipulate independent variables and the degree to which they test hypotheses. Good work in biology need not be experimental in either sense. And good hypothesis testing need not be manipulative in either of the senses mentioned at the beginning of this paragraph. This at least opens up space for the recognition that good descriptive work, like good manipulative experimentation, can serve as an important means of testing theoretical claims—a position that has often been denied in this age of high technology experimentation. And it is precisely the role of a model organism, such as *Polistes*, to serve as the vehicle for such descriptive work.

Polistes again

By combining the things just said about so-called basic research and hypothesis testing with Darden's ideas about the uses of anomalies, one can make some useful observations about the literature on *Polistes*. *Polistes* workers will be far more qualified than I to draw out the morals, for my acquaintance with that literature is relatively recent and slight. Nonetheless, I will hazard some tentative observations.

1. The use of *Polistes* as one of a large number of model organisms for the study of eusociality is best understood as belonging with attempts to convert the existence of sterile castes into a *model anomaly* for the classical theory of evolution by means of natural selection. That is, it belongs with attempts to show that the *normal* processes of selection are better interpreted via *inclusive fitnesses* or by *selection acting at multiple hierarchical levels* (not discussed in this chapter) rather than by *individual fitnesses*, and that by such changes of theory one can provide a natural and comprehensive account of the *origin* and *maintenance* of sterile castes. The elaboration of a number of models of kin selection allowed the use of 'descriptive' work with *Polistes* (but by no means only *Polistes*) to test a number of hypotheses bearing on the effectiveness of selection, calculated in terms of inclusive fitness, to produce or maintain sterile castes and various social behaviours.

2. As is shown by the diversity of topics taken up in this volume, much very interesting work on *Polistes* is not directly relevant to this particular enterprise. Yet, even from the monomaniacal perspective of someone interested primarily in kin selection theory or in the origin of eusociality and only incidentally in *Polistes*, it is important to foster work that does not appear to bear directly on these focal problems. We must, therefore, encourage and appreciate a wide range of such work. First and foremost, it is intrinsically interesting. But also, quite generally, a thorough understanding of all aspects of the life history of one's model organisms is absolutely essential as a basis for evaluating the importance of various aspects of

their biology for whichever problem one chooses to address. For example, as Hansell (this volume) shows, something like the study of nest architecture can reveal surprising connections between factors hitherto ignored and the conditions under which social life is feasible or sustainable. Nests reinforce and constrain social life, and they are a resource that can require group living and co-operation to exploit. Indeed, not to understand *Polistes* nests would be not fully to understand the organism. Such work, to repeat, is intrinsically important. And even for the monomaniac, as Hansell makes clear, such fundamental properties of *Polistes* biology as the constraints imposed by nest materials and nest building may need to enter into models of the resources that constrain society formation.[17] Beyond this, we can well imagine a special case anomaly arising because, under the constraint of restriction to some particular nest material—perhaps fungus-infected plant matter—the social system of the wasp must be changed in particular ways.

3. From the perspective of someone interested in inclusive fitness and the origin and maintenance of eusociality, at least three papers here attempt to turn anomalies for kin selection theory into model anomalies—those of Jacques Gervet *et al.*, West-Eberhard, and Gadagkar (using *Ropalidia* rather than *Polistes*). If 'demographic predisposition', 'developmental switch', or 'pal selection' models, to mention just three, should prove as widely applicable as their authors hope, some of the findings presented in this volume will be turned into model anomalies. Note that this does not imply that any of these mechanisms contradict kin selection theory; rather, if they prove to be of general importance, they will count as normal processes contributing to the evolution of sociality and of specific social behaviours. As such they would be classified as 'normal' process in much the way that linkage is a normal Mendelian process, even though nothing like it was (apparently) envisaged by Mendel. Once again, we see the delicacy of estimating the relative importance of interacting causes and mechanisms and the difficulty of ascertaining or predicting their relative importance.

4. Many papers in this volume are not directly motivated by the issues of kin selection theory or are concerned with what look, initially, like special case anomalies. Thus one *might* look at the *congeneric* parasites like *Polistes sulficer*, discussed by Cervo and Dani, as special cases. Nest parasitism by another *Polistes* species poses a particular problem of recognition of conspecifics and/or of defence of the nest which may (or may not) call forth a particular solution on the part of the host species. *If* it turns out that this is what is at stake here, understanding the dynamics of nest parasitism and the defences against it *may* prove to give us an account of what is 'normal' under special circumstances—i.e. we may have a special case anomaly.

5. However, both in this instance and in general, it is not easy to assess the depth and importance of unexpected findings. In most instances, it takes several years (and sometimes several scientific generations) to find out whether a new finding should be dismissed as an artifact, should count as a special case, or should be taken to shed light on normal cases. It may take equally long to learn whether a major family of models (e.g. those of kin selection theory) can be correctly applied to a

particular range of cases or anomalies. One modest example illustrates how much it may take to solve a middle-sized problem, namely, the development of the problem of kin recognition as part of the elaboration of inclusive fitness theory (see Gamboa this volume, for recent work on kin recognition in *Polistes*). Part of what it means to say that basic research is an attempt to find what is not yet known is that *the normal position of scientific researchers is not to know where their findings will lead*. One of the joys and difficulties of the enterprise of real research is the continuing effort to evaluate which findings are significant—and just how significant they are.

6. For the class of problems surrounding the problem of the scope and importance of inclusive fitness in evolutionary biology, *Polistes* is by no means the only suitable model organism. Indeed, it is crucial to use many organisms across a wide phylogenetic range for this purpose—and it may well prove that some other organism serves as a better general model than *Polistes*.

Some concluding remarks

In a letter responding to the presentation on which this chapter is based, Mary Jane West-Eberhard (pers. comm.) objects that I over-emphasize the importance of testing kin selection theory and related matters in this chapter. Perhaps so. She suggests that the main significance of wasps is 'as a wellspring of theoretical ideas on the evolution of social behaviour and, especially, on the importance of individuality and flexibility in the evolution of behaviour', a point also stressed in the editorial introduction to this volume. She connects this to a general point about comparative biology, which she contrasts with such fields as genetics and physiology. In this context she says: 'Wasps are not ... important as a 'test organism' like *Drosophila*, phage, garden peas, or the white rat, although they have sometimes been used that way. They are important as an idea generator, and I think there are things about wasps that make this so' (see also the editorial introduction, p. viii).

I have no desire to deny the power of the ideas generated in the course of observing and studying wasps. On the contrary, West-Eberhard is surely right to insist on the extraordinary conceptual fecundity of those biologists who have immersed themselves in wasps and tried to go where the organisms led them. But I believe that the apparatus of 'anomaly driven theory redesign', set forth above goes some considerable distance toward recognition of the sort of thing she has in mind. It certainly goes beyond a classical account of the use of a model organism as a 'test organism', e.g. in providing heuristics for localizing difficulties and thus suggesting where novel hypotheses might be fruitful. In particular, the key innovation is the use of a set of iterated steps, with corrective feedback, combined with the behaviour of the model organism, to generate new models and working hypotheses. Without these new models and some procedures for evaluation of the power of the ideas generated, our working estimates of the worth of those ideas and of the strength of the tests that can be performed to evaluate them would be far weaker than they are. With them, one is empowered to challenge the prior models with which the investigations in question began. Model testing in accordance with anomaly driven theory redesign is comparative, like much of biology. And one of the most remarkable uses of a model organism like *Polistes* is to assist us in generating the new models that make comparative testing possible.

I shall close with a final reflection, based on the conclusion of Malte Andersson's (1984) review, 'The Evolution of Eusociality'. Andersson surveys the evidence for and against accounts of the origin and maintenance of eusociality based on kin selection. He emphasizes that the fit between kin selection models and the findings in a great variety of organisms is, at best, partial. Among the many additional factors he suggests as entering into the origin of eusociality are mutualism *à la* Michener, West-Eberhard, and Evans; parental manipulation *à la* Alexander and others; various modes of inbreeding and symbiont transmission such as are found in termites; and various ecological factors connected with provisioning a nest or den. For these reasons, but also for others, I concur with his concluding admonition, repeated below, which is a plea for explanatory pluralism.

This chapter started with acknowledgement of the importance of the fact that evolution is a branching process. Given this, I believe it would be a mistake to expect a unified theory of the origin or of the maintenance of social systems. Mosaic evolution is common. In her chapter, West-Eberhard stresses the way in which developmental switches allow organisms to achieve phenotypic flexibility and to pass from one constellation of co-adapted traits to another. If such mosaicism is taken seriously at both organismic and evolutionary scales, we should expect a kind of theoretical pluralism—that is, we should expect to find different models of (for example) the origin of sociality applying in different cases and contributing in different degrees to our understanding of sociality in different clades. These reflections, drawn from a different starting point than Andersson's, support the views expressed in his concluding paragraph, the first half of which reads as follows:

[T]here has been an emphasis in sociobiology on 'strong inference' based on alternative rather than complementary hypotheses. When feasible, this approach has much to say for it, but clear-cut answers in terms of a single hypothesis often are not possible in evolutionary biology. Strong inference works excellently when the hypotheses are qualitative and mutually exclusive [as is often found in physics]. . . . But the situation is very different for biologists who try to explain the evolution of a trait such as eusociality. Here, different hypotheses usually are not mutually exclusive, and the main problem is to identify and estimate the relative importance of each factor. The time scale over which evolution occurs often precludes experimental tests, and the comparative method—with all its pitfalls—may be the only tool available. (Andersson 1984, p. 182, references omitted.)

We are after all, in the realm of natural history. At the heart of natural history we find, and will always find, arguments over the relative frequency and relative importance of interacting factors and multiple causes acting jointly. Let us do honour to this fact by allowing a healthy theoretical pluralism and a very broad comparative base covering a wide range of organisms. This, I believe, is necessary to foster the understanding of complex phenomena, such as eusociality, that result from the joint operation of many contributing causes.

Acknowledgements

I have profited greatly from correspondence with Mary Jane West-Eberhard, from her criticism of a draft of this paper (with which she has substantial constructive disagreements), and from discussion at the conference on which this volume is based. I have improved the text thanks to suggestions from Lindley Darden, Marjorie Grene, and Maxwell Palmer. I am grateful to them and to numerous colleagues who have sharpened my thinking about model organisms.

Notes

1. Beatty (1995) argues that regularities like those exemplified in the genetic code and the Krebs cycle (to cite controversial examples) are typical in being historically contingent regularities. Beatty, R. Brandon, and I plan a book on evolutionary contingency arguing for the strong thesis that all distinctively biological regularities are 'historical', i.e. they are evolutionary contingencies.

2. This term is understood here according to what I take to be standard usage: '*Eusociality* is defined as the presence of overlapping adult generations, co-operative brood care, and extreme reproductive division of labour—namely, an essentially sterile caste' (Carpenter 1991, p. 7; see also Cowan 1991, p. 34; Michener 1974, p. 38; Wilson 1971, pp. 4–5).

3. At the end of this chapter I will discuss M. J. West-Eberhard's suggestion that an alternative 'use' of *Polistes* species may be more important than their use as model organisms.

4. Zallen (1993), referring to Sturtevant (1971), reminds us of his distinction between wise (unwise) and lucky (unlucky) choices of experimental organisms. Advance considerations of the reasons for expecting an organism to be especially suited to a particular investigation (as Warburg did for *Chlorella*) is part of what goes into making a wise choice. But there are, as Sturtevant and many others have pointed out, lucky (and unlucky) as well as wise (and unwise) choices. Thus Mendel's choice of *Hieracium* to meet some objections of Nägeli proved unlucky because of that organism's cryptic and aberrant failure to cross regularly, and Morgan's choice of *Drosophila* as a genetic organism was lucky, since he did not know at the time (1910) of the small number of linkage groups and their correspondence with the chromosomes, and he could not anticipate the discovery nearly twenty years later of salivary gland chromosomes. For more on the choice of *Drosophila* and the Morgan group's development of a production system for genetic knowledge by use of this little dipteran, see Kohler (1993, 1994).

5. Note, however, that physiological features *may* be crucial to the origin of eusociality. See, e.g. the role that M. J. West-Eberhard (building on her own work and that of many others) ascribes to ovarian regression of dominated females in the formation of semi-social groups in *Polistes*. There is a strong connection here to work on other Hymenoptera, e.g. on *Ceratina* by Sakagami and Maeta (e.g. 1987*a*, 1987*b*), that suggests that this physiological feature plays an important role in evolution of sociality across a wide range of organisms. Judgements of relevance depend on background knowledge and are corrected, sometimes radically, as we learn more about the system or problem under investigation.

6. Note that these findings do not contradict kin selection theory; rather, they require either a non-kin-theoretic explanation or a kin-theoretic explanation in which the cost/benefit term is powerful enough to compensate for the surprisingly low relatedness of *Ropalidia* nestmates. Thus, if Gadagkar's findings prove robust, it will still be necessary to evaluate their theoretical import.

7. In fairness to Kuhn, it is important to stress that he views 'normal science', as opposed to science in crisis or revolutionary science, as a puzzle-solving activity *within* the confines of a single world view, instantiated in a 'disciplinary matrix'. Thus the claim in the text speaks to difficulties with the handling of anomalies in the context of crises and scientific revolutions, not of 'puzzles' within a stable theoretical context. The latter sort of work belongs to Kuhn's 'normal science' and it does allow a great deal of piecemeal progress. However, it is restricted to work within the secure and unchallenged assumptions of a stable master theory.

8. A condition which, in spite of much mythology to the contrary, did not really arise until the middle of the twentieth century.

9. I was reminded by Maxwell Palmer of a second candidate for the status of a Kuhnian anomaly, this time affecting vitalist biology in the early nineteenth century. It is the synthesis of urea by Wöhler in 1828.

10. Contingently because if cometary impacts occurred every few generations, their effects would be among the environmental excursions on which selective processes would operate.

11. It is likely that Darwin himself would have preferred to treat family selection yielding neuter castes and selection between tribes of humans (which provided him with a mechanism for the production of human moral sentiments and altruism) as special cases, explicable by unusual rather than generally applicable mechanisms. If so, in the current classification, he would tend to view the phenomena involved as special case anomalies rather than model anomalies. An additional puzzle in understanding Darwin is whether his treatment of neuter castes fits more naturally with models of *inclusive fitness* or of *group selection*.

12. An additional problem arises at this point: the actual proportion of mice carrying t-alleles in the population is lower than the proportion in gametes and pups would imply, given all the relevant selection coefficients. Lewontin and Dunn (1960) argue that population structure and group selection lower the population equilibrium from what it would be in the presence only of meiotic drive and selection against the t-allele in a panmictic population.

13. M. J. West-Eberhard (pers. comm.) has emphasized the importance of giving pre-existing theory its due. One reason for this is that it is all too common (and in the investigator's narrow self interest) to portray interesting findings as more powerful or important than they are by making it appear that they contradict or cannot be fit with existing theory when, in fact, the resources of that theory have been misconstrued or have not been properly exploited. In Darden's terms, this speaks to the difficulty of establishing that a puzzling finding ought genuinely to count as an anomaly.

14. An aspect of Darden's and others' treatment of these issues that I have no space to discuss here is the use of exemplary instances (Kuhn's 'exemplars') as models for a wider class of cases. Sometimes the redesign is best accomplished by use of a new or revised exemplar rather than by explicit construction of a new model or theory.

15. With a specific account of why the remainder should be dismissed as artifacts or as explicable on the basis of factors belonging to some other domain.

16. Of course, these two are interactive; no description of a fossil is feasible without some rudimentary interpretation of the nature and allegiances of the fossil.

17. Indeed, for those pursuing the program of explaining features of social organism in terms of selection acting at several hierarchical levels, a crucial problem is the relation of individuals to the group. Constraints imposed by a common nest are a crucial part of the work on this theme which, as M. J. West-Eberhard (pers. comm.) emphasized to me, has been a continuing preoccupation in the study of *Polistes* for at least fifty years.

References

Agassiz, L. (1845). *Nomenclator zoologicus, continens nomina systematica generum animalium tam viventium quam fossilium.* Fasc. VII et VIII, Hymenoptera, Vol. 27, Soloduri.

Akre, R. D. (1982). Social wasps. In *The social insects*, Vol. 4 (ed. H. Hermann), pp. 1–105. Academic Press, New York.

Akre, R. D. and Reed, H. C. (1983). Evidence for a queen pheromone in *Vespula* (Hymenoptera: Vespidae). *Canadian Entomologist*, **115**, 371–7.

Akre, R. D., Hill, W. B., Mac Donald, J. F., and Garnett, W. B. (1975). Foraging distances of *Vespula pensylvanica* workers (Hymenoptera; Vespidae). *Journal of the Kansas Entomological Society*, **48**, 12–6.

Alam, S. M. (1959). Some interesting revelations about the nest of *Polistes hebroeus* Fabr. (Vespidae, Hymenoptera)—the common yellow wasps of India. *Proceedings of the Zoological Society, Calcutta*, **11**, 113–22.

Alcock, J. (1987). Leks and hilltopping in insects. *Journal of Natural History*, **21**, 319–28.

Aldrovandi, U. (1638). *De animalibus insectis libri septem cum singulorum iconibus ad nivum expressis.* Clementen Ferronium, Bononiae.

Alexander, B. (1987). Eusociality and parasitism in nest-provisioning insects. In *Chemistry and biology of social insects*, (ed. J. Eder and H. Rembold), p. 387. Verlag J. Peperny, München.

Alexander, R. D. (1974). The evolution of social behavior. *Annual Review of Ecology and Systematics*, **4**, 325–83.

Alexander, R. D., Noonan, K. M., and Crespi, B. J. (1991). The evolution of eusociality. In *The biology of the naked mole rat*, (ed. P. W. Sherman, J. U. M. Jarvis, and R. D. Alexander), pp. 3–44. Princeton University Press.

Alford, R. D. (1975). *Bumblebees.* Davies-Poynter, London.

Allen, J. L., Schulze-Kellman, K., and Gamboa, G. J. (1982). Clumping patterns during overwintering in the paper wasp, *Polistes exclamans*: effects on relatedness. *Journal of the Kansas Entomological Society*, **55**, 97–100.

Anderson, R. H. (1963). The laying worker in the Cape honey bee, *Apis mellifera capensis*. *Journal of Apiculture Research*, **2**, 85–92.

Andersson, M. (1984). The evolution of eusociality. *Annual Review of Ecology and Systematics*, **15**, 165–90.

Aoki, S. (1977). *Colophina clematis* (Homoptera, Pemphigidae), an aphid species with "soldiers". *Kontyû*, **45**, 276–82.

Arathi, H. S. and Gadagkar, R. (1993). Can genetically unrelated individuals join colonies of *Ropalidia marginata*? In *Proceedings of the 21st Annual Conference of the Ethological Society of India, Tirupati* (In press).

Ashmead, W. L. (1904). Descriptions of new genera and species of Hymenoptera from the Philippine Islands. *Proceedings of the United States National Museum*, **28**, 127–58.

Baerends, G. P. (1941). Fortpflanzungsverhalten und Orientierung der Grabwespe *Ammophila campestris* Jur. *Tijdschrift voor Entomologie*, **84**, 68–275.

Baerends, G. P. (1959). Ethological studies of insect behavior. *Annual Review of Entomology*, **4**, 207–34.

Baerends, G. P. (1976.) The functional organization of behaviour. *Animal Behaviour*, **24**, 726–38.

Bagnères, A. G., Lange, C., Clément, J. L., and Joulie, C. (1988). Les hydrocarbures cuticulaires des *Reticulitermes* français: variations specifiques et coloniales. *Actes Colloques Insectes Sociaux*, **4**, 34–42.

Bagnères, A. G., Clément, J. L., Lange, C., and Blum, M. S. (1990). Cuticular compounds in *Reticulitermes* termites: species, caste and colonial signature. In *Proceedings of the 11th International Congress IUSSI*, pp. 423–4. Bangalore, India.

Bagnères, A. G., Errard, C., Mulheim, C., Joulie, C., and Lange, C. (1991a). Induced mimicry of colony odors in ants. *Journal of Chemical Ecology*, **17**, 1641–64.

Bagnères, A. G., Killian, A., Clément, J. L., and Lange, C. (1991b). Interspecific recognition among termites of the genus *Reticulitermes*: Evidence for a role for the cuticular hydrocarbons. *Journal of Chemical Ecology*, **17**, 2397–420.

Bagnères, A. G., Provost, E., Clément, J. L., Plateaux, L., and Morgan, E. D. (1991c). Chemosystematics in Leptothoracine ants. In *Proceedings of the 1st European Congress of Social Insects*, p. 46. Leuven, Belgium.

Bagnères, A. G., Lorenzi, M. C., Clément, J. L., Dusticier, G., and Turillazzi, S. (1994). Fluctuations of the chemical signature of two species of paper wasps (*Polistes biglumis bimaculatus* and *P. atrimandibularis*) during a cycle of parasitism. In *Les insectes sociaux* (ed. A. Lenoir, G. Arnold, and M. Lepage), p. 346. Université Paris Nord.

Balduf, W. V. (1961). A large population of *Polistes annularis* (Linn.). *Entomological News*, **72**, 259–60.

Barth, R. H., Lester, L. J., Sroka, P., Kessler, T., and Hearn, R. (1975). Juvenile hormone promotes dominance behavior and ovarian development in social wasps (*Polistes annularis*). *Experientia*, **31**, 691–2.

Bateson, P. (1988). The active role of behaviour in evolution. In *Evolutionary processes and metamorphs* (ed. M.-W. Ho and S. W. Fox), pp. 191–207. Wiley, Chichester.

Bateson, W., Saunders, E., and Punnett, R. (1905). Further experiments on inheritance in sweet peas and stocks: Preliminary account. (*Proceedings of the Royal Society*, London, **77**, 236–8). (Reprinted) in *Scientific papers of William Bateson* (1928), Vol. 2, (ed. R. Punnett), pp. 139–41. Cambridge University Press.

Batra, S. W. T. (1977). Bees of India (Apoidea), their behaviour, management and a key to the genera. *Oriental Insects*, **11**, 289–324.

Batschelet, E. (1981). *Circular statistic in biology*. Academic Press, New York.

Beani, L. and Calloni, C. (1991a). Leg tegumental glands and male rubbing behavior at leks in *Polistes dominulus* (Hymenoptera: Vespidae). *Journal of Insect Behavior*, **4**, 449–62.

Beani, L. and Calloni, C. (1991b). Male rubbing behavior and the hypothesis of pheromonal release in polistine wasps (Hymenoptera Vespidae). *Ethology Ecology & Evolution (Special Issue)*, **1**, 51–4.

Beani, L. and Lorenzi, M. C. (1992). Different tactics of mate searching by *Polistes biglumis bimaculatus* males (Hymenoptera Vespidae). *Ethology Ecology & Evolution (Special Issue)*, **2**, 43–5.

Beani, L. and Turillazzi, S. (1988a). Alternative mating tactics in males of *Polistes dominulus* (Hymenoptera: Vespidae). *Behavioral Ecology and Sociobiology*, **22**, 257–64.

Beani, L. and Turillazzi, S. (1988b). An experiment on the relationship between spatial behaviour and mating success in the male *Polistes gallicus* (L.) (Hymenoptera: Vespidae). *Monitore Zoologico Italiano (Nuova Serie)*, **22**, 323–30.

Beani, L. and Turillazzi, S. (1990a). Male swarms at landmarks and scramble competition polygyny in *Polistes gallicus* (Hymenoptera: Vespidae). *Journal of Insect Behavior*, **3**, 545–56.

Beani, L. and Turillazzi, S. (1990b). Overlap at landmarks by lek-territorial and swarming males of two sympatric polistine wasps (Hymenoptera Vespidae). *Ethology Ecology & Evolution*, **2**, 419–31.

Beani, L. and Turillazzi, S. (1994). Aerial patrolling and stripes display in males of *Parischnogaster mellyi* (Hymenoptera Stenogastrinae). *Ethology Ecology & Evolution (Special Issue)*, **3**, 43–6.

Beani, L., Cervo, R., Lorenzi, M. C., and Turillazzi, S. (1992). Landmark-based mating systems in four *Polistes* species (Hymenoptera: Vespidae). *Journal of the Kansas Entomological Society*, **65**, 211–17.

Beattie, A. J., Turnbull, C. L., Hough, T., and Knox, R. B. (1986). Antibiotic production: a possible function for the metapleural glands of ants (Hymenoptera: Formicidae). *Annals of the Entomological Society of America*, **79**, 448–50.

Beatty, J. (1995). The evolutionary contingency thesis. In *Concepts, theories and rationality in the biological sciences*, (ed. G. Wolters, J. G. Lennox, and P. McLaughlin), pp. 45–81. UVK-Universitätsverlag Konstanz/University of Pittsburgh Press.

Beaumont, J. de and Matthey, R. (1945). Observations sur les *Polistes* parasites de la Suisse. *Bulletin de la Société Vaudoise des Sciences Naturelles*, **62**, 30–47.

Bechtel, W. and Richardson, R. (1992). *Discovering complexity: Decomposition and localization as strategies in scientific research*. Princeton University Press.

Bell, W. J. (1973). Factors controlling initiation of vitellogenesis in a primitively social bee, *Lasioglossum zephyrum* (Hymenoptera: Halictidae). *Insectes Sociaux*, **20**, 253–60.

Bell, W. J. and Barth, R. H. (1971). Initiation of yolk deposition by juvenile hormone. *Nature New Biology*, **230**, 220–1.

Bell, W. J. and Bohm, M. K. (1975). Oosorption in insects. *Biological Review*, **50**, 373–96.

Bennett, N. C., Jarvis, J. U. M., and Davies, K. C. (1988). Daily and seasonal temperatures in the burrows of African rodent moles. *South African Journal of Zoology*, **23**, 189–95.

Benton, T. G. and Foster, W. A. (1992). Altruistic housekeeping in a social aphid. *Proceedings of the Royal Society of London*, **247**, 199–202.

Bequaert, J. (1918). A revision of the Vespidae of the Belgian Congo based on the collection of the American Museum Congo Expedition with a list of Ethiopian diplopterous wasps. *Bulletin of the American Museum of Natural History*, **39**, 1–384.

Bequaert, J. (1930). Ashmead's genus *Polistella* (Hymenoptera, Vespidae). *Pan-Pacific Entomologist*, **7**, 91–3.

Bequaert, J. (1934). Les races de coloration de *Vespa luctuosa* de Saussure et de *Polistes tenebricosus* Lepeletier. *Bulletin du Musée Royal d'Histoire Naturelle de Belgique*, **10**, 1–11.

Bequaert, J. (1937). The American *Polistes* with prepectal suture. Their structural characters, distribution and variation. *Archivos do Instituto de Biologia Vegetal, Rio de Janeiro*, **3**, 171–205.

Bequaert, J. (1938). Études sur les Hyménoptères Diploptères d'Afrique. II. Quelques *Polistes* nouveaux pour la faune du Congo Belge avec un aperçu des espèces ethiopiennes et malgaches du genre. *Revue de Zoologie et de Botanique Africaine*, **31**, 129–52.

Bequaert, J. (1940). An introductory study of *Polistes* in the United States and Canada with descriptions of some new North and South American forms (Hymenoptera: Vespidae). *Journal of the New York Entomological Society*, **48**, 1–31.

Bernstein, I. S. (1991). The correlation between kinship and behaviour in non-human primates. In *Kin recognition*, (ed. P. W. Hepper), pp. 6–29. Cambridge University Press.

Billen, J. (1982). Ovariole development in workers of *Formica sanguinea* Latr. (Hymenoptera: Formicidae). *Insectes Sociaux*, **29**, 86–94.

Billen, J. (1986). Étude morphologique des glandes tarsales chez la guêpe *Polistes annularis* (L.) (Vespidae, Polistinae). *Actes des Colloques Insectes Sociaux*, **3**, 51–60.

Blaustein, A. R. and O'Hara, R. K. (1982). Kin recognition in *Rana cascadae* tadpoles: maternal and paternal effects. *Animal Behaviour*, **30**, 1151–57.

Blom, J. van der and Velthuis, H. H. W. (1988). Social behaviour of the carpenter bee

Xylocopa pubescens (Spinola). *Ethology*, **79**, 281–94.

Blomquist, J. G. and Dillwith, J. W. (1985). Cuticular lipids. In *Comprehensive insect physiology, biochemistry and pharmacology*, Vol. 3 (ed. G. A. Kerkut and L. I. Gilbert), pp. 117–54. Pergamon Press, Oxford.

Blomquist, J. G., Howard, R. W., and McDaniel, C. A. (1979). Structure of the cuticular hydrocarbons of the termite *Zootermopsis angusticollis* (Hagen). *Insect Biochemistry*, **9**, 365–70.

Blum, M. S. (1981). *Chemical defenses of arthropods*. Academic Press, New York.

Blüthgen, P. (1938). Systematisches Verzeichnis der Faltenwespen Mitteleuropas, Skandinaviens und Englands. *Konowia*, **16**, 270–95.

Blüthgen, P. (1943). Die europäischen Polistinen. *Archiv für Naturgeschichte, N. F.*, **12**, 94–129.

Bohm, M. K. (1972). Effects of environment and juvenile hormone on ovaries of the wasp, *Polistes metricus*. *Journal of Insect Physiology*, **18**, 1875–83.

Bonavita-Cougourdan, A., Clément, J. L., and Lange, C. (1987). Nestmate recognition: cuticular hydrocarbons and colony odor in the ant *Camponotus vagus* Scop. *Journal of Entomological Science*, **22**, 1–10.

Bonavita-Cougourdan, A., Clément, J. L., and Povéda, A. (1990). Les hydrocarbures cuticulaires et les processus de reconnaissance chez les fourmis: le code d'information complexe de *Camponotus vagus*. *Actes des Colloques Insectes Sociaux*, **6**, 273–80.

Bonavita-Cougourdan, A., Théraulaz, G., Bagnères, A. G., Roux, M., Pratte, M., Provost, E., and Clément, J. L. (1991). Cuticular hydrocarbons, social organization and ovarian development in a polistine wasp: *Polistes dominulus* Christ. *Comparative Biochemistry and Physiology*, **100B**, 667–80.

Bonavita-Cougourdan, A., Clément, J. L., and Lange, C. (1993). Functional subcaste discrimination (forager and brood-tends) in the ant *Camponotus vagus*. Polymorphism of cuticular hydrocarbon patterns. *Journal of Chemical Ecology*, **19**, 1461–77.

Boomsma, J. J. and Grafen, A. (1991). Colony-level sex ratio selection in the eusocial Hymenoptera. *Journal of Evolutionary Biology*, **3**, 383–407.

Bordas, L. (1908). Les glandes cutanées de quelques Vespides. *Bulletin de la Société Zoologique de France*, **33**, 59–64.

Bornais, K. M., Larch, C. M., Gamboa, G. J., and Daily, R. B. (1983). Nestmate discrimination among laboratory overwintered foundresses of the paper wasp, *Polistes fuscatus* (Hymenoptera: Vespidae). *Canadian Entomologist*, **115**, 655–8.

Borsato, W. (1992). Le vespe dei monti Lessini (Verona) (Hymenoptera, Vespidae). *Bollettino del Museo Civico di Storia Naturale, Verona*, **16**, 333–46.

Bradbury, J. W. (1985). Contrasts between insects and vertebrates in the evolution of male display, female choice, and lek mating. In *Experimental behavioral ecology and sociobiology* (ed. B. Hölldobler and M. Lindauer), pp. 289–93. Fischer Press, New York.

Bradbury, J. W. and Davies, N. B. (1987). Relative roles of intra and intersexual selection. In *Sexual selection: testing the alternatives* (ed. J. W. Bradbury and M. B. Andersson), pp. 143–63. Wiley, Chichester.

Bradbury, J. W. and Gibson, R. M. (1983). Leks and mate choice. In *Mate choice* (ed. P. B. Bateson), pp. 109–38. Cambridge University Press.

Brandon, R. (1994). Theory and experiment in evolutionary biology. *Synthese*, **99**, (In press).

Breed, M. D. and Julian, G. E. (1992). Do simple rules apply in honey-bee nestmate discrimination? *Nature*, **357**, 685–6.

Breed, M. D. and Stiller, T. M. (1992). Honey bee, *Apis mellifera*, nestmate discrimination: hydrocarbon effects and the evolutionary implications of comb choice. *Animal Behaviour*, **43**, 875–83.

Breed, M. D., Fewell, J. H., and Williams, K. R. (1988). Comb wax mediates the acquisition of nest-mate recognition cues in honey bees. *Proceedings of the National Academy of Sciences of USA*, **85**, 8766–9.

Breed, M. D., Snyder, L. E., Lynn, T. L., and Morhart, J. A. (1992). Acquired chemical camouflage in a tropical ant. *Animal Behaviour*, **44**, 519–23.

Bremer, K. (1992). Ancestral areas: A cladistic reinterpretation of the center of origin concept. *Systematic Biology*, **41**, 436–45.

Brèthes, J. (1903). Contribución al estudio de los véspidos sudamericanos y especialmente argentinos. *Anales del Museo Nacional de Buenos Aires*, (3), **9**, 15–39.

Brian, M. V. (1956). Studies on caste differentiation in *Myrmica rubra* L. 4. Controlled larval nutrition. *Insectes Sociaux*, **3**, 369–94.

Brockmann, H. J. (1984). The evolution of social behaviour in insects. In *Behavioural ecology: an evolutionary approach* (ed. J. R. Krebs and N. B. Davies), pp. 340–61. Blackwell, Oxford.

Brockmann, H. J. (1993). Parasitizing conspecifics: comparisons between Hymenoptera and birds. *Tree*, **8**, 2–4.

Brockmann, H. J. and Dawkins, R. (1979). Joint nesting in a digger wasp as an evolutionarily stable preadaptation to social life. *Behaviour*, **71**, 3–4.

Brothers, D. J. (1992). The first Mesozoic Vespidae (Hymenoptera) from the Southern Hemisphere, Botswana. *Journal of Hymenoptera Research*, **1**, 119–24.

Brown, J. L. (1975). *The evolution of behavior*, pp. 1–761. W. W. Norton and Company, Inc., New York.

Brown, J. L. (1987). *Helping and communal breeding in birds*. Princeton University Press.

Brown, W. V., Spradbery, J. P., and Lacey, M. J. (1991). Changes in the cuticular hydrocarbon composition during development of the social wasp, *Vespula germanica* (F.) (Hymenoptera: Vespidae). *Comparative Biochemistry and Physiology*, **99B**, 553–62.

Brown, W. V., Watson, J. A. L., Carter, F. L., Lacey, M. J., Barrett, R. A., and McDaniel, C. A. (1990). Preliminary examination of cuticular hydrocarbons of worker termites as chemotaxonomic characters for some Australian species of *Coptotermes* (Isoptera: Rhinotermitidae). *Sociobiology*, **16**, 305–28.

Bura, E. A. and Gamboa, G. J. (1994). Kin recognition by social wasps: asymmetric tolerance between aunts and nieces. *Animal Behaviour*, **47**, 977–9.

Burian, R. (1992). How the choice of experimental organism matters: Biological practices and discipline boundaries. *Synthese*, **92**, 151–66.

Burian, R. (1993). How the choice of experimental organism matters: Reflections on an epistemological aspect of biological practice. *Journal of the History of Biology*, **26**, 351–67.

Burk, T. (1988). Acoustic signals, arms races and the costs of honest signalling. *Florida Entomologist*, **71**, 400–9.

Butler, C. G., Fletcher, D. J. C., and Watler, D. (1969). Nest-entrance marking with pheromones by the honeybee, *Apis mellifera* L. and by a wasp, *Vespula vulgaris* L. *Animal Behaviour*, **17**, 142–7.

Butts, D. P., Espelie, K. E., and Hermann, H. R. (1991). Cuticular hydrocarbons of four species of social wasps in the subfamily Vespinae: *Vespa crabro* (L.), *Dolichovespula maculata* (L.), *Vespula squamosa* (Drury), and *Vespula maculifrons* (Buysson). *Comparative Biochemistry and Physiology*, **99B**, 87–91.

Butts, D. P., Camann, M. A., and Espelie, K. E. (1993). Discriminant analysis of cuticular hydrocarbons of the baldfaced hornet, *Dolichovespula maculata* (Hymenoptera: Vespidae). *Sociobiology*, **21**, 193–201.

Buysson, R. du (1905). Descriptions d'Hyménoptères nouveaux. *Bulletin de la Société Entomologique de France*, **1905**, 256–8.

Buysson, R. du (1909). Monographie des Vespides du genre *Belonogaster*. *Annales de la Société Entomologique de France*, **78**, 199–270.

Byers, J. A. and Bekoff, M. (1986). What does "kin recognition" mean? *Ethology,* **72**, 342–5.

Camazine, S. (1991). Self-organizing pattern formation on the combs of honey bee colonies. *Behavioral Ecology and Sociobiology,* **28**, 61–76.

Cameron, S. A. and Robinson, G. E. (1990). Juvenile hormone does not affect division of labor in bumble bee colonies (Hymenoptera: Apidae). *Annals of Entomological Society of America,* **83**, 626–31.

Cancello, E. M. (1991). Two different mounds of *Cornitermes bequaerti* (Termitidae, Nasutitermitinae): An example of the plasticity in termite nest architecture in the tropics. *Revista Brasileira de Entomologia,* **35**, 603–6.

Cane, J. H. (1983). Chemical evolution and chemosystematics of the Dufour's gland secretions of the lactone-producing bees (Hymenoptera: Colletidae, Halictidae, and Oxaeidae). *Evolution,* **37**, 657–74.

Cann, R. L., Stoneking, M., and Wilson, A. C. (1987). Mitochondrial DNA and human evolution. *Nature,* **325**, 31–6.

Carpenter, J. M. (1981). The phylogenetic relationships and natural classification of the Vespoidea (Hymenoptera). *Systematic Entomology,* **7**, 11–38, (1982).

Carpenter, J. M. (1987). Phylogenetic relationships and classification of the Vespinae (Hymenoptera: Vespidae). *Systematic Entomology,* **12**, 413–31.

Carpenter, J. M. (1988*a*). The phylogenetic system of the Stenogastrinae (Hymenoptera: Vespidae). *Journal of the New York Entomological Society,* **96**, 140–75.

Carpenter, J. M. (1988*b*). Choosing among multiple equally parsimonious cladograms. *Cladistics,* **4**, 291–6.

Carpenter, J. M. (1989*a*). Testing scenarios: wasp social behavior. *Cladistics,* **5**, 131–44.

Carpenter, J. M. (1989*b*). The phylogenetic system of the Gayellini (Hymenoptera: Vespidae; Masarinae). *Psyche,* **95**, 211–41.

Carpenter, J. M. (1990). Of genetic distances and social wasps. *Systematic Zoology,* **39**, 391–7.

Carpenter, J. M. (1991). Phylogenetic relationships and the origin of social behavior in the Vespidae. In *The social biology of wasps* (ed. K. G. Ross and R. W. Matthews), pp. 7–32. Comstock, Ithaca.

Carpenter, J. M. (1992). Incidit in Scyllam qui vult vitare Charybdim. Review of Brooks, D. R. and D. A. McLennan, Phylogeny, ecology and behavior (University of Chicago Press, Chicago, 1991). *Cladistics,* **8**, 100–2.

Carpenter, J. M. (1993). Biogeographic patterns in the Vespidae (Hymenoptera): Two views of Africa and South America. In *Biological relationships between Africa and South America* (ed. P. Goldblatt), pp. 139–55. Yale University Press, New Haven.

Carpenter, J. M. and Cumming, J. M. (1985). A character analysis of the North American potter wasps (Hymenoptera:Vespidae; Eumeninae). *Journal of Natural History,* **19**, 877–916.

Carpenter, J. M. and Rasnitsyn, A. P. (1990). Mesozoic Vespidae. *Psyche,* **97**, 1–20.

Carpenter, J. M., Strassmann, J. E., Turillazzi, S., Hughes, C. R., Solìs, C. R., and Cervo, R. (1993). Phylogenetic relationships among paper wasp social parasites and their hosts (Hymenoptera: Vespidae; Polistinae). *Cladistics,* **9**, 129–46.

Cervo, R. (1990). Il parassitismo sociale nei *Polistes.* Unpublished D. Phil. thesis. University of Florence, Italy.

Cervo, R. (1994). Morphological adaptations to the parasitic life in *Polistes sulcifer* and *Polistes atrimandibularis* (Hymenoptera Vespidae). *Ethology Ecology & Evolution, (Special Issue),* **3**, 61–6.

Cervo, R. and Lorenzi, M. C. (1994). The control of host reproductive capacity by the social parasite *Polistes atrimandibularis* (Hymenoptera Vespidae). *Ethology Ecology &*

Evolution, **6**, 415–6.

Cervo, R. and Turillazzi, S. (1985). Associative foundation and nesting sites in *Polistes nimpha*. *Naturwissenschaften*, **72**, 48–9.

Cervo, R. and Turillazzi, S. (1989). Nest exchange experiments in *Polistes gallicus* (L.) (Hymenoptera Vespidae). *Ethology Ecology & Evolution*, **1**, 185–93.

Cervo, R. and Turillazzi, S. (1990). On the evolution of social parasitism in *Polistes* wasps (Hymenoptera Vespidae). In *Social insects and the enviroment* (ed. G. K. Veeresh, B. Mallik, and C. A. Viraktamath), pp. 158–9. Oxford and IBH Publishing Co., New Dehli.

Cervo, R., Bertocci, F., and Turillazzi, S. (1993). Factors influencing choice of host nest in the social parasite *Polistes sulcifer* (Hymenoptera Vespidae). *Ethology Ecology & Evolution*, **5**, 385–6.

Cervo, R., Dani, F. R., and Turillazzi, S. (1992). Preliminary note on *Polistes atrimandibularis*, social parasite of *Polistes gallicus* (Hymenoptera Vespidae). *Ethology Ecology & Evolution, (Special Issue)*, **2**, 49–53.

Cervo, R., Lorenzi, M. C., and Turillazzi, S. (1990a). Non-aggressive usurpation of the nest of *Polistes biglumis bimaculatus* by the social parasite *Sulcopolistes atrimandibularis* (Hymenoptera, Vespidae). *Insectes Sociaux*, **37**, 333–47.

Cervo, R., Lorenzi, M. C., and Turillazzi, S. (1990b). On the strategies of host nest invasion in three species of *Sulcopolistes*, social parasites of *Polistes* wasps. *Actes des Colloques Insectes Sociaux*, **6**, 69–74.

Cervo, R., Lorenzi, M. C., and Turillazzi, S. (1990c). *Sulcopolistes atrimandibularis*, social parasite and predator of an Alpine *Polistes* (Hymenoptera, Vespidae). *Ethology*, **86**, 71–8.

Chadab, R. (1979). Army-ant predation on social wasps. Unpublished D. Phil. thesis. University of Connecticut.

Chandrashekara, K. and Gadagkar, R. (1990). Behavioural castes and their correlates in the primitively eusocial wasp, *Ropalidia marginata* (Lep.) (Hymenoptera: Vespidae). In *Social insects: an Indian perspective*, (ed. G. K. Veeresh, A. R. V. Kumar and T. Shivashankar), pp. 153–60. IUSSI-Indian Chapter, Bangalore.

Chandrashekara, K. and Gadagkar, R. (1991). Unmated queens in the primitively eusocial wasp *Ropalidia marginata* (Lep.) (Hymenoptera: Vespidae). *Insectes Sociaux*, **38**, 213–16.

Chandrashekara, K. and Gadagkar, R. (1992). Queen succession in the primitively eusocial tropical wasp *Ropalidia marginata* (Lep.) (Hymenoptera: Vespidae). *Journal of Insect Behavior*, **5**, 193–209.

Charnley, H. W. (1973). The value of the propodeal orifice and the phallic capsule in vespid taxonomy (Hymenoptera, Vespidae). *Bulletin of the Buffalo Society of Natural Sciences*, **26**, 1–79.

Charnov, E. L. (1978). Evolution of eusocial behavior, offspring choice or parental parasitism? *Journal of Theoretical Biology*, **75**, 451–65.

Cheesman, L. E. (1951). A collection of *Polistes* from Papuasia in the British Museum. *Annals and Magazine of Natural History*, **4**, 982–93.

Choudhary, M., Strassmann, J. E., Solìs, C. R., and Queller, D. C. (1993). Microsatellite variation in a social insect. *Biochemical Genetics*, **31**, 87–96.

Choudhary M., Strassmann, J. E., Queller, D. C., Turillazzi, S., and Cervo, R. (1994). Social parasites in polistine wasps are monophyletic: implications for sympatric speciation. *Proceedings of the Royal Society of London, B*, **257**, 31–5.

Christensen, T. A., Itagaki, H., Teal, P. E. A., Jasensky, R. D., Tumlinson, J. H., and Hildebrand, J. G. (1991). Innervation and neural regulation of the sex pheromone gland in female *Heliothis* moths. *Proceedings of National Academy of Science USA*, **88**, 4971–5.

Clarke, B. C. (1966). The evolution of morph-ratio clines. *American Naturalist*, **100**, 389–402.

Cornell, T. J., Berven, K. A., and Gamboa, G. J. (1989). Kin recognition by tadpoles and froglets of the wood frog, *Rana sylvatica*. *Oecologia*, **78**, 312–16.

Coster-Longman, C. (1994). Laboratory observations on the social behaviour of *Parischnogaster alternata* (Vespidae Stenogastrinae). *Ethology Ecology & Evolution (Special Issue)*, **3**, 31–6.

Coster-Longman, C. and Turillazzi, S. (1995). Nest architecture in *Parischnogaster alternata* Sakagami (Stenogastrinae, Vespidae) intra-specific variability in building strategies. *Insectes Sociaux*, **42**, 1–16.

Cowan, D. P. (1991). The solitary and presocial Vespidae. In *The social biology of wasps* (ed. K. Ross and R. Matthews), pp. 33–73. Comstock, Ithaca.

Craig, R. (1983). Subfertility and the evolution of eusociality by kin selection. *Journal of Theoretical Biology*, **100**, 379–97.

Cramp, S. and Simmons, K. E. L. (1980). *The birds of the Western Palearctic. Hawks and Bustards,* Vol. 2. Oxford University Press.

Crespi, B. J. (1992). Eusociality in Australian gall thrips. *Nature*, **359**, 724–6.

Croizat, L., Nelson, G., and Rosen, D. E. (1974). Centers of origin and related concepts. *Systematic Zoology*, **23**, 265–87.

Cronin, H. (1991). *The ant and the peacock: Altruism and sexual selection from Darwin to today*. Cambridge University Press.

Crozier, R. H. (1970). Coefficients of relationship and the identity of genes by descent in the Hymenoptera. *American Naturalist*, **104**, 216–17.

Crozier, R. H. (1977). Evolutionary genetics of the Hymenoptera. *Annual Review of Entomology*, **22**, 263–88.

Dalla Torre, K. W. (1904). Vespidae. *Genera Insectorum*, **19**, 1–108.

Dani, F. R. and Cervo, R. (1992). Reproductive strategies following nest loss in *Polistes gallicus* (Hymenoptera Vespidae). *Ethology Ecology & Evolution, (Special Issue)*, **2**, 49–53.

Dani, F. R., Cervo, R., and Turillazzi, S. (1992*a*). Preliminary observations on abdominal stroking behaviour in *Polistes dominulus* (Christ) (Hymenoptera: Vespidae). In *Biology and evolution of social insects*, (ed. J. Billen), pp. 281–5, Leuven University Press, Leuven, Belgium.

Dani, F. R., Cervo, R., and Turillazzi, S. (1992*b*). Abdomen stroking behaviour and its possible functions in *Polistes dominulus* (Christ) (Hymenoptera, Vespidae). *Behavioural Processes*, **28**, 51–8.

Dani, F. R., Turillazzi, S., and Morgan, E. D. (1994*a*). Analisi della secrezione delle ghiandole sternali di *Polistes dominulus* (Christ) (Hymenoptera: Vespidae). *Atti del XVII Congresso Nazionale Italiano di Entomologia*, 365.

Dani, F. R., Oldham, N. J., Turillazzi, S., and Morgan, E. D. (1994*b*). Preliminary data on the Dufour's gland secretion of *Polistes dominulus* and its social parasite *Polistes sulcifer* (Hymenoptera: Vespidae). In *Les insectes sociaux* (ed. A. Lenoir, G. Arnold, and M. Lepage), p. 407. Université Paris Nord.

Darchen, R. (1976*a*). La formation d'une nouvelle colonie de *Polybiodes tabida* (Fab.) (Vespidae, Polybiini). *Comptes Rendus de l'Academie des Sciences, Paris (D)*, **282**, 457–9.

Darchen, R. (1976*b*). *Ropalidia cincta*, guêpe sociale de la savane de Lampto (Côte d'Ivoire) (Hymenoptera Vespidae). *Annales de la Société Entomologique de France*, **12**, 579–601.

Darden, L. (1991). *Theory change in science: Strategies from Mendelian genetics*. Oxford University Press, New York.

Darden, L. (1992). Strategies for anomaly resolution. In *Minnesota studies in the philosophy of science. Cognitive models of science*, Vol. XV, (ed. R. Giere), pp. 251–73. University of Minnesota Press, Minneapolis.

Darden, L. (1995). Exemplars, abstractions, and anomalies: representations and theory change in mendelian and molecular genetics. In *Concepts, theories, and rationality in the biological sciences*, (ed. G. Wolters, J. G. Lennox, and P. McLaughlin), pp. 137–58. UVK-Universitätsverlag Konstanz/University of Pittsburgh Press.

Darlington, J. P. E. C. (1984). Two types of mound built by the termite *Macrotermes subhyalinus* in Kenya. *Insect Science and its Application*, 5, 481–92.

Darlington, P. J. (1957). *Zoogeography*. Wiley, New York.

Darwin, C. (1859). *The origin of species*, (facsimile of first edn) [1966]. Harvard University Press, Cambridge.

Darwin, C. (1872). *The origin of species*, (6th edn). The Modern Library, New York.

Darwin, C. (1868). *Variation in animals and plants under domestication*, (1896 edn) (reprinted 1972). AMS Press, New York.

Darwin, C. (1871). *The descent of man and selection in relation to sex*. John Murray, London.

Das, B. P. and Gupta, V. K. (1984). A catalogue of the families Stenogastridae [sic!] and Vespidae from the Indian subregion (Hymenoptera: Vespoidea). *Oriental Insects*, 17, 395–464.

Das, B. P. and Gupta, V. K. (1989). The social wasps of India and the adjacent countries (Hymenoptera: Vespidae). *Oriental Insects Monograph*, 11, 1–292.

Davidson, E. H. (1986). *Gene activity in early development*. Academic Press, New York.

Dawkins, R. (1976). *The selfish gene*. Oxford University Press.

Day, M. C. (1979). The species of Hymenoptera described by Linnaeus in the genera *Sphex, Chrysis, Vespa, Apis* and *Mutilla*. *Biological Journal of the Linnean Society, London*, 12, 45–84.

Deleurance, E. P. (1950). Sur le mécanisme de la monogynie fonctionnelle chez les *Polistes* (Hyménoptères-Vespides). *Compte Rendus de l'Academie des Sciences, Paris*, 230, 782–4.

Deleurance, E. P. (1952a). Le polymorphisme social et son déterminisme chez les guêpes. In *Structure et physiologie des sociétés animales*, (ed. P.-P. Grassé), Publications du Centre Nationale de la Recherche Scientifique, Colloques Int, 34, 141–55.

Deleurance, E. P. (1952b). Étude du cycle biologique du couvain chez *Polistes*. Les phases "couvain normal" et "couvain abortif". *Behaviour*, 4, 104–15.

Deleurance, E. P. (1955). Contribution à l'étude biologique des *Polistes* (Hyménoptères-Vespides). II. Le cycle évolutif du couvain. *Insectes Sociaux*, 2, 285–302.

Deleurance, E. P. (1957). Contribution à l'étude biologique des *Polistes* (Hyménoptères-Vespides). I. L'activité de construction. *Behaviour*, 11, 67–84.

Demolin, G. and Martin, J. C. (1980). Biologie de *Sulcopolistes semenowi* (Morawitz) parasite de *Polistes nimpha* (Christ), Hymenoptera: Vespidae. *Biologie-Ecologie Méditerranéenne*, 8, 181–2.

Deneubourg, J. L., Goss, S., Franks, N., and Pasteels, J. M. (1989). The blind leading the blind: modelling chemically mediated army ant raid patterns. *Journal of Insect Behavior*, 2, 719–25.

Deneubourg, J. L., Theraulaz, G., and Beckers, R. (1992). Swarm made architecture. In *Proceedings of the ECAL 1991 Conference*. MIT Press, Cambridge, USA.

Dettner, K. and Liepert, C. (1994). Chemical mimicry and camouflage. *Annual Review of Entomology*, 39, 129–54.

Dew, H. E. and Michener, C. D. (1978). Foraging flights of two species of *Polistes* wasps (Hymenoptera, Vespidae). *Journal of the Kansas Entomological Society*, 51, 380–5.

Disderi, G. S. (1816). Vespae Gallicae Historia. *Mémoires de l'Académie Royale des Sciences de Turin, pour les années* 1813–14, 1–19.

Distefano, S. L. (1968). Ricerche sulla fauna e sulla zoogeografia della Sicilia. 43. Osservazioni su *Sulcopolistes sulcifer* (Zimmermann) parassita sociale di *Polistes gallicus*. *Bollettino delle Sedute dell'Accademia Gioenia di Scienze Naturali di Catania*,

9, 662–78.

Distefano, S. L. (1971). Eccezionale riutilizzazione di un vecchio favo per una nuova colonia di *Polistes gallicus* (L.). In *Atti del XVII Congresso Nazionale Italiano di Entomologia, Firenze*, 132–3.

Distefano, S. L. (1972). Inconsuete sciamature autunnali di maschi di *Polistes* (Hymenoptera, Vespidae). In *Atti del XIX Congresso Nazionale Italiano di Entomologia, Siena*, 59–64.

Downing, H. A. (1991*a*). A role of the Dufour's gland in the dominance interactions of the paper wasp, *Polistes fuscatus* (Hymenoptera: Vespidae). *Journal of Insect Behavior*, **4**, 557–65.

Downing, H. A. (1991*b*). The function and evolution of exocrine glands. In *The social biology of wasps* (ed. K. G. Ross and R. W. Matthews), pp. 540–69. Comstock, Ithaca.

Downing, H. A. and Jeanne, R. L. (1982). A description of the ectal mandibular gland in the paper wasp *Polistes fuscatus* (Hymenoptera: Vespidae). *Psyche*, **89**, 317–20.

Downing, H. A. and Jeanne, R. L. (1983). Correlation of season and dominance status with activity of exocrine glands in *Polistes fuscatus* (Hymenoptera: Vespidae). *Journal of the Kansas Entomological Society*, **56**, 387–97.

Downing, H. A. and Jeanne, R. L. (1986). Intra- and interspecific variation in nest architecture in the paper wasp *Polistes* (Hymenoptera, Vespidae). *Insectes Sociaux*, **33**, 422–43.

Downing, H. A. and Jeanne, R. L. (1987). A comparison of nest construction behavior in two species of *Polistes* paper wasps (Insecta, Hymenoptera: Vespidae). *Journal of Ethology*, **5**, 53–66.

Downing, H. A. and Jeanne, R. L. (1988). Nest construction by the paper wasp *Polistes*: a test of stigmergy theory. *Animal Behaviour*, **36**, 1729–39.

Downing, H. A. and Jeanne, R. L. (1990). The regulation of complex building behavior in the paper wasp, *Polistes fuscatus* (Insecta, Hymenoptera, Vespidae). *Animal Behaviour*, **39**, 105–24.

Downing, H. A., Post, D. C., and Jeanne, R. L. (1985). Morphology of sternal glands in male polistine wasps (Hymenoptera: Vespidae). *Insectes Sociaux*, **32**, 186–97.

Dropkin, J. A. and Gamboa, G. J. (1981). Physical comparisons of foundresses of the paper wasp *Polistes metricus* (Hymenoptera: Vespidae). *The Canadian Entomologist*, **113**, 457–61.

Ducke, A. (1904). Sobre as vespidas sociaes do Pará. *Boletim do Museu Goeldi, Pará*, **4**, 317–74.

Ducke, A. (1910). Révision des guêpes sociales polygames d'Amérique. *Annales Musei Nationalis Hungarici*, **8**, 449–544.

Dyer, F. C. (1991). Bees acquire route-based memories but not cognitive maps in a familiar landscape. *Animal Behaviour*, **41**, 239–46.

Eberhard, W. G. (1985). *Sexual selection and animal genitalia*. Harvard University Press, Cambridge.

Eberhard, W. G. and Gutierrez, E. E. (1991). Male dimorphisms in beetles and earwigs and the question of developmental constraints. *Evolution*, **45**, 18–28.

Edwards, R. (1980). *Social wasps*. The Rentokil Library, Felcourt, West Sussex.

Eickwort, G. C. (1981). Presocial insects. In *Social insects*, Vol. 2, (ed. H. R. Hermann), pp. 199–280. Academic Press, New York.

Eickwort, K. (1969). Separation of the castes of *Polistes exclamans* and notes on its biology (Hym. Vespidae). *Insectes Sociaux*, **16**, 67–72.

Eisner, T., Hicks, K., Eisner, M., and Robson, D. S. (1978). 'Wolf-in-sheep's-clothing' strategy of a predaceous insect larva. *Science*, **199**, 790–4.

Elliott, N. B. and Elliott, W. M. (1987). Nest usurpation by females of *Cerceris cribrosa*.

Journal of Kansas Entomological Society, **60**, 397–402.

Emlen, S. T. and Oring, L. W. (1977). Ecology, sexual selection, and the evolution of mating systems. *Science*, **197**, 215–23.

Endler, J. A. (1977). *Geographic variation, speciation, and clines*. Princeton University Press.

Esch, H. E. and Burns, J. E. (1995). Honeybees use optic flow to measure the distance of a food source. *Naturwissenschaften*, **82**, 38–40.

Eshel, I. and Cavalli-Sforza, L. L. (1982). Assortment of encounters and evolution of cooperativeness. *Proceedings of the National Academy of Sciences USA*, **79**, 1331–5.

Espelie, K. E. and Hermann, H. R. (1988). Congruent cuticular hydrocarbons: biochemical convergence of a social wasp, an ant and a host plant. *Biochemical Systematics and Ecology*, **16**, 505–8.

Espelie, K. E. and Hermann, H. R. (1990). Surface lipids of the social wasps *Polistes annularis* (L.) and its nest and nest pedicel. *Journal of Chemical Ecology*, **16**, 1841–52.

Espelie, K. E. and Himmelsbach, D. S. (1990). Characterization of pedicel, paper, and larval silk from nest of *Polistes annularis* (L.). *Journal of Chemical Ecology*, **16**, 3467–77.

Espelie, K. E., Wenzel, J. W., and Chang, G. (1990). Surface lipids of social wasps *Polistes metricus* Say and its nest and nest pedicel and their relation to nestmate recognition. *Journal of Chemical Ecology*, **16**, 2229–41.

Espelie, K. E., Gamboa, G. J., Grudzien, T. A., and Bura, E. A. (1994). Cuticular hydrocarbons of the paper wasp, *Polistes fuscatus*: a search for recognition pheromones. *Journal of Chemical Ecology*, **20**, 1677–87.

Evans, H. E. (1966a). *The comparative ethology and evolution of the sand wasps*. Harvard University Press, Cambridge.

Evans, H. E. (1966b). The behavior patterns of solitary wasps. *Annual Review of Entomology*, **11**, 123–54.

Evans, H. E. (1973). Burrow sharing and nest transfer in the digger wasp *Philanthus gibbosus* (Fabricius). *Animal Behaviour*, **21**, 302–8.

Evans, H. E. (1977). Extrinsic versus intrinsic factors in the evolution of insect sociality. *Bioscience*, **27**, 613–17.

Evans, H. E. and West-Eberhard, M. J. (1970). *The wasps*. University of Michigan Press, Ann Arbor, Michigan.

Evans, J. D. (1993). Parentage analyses in ant colonies using simple sequence repeat loci. *Molecular Ecology*, **2**, 393–7.

Farris, J. S. (1969). A successive approximations approach to character weighting. *Systematic Zoology*, **18**, 374–85.

Farris, J. S. (1979). The information content of the phylogenetic system. *Systematic Zoology*, **28**, 483–519.

Farris, J. S. (1988). *Hennig86*, version 1.5. Program and documentation. Port Jefferson, New York.

Farris, J. S. (1989). The retention index and the rescaled consistency index. *Cladistics*, **5**, 417–19.

Ferguson, I. D., Gamboa, G. J., and Jones, J. K. (1987). Discrimination between natal and non-natal nests by the social wasps *Dolichovespula maculata* and *Polistes fuscatus* (Hymenoptera: Vespidae). *Journal of the Kansas Entomological Society*, **60**, 65–9.

Ferton, C. (1921). Notes détachées sur l'instinct des Hyménoptères mellifères et ravisseurs. *Annales de la Société Entomologique de France*, **89**, 329–75.

Field, J. (1989). Intraspecific parasitism and nesting success in the solitary wasp *Ammophila sabulosa*. *Behaviour*, **110**, 23–46.

Field, J. (1992). Intraspecific parasitism as an alternative reproductive tactic in nest-building wasps and bees. *Biological Reviews*, **67**, 79–126.

Fisher, R. M. (1983). Recognition of host nest odour by the bumblebee social parasite

Psithyrus ashtoni (Hymenoptera: Apidae). *Journal of New York Entomological Society*, **91**, 503–7.

Fisher, R. M. (1984). Evolution and host specificity: a study of the invasion success of a specialized bumblebee social parasite. *Canadian Journal of Zoology*, **62**, 1641–4.

Fisher, R. M. (1985). Evolution and host specificity: dichotomous invasion success of *Psithyrus citrinus* (Hymenoptera: Apidae), a bumblebee social parasite in colonies of its two hosts. *Canadian Journal of Zoology*, **63**, 977–81.

Fisher, R. M. (1987). Temporal dynamics of facultative social parasitism in bumble bees (Hymenoptera: Vespidae). *Animal Behaviour*, **35**, 1628–36.

Fisher, R. M. (1993). How important is the sting in insect social evolution? *Ethology Ecology & Evolution*, **5**, 157–68.

Fishwild, T. G. and Gamboa, G. J. (1992). Colony defence against conspecifics: caste specific differences in kin recognition by paper wasps, *Polistes fuscatus*. *Animal Behaviour*, **43**, 95–102.

Flanders, S. E. (1942). Oosorption and ovulation in relation to oviposition in the parasitic Hymenoptera. *Annals of the Entomological Society of America*, **35**, 251–66.

Flanders, S. E. (1969). Social aspects of facultative gravidity and agravidity in Hymenoptera. In *Proceedings of the 6th Congress of the IUSSI, Bern*, 47–53.

Fletcher, D. J. C. and Michener, C. D. (1987). *Kin recognition in animals*. Wiley, New York.

Fluri, P., Luscher, M., Wille, H., and Gerig, L. (1982). Changes in weight of the pharyngeal gland and haemolymph titres of juvenile hormone, protein and vitellogenin in worker honey bees. *Journal of Insect Physiology*, **28**, 61–8.

Formigoni, A. (1956). Neurosecretion et organes endocrines chez *Apis mellifica* L. *Annales des Sciences naturelles (Zoologie)*, **18**, 283–91.

Francescato, E., Tindo, M., Turillazzi, S., and Dejean, A. (1994). Nest recognition in *Belonogaster juncea* (Hymenoptera Vespidae). *Ethology Ecology & Evolution (Special Issue)*, **3**, 53–6.

Francescato, E., Turillazzi, S., and Dejean, A. (1993). Swarming behaviour in *Polybioides tabida* (Hymenoptera, Vespidae). *Actes des Colloques Insectes Sociaux*, **8**, 121–6.

Francke, W., Hindorf, G., and Reith, W. (1978). Methyl-1,6-dioxaspiro [4,5] decane als Duftstoffe von *Paravespula vulgaris* (L.). *Angewandte Chemie*, **90**, 915.

Franks, N., Blum, M., Smith, R. K., and Allies, A. B. (1990). Behavior and chemical disguise of cuckoo ant *Leptothorax kutteri* in relation to its host *Leptothorax acervorum*. *Journal of Chemical Ecology*, **16**, 1431–44.

Free, J. B. (1955). The behaviour of egg-laying workers of Bumble-bees colonies. *British Journal of Animal Behaviour*, **3**, 147–53.

Freisling, J. (1943). Zur Psychologie der Feldwespe. *Zeitschrift für Tierpsychologie*, **5**, 438–63.

Frumhoff, P. C. and Reeve, H. K. (1994). Using phylogenies to test hypotheses of adaptation: a critique of some current proposals. *Evolution*, **48**, 172–80.

Furey, R. E. (1992). Division of labour can be morphological and/or temporal: a reply to Tsuji. *Animal Behaviour*, **44**, 571.

Gadagkar, R. (1985). Evolution of insect sociality: A review of some attempts to test modern theories. *Proceedings of the Indian Academy of Sciences (Animal Sciences)*, **94**, 309–24.

Gadagkar, R. (1990a). Social biology of *Ropalidia*: Investigations into the origins of eusociality. In *Social insects and the environment* (ed. G. K. Veeresh, B. Mallik, and C. A. Viraktamath), pp. 9–11. Proceedings of the XI International Congress of the International Union for the Study of Social Insects, Bangalore. Oxford and India Book House, New Delhi.

Gadagkar, R. (1990b). Evolution of eusociality: The advantage of assured fitness returns. *Philosophical Transactions of the Royal Society of London, B*, **329**, 17–25.

Gadagkar, R. (1990c). The haplodiploidy threshold and social evolution. *Current Science*, **59**, 374–6.

Gadagkar, R. (1990d). Evolution of insect societies: Some insights from studying tropical wasps. In *Social insects: An Indian perspective*. (ed. G. K. Veeresh, A. R. V. Kumar and T. Shivashankar), pp. 129–52. IUSSI – Indian Chapter, Bangalore.

Gadagkar, R. (1991a). On testing the role of genetic asymmetries created by haplodiploidy in the evolution of eusociality in the Hymenoptera. *Journal of Genetics*, **70**, 1–31.

Gadagkar, R. (1991b). Demographic predisposition to the evolution of eusociality – A hierarchy of models. *Proceedings of the National Academy of Science, USA*, **88**, 10993–7.

Gadagkar, R. (1991c) *Belonogaster, Mischocyttarus, Parapolybia*, and independent founding *Ropalidia*. In *The social biology of wasps*, (ed. K. R. Ross and R. W. Matthews), pp. 149–90. Comstock, Ithaca, N. Y.

Gadagkar, R. (1994a). Why the definition of eusociality is not helpful to understand its evolution and what should we do about it. *Oikos*, (In press).

Gadagkar, R. and Joshi, N. V. (1983). Quantitative ethology of social wasps: time-activity budgets and caste differentiation in *Ropalidia marginata* (Lep.) (Hymenoptera: Vespidae). *Animal Behaviour*, **31**, 26–31.

Gadagkar, R., Vinutha, C., Shanubhogue, A., and Gore, A. P. (1988). Pre-imaginal biasing of caste in a primitively eusocial insect. *Proceedings of the Royal Society of London, B*, **233**, 175–89.

Gadagkar, R., Bhagavan, S., Malpe, R., and Vinutha, C. (1990). On reconfirming the evidence for pre-imaginal caste bias in a primitively eusocial wasp. *Proceedings of the Indian Academy of Sciences (Animal Sciences)*, **99**, 141–50.

Gadagkar, R., Bhagavan, S., Chandrashekara, K., and Vinutha, C. (1991a). The role of larval nutrition in a pre-imaginal biasing of caste in a primitively eusocial wasp *Ropalidia marginata* (Lep.) (Hymenoptera: Vespidae). *Ecological Entomology*, **16**, 435–40.

Gadagkar, R., Chandrashekara, K., Chandran, S., and Bhagavan, S. (1991b). Worker-brood genetic relatedness in a primitively eusocial wasp – A pedigree analysis. *Naturwissenschaften*, **78**, 523–6.

Gadagkar, R., Chandrashekara, K., Chandran, S., and Bhagavan, S. (1993). Serial polygyny in a primitively eusocial wasp: implications for the evolution of eusociality. In *Queen number and sociality in insects* (ed. L. Keller), pp. 187–214. Oxford University Press.

Gambino, P. (1993). Antibiotic activity of larval saliva of *Vespula* wasps. *Journal of Invertebrate Pathology*, **61**, 110.

Gamboa, G. J. (1978). Intraspecific defence: advantage of social cooperation among paper wasp foundresses. *Science*, **199**, 1463–5.

Gamboa, G. J. (1981). Nest sharing and maintenance of multiple nests by the paper wasp *Polistes metricus* (Hymenoptera: Vespidae). *Journal of the Kansas Entomological Society*, **54**, 153–5.

Gamboa, G. J. (1988). Sister, aunt-niece, and cousin recognition by social wasps. *Behavior Genetics*, **18**, 409–23.

Gamboa, G. J. (1992). Recognizing relatives. *Bioscience*, **42**, 727–8.

Gamboa, G. J. and Dew, H. E. (1981). Intracolonial communication by body oscillations in the paper wasp, *Polistes metricus*. *Insectes Sociaux*, **28**, 13–26.

Gamboa, G. J., Heacock, B. D., and Wiltjer, S. L. (1978). Division of labor and subordinate longevity in foundresses associations of the paper wasp *Polistes metricus* (Hymenoptera-Vespidae). *Journal of the Kansas Entomological Society*, **51**, 343–52.

Gamboa, G. J., Reeve, H. K., and Pfennig, D. W. (1986a). The evolution and ontogeny of nestmate recognition in social wasps. *Annual Review of Entomology*, **31**, 431–54.

Gamboa, G. J., Reeve, H. K., Ferguson, I. D., and Wacker, T. L. (1986b). Nestmate recognition in social wasps: the origin and acquisition of recognition odours. *Animal*

Behaviour, **34**, 685–95.

Gamboa, G. J., Klahn, J. E., Parman, A. O., and Ryan, R. E. (1987). Discrimination between nestmate and non-nestmate kin by social wasps (*Polistes fuscatus*, Hymenoptera: Vespidae). *Behavioral Ecology and Sociobiology,* **21**, 125–8.

Gamboa, G. J., Berven, K. A., Schemidt, R. A., Fishwild, T. G., and Jankens, K. M. (1991a). Kin recognition by larval wood frogs (*Rana sylvatica*): effects of diet and prior exposure to conspecifics. *Oecologia,* **86**, 319–24.

Gamboa, G. J., Foster, R. L., Scope, J. A., and Bitterman, A. M. (1991b). Effects of stage of colony cycle, context, and intercolony distance on conspecific tolerance by paper wasps (*Polistes fuscatus*). *Behavioral Ecology and Sociobiology,* **29**, 87–94.

Gamboa, G. J., Reeve, H. K., and Holmes, W. G. (1991c). Conceptual issues and methodology in kin recognition research: a critical discussion. *Ethology,* **88**, 109–27.

Gamboa, G. J., Wacker, T. L., Duffy, K. G., Dobson, S. W., and Fishwild, T. G. (1992). Defence against intraspecific usurpation by paper wasp cofoundresses (*Polistes fuscatus*, Hymenoptera: Vespidae). *Canadian Journal of Zoology,* **70**, 2369–72.

Gary, N. E. (1975). Activities and behavior of honey bees. In *The hive and the honey bee,* (ed. Dadant & Sons), pp.185–264. Dadant & Sons, Hamilton.

Gastreich, K. R., Queller, D. C., Hughes, C. R., and Strassmann, J. E. (1990). Kin discrimination in a tropical swarm-founding wasp, *Parachartergus colobopterus. Animal Behaviour,* **40**, 598–601.

Gayon, J. (1995). Neo-Darwinism. In *Concepts, theories, and rationality in the biological sciences,* (ed. G. Wolters, J. G. Lennox, and P. McLaughlin), pp. 1–25. UVK-Universitätsverlag Konstanz/University of Pittsburgh Press.

Geiger, K., Kratzsch, D., and Menzel, R. (1994). Bees do not use landmark cues seen during displacement for displacement compensation. *Naturwissenschaften,* **81**, 415–17.

Gerace, L. and Turillazzi, S. (1992). Size and characteristics of colonies of *Eustenogaster fraterna* (Hymenoptera Vespidae Stenogastrinae). *Ethology Ecology & Evolution (Special Issue),* **2**, 65–7.

Gervet, J. (1962). Étude de l'effet de groupe sur la ponte dans la société polygyne de *P. gallicus* L. (Hymen. Vesp.). *Insectes Sociaux,* **9**, 231–63.

Gervet, J. (1964). La ponte et sa régulation dans la société polygyne de *Polistes gallicus* L. *Annales de Sciences Naturelles, Zoologie, 12ème série,* **6**, 601–778.

Gervet, J. (1986). La famille ou les copains: quel mode de sélection chez les Guêpes? *Bulletin de la SFECA,* **1**, 61–8.

Gervet, J. and Theraulaz, G. (1991). Behavioural screening and selection through affinity: the case of polygyny in paper wasps (*Polistes dominulus*). *International Journal of Comparative Psychology,* **4**, 253–71.

Getz, W. M. (1991). The honey bee as a model kin recognition system. In *Kin recognition,* (ed. P. G. Hepper), pp. 358–412. Cambridge University Press.

Giannotti, E. (1992). Estudos biológicos e etológicos da vespa social neotropical *Polistes (Aphanilopterus) lanio lanio* (Fabricius 1775) (Hymenoptera, Vespidae). Unpublished D. Phil. thesis. University of São Paulo, Rio Claro, Brazil.

Gibo, D. L. (1978). The selective advantage of foundress associations in *Polistes fuscatus* (Hymenoptera: Vespidae): a field study of the effects of predation on productivity. *Canadian Entomologist,* **110**, 519–40.

Gibo, D. L. and Metcalf, R. L. (1978). Early survival of *Polistes apachus* (Hymenoptera Vespidae) colonies in California: a field study of an introduced species. *Canadian Entomologist,* **110**, 1339–43.

Gibson, R. M. and Bradbury, J. W. (1985). Sexual selection in lekking sage grouse: phenotypic correlates of male mating success. *Behavioral Ecology and Sociobiology,* **18**, 117–23.

Gittleman, J. L. (1989). *Carnivore behavior, ecology, and evolution.* Chapman and Hall,

London.

Godfray, H. C. J. and Grafen, A. (1988). Unmatedness and the evolution of eusociality. *American Naturalist*, **131**, 303–5.

Goloboff, P. A. (1993). *NONA*, version 2. Program and documentation. New York.

Gould, J. L. (1986). The locale map of honey bees: do insects have cognitive maps? *Science*, **232**, 861–3.

Gould, S. J. and Eldredge, N. (1977). Punctuated equilibria: the tempo and mode of evolution reconsidered. *Paleobiology*, **3**, 115–51.

Grafen, A. (1986). Split sex ratios and the evolutionary origins of eusociality. *Journal of Theoretical Biology*, **122**, 95–121.

Grafen, A. (1991). Modelling in behavioural ecology. In *Behavioural ecology – An evolutionary approach*, (ed. J. R. Krebs and N. B. Davies), pp. 5–31. Blackwell Scientific Publications, Oxford.

Grassé, P. (1959). La reconstruction du nid et les coordinations interindividuelles chez *Bellicositermes natalensis* et *Cubitermes* sp. La theorie de la stigmergie: essai d'interpretation du comportement des termites constructeurs. *Insectes Sociaux*, **6**, 41–81.

Greene, A. (1991). *Dolichovespula* and *Vespula*. In *The social biology of wasps*, (ed. K. G. Ross and R. W. Matthews), pp. 263–305. Comstock, Ithaca.

Greene, A., Akre, R. D., and Landolt, P. (1976). The aerial yellowjacket, *Dolichovespula arenaria* (Fab.): nesting biology, reproductive production, and behavior (Hymenoptera: Vespidae). *Melanderia*, **26**, 1–34.

Greene, A., Akre, R. D., and Landolt, P. J. (1978). Behavior of the yellowjacket social parasite, *Dolichovespula arctica* (Rohwer) (Hymenoptera: Vespidae). *Melanderia*, **29**, 1–28.

Grinfel'd, E. K. (1977). The feeding of the social wasp *Polistes gallicus* (Hymenoptera, Vespidae). *Entomological Review, Washington*, **56**, 24–9.

Grosberg, R. K. and Quinn, J. F. (1986). The genetic control and consequences of kin recognition by the larvae of a colonial marine invertebrate. *Nature*, **322**, 456–59.

Guiglia, D. (1971). *Les guêpes sociales (Hymenoptera Vespidae) d'Europe occidentale et septentrionale*, Faune de l'Europe et du Bassin Mediterranéen 6. Masson et Cie, Paris.

Habersetzer, C. and Bonavita-Cougourdan, A. (1993). Cuticular spectra in the slave making ant *Polyergus rufescens* and the slave species *Formica rufibarbis*. *Physiological Entomology*, **18**, 160–6.

Halffter, G. and Edmonds, W. D. (1982). *The nesting behaviour of dung beetles* (Scarabaeinae). Instituto de Ecologia, Mexico.

Halliday, T. R. (1983). The study of mate choice. In *Mate choice* (ed. P. B. Bateson), pp. 3–32. Cambridge University Press.

Halliday, T. R. (1987). Physiological constraints on sexual selection. In *Sexual selection: testing the alternatives* (ed. J. W. Bradbury and M. B. Andersson), pp. 247–64. Wiley, Chichester.

Halliday, T. R. (1992). Sexual selection in amphibians and reptiles: theoretical issues and new directions. In *Herpetology: current research on the biology of amphibians and reptiles* (ed. K. Adler), pp. 81–95. Proceedings of the First World Congress of Herpetology. Society for the Study of Amphibians and Reptiles, Oxford (Ohio).

Hamilton, W. D. (1964a). The genetical evolution of social behaviour. I. *Journal of Theoretical Biology*, **7**, 1–16.

Hamilton, W. D. (1964b). The genetical evolution of social behaviour. II. *Journal of Theoretical Biology*, **7**, 17–52.

Hamilton, W. D. (1967). Extraordinary sex ratios. *Sciences*, **156**, 477–8.

Hamilton, W. D. (1972). Altruism and related phenomena, mainly in social insects. *Annual Review of Ecology and Systematics*, **3**, 193–232.

Hamilton, W. D. and May, R. M. (1977). Dispersal in stable habitats. *Nature*, **269**,

578–81.

Hansell, M. H. (1977). Social behaviour of a three wasp colony: Stenogastrinae Vespidae. In *Abstracts of the 15th International Ethological Conference, Bielefeld, W. Germany*, 30.

Hansell, M. H. (1981). Nest construction in the subsocial wasp *Parischnogaster mellyi* (Saussure) (Stenogastrinae Hymenoptera). *Insectes Sociaux*, **28**, 208–16.

Hansell, M. H. (1982). Brood development in the subsocial wasp *Parischnogaster mellyi* (Saussure) (Stenogastrinae Hymenoptera). *Insectes Sociaux*, **29**, 3–14.

Hansell, M. H. (1983). Social behaviour and colony size in the wasp *Parischnogaster mellyi* (Saussure). *Proceedings of the Koninklijke Nederlandse Akademie van Wetenschappen, Series C*, **86**, 167–78.

Hansell, M. H. (1986). Colony biology of the stenogastrine wasp *Holischnogaster gracilipes* (Van der Vecht) on Mount Kinabalu. *Entomologist's Monthly Magazine*, **122**, 31–6.

Hansell, M. H. (1987a). Nest building as a facilitating and limiting factor in the evolution of eusociality in the Hymenoptera. *Oxford Surveys in Evolutionary Biology*, Vol. 4, (ed. P. H. Harvey and L. Partridge), pp. 155–81. Oxford University Press, Oxford.

Hansell, M. H. (1987b). Elements of eusociality in colonies of *Eustenogaster calyptodoma* (Sakagami and Yoshikawa) (Stenogastrinae, Vespidae). *Animal Behaviour*, **35**, 131–41.

Hansell, M. H. (1989). Les nids des insectes sociaux. *La Recherche*, **20**, 14–22.

Hansell, M. H. (1993). The ecological impact of animal nests and burrows. *Functional Ecology*, **7**, 5–12.

Hansell, M. H. and Turillazzi, S. (1995). Nest structure and building material of three species of *Anischnogaster* (Vespidae, Stenogastrinae) from Papua New Guinea. *Tropical Zoology*, **8**, 203–19.

Hansell, M. H., Samuel, C., and Furtado, J. H. (1982). *Liostenogaster flavolineata*: social life in the small colonies of an Asian tropical wasp. In *The biology of social insects*, (ed. M. Breed, C. D. Michener, and H. E. Evans), pp. 192–5. Westview Press, Boulder.

Haverty, M. I., Nelson, L. J., and Page, M. (1990). Cuticular hydrocarbons of four populations of *Coptotermes formosanus* Shiraki in the United States. Similarities and origins of introductions. *Journal of Chemical Ecology*, **16**, 1635–47.

Haydak, M. H. (1957). Changes with age in the appearance of some internal organs of the honeybee. *Bee World*, **38**, (8), 197–207.

Heath, R. R. and Landolt, P. J. (1988). The isolation, identification and synthesis of the alarm pheromone of *Vespula squamosa* (Drury) (Hymenoptera: Vespidae) and associated behavior. *Experientia*, **44**, 82–3.

Hedrick, A. V. (1986). Female preference for male calling bout duration in a field cricket. *Behavioral Ecology and Sociobiology*, **19**, 73–7.

Heldmann, G. (1936). Über das Leben auf Waben mit mehreren überwinterten Weibchen von *Polistes gallica* L. *Biologisches Zentralblatt*, **56**, 389–400.

Henderson, G., Andersen, J. F., Phillips, J. K., and Jeanne, R. L. (1990). Internest aggression and identification of possible nestmate discrimination pheromones in polygynous ant *Formica montana*. *Journal of Chemical Ecology*, **16**, 2217–27.

Hennig, W. (1966). *Phylogenetic systematics*. University of Illinois Press, Urbana.

Hepburn, H. R., Magnuson, P., Herbert, L., and Whiffler, L. A. (1991). The development of laying workers in field colonies of the Cape honey bee. *Journal of Apicultural Research*, **30**, 107–12.

Hepper, P. G. (ed.) (1991). *Kin recognition*. Cambridge University Press.

Hermann, H. R. and Blum, M. S. (1981). Defensive mechanisms in the social Hymenoptera. In *Social insects*, Vol. 2 (ed. H. R. Hermann), pp. 78–197. Academic Press, New York.

Hermann, H. R. and Dirks, T. F. (1974). Sternal glands in polistine wasps: Morphology and associated behavior. *Journal of the Georgia Entomological Society*, **9**, 1–8.

Hermann, H. R. and Dirks, T. F. (1975). Biology of *Polistes annularis* (Hymenoptera:

Vespidae). I. Spring behavior. *Psyche*, **82**, 97–108.

Hermann, H. R. and Gerling, D. (1974). The cohibernation and mating activity of five Polistine wasp species (Hymenoptera: Vespidae: Polistinae). *Journal of the Georgia Entomological Society*, **9**, 203–4.

Heselhaus, F. (1922). Die Hautdrüsen der Apiden und verwandter Formen. *Zoologische Jahrbücher, Anatomie und Ontogenie der Tiere*, **43**, 369–464.

Hibino, Y. (1980). On the survival of nests and foraging behaviour of the Japanese paper wasp, *Polistes jadwigae*. In *Abstracts of the 16th International Congress of Entomology*, Kyoto, 436.

Highnam, K. C. (1964). Endocrine relationships in insect reproduction. *Symposia of the Royal Entomological Society of London*, **2**, 26–42.

Hildebrandt, H. H. and Kaatz, H. H. (1990). Impact of queen pheromone on the physiological status of worker honey bees (*Apis mellifera* L.). In *Social insects and the environment* (ed. G. K. Veeresh, B. Mallik, and C. A. Viraktamath), pp. 740–1 (Proceedings of the 11th International Congress IUSSI, Bangalore, India). Oxford and IBH, New Delhi.

Hinton, G. E. and Nowlan, S. J. (1987). How learning can guide evolution. *Complex Systems*, **1**, 495–502.

Hirose, Y. and Yamasaki, M. (1984). Dispersal of females for colony founding in *Polistes jadwigae* Dalla Torre (Hymenoptera, Vespidae). *Kontyû*, **52**, 65–71.

Hölldobler, B. (1977). Communication in social Hymenoptera. In *How animals communicate*, (1st edn) (ed. T. A. Sebeok), pp. 418–71. Indiana University Press, Bloomington.

Hölldobler, B. and Michener, C. D. (1980). Mechanism of identification and discrimination in social Hymenoptera. In *Evolution of social behaviour: hypotheses and empirical tests* (ed. H. Markl), pp. 35–58. Dahlem Konferenzen 1980, Verlag Chemie, Weinheim.

Hölldobler, B. and Wilson, E. O. (1983). The evolution of communal nest-weaving in ants. *American Scientist*, **71**, 490–9.

Hölldobler, B. and Wilson, E. O. (1990). *The ants*. Harvard University Press, Cambridge.

Holmes, F. L. (1993). The old martyr of science: The frog in experimental physiology. *Journal of the History of Biology*, **26**, 311–28.

Hook, A. (1982). Observations on a declining nest of *Polistes tepidus* (F.) (Hymenoptera: Vespidae). *Journal of the Australian Entomological Society*, **21**, 277–8.

Howard, R. W. and Blomquist, G. J. (1982). Chemical ecology and biochemistry of insect hydrocarbons. *Annual Review of Entomology*, **27**, 149–72.

Howard, R. W., McDaniel, C. A., and Blomquist, G. J. (1978). Cuticular hydrocarbons of the eastern subterranean termite, *Reticulitermes flavipes* (Kollar) (Isoptera: Rhinotermitidae). *Journal of Chemical Ecology*, **4**, 233–45.

Howard, R. W., McDaniel, C. A., and Blomquist, G. J. (1980). Chemical mimicry as an integrating mechanism: cuticular hydrocarbons of a termitophile and its host. *Science*, **210**, 431–3.

Howard, R. W., McDaniel, C. A., and Blomquist, G. J. (1982a). Chemical mimicry as an integrating mechanism for three termitophiles associated with *Reticulitermes virginicus* (Banks). *Psyche*, **89**, 157–67.

Howard, R. W., McDaniel, C. A., Nelson, D. R., Blomquist, G. J., Gelbaum, L. T., and Zalkow, L. H. (1982b). Cuticular hydrocarbons of *Reticulitermes virginicus* (Banks) and their role as potential species and caste-recognition cues. *Journal of Chemical Ecology*, **8**, 1227–39.

Howard, R. W., Thorne, B. L., Levings, S. C., and McDaniel, C. A. (1988). Cuticular hydrocarbons as chemotaxonomic characters for *Nasutitermes corniger* (Motschulsky) and *N. ephratae* (Holmgren) (Isoptera: Termitidae). *Annals of the Entomological Society of America,* **81**, 395–9.

Howard, R. W., Akre, R. D., and Garnett, W. B. (1990). Chemical mimicry in an obligate predator of carpenter ants (Hymenoptera: Formicidae). *Annals of the Entomological Society of America,* **83**, 607–16.

Huang, Z.-Y., Robinson, G. E., Tobe, S. S., Yagi, K. J., Strambi, C., Strambi, A., and Stay, B. (1991). Hormonal regulation of behavioural development in the honey bee is based on changes in the rate of juvenile hormone biosynthesis. *Journal of Insect Physiology,* **37**, 733–41.

Hughes, C. R. (1987). Group nesting and reproductive conflict in primitively eusocial wasps. Unpublished D. Phil. thesis. Rice University. Ann Arbor Microfilms.

Hughes, C. R. and Queller, D. C. (1993). Detection of highly polymorphic microsatellite loci in a species with a little allozyme polymorphism. *Molecular Ecology,* **2**, 131–7.

Hughes, C. R. and Strassmann, J. E. (1988). Foundress mortality after worker emergence in social wasps (*Polistes*). *Ethology,* **79**, 265–80.

Hughes, C. R., Beck, M. O., and Strassmann, J. E. (1987). Queen succession in the social wasp *Polistes annularis. Ethology,* **76**, 124–32.

Hughes, C. R., Queller, D. C., Strassmann, J. E., and Davis, S. K. (1993). Relatedness and altruism in *Polistes* wasps. *Behavioral Ecology,* **4**, 128–37.

Hungerford, H. B. and Williams, F. X. (1912). Biological notes on some Kansas Hymenoptera. *Entomological News,* **23**, 241–60.

Hunt, J. H. (1994). Nourishment and evolution in wasps *sensu lato.* In *Nourishment and evolution in insect societies,* (ed. J. H. Hunt and C. A. Nalepa), pp. 211–44. Westview Press, Boulder.

Hunt, J. H. and Gamboa, G. J. (1978). Joint nest use by two paper wasp species. *Insectes Sociaux,* **25**, 373–4.

Hunt, J. H. and Noonan, K. C. (1979). Larval feeding by male *Polistes fuscatus* and *Polistes metricus* (Hymenoptera: Vespidae). *Insectes Sociaux,* **26**, 247–51.

Ihering, R. von (1904). As vespas sociaes do Brazil. *Revista do Museu Paulista,* **6**, 97–309.

Ikan, R., Gottlieb, R., Bergmann, E. D., and Ishay, J. (1969). The pheromone of the queen of the oriental hornet, *Vespa orientalis. Journal of Insect Physiology,* **15**, 1709–12.

Ishay, J. (1972). Thermoregulatory pheromones in wasps. *Experientia,* **28**, 1185–7.

Ishay, J. (1973). The influence of cooling and queen pheromone on cell building and nest architecture by *Vespa orientalis* (Vespinae, Hymenoptera). *Insectes Sociaux,* **20**, 243–52.

Ishay, J. and Perna, B. (1979). Building pheromones of *Vespa orientalis* and *Polistes foederatus. Journal of Chemical Ecology,* **5**, 259–72.

Ishay, J., Ikan, R., and Bergmann, E. D. (1965). The presence of pheromones in the oriental hornet, *Vespa orientalis* F. *Journal of Insect Physiology,* **11**, 1307–9.

Itô, Y. (1985). A comparison of frequency on intra-colony aggressive behaviours among five species of polistine wasps (Hymenoptera Vespidae). *Zeitschrift für Tierpsychologie,* **68**, 152–67.

Itô, Y. (1986a). Observations on the social behaviour of three polistine wasps (Hymenoptera: Vespidae). *Journal of the Australian Entomological Society,* **25**, 309–14.

Itô, Y. (1986b). Spring behaviour of an Australian paper wasp *Polistes humilis synoecus*: Colony founding by haplometrosis and utilization of old nests. *Kontyû,* **54**, 191–202.

Itô, Y. (1986c). On the pleometrotic route of social evolution in the Vespidae. *Monitore Zoologico Italiano (Nuova Serie),* **20**, 241–62.

Itô, Y. (1989). The evolutionary biology of sterile soldiers in aphids. *Trends in Ecology and Evolution,* **4**, 69–73.

Itô, Y. (1993). *Behaviour and social evolution of wasps. The communal aggregation hypothesis.* Oxford University Press.

Iwata, K. (1955). The comparative anatomy of the ovary in Hymenoptera. Part I. Aculeata.

Mushi, **29**, 17–34.

Iwata, K. (1966). Description of the nest of so called *Belonogaster griseus* var. *meneliki* Gribodo collected by Dr. K. Yamashita in Ethiopia, with a general consideration of the life of the genus (Hymenoptera, Vespidae). *Mushi*, **39**, 57–64.

Jacklyn, P. (1991). Evidence of adaptive variation in the orientation of *Amitermes* (Isoptera: Termitinae) mounds from Northern Australia. *Australian Journal of Zoology*, **39**, 569–77.

Jacobs-Jensen, U. F. (1959). Zur Orientierung der Hummeln und einiger anderer Hymenopteren. *Zeitschrift für Vergleichende Physiologie*, **41**, 597–641.

Jaffé, K. (1983) Chemical communication systems in the ant *Atta cephalotes*. In *Social insects in the tropics*, Vol. 2 (ed. P. Jaisson), pp. 165–80. Université Paris Nord.

Jaisson, P. (1991). Kinship and fellowship in ants and social wasps. In *Kin recognition*, (ed. P. G. Hepper), pp. 60–93. Cambridge University Press.

Jamon, M. (1987). Effectiveness and limitation of random search in homing behaviour. In *Cognitive processes and spatial orientation in animal and man* (ed. P. Ellen and C. Thinus-Blanc), pp. 284–94. NATO-ASI Series, M. Nijhoff Publ., Dordrecht.

Janet, C. (1895). Études sur les fourmis, les guêpes et les abeilles. Note 9. Sur *Vespa crabro* L. Histoire d'un nid depuis son origine. *Mémoires de la Société Zoologique de France*, **8**, 1–140.

Janet, C. (1903). *Observations sur les guêpes*. C. Naud, Paris.

Jarvis, J. U. M. (1981). Eusociality in a mammal: cooperative breeding in naked mole rat colonies. *Science*, **212**, 571–3.

Jarvis, J. U. M., O'Riain, M. J., Bennett, N. C., and Sherman, P. W. (1994). Mammalian eusociality: a family affair. *Nature*, **9**, 47–51.

Jeanne, R. L. (1970). Chemical defense of brood by a social wasp. *Science*, **168**, 1465–6.

Jeanne, R. L. (1972). Social biology of the Neotropical wasp *Mischocyttarus drewseni*. *Bulletin of the Museum of Comparative Zoology, Harvard University*, **144**, 63–150.

Jeanne, R. L. (1975). The adaptiveness of social wasp nest architecture. *Quarterly Review of Biology*, **50**, 267–87.

Jeanne, R. L. (1977*a*). A specialization in nest petiole construction by queens of *Vespula* spp. (Hymenoptera: Vespidae). *Journal of the New York Entomological Society*, **85**, 127–9.

Jeanne, R. L. (1977*b*). Behavior of the obligate social parasite *Vespula artica* (Hymenoptera: Vespidae). *Journal of the Kansas Entomological Society*, **50**, 541–57.

Jeanne, R. L. (1979*a*). Construction and utilization of multiple combs in *Polistes canadensis* in relation to the biology of a predaceous moth. *Behavioral Ecology and Sociobiology*, **4**, 293–310.

Jeanne, R. L. (1979*b*). A latitudinal gradient in rates of ant predation. *Ecology*, **60**, 1211–24.

Jeanne, R. L. (1980). Evolution of social behavior in the Vespidae. *Annual Review of Entomology*, **25**, 371–96.

Jeanne, R. L. (1981*a*). Alarm recruitment, attack behavior, and the role of the alarm pheromone in *Polybia occidentalis* (Hymenoptera: Vespidae). *Behavioral Ecology and Sociobiology*, **9**, 143–8.

Jeanne, R. L. (1981*b*). Chemical communication during swarm emigration in the social wasp *Polybia sericea* (Olivier). *Animal Behaviour*, **29**, 102–13.

Jeanne, R. L. (1982). Evidence for an alarm substance in *Polistes canadensis*. *Experientia*, **38**, 329–30.

Jeanne, R. L. (1986*a*). The organization of work in *Polybia occidentalis*: costs and benefits of specialization in a social wasp. *Behavioral Ecology and Sociobiology*, **19**, 333–41.

Jeanne, R. L. (1986*b*). The evolution of the organization of work in social insects. *Monitore Zoologico Italiano (Nuova Serie)*, **20**, 119–33.

Jeanne, R. L. (1991*a*). Polyethism. In *The social biology of wasps* (ed. K. G. Ross and R. W. Matthews), pp. 389–425. Comstock, Ithaca.

Jeanne, R. L. (1991b). The swarm-founding Polistinae. In *The social biology of wasps* (ed. K. G. Ross and R. W. Matthews), pp. 191–231. Comstock, Ithaca.

Jeanne, R. L. and Fagen, R. (1974). Polymorphism in *Stelopolybia areata* (Hymenoptera, Vespidae). *Psyche,* **81**, 155–66.

Jeanne, R. L., Downing, H. A., and Post, D. C. (1983). Morphology and function of sternal glands in polistine wasps (Hymenoptera: Vespidae). *Zoomorphology,* **103**, 149–64.

Jeanne, R. L., Downing, H. A., and Post, D. C. (1988). Age polyethism and individual variation in *Polybia occidentalis,* an advanced eusocial wasp. In *Interindividual behavioral variability in social insects,* (ed. R. L. Jeanne), pp. 323–57. Westview Press, Boulder and London.

Jerne, N. (1966). The natural selection theory of antibody formation; ten years later. In *Phage and the origins of molecular biology* (ed. J. Cairns, G. Stent, and J. Watson), pp. 301–12. Cold Spring Harbor Laboratory of Quantitative Biology, Cold Spring Harbor.

Karsai, I. and Pénzes, Z. (1993). Comb building in social wasps: self-organization and stigmergic script. *Journal of Theoretical Biology,* **161**, 505–25.

Karsai, I. and Wenzel, J. W. (1995). Nests build on the dorsum of conspecific in *Polistes*: the value of anomalous behaviour. *Animal Behaviour,* **49**, (in press).

Kasuya, E. (1980). Behavioral ecology of Japanese paper wasps, *Polistes* spp. (Hymenoptera, Vespidae). I. Extranidal activities of *Polistes chinensis antennalis.* *Researches on Population Ecology,* **22**, 242–54.

Kasuya, E. (1981a). Male mating territory in a Japanese paper wasp, *Polistes jadwigae* Dalla Torre (Hymenoptera: Vespidae). *Kontyû,* **49**, 607–14.

Kasuya, E. (1981b). Polygyny in the Japanese paper wasp *Polistes jadwigae* Dalla Torre (Hymen. Vesp.). *Kontyû,* **49**, 306–13.

Kasuya, E. (1982). Take-over of nests in a Japanese paper wasp, *Polistes chinensis antennalis* (Hymenoptera: Vespidae). *Applied Entomology and Zoology.,* **17**, 427–31.

Kasuya, E., Hibino, Y., and Itô, Y. (1980). On "intercolonial" cannibalism in Japanese paper wasps, *Polistes chinensis antennalis* Pérez and *P. jadwigae* Dalla Torre (Hymenoptera: Vespidae). *Researches in Population Ecology,* **22**, 255–62.

Keegans, S. J., Morgan, E. D., Agosti, D., and Wehner, R. (1992a). What do glands tell us about species? A chemical case study of *Cataglyphis* ants. *Biochemical Systematics and Ecology,* **20**, 559–72.

Keegans, S. J., Morgan, E. D., Jackson, B. D., Berresford, G., Turillazzi, S., and Billen, J. (1992b). The chemical nature and origin of the abdominal substance of the wasp *Liostenogaster flavolineata* (Vespidae: Stenogastrinae). In *Biology and evolution of social insects,* (ed. J. Billen), pp. 217–21. Leuven University Press, Leuven.

Keegans, S. J., Morgan, E. D., Turillazzi, S., Jackson, B. D., and Billen, J. (1993). The Dufour gland and the secretion placed on eggs of two species of social wasps, *Liostenogaster flavolineata* and *Parischnogaster jacobsoni* (Vespidae: Stenogastrinae). *Journal of Chemical Ecology,* **19**, 279–90.

Keeping, M. G. (1989). Social biology and colony dynamics of the polistine wasp *Belonogaster petiolata* (Hymenoptera: Vespidae). Unpublished D. Phil. thesis. University of the Witwatersrand, Johannesburg, South Africa.

Keeping, M. G. (1990a). Colony foundation and nestmate recognition in the social wasp *Belonogaster petiolata. Ethology,* **85**, 1–12.

Keeping, M. G. (1990b). Rubbing behavior and morphology of van der Vecht's gland in *Belonogaster petiolata* (Hymenoptera: Vespidae). *Journal of Insect Behavior,* **3**, 85–104.

Keeping, M. G. (1991). Nest construction in the social wasp, *Belonogaster petiolata* (Degeer) (Hymenoptera: Vespidae). *Journal of the Entomological Society of South Africa,* **54**, 17–28.

Keeping, M. G. (1992). Social organization and division of labour in colonies of the polistine wasp, *Belonogaster petiolata*. *Behavioral Ecolology and Sociobiology*, **31**, 211–24.

Keeping, M. G. and Crewe, R. M. (1983). Parasitoids, commensals and colony size in nests of *Belonogaster* (Hymenoptera: Vespidae). *Journal of the Entomological Society of South Africa*, **46**, 309–23.

Keeping, M. G. and Crewe, R. M. (1987). The ontogeny and evolution of foundress associations in *Belonogaster petiolata* (Hymenoptera: Vespidae). In *Chemistry and biology of social insects* (ed. J. Eder and H. Rembold), pp. 383–4. Verlag J. Peperny, München.

Keeping, M. G., Lipschitz, L., and Crewe, R. M. (1986). Chemical mate recognition and release of male sexual behavior in polybiine wasp, *Belonogaster petiolata* (Degeer) (Hymenoptera: Vespidae). *Journal of Chemical Ecology*, **12**, 773–9.

Keller, L. and Reeve, H. K. (1994). Partitioning of reproduction in animal societies. *Tree*, **9**, (3), 98–102.

Kent, D. S. and Simpson, J. A. (1992). Eusociality in the beetle *Austroplatypus incompertus* (Coleoptera: Curculionidae). *Naturwissenschaften*, **79**, 86–7.

Khoo, S. G. and Yong, H. S. (1987). Nest structure and colony defence in the stingless bee *Trigona terminator*. *Nature Malaysiana*, **12**, 4–15.

Kilgore, D. L. and Knudsen, K. L. (1977). Analysis of materials in cliff and barn swallow nests: Relationship between mud selection and nest architecture. *Wilson Bulletin*, **89**, 562–71.

Kirby, W. and Spence, W. (1826). *Introduction to entomology*, (1st edn), Vol. 3. London.

Kirby, W. and Spence, W. (1828). *Introduction to entomology*, (5th edn). London.

Kistner, D. H. (1982). *The social insect's bestiary*. In *Social insects III* (ed. H. R. Hermann), pp. 1–244. Academic Press, New York.

Klahn, J. E. (1979). Philopatric and nonphilopatric foundress associations in the social wasp *Polistes fuscatus*. *Behavioral Ecology and Sociobiology*, **5**, 417–24.

Klahn, J. E. (1981). Alternative reproductive tactics of single foundresses of a social wasp, *Polistes fuscatus*. Unpublished D. Phil. thesis. University of Iowa. Ann Arbor Microfilms.

Klahn, J. E. (1988). Intraspecific comb usurpation in the social wasp, *Polistes fuscatus*. *Behavioral Ecology and Sociobiology*, **23**, 1–8.

Klahn, J. E. and Gamboa, G. J. (1983). Social wasps: discrimination between kin and nonkin brood. *Science*, **221**, 482–4.

Knisley, C. B. (1985). Utilisation of tiger beetle larval burrows by a nest-provisioning wasp, *Leucodynerus russatus* (Bohart) (Hymenoptera: Eumenidae). *Proceedings of the Entomological Society of Washington*, **87**, 481.

Kohler, R. (1993). *Drosophila*: A life in the laboratory. *Journal of the History of Biology*, **26**, 281–310.

Kohler, R. (1994). *Lords of the fly:* Drosophila *genetics and the experimental life*. University of Chicago Press, Chicago.

Kojima, J. (1982). Notes on rubbing behavior in *Ropalidia gregaria* (Hymenoptera, Vespidae). *New Entomologist*, **31**, 17–19.

Kojima, J. (1983a). Defense of the pre-emergence colony against ants by means of a chemical barrier in *Ropalidia fasciata* (Hymenoptera, Vespidae). *Japanese Journal of Ecology*, **33**, 213–23.

Kojima, J. (1983b). Occurrence of the rubbing behavior in a paper wasp, *Parapolybia indica* (Hymenoptera, Vespidae). *Kontyû*, **51**, 158–9.

Kojima, J. (1992). The ant repellent function of the rubbing substance in an Old World polistine, *Parapolybia indica* (Hymenoptera Vespidae). *Ethology Ecology & Evolution*, **4**, 183–5.

Kojima, J. and Keeping, M. G. (1988). Description of mature larvae of *Belonogaster dubia*

Kohl (Hymenoptera, Vespidae), with notes on the larval characters of the Polistinae. *Kontyû*, **56**, 817–23.

Kojima, J. and Kojima, K. (1988). Three new species of *Polistes* Latreille (Hymenoptera: Vespidae) from Papua New Guinea, with notes on the taxonomic status of the subgenus *Stenopolistes* van der Vecht. *Journal of the Australian Entomological Society*, **27**, 69–80.

Kojima, J. and Suzuki, T. (1986). Timing of mating in five Japanese polistine wasps (Hymenoptera: Vespidae): Anatomy of fall females. *Journal of the Kansas Entomological Society*, **59**, 401–4.

Kort, C. A. D. de and Granger, N. A. (1981). Regulation of the juvenile hormone titer. *Annual Review of Entomology*, **26**, 1–28.

Kramer, S. and Wigglesworth, V. B. (1950). The outer layers of the cuticle in the cockroach *Periplaneta americana* and the function of the oenocytes. *Quarterly Journal of Microscopical Science*, **91**, 63–72.

Krebs, H. (1975). The August Krogh principle: 'For many problems there is an animal on which it can be most conveniently studied'. *Journal of Experimental Zoology*, **194**, 221–6.

Krogh, A. (1929). Progress of physiology. *American Journal of Physiology*, **90**, 243–51.

Krombein, K. V. (1976). *Eustenogaster*, a primitive social Sinhalese wasp. *Loris*, **14**, 303–6.

Krombein, K. V. (1991). Biosystematic studies of Ceylonese wasps, XIX: Natural history notes in several families (Hymenoptera: Eumenidae, Vespidae, Pompilidae, and Crabronidae). *Smithsonian Contributions to Zoology*, **515**, 1–41.

Kuhn, T. (1962). *The structure of scientific revolutions*, (Revised edition, 1970). University of Chicago Press.

Kundu, H. L. (1965). Notes on observations of the behaviour of a social wasp *Polistes hebraeus* (Hymenoptera). In *Proceedings of the 12th International Congress of Entomology, London*, 304.

Kundu, H. L. (1967). Observations on *Polistes hebraeus* (Hymenoptera). *Journal of Birla Institute of Technology and Science*, **1**, 152–61.

Kurczewski, F. E. and Miller, R. C. (1983). Nesting behavior of *Philanthus sanbornii* in Florida (Hymenoptera: Sphecidae). *The Florida Entomologist*, **66**, 199–206.

Landolt, P. J. and Akre, R. D. (1979a). Occurrence and location of exocrine glands in some social Vespidae (Hymenoptera). *Annals of the Entomological Society of America*, **72**, 141–8.

Landolt, P. J. and Akre, R. D. (1979b). Ultrastructure of the thoracic gland of queens of the western yellowjacket *Vespula pensylvanica* (Hymenoptera: Vespidae). *Annals of the Entomological Society of America*, **72**, 586–90.

Landolt, P. J. and Heath, R. R. (1987). Alarm pheromone behavior of *Vespula squamosa* (Hymenoptera: Vespidae). *The Florida Entomologist*, **70**, 221–5.

Landolt, P. J., Akre, R. D., and Greene, A. (1977). Effects of colony division on *Vespula atropilosa* (Sladen) (Hymenoptera: Vespidae). *Journal of the Kansas Entomological Society*, **50**, 135–47.

Larch, C. M. and Gamboa, G. J. (1981). Investigation of mating preference for nestmates in the paper wasp *Polistes fuscatus* (Hymenoptera: Vespidae). *Journal of the Kansas Entomological Society*, **54**, 811–14.

Latreille, P. A. (1802). *Histoire naturelle, générale et particulière des Crustacés et des Insectes*, Vol. 3. Paris.

Latreille, P. A. (1804). *Historia naturalis Insectorum*, Vol. 13. Paris.

Layton, J. M., Camann, M. A., and Espelie, K. E. (1994). Cuticular lipid profiles of queens, workers, and males of social wasp *Polistes metricus* Say are colony-specific. *Journal of Chemical Ecology*, **20**, 2307–21.

Lemaire, M., Lange, C., Lefebvre, J., and Clément, J. L. (1986). Strategie de camouflage du predateur *Hypoponera eduardi* dans les sociétés de *Reticulitermes* européens. *Actes des*

Colloques Insectes Sociaux, **3**, 97–101.

Lenoir, A., Yamaoka, R., Francoeur, A., and Cerda, X. (1993). On the origin of chemical mimicry between a parasitic ant (*Formicoxenus provancheri*) and its host (*Myrmica incompleta*). In *Abstracts of the XXIII International Ethological Conference, Torremolinos, Spain,* 61.

Lester, L. J. and Selander, R. K. (1981). Genetic relatedness and the social organization of *Polistes* colonies. *American Naturalist,* **117**, 147–66.

Levins, R. (1968). *Evolution in changing environments.* Princeton University Press.

Lewontin, R. and Dunn, L. (1960). The evolutionary dynamics of a polymorphism in the house mouse. *Genetics,* **45**, 705–22.

Lillie, F. R. (1927). The gene and the ontogenetic process. *Science,* **64**, 361–8.

Lin, N. (1972). Territorial behavior among males of the social wasp *Polistes exclamans* Viereck. *Proceedings of the Entomological Society of Washington,* **74**, 148–55.

Lin, N. and Michener, C. D. (1972). Evolution of sociality in insects. *Quarterly Review of Biology,* **47**, 131–59.

Lindauer, M. (1961). *Communication among social bees.* Harvard University Press, Cambridge.

Litt, M. and Luty, J. (1989). A hypervariable microsatellite revealed by in vitro amplification of a dinucleotide repeat within the cardiac muscle actin gene. *American Journal of Human Genetics,* **44**, 397–401.

Litte, M. (1977). Behavioral ecology of the social wasp, *Mischocyttarus mexicanus. Behavioral Ecology and Sociobiology,* **2**, 229–46.

Litte, M. (1979). *Mischocyttarus flavitarsis* in Arizona: social and nesting biology of a polistine wasp. *Zeitschrift für Tierpsychologie,* **50**, 282–312.

Litte, M. (1981). Social biology of the polistine wasp *Mischocyttarus labiatus:* survival in a Colombian rain forest. *Smithsonian Contributions to Zoology,* **327**, 1–27.

Lloyd, J. E. (1981). Sexual selection: individuality, identification, and recognition in a bumblebee and other insects. *Florida Entomologist,* **64**, 89–107.

Lloyd, J. E. (1983). Bioluminescence and communication in insects. *Annual Review of Entomology,* **28**, 131–60.

Lockey, K. H. (1988). Lipids of the insect cuticle: origin, composition and function. *Comparative Biochemistry and Physiology,* **89B**, 595–645.

Lorenzi, M. C. (1992). Epicuticular hydrocarbons of *Polistes biglumis bimaculatus* (Hymenoptera Vespidae): preliminary results. *Ethology Ecology & Evolution (Special Issue),* **2**, 61–3.

Lorenzi, M. C. and Cervo, R. (1992). Behaviour of *Polistes biglumis bimaculatus* (Hymenoptera: Vespidae) foundresses on alien conspecific nests. In *Biology and evolution of social insects* (ed. J. Billen), pp. 273–9. Leuven University Press, Leuven.

Lorenzi, M. C. and Cervo, R. (1993). Intraspecific usurpation in *Polistes biglumis bimaculatus* (Hymenoptera Vespidae). *Ethology Ecology & Evolution,* **3**, 397–8.

Lorenzi, M. C. and Cervo, R. (1995). Usurpation and late associations in the solitary founding social wasp, *Polistes biglumis bimaculatus* (Hymenoptera: Vespidae). *Journal of Insect Behavior,* **8**, 443–51.

Lorenzi, M. C. and Turillazzi, S. (1986). Behavioural and ecological adaptations to the high mountain enviroment of *Polistes biglumis bimaculatus. Ecological Entomology,* **11**, 199–204.

Lorenzi, M. C., Cervo, R., and Turillazzi, S. (1992). Effects of social parasitism of *Polistes atrimandibularis* on the colony cycle and brood production of *Polistes biglumis bimaculatus* (Hymenoptera, Vespidae). *Bollettino di Zoologia,* **59**, 267–71.

Lorenzi, M. C., Bertolino, F., and Beani, L. (1994*a*). Nuptial system of a social parasite wasp, *Polistes semenowi* (Hymenoptera Vespidae). *Ethology Ecology & Evolution (Special Issue),* **3**, 57–60.

Lorenzi, M. C., Bagnères, A. G., Clément, J. L., and Turillazzi, S. (1994*b*). Mechanisms of colony recognition: features of the epicuticular hydrocarbons of *Polistes biglumis bimaculatus* (Hymenoptera Vespidae). *Ethology Ecology & Evolution,* **6**, 428–9.

MacDonald, J. F. and Matthews, R. W. (1975). *Vespula squamosa*: a yellowjacket wasp evolving toward parasitism. *Science,* **190**, 1003–4.

MacCormack, A. T. (1982). Foundress associations and early colony failure in *Polistes exclamans*. Unpublished D. Phil. thesis, University of North Carolina at Chapel Hill, Chapel Hill. Ann Arbor Microfilms.

Maddison, D. R. (1991). African origin of human mitochondrial DNA reexamined. *Systematic Zoology,* **40**, 355–63.

Maddison, D. R., Ruvolo, M., and Swofford, D. L. (1992). Geographic origins of human mitochondrial DNA: Phylogenetic evidence from control region sequences. *Systematic Biology,* **41**, 111–24.

Makino, S. (1983). Biology of *Latibulus argiolus* (Hymenoptera, Ichneumonidae), a parasitoid of the paper wasp *Polistes biglumis* (Hymenoptera, Vespidae). *Kontyû,* **51**, 426–34.

Makino, S. (1985*a*). Foundress-replacement on nests of the monogynic paper wasp *Polistes biglumis* in Japan (Hymenoptera: Vespidae). *Kontyû,* **53**, 143–9.

Makino, S. (1985*b*). List of parasitoids of polistine wasps. *Sphecos,* **10**, 19–25.

Makino, S. (1989*a*). Usurpation and nest rebuilding in *Polistes riparius*: two ways to reproduce after the loss of original nest (Hymenoptera: Vespidae). *Insectes Sociaux,* **36**, 116–28.

Makino, S. (1989*b*). Switching of behavioral option from renesting to nest usurpation after nest loss by the foundress of a paper wasp, *Polistes riparius*: a field test. *Journal of Ethology,* **7**, 62–4.

Makino, S. (1991). Nest usurpation as a means of reproduction for paper wasp foundresses and conflicts between usurpers and workers. In *Abstracts of the 22nd International Ethological Conference,* 34.

Makino, S. and Aoki, S. (1982). Observations on two polygynic colonies of *Polistes biglumis* in Hokkaido, northern Japan (Hymenoptera, Vespidae). *Kontyû,* **50**, 175–82.

Makino, S. and Sayama, K. (1991). Comparison of intraspecific nest usurpation between two haplometrotic paper wasp species (Hymenoptera: Vespidae: *Polistes*). *Journal of Ethology,* **9**, 121–8.

Malyshev, S. I. (1968). *Genesis of the Hymenoptera and the phases of their evolution,* (trans. B. Haigh; ed. O. W. Richards and B. Uvarov). Methuen, London.

Mar, T., Brill, J., Bertsch, W., Fletcher, D. J. C., and Crewe, R. (1987). Investigation of cuticular hydrocarbons from selected honeybee populations by gas chromatography with pattern recognition. *Journal of Chromatography,* **399**, 277–90.

Marchal, P. (1896). Observation sur les *Polistes*. Cellule primitive et première cellule du nid. Provision de miel. Association de reines fondatrices. *Bulletin de la Société Zoologique de France,* **21**, 15–21.

Marino Piccioli, M. T. (1968). The extraction of the larval peritrophic sac by the adults in *Belonogaster* (Hymenoptera Vespoidea). *Monitore Zoologico Italiano (Nuova Serie) (Supplemento),* **2**, 203–6.

Marino Piccioli, M. T. and Pardi, L. (1970). Studi sulla biologia di *Belonogaster* (Hymenoptera, Vespidae). 1. Sull'etogramma di *Belonogaster griseus* (Fab). *Monitore Zoologico Italiano (Nuova Serie) (Supplemento),* **3**, 197–225.

Marino Piccioli, M. T. and Pardi, L. (1978). Studies on the biology of *Belonogaster* (Hymenoptera, Vespidae). 3. The nest of *Belonogaster griseus* (Fab.). *Monitore Zoologico Italiano (Nuova Serie) (Supplemento),* **10**, 179–228.

Marino Piccioli, M. T. and Pardi, L. (1980). Social dominance and trophallaxis in biginic societies of *Polistes gallicus* (L.). *Rendiconti della Classe di Scienze Fisiche, Matematiche e Naturali dell'Accademia Nazionale dei Lincei,* **68**, 443–8.

Maschwitz, U. (1964). Gefahrenalarmstoffe und Gefahrenalarmierung by sozialien Hymenopteren. *Zeitschrift für Vergleichende Physiologie,* **47**, 596–655.

Maschwitz, U. (1984). Alarm pheromone in the long-cheeked wasp *Dolichovespula saxonica* (Hym. Vespidae). *Deutsche Entomologische Zeitschrift, N. F.,* **31**, 33–4.

Maschwitz, U., Koob, K., and Schildknecht, H. (1970). Ein Beitrag zur Funktion der Metathoracaldrüse der Ameisen. *Journal of Insect Physiology,* **16**, 387–404.

Maschwitz, U., Beier, W., Dietrich, I., and Keidel, W. (1974). Futterverständigung bei Wespen der Gattung *Paravespula. Naturwissenschaften,* **61**, 506.

Maschwitz, U., Dorow, W. H. O., and Botz, T. (1990). Chemical composition of the nest walls, and nesting behaviour, of *Ropalidia (Icarielia) opifex* van der Vecht, 1962 (Hymenoptera: Vespidae), a Southeast Asian social wasp with translucent nests. *Journal of Natural History,* **24**, 1311–19.

Mather, K. (1973). *Genetical structure of populations,* pp. 1–197. Halsted Press, New York.

Mathis, A. and Smith, J. F. (1993). Fathead minnows, *Pimephales promelas,* learn to recognize northern pike, *Esox lucius,* as predators on the basis of chemical stimuli from minnows in the pike's diet. *Animal Behaviour,* **46**, 645–56.

Matsuda, R. (1982). The evolutionary process in talitrid amphipods and salamanders in changing environments, with a discussion of 'genetic assimilation' and some other evolutionary concepts. *Canadian Journal of Zoology,* **60**, 733–49.

Matsuda, R. (1987). *Animal evolution in changing environments.* John Wiley, New York.

Matsuura, M. (1977). The life of the paper wasps. *Shizen,* **32**, 26–36 (in Japanese).

Matsuura, M. (1984). Comparative biology of the five Japanese species of the genus *Vespa* (Hymenoptera, Vespidae). *Bulletin of the Faculty of Agriculture, Mie University,* **69**, 1–131.

Matsuura, M. (1990). Biology of three *Vespa* species in Central Sumatra (Hymenoptera, Vespidae). In *Natural history of social wasps and bees in Equatorial Sumatra* (ed. S. F. Sakagami, R. Ohgushi, and D. W. Roubik), pp. 113–24. Hokkaido University Press, Sapporo.

Matsuura, M. (1991). *Vespa* and *Provespa.* In *The social biology of wasps* (ed. K. G. Ross and R. W. Matthews), pp. 232–62. Comstock, Ithaca.

Matsuura, M. and Yamane, Sk. (1990). *Biology of the vespine wasps.* Springer-Verlag, Berlin.

Matthes-Sears, W. and Alcock, J. (1986). Hilltopping behavior of *Polistes commanchus navajoe* (Hymenoptera: Vespidae). *Ethology,* **71**, 42–53.

Matthews, R. W. (1968). *Microstigmus comes:* Sociality in a sphecid wasp. *Science,* **160**, 787–8.

Matthews, R. W. and Naumann, I. D. (1988). Nesting biology and taxonomy of *Arpactophilus mimi,* a new species of social sphecid (Hymenoptera: Sphecidae). *Australian Journal of Zoology,* **36**, 585–97.

Matthews, R. W. and Starr, C. K. (1984). *Microstigmus comes* wasps have a method of nest construction unique among social insects. *Biotropica,* **16**, 55–8.

Maynard Smith, J. (1976). Group selection. *Quarterly Review of Biology,* **51**, 277–83.

Maynard Smith, J. (1987). When learning guides evolution. *Nature,* **329**, 761–2.

Maynard Smith, J. (1991). Theories of sexual selection. *Trends in Ecology and Evolution,* **6**, 146–51.

Maynard Smith, J., Burian, R., Kaufman, S., Alberch, P., Campbell, J., Goodwin, B., Lind, R., Raup, D., and Wolpert, L. (1985). Developmental constraints and evolution. *Quarterly Review of Biology,* **60**, 265–87.

McComb, K. (1987). Roaring by red deer stags advances the date of oestrous in hinds.

Nature, (London), **330**, 648–9.

McCorquodale, D. B. (1989). Soil softness, nest initiation and nest sharing in the wasp, *Cerceris antipodes* (Hymenoptera: Sphecidae). *Ecological Entomology,* **14**, 191–6.

McCorquodale, D. B. (1990). Oocyte development in the primitively social wasp, *Cerceris antipodes* (Hymenoptera Sphecidae). *Ethology Ecology & Evolution,* **2**, 345–61.

McDaniel, C. A. (1990). Cuticular hydrocarbons of the formosan termite *Coptotermes formosanus. Sociobiology,* **16**, 265–73.

McGovern, J. N., Jeanne, R. L., and Effland, M. J. (1988). The nature of wasp nest paper. *Tappi Journal,* **71**, 133–9.

Mead, F. (1991). Social parasitism of a *Polistes dominulus* Christ colony by *Sulcopolistes semenowi* Morawitz: Changes in social activity among the queens and development of the usurped colony. *Journal of Ethology,* **9**, 37–40.

Mead, F. and Gabouriaut, D. (1993). Post-eclosion sensitivity to social context in *Polistes dominulus* Christ females (Hymenoptera, Vespidae). *Insectes Sociaux,* **40**, 11–20.

Meinwald, J., Smollanoff, J., Chibnall, A. C., and Eisner, T. (1975). Characterisation and synthesis of waxes from homopterous insects. *Journal of Chemical Ecology,* **2**, 269–74.

Metcalf, R. A. (1980). Sex-ratios, parent-offspring conflict, and local competition for mates in the social wasp *Polistes metricus* and *Polistes variatus. American Naturalist,* **116**, 642–54.

Metcalf, R. A. and Whitt, G. S. (1977*a*). Intra-nest relatedness in the social wasp *Polistes metricus.* A genetic analysis. *Behavioral Ecology and Sociobiology,* **2**, 339–51.

Metcalf, R. A. and Whitt, G. S. (1977*b*). Relative inclusive fitness in the social wasp, *Polistes metricus. Behavioral Ecology and Sociobiology,* **2**, 353–60.

Michener, C. D. (1964). Reproductive efficiency in relation to colony size in hymenopterous societies. *Insectes Sociaux,* **11**, 317–41.

Michener, C. D. (1969). Comparative social behaviour of bees. *Annual Review of Entomology,* **14**, 299–342.

Michener, C. D. (1974). *The social behavior of the bees: A comparative study.* Harvard University Press, Cambridge.

Michener, C. D. (1990). Wasps and our knowledge of insect social behaviour. In *Social insects and the environment* (ed. G. K. Veeresh, B. Mallik, and C. A. Viraktamath), pp. 61–62 (*Proceedings, 11th International Congress IUSSI, Bangalore, India*), Oxford and IBH, New Delhi.

Michener, C. D. and Grimaldi, D. A. (1988). The oldest fossil bee: Apoid history, evolutionary stasis, and antiquity of social behavior. *Proceedings of the National Academy of Sciences of USA,* **85**, 6424–6.

Miller, E. H. (1988). Description of bird behavior for comparative purposes. *Current Ornithology,* **5**, 347–94.

Mintzer, A. C., Williams, H. J., and Vinson, S. B. (1987). Identity and variation of hexane soluble cuticular components produced by the acacia ant *Pseudomyrmex ferruginea. Comparative Biochemistry and Physiology,* **86B**, 27–30.

Miyano, S. (1980). Life tables of colonies and workers in a paper wasp, *Polistes chinensis antennalis,* in central Japan (Hymenoptera: Vespidae). *Researches in Population Ecology,* **22**, 69–88.

Miyano, S. (1983). Number of offspring and seasonal changes of their body weight in a paper wasp, *Polistes chinensis antennalis* Perez (Hymenoptera: Vespidae), with reference to male production by workers. *Researches in Population Ecology,* **25**, 198–209.

Miyano, S. (1986). Colony development, worker behavior and male production in orphan colonies of a Japanese paper wasp, *Polistes chinensis antennalis* Perez (Hymenoptera: Vespidae). *Researches in Population Ecology,* **28**, 347–61.

Miyano, S. (1991). Worker reproduction and related behavior in orphan colonies of a Japanese paper wasp, *Polistes jadwigae* (Hymenoptera, Vespidae). *Journal of Ethology,*

9, 135–46.

Molitor, A. (1937). Versuche betreffend das "Orts-(bzw. Gegestands-) Gedachtnis" von Apiden, Sphegiden und Vespiden. *Zoologischer Anzeiger*, **117**, 110–12.

Molitor, A. (1939). Beobachtungen, den "Ortsinn" und Netsbau der Vespiden betreffend. *Zoologischer Anzeiger*, **126**, 239–45.

Montagner, M. (1966). Le mécanisme et le conséquences des comportement trophallactiques chez le guêpes du genre *Vespa. Bulletin Biologique de la France et de la Belgique*, **166**, 187–323.

Mora, G. (1990). Paternal care in a neotropical harvestman, *Zygopachylus albomarginis* (Arachnida, Opiliones: Gonyleptidae). *Animal Behaviour*, **39**, 582–93.

Moran, N. A. and Whitham, T. G. (1988). Evolutionary reduction of complex life cycles: loss of host-alternation in *Pemphigus* (Homoptera: Aphididae). *Evolution*, **42**, 717–28.

Morel, L. and Vander Meer, R. K. (1987). Nestmate recognition in *Camponotus floridanus*: behavioral and chemical evidence for the role of age and social experience. In *Chemistry and biology of social insects* (ed. J. Eder and H. Rembold), pp. 471–2. Verlag J. Peperny, München.

Morel, L., Vander Meer, R. K., and Lavine, B. K. (1988). Ontogeny of nestmate recognition cues in the red carpenter ant (*Camponotus floridanus*) *Behavioral Ecology and Sociobiology*, **22**, 175–83.

Morimoto, R. (1954a). On the nest development of *Polistes chinensis antennalis* Perez. (Studies on the social Hymenoptera of Japan. III). *Science Bulletin of the Faculty of Agriculture, Kyushu University*, **14**, 337–53.

Morimoto, R. (1954b). On the nest development of *Polistes chinensis antennalis* Perez. II. (Studies on the social Hymenoptera of Japan. IV). *Science Bulletin of the Faculty of Agriculture, Kyushu University*, **14**, 511–22.

Morimoto, R. (1954c). On the nest development of *Polistes chinensis antennalis* Perez. III. Relation between the removal of eggs and larvae from the nest and the oviposition of the founding female (Studies on the social Hymenoptera of Japan. V). *Science Bulletin of the Faculty of Agriculture, Kyushu University*, **14**, 523–33.

Morimoto, R. (1959). On the nesting activity of the founding female of *Polistes chinensis antennalis* Perez. I-II. *Science Bulletin of the Faculty of Agriculture, Kyushu University*, **17**, 99–113, 115–28.

Morimoto, R. (1961). On the dominance order in *Polistes* wasps. I. (Studies on the social Hymenoptera of Japan. XII). *Scientific Bulletin of the Faculty of Agriculture Kyushu University*, **18**, 339–51.

Morrone, J. J. and Carpenter, J. M. (1994). In search of a method for cladistic biogeography: An empirical comparison of component analysis, Brooks parsimony analysis, and three-area statements. *Cladistics*, **10**, 99–153.

Motro, M., Motro, U., Ishay, J. S., and Kugler, J. (1979). Some social and dietary prerequisites of oocyte development in *Vespa orientalis* L. workers. *Insectes Sociaux*, **26**, 155–64.

Mueller, U. G. (1991). Haplodiploidy and the evolution of facultative sex ratios in a primitively eusocial bee. *Science*, **254**, 442–4.

Muralidharan, K., Shaila, M. S., and Gadagkar, R. (1986). Evidence for multiple mating in the primitively eusocial wasp *Ropalidia marginata* (Lep.) (Hymenoptera: Vespidae). *Journal of Genetics*, **65**, 153–8.

Myles, T. G. (1988). Resource inheritance in social evolution from termites to man. In *The Ecology of social behaviour* (ed. C. N. Slobodchikoff), pp. 379–422. Academic Press, San Diego.

Naumann, M. G. (1970). The nesting behavior of *Protopolybia pumila* in Panama (Hymenoptera: Vespidae). Unpublished D. Phil. thesis. University of Kansas.

Naumann, M. G. (1975). Swarming behavior: evidence for communication in social wasps. *Science*, **189**, 642–4.

Nelson, G. (1979). Cladistic analysis and synthesis: principles and definitions, with a historical note on Adanson's *Familles des Plantes* (1763–1764). *Systematic Zoology*, **28**, 1–21.

Nelson, G. and Ladiges, P. Y. (1991*a*). Standard assumptions for biogeographic analysis. *Australian Systematic Botany*, **4**, 41–58.

Nelson, G. and Ladiges, P. Y. (1991*b*). Three-area statements: Standard assumptions for biogeographic analysis. *Systematic Zoology*, **40**, 470–84.

Nelson, G. and Ladiges, P. Y. (1992). *TAS*. Program and documentation. New York and Melbourne.

Nelson, G. and Platnick, N. I. (1981). *Systematics and biogeography: Cladistics and vicariance*. Columbia University Press, New York.

Nelson, J. M. (1968). Parasites and symbionts of nests of *Polistes* wasps. *Annals of the Entomological Society of America*, **61**, 1528–39.

Nelson, J. M. (1982). External morphology of *Polistes* (paper wasp) larvae in the United States. *Melanderia*, **38**, 1–29.

Nessov, L. A. (1988). Traces of life activity of organisms of late Mesozoic-Paleocene of middle Asia and Kazakhstan as the indicators of the paleoenvironment of the vertebrates. In *Fossil traces of vital activity and dynamics of the environment in ancient biotopes*. Transaction of the XXX Session All-Union Paleontological Society, (ed. T. N. Bogdanova, L. I. Khozatsky, and A. A. Istchenko), pp. 76–90. Kiev.

Nevo, E. (1979). Adaptive convergence and divergence of subterranean mammals. *Annual Reviews of Ecology and Systematics*, **10**, 269–308.

Nickell, W. P. (1958). Variations in engineering features of the nests of several species of birds in relation to nest sites and nesting materials. *Butler University Botanical Studies*, **13**, 121–40.

Nijhout, F. (1994). *Insect hormones*. Princeton University Press.

Nijhout, H. F. and Wheeler, D. E. (1982). Juvenile hormone and the physiological basis of insect polymorphisms. *The Quarterly Review of Biology*, **57**, 109–33.

Nixon, K. C. (1992). *CLADOS*, version 1.2. Program and documentation. Trumansburg, NY.

Nixon, K. C. and Davis, J. I. (1991). Polymorphic taxa, missing values and cladistic analysis. *Cladistics*, **7**, 233–41.

Noirot, C. and Quennedey, A. (1991). Glands, gland cells, glandular units: some comments on terminology and classification. *Annales de la Société Entomologique de France (Nouvelle Série)*, **27**, 123–8.

Nonacs, P. (1986). Ant reproductive strategies. *Quarterly Review of Biology*, **61**, 1–21.

Nonacs, P. (1991). Alloparental care and eusocial evolution: the limits of Queller's head-start advantage. *Oikos*, **61**, 122–5.

Nonacs, P. and Reeve, H. K. (1993). Opportunistic adoption of orphaned nests in paper wasps as an alternative reproductive strategy. *Behavioural Processes*, **30**, 47–60.

Noonan, K. M. (1978). Sex ratio of parental investment in colonies of the social wasp *Polistes fuscatus*. *Science*, **199**, 1354–6.

Noonan, K. M. (1979). Individual strategies of inclusive fitness maximizing in the social wasp, *Polistes fuscatus* (Hymenoptera: Vespidae). Unpublished D. Phil. thesis. University of Michigan. Ann Arbor Microfilms.

Noonan, K. M. (1981). Individual strategies of inclusive-fitness-maximizing in *Polistes fuscatus* foundresses. In *Natural selection and social behavior* (ed. R. D. Alexander and D. W. Tinkle), pp. 18–44. Chiron, New York.

Nowbahari, E., Lenoir, A., Clément, J. L., Lange, C., Bagnères, A. G., and Joulie, C. (1990). Individual geographical and experimental variation of cuticular hydrocarbons of the ant

Cataglyphis cursor (Hymenoptera: Formicidae): their use in nest and subspecies recognition. *Biochemical Systematics and Ecology*, **18**, 63–73.

Nozawa, K. and Itô, Y. (1989). Biochemical-genetic differentiation among nine species of polistine wasps from Japan. *Insectes Sociaux*, **36**, 183–96.

O'Donnell, S. (1992). Off-nest gastral rubbing observed in *Mischocyttarus immarginatus* (Hymenoptera: Vespidae) in Costa Rica. *Sphecos*, **23**, 5.

O'Donnell, S. and Jeanne, R. L. (1990). Forager specialization and the control of nest repair in *Polybia occidentalis* Olivier (Hymenoptera: Vespidae). *Behavioral Ecology and Sociobiology*, **27**, 359–64.

O'Donnell, S. and Jeanne, R. L. (1991). Interspecific occupation of a tropical social wasp colony (Hymenoptera: Vespidae: *Polistes*). *Journal of Insect Behavior*, **4**, 397–400.

O'Donnell, S. and Jeanne, R. L. (1993). Methoprene accelerates age polyethism in workers of a social wasp (*Polybia occidentalis*). *Physiological Entomology*, **18**, 189–94.

Obin, M. S. (1986). Nestmate recognition cues in laboratory and field colonies of *Solenopsis invicta* Buren (Hymenoptera: Formicidae): effect of environment and role of cuticular hydrocarbons. *Journal of Chemical Ecology*, **12**, 1965–75.

Ohgushi, R., Yamane, Sô., and Abbas, N. D. (1985). Descriptions and re-descriptions of 5 types of stenogastrine nests collected in Sumatera Barat, Indonesia, with some biological notes (Hymenoptera, Vespidae). *Sumatra Nature Study (Entomology), Kunazawa University*, 1–12.

Ohgushi, R., Sakagami, S. F. and Yamane, Sô. (1990). Nest architecture of the stenogastrine wasps: Diversity and evolution (Hymenoptera, Vespidae). A comparative review. In *Natural history of social wasps and bees in Equatorial Sumatra* (ed. S. F. Sakagami, R. Ohgushi, and D. W. Roubik), pp. 41–72. Hokkaido University Press, Sapporo.

Ondricek-Fallscheer, R. L. (1992). A morphological comparison of the sting apparatuses of socially parasitic and nonparasitic species of yellowjackets (Hymenoptera: Vespidae). *Sociobiology*, **20**, 245–300.

Ordway, E. (1965). Caste differentiation in *Augochlorella*. *Insectes Sociaux*, **12**, 291–308.

Ormerod, E. L. (1868). *British social wasps: an introduction to their anatomy and physiology, architecture, and general natural history, with illustrations of the different species and their nests.* Longmans, Green, Reader, and Dyer, London.

Oster, G. F. and Wilson, E. O. (1978). *Caste and ecology in the social insects.* Princeton University Press.

Pagden, H. T. (1962). More about *Stenogaster*. *Malayan Nature Journal*, **16**, 95–102.

Page, R. D. M. (1987). Graphs and generalized tracks: Quantifying Croizat's panbiogeography. *Systematic Zoology*, **36**, 1–17.

Page, R. D. M. (1988). Quantitative cladistic biogeography: Constructing and comparing area cladograms. *Systematic Zoology*, **37**, 254–70.

Page, R. D. M. (1989a). Comments on component-compatibility in historical biogeography. *Cladistics*, **5**, 167–82.

Page, R. D. M. (1989b). *COMPONENT*, release 1.5. Program and documentation. Auckland, New Zealand.

Page, R. D. M. (1990). Component analysis: A valiant failure? *Cladistics*, **6**, 119–36.

Page, R. E. and Metcalf, R. A. (1982). Multiple mating, sperm utilization, and social evolution. *American Naturalist*, **119**, 263–81.

Page, R. E., Metcalf, R. A., Metcalf, R. L., Erickson, E. H., and Lampman, R. L. (1991). Extractable hydrocarbons and kin recognition in honeybee (*Apis mellifera* L.). *Journal of Chemical Ecology*, **17**, 745–56.

Page, R. E., Robinson, G. E., and Fondrk, M. K. (1989). Genetic specialists, kin recognition and nepotism in honey-bee colonies. *Nature*, **338**, 576–9.

Pamilo, P. (1984). Genetic correlation and regression in social groups: multiple alleles, multiple loci, and subdivided populations. *Genetics*, **107**, 307–20.

Pamilo, P. and Crozier, R. H. (1982). Measuring genetic relatedness in natural populations: methodology. *Theorethical Population Biology*, **21**, 171–93.

Papi, F. (1992). General aspects. In *Animal homing* (ed. F. Papi), pp. 1–18. Chapman and Hall, London.

Pardi, L. (1938). Origine e comportamento del glicogeno nei corpi grassi di *Polistes gallicus* L. (Hymenoptera-Vespidae). *Archivio Italiano di Anatomia*, **60**, 281–300.

Pardi, L. (1939). I corpi grassi degli Insetti. *Redia*, **25**, 87–128.

Pardi, L. (1940). Ricerche sui Polistini, I. Poliginia vera e apparente in *Polistes gallicus* (L.). *Atti della Società Toscana di Scienze Naturali*, **49**, 1–9.

Pardi, L. (1942). Ricerche sui Polistini. V. La poliginia iniziale in *Polistes gallicus* (L.). *Bollettino dell'Istituto di Entomologia dell'Università di Bologna*, **14**, 1–106.

Pardi, L. (1946). Ricerche sui Polistini. VI. La "dominazione" e il ciclo ovarico annuale in *Polistes gallicus* (L.). *Bollettino dell'Istituto di Entomologia dell'Università di Bologna*, **15**, 25–84.

Pardi, L. (1948*a*). Dominance order in *Polistes* wasps. *Physiological Zoology*, **21**, 1–13.

Pardi, L. (1948*b*). Ricerche sui Polistini. 11. Sulla durata della permanenza delle femmine nel nido e sull'accrescimento della società in *Polistes gallicus* (L.). *Memorie della Società Toscana di Scienze Naturali (Serie B)*, **55**, 3–15.

Pardi, L. (1951). Studio delle attività e della divisione di lavoro in una società di *Polistes gallicus* (L.) dopo la comparsa delle operaie. (Ricerche sui Polistini, XII). *Archivio Zoologico Italiano*, **36**, 363–431.

Pardi, L. (1952). Dominazione e gerarchia in alcuni Invertebrati. In *Structure et physiologie des sociétés animales*. Publications du Centre Nationale de la Recherche Scientifique, Colloques Int., No. 34, (ed. P.-P. Grassé), pp. 183–97.

Pardi, L. (1974). Polymorphismus bei sozialen Faltenwespen. In *Sozial Polymorphismus bei Insekten*, (ed. G. H. Schmidt). Wissenschaftl, Verlagsgesell, MBH, Stuttgart.

Pardi, L. (1977). Su alcuni aspetti della biologia di *Belonogaster* (Hymenoptera, Vespidae). *Bollettino dell'Istituto di Entomologia dell'Università di Bologna*, **33**, 281–99.

Pardi, L. (1980). Le vespe sociali: biologia ed evoluzione del comportamento. *Contributi del Centro Linceo Interdisciplinare di Scienze Matematiche e Loro Applicazioni*, **51**, 161–221.

Pardi, L. and Cavalcanti, M. (1951). Esperienze sul meccanismo della monoginia funzionale in *Polistes gallicus* (L.) (Hymenoptera Vespidae). *Bollettino di Zoologia*, **18**, 247–52.

Pardi, L. and Marino Piccioli, M. T. (1970). Studi sulla biologia di *Belonogaster* (Hymenoptera, Vespidae). 2. Differenziamento castale incipiente in *B. griseus* (Fab.). *Monitore Zoologico Italiano (Nuova Serie) (Supplemento)*, **3**, 235–65.

Pardi, L. and Marino Piccioli, M. T. (1981). Studies on the biology of *Belonogaster* (Hymenoptera, Vespidae). 4. On caste differences in *Belonogaster griseus* (Fab.) and the position of this genus among social wasps. *Monitore Zoologico Italiano (Nuova Serie) (Supplemento)*, **14**, 131–46.

Pardi, L. and Turillazzi, S. (1982). Biologia delle Stenogastrinae (Hymenoptera, Vespoidea). *Atti dell'Accademia Nazionale Italiana di Entomologia, Rendiconti*, **30**, 1–21.

Parker, F. D. (1966). A revision of the North American species in the genus *Leptochilus* (Hymenoptera: Eumenidae). *Miscellaneous Publications of the Entomological Society of America*, **5**, 151–229.

Payne, R. (1978). The ecology of brood parasitism in birds. *Annual Review of Ecology and Systematics*, **8**, 1–28.

Peckham, G. W. and Peckham, E. G. (1889). Observations on sexual selection in spiders of the family Attidae. *Occasional Papers of the Natural History Society of Wisconsin*, **1**,

1–60.

Peeters, C. (1987). The diversity of reproductive systems in ponerine ants. In *Chemistry and biology of social insects* (ed. J. Eder and H. Rembold), pp. 253–354. Verlag J. Peperny, München.

Peeters, C. and Crewe, R. (1985). Worker reproduction in the ponerine ant *Ophthalmopone berthoudi*: an alternative form of eusocial organization. *Behavioral Ecology and Sociobiology*, **18**, 29–37.

Peeters, C. and Crewe, R. (1986). Queenright and queenless breeding systems within the genus *Pachycondyla* (Hymenoptera: Formicidae). *Journal of Entomological Society, South Africa*, **49**, 251–5.

Perna, B., Marino Piccioli, M. T., and Turillazzi, S. (1978). Osservazioni sulla poliginia di *Polistes foederatus* (Kohl) (Hymenoptera, Vespidae) indotta in cattività. *Bollettino dell'Istituto di Entomologia dell'Università di Bologna*, **34**, 55–63.

Petersen, B. (1987). Subspecies of the Indo-Australian *Polistes stigma* (Fabricius) (Hymenoptera: Vespidae). *Entomologica Scandinavica*, **18**, 227–59.

Petersen, B. (1990). Polistine wasps of the subgenera *Stenopolistes* and *Megapolistes* from the Philippine, Moluccan and Bismarck Islands (Hymenoptera: Vespidae). *Entomologica Scandinavica*, **21**, 53–66.

Pfennig, D. W. (1990). Nestmate and nest discrimination among workers from neighboring colonies of social wasps *Polistes exclamans*. *Canadian Journal of Zoology*, **68**, 268–71.

Pfennig, D. W. and Reeve, H. K. (1989). Neighbor recognition and context-dependent aggression in a solitary wasp, *Sphecius speciosus* (Hymenoptera: Sphecidae). *Ethology*, **80**, 1–18.

Pfennig, D. W. and Reeve, H. K. (1993). Nepotism in a solitary wasp as revealed by DNA fingerprinting. *Evolution*, **47**, 700–4.

Pfennig, D. W., Gamboa, G. J., Reeve, H. K., Shellman-Reeve, J., and Ferguson, D. (1983*a*). The mechanism of nestmate discrimination in social wasps (*Polistes*, Hymenoptera: Vespidae). *Behavioral Ecology and Sociobiology*, **13**, 299–305.

Pfennig, D. W., Reeve, H. K., and Shellman, J. S. (1983*b*). Learned component of nestmate discrimination in workers of a social wasp, *Polistes fuscatus* (Hymenoptera: Vespidae). *Animal Behaviour*, **31**, 412–16.

Phelan, P. L., Smith, A. W., and Needham, G. R. (1991). Mediation of host selection by cuticular hydrocarbons in the honeybee tracheal mite *Acarapis woodi* (Rennie). *Journal of Chemical Ecology*, **17**, 463–73.

Pickering, J. (1980). Sex ratio, social behavior and ecology in *Polistes* (Hymenoptera, Vespidae), *Pachysomoides* (Hymenoptera, Ichneumonidae) and *Plasmodium* (Protozoa, Haemosporida). Unpublished D. Phil. thesis. Harvard University.

Pierce, W. D. (1909). A monographic revision of the twisted winged insects comprising the order Strepsiptera Kirby. *Bulletin of the United States National Museum*, **66**, 1–232.

Platnick, N. I. (1987). An empirical comparison of microcomputer parsimony programs. *Cladistics*, **3**, 121–44.

Platnick, N. I., Griswold, C. E., and Coddington, J. A. (1991). On missing entries in cladistic analysis. *Cladistics*, **7**, 337–43.

Polak, M. (1992). Distribution of virgin females influences mate-searching behavior of male *Polistes canadensis* (L.) (Hymenoptera: Vespidae). *Journal of Insect Behavior*, **5**, 531–5.

Polak, M. (1993*a*). Competition for landmark territories among male *Polistes canadensis* (L.) (Hymenoptera: Vespidae): large-size advantage and alternative mate-acquisition tactics. *Behavioral Ecology*, **4**, 325–31.

Polak, M. (1993*b*). Landmark territoriality in the Neotropical paper wasps *Polistes canadensis* (L.) and *P. carnifex* (F.) (Hymenoptera: Vespidae). *Ethology*, **95**, 278–90.

Pope, R. D. (1983). Some aphid waxes, their form and function. *Journal of Natural History*,

17, 489–506.

Post, D. C. and Jeanne, R. L. (1981). Colony defense against ants by *Polistes fuscatus* (Hymenoptera: Vespidae) in Wisconsin. *Journal of the Kansas Entomological Society*, **54**, 599–615.

Post, D. C. and Jeanne, R. L. (1982*a*). Recognition of former nestmates during colony founding by the social wasp *Polistes fuscatus* (Hymenoptera: Vespidae). *Behavioral Ecology and Sociobiology*, **11**, 283–5.

Post, D. C. and Jeanne, R. L. (1982*b*). Sternal glands in three species of male social wasps of the genus *Mischocyttarus* (Hymenoptera: Vespidae). *Journal of the New York Entomological Society*, **90**, 8–15.

Post, D. C. and Jeanne, R. L. (1983*a*). Male reproductive behaviour of the social wasp *Polistes fuscatus* (Hymenoptera: Vespidae). *Zeitschrift für Tierpsychologie*, **62**, 157–71.

Post, D. C. and Jeanne, R. L. (1983*b*). Relatedness and mate selection in *Polistes fuscatus* (Hymenoptera: Vespidae). *Animal Behaviour*, **31**, 1260–1.

Post, D. C. and Jeanne, R. L. (1983*c*). Sternal glands in males of six species of *Polistes (Fuscopolistes)* (Hymenoptera: Vespidae). *Journal of the Kansas Entomological Society*, **56**, 32–9.

Post, D. C. and Jeanne, R. L. (1983*d*). Venom: source of a sex pheromone in the social wasp *Polistes fuscatus* (Hymenoptera: Vespidae). *Journal of Chemical Ecology*, **9**, 259–66.

Post, D. C. and Jeanne, R. L. (1984). Venom as an interspecific sex pheromone, and species recognition by a cuticular pheromone in paper wasps (*Polistes*, Hymenoptera: Vespidae). *Physiological Entomology*, **9**, 65–75.

Post, D. C., Downing, H. A., and Jeanne, R. L. (1984*a*). Alarm response to venom by social wasps *Polistes exclamans* and *P. fuscatus* (Hymenoptera: Vespidae). *Journal of Chemical Ecology*, **10**, 1425–33.

Post, D. C., Mohamed, M. A., Coppel, H. C., and Jeanne R. L. (1984*b*). Identification of ant repellent allomone produced by social wasp *Polistes fuscatus* (Hymenoptera, Vespidae). *Journal of Chemical Ecology*, **10**, 1799–807.

Post, D.C., Jeanne, R. L., and Erickson, Jr., E. H. (1988). Variation in behavior among workers of the primitively social wasp *Polistes fuscatus variatus*. In *Interindividual behavioral variability in social insects*, (ed. R.L. Jeanne), pp. 283–321. Westview Press, Boulder.

Potter, N. B. (1965). Some aspects of the biology of *Vespula vulgaris* L. Unpublished D. Phil. thesis. University of Bristol.

Prance, G. T. (1992). Ant association with *Parinari excelsa* (Chrysobalanaceae) in Marajo, Brazil. *Biotropica*, **24**, 102–4.

Pratte, M. (1982). Relations antérieures et association de fondation chez *Polistes gallicus* L. *Insectes Sociaux*, **29**, 352–7.

Pratte, M. (1990*a*). Foundress association in the paper wasp *Polistes dominulus* Christ. Effects of dominance hierarchy on the division of labour. *Behaviour*, **111**, 208–19.

Pratte, M. (1990*b*). Effects of changes in brood composition on the activities of three associated foundresses of the paper wasp *Polistes dominulus* (Christ). *Behavioural Processes*, **22**, 187–95.

Pratte, M. and Gervet, J. (1980). Le modèle sociobiologique, ses conditions de validité dans le cas des sociétés d'Hyménoptères. *Année Biologique*, **19**, 163–201.

Pratte, M. and Gervet, J. (1992). Effects of prior residence and previous cohabitation on the *Polistes dominulus* Christ dominance hierarchy. *Ethology*, **90**, 72–80.

Pratte, M., Strambi, C., Gervet, J., and Strambi, A. (1982). Parametres physiologiques et ethologiques dans un guepier de *Polistes gallicus* L. *Insectes Sociaux*, **29**, 383–401.

Provost, E., Rivière, G., Roux, M., Morgan, E. D., and Bagnères, A. G. (1993). Change in the chemical signature of the ant *Leptothorax lichtensteini* Bondroit with time. *Insect*

Biochemistry and Molecular Biology, **23**, 945–57.

Pulich, W. M. (1969). Unusual feeding behavior of three species of birds. *Wilson Bulletin,* **81**, 472.

Queller, D. C. (1989). The evolution of eusociality: Reproductive head starts of workers. *Proceedings of the National Academy of Science, USA,* **86**, 3224–6.

Queller, D. C. (1994). Extended parental care and the origin of eusociality. *Proceedings of the Royal Society of London, B,* (In press).

Queller, D. C. and Goodnight, K. F. (1989). Estimating relatedness using genetic markers. *Evolution,* **43**, 258–75.

Queller, D. C. and Strassmann, J. E. (1988). Reproductive success and group nesting in the paper wasp *Polistes annularis.* In *Reproductive success: Studies of individual variation in contrasting breeding systems* (ed. T. H. Clutton-Brock), pp. 76–96. University of Chicago Press.

Queller, D. C. and Strassmann, J. E. (1989). Measuring inclusive fitness in social wasps. In *The genetics of social evolution,* (ed. M. D. Breed and R. E. Page), pp. 103–22. Westview, Boulder, Colorado.

Queller, D. C., Strassmann, J. E., and Hughes, C. R. (1988). Genetic relatedness in colonies of tropical wasps with multiple queens. *Science,* **242**, 1155–7.

Queller, D. C., Hughes, C. R., and Strassmann, J. E. (1990). Wasps fail to make distinction. *Nature,* **344**, 388.

Queller, D. C., Negrón-Sotomayor, J., Strassmann, J. E., and Hughes, C. R. (1993*a*). Queen number and genetic relatedness in a neotropical wasp, *Polybia occidentalis. Behavioral Ecology,* **4**, 7–13.

Queller, D. C., Strassmann, J. E., and Hughes, C. R. (1993*b*). Microsatellites and kinship. *Trends in Ecology and Evolution,* **8**, 285–8.

Rabaud, E. (1924). Le retour au nid de *Vespa sylvestris. La Feuille des Jeunes Naturalistes (Nouvelle Série),* **1**, 7–11.

Rau, P. (1928). Autumn and spring in the life of *Polistes annularis* and *P. pallipes. Bulletin of the Brooklyn Entomological Society,* **23**, 230–5.

Rau, P. (1929*a*). At the end of the season with *Polistes rubiginosus* (Hymenoptera: Vespidae). *Entomological News,* **40**, 7–13.

Rau, P. (1929*b*). The habitat and dissemination of four species of *Polistes* wasps. *Ecology,* **10**, 191–200.

Rau, P. (1933). *The jungle bees and wasps of Barro Colorado Island,* pp. 1–324. Phil Rau, Kirkwood.

Rau, P. (1935). The duties of a wasp queen *Polistes pallipes. Entomological News,* **46**, 25–7.

Rau, P. (1940). Co-operative nest-founding by the wasp, *Polistes annularis* Linn. *Annals of the Entomological Society of America,* **33**, 617–20.

Rau, P. (1941). The swarming of *Polistes* wasps in the temperate regions. *Annals of the Entomological Society of America,* **34**, 580–4.

Rau, P. and Rau, N. (1918). *Wasps studies afield.* Princeton University Press, Princeton. (Republished by Dover Publications Inc., New York 1970).

Raup, D. (1991). *Extinctions: Bad genes or bad luck?* Norton, New York.

Raveret Richter, M., Downing, H., and Richter, W. (1987). A novel social wasp behavior: worker mouthing and rubbing of teneral *Polistes pacificus* (Hymenoptera: Vespidae). *Journal of the Kansas Entomological Society,* **60**, (2), 347–9.

Ray, J. (1691). *The wisdom of God manifested in the works of the Creation.* London.

Reed, H. C. and Akre, R. D. (1982). Morphological comparison between the obligate social parasite, *Vespula austriaca* (Panzer), and its host, *Vespula acadica* (Sladen) (Hymenoptera: Vespidae). *Psyche,* **89**, 183–95.

Reed, H. C. and Akre, R. D. (1983). Usurpation behaviour of the yellowjacket social parasite

Vespula austriaca (Panzer) (Hymenoptera, Vespidae). *American Midland Naturalist*, **30**, 259–73.

Reed, H. C. and Akre, R. D. (1990). Evolution of social parasitism in yellow jackets and hornets (Hymenoptera: Vespidae: Vespinae). In *Social insects and the environment* (ed. G. K. Veeresh, B. Mallik, and C. A. Viraktamath), pp. 160–1. Oxford and IBH Publishing Co., New Delhi.

Reed, H. C. and Landolt, P. J. (1990). Sex attraction in paper wasp, *Polistes exclamans* Viereck (Hymenoptera: Vespidae), in a wind tunnel. *Journal of Chemical Ecology*, **16**, 1277–87.

Reed, H. C. and Landolt, P. J. (1991). Swarming of paper wasp sexuals at towers in Florida (Hymenoptera: Vespidae: *Polistes*). *Annals of the Entomological Society of America*, **84**, 628–35.

Reeve, H. K. (1991). *Polistes*. In *The social biology of wasps* (ed. K. G. Ross and R. W. Matthews), pp. 99–148. Comstock, Ithaca.

Reeve, H. K. and Gamboa, G. J. (1987). Queen regulation of worker foraging in paper wasps: a social feedback control system (*Polistes fuscatus*, Hymenoptera: Vespidae). *Behaviour*, **102**, 147–67.

Reeve, H. K. and Ratnieks, F. L. W. (1993). Queen–queen conflicts in polygynous societies: mutual tolerance and reproductive skew. In *Queen number and sociality in insects* (ed. L. Keller), pp. 45–85. Oxford University Press.

Reeve, H. K. and Sherman, P. W. (1991). Intracolonial aggression and nepotism by the breeding female naked mole-rat. In *The biology of the naked mole-rat* (ed. P. W. Sherman, J. U. M. Jarvis, and R. D. Alexander), pp. 337–57. Princeton University Press.

Rensch, B. (1960). *Evolution above the species level.* Columbia University Press, New York.

Rheinberger H. J. (1995). From experimental systems to cultures of experimentation. In *Concepts, theories, and rationality in the biological sciences*, (ed. G. Wolters, J. G. Lennox, and P. McLaughlin), pp. 107–22. UVK-Universitätsverlag Konstanz/University of Pittsburgh Press.

Richards, J. G. and King, P. E. (1967). Chorion and vitelline membranes and their role in resorbing eggs of the Hymenoptera. *Nature*, **214**, 601–2.

Richards, O. W. (1962). *A revisional study of the masarid wasps.* British Museum (Natural History), London.

Richards, O. W. (1969). The biology of some W. African social wasps (Hymenoptera: Vespidae, Polistinae). *Memorie della Società Entomologica Italiana*, **48**, 79–93.

Richards, O. W. (1971). The biology of the social wasps (Hymenoptera, Vespidae). *Biological Reviews*, **46**, 483–528.

Richards, O. W. (1973). The subgenera of *Polistes* Latreille (Hymenoptera, Vespidae). *Revista Brasileira de Entomologia*, **17**, 85–104.

Richards, O. W. (1978a). *The social wasps of the Americas, excluding the Vespinae.* British Museum (Natural History), London.

Richards, O. W. (1978b). The Australian social wasps (Hymenoptera: Vespidae). *Australian Journal of Zoology, Supplementary Series*, **61**, 1–132.

Richards, O. W. (1982). A revision of the genus *Belonogaster* de Saussure (Hymenoptera: Vespidae). *Bulletin of the British Museum of Natural History (Entomology)*, **44**, 31–114.

Richards, O. W. and Richards, M. J. (1951). Observations on the social wasps of South America (Hymenoptera Vespidae). *Transactions of the Royal Entomological Society of London*, **102**, 1–170.

Robinson, G. E. (1992). Regulation of division of labor in insect societies. *Annual Review Entomology*, **37**, 637–65.

Robinson, G. E., Strambi, A., Strambi, C., Paulino-Simoes, Z. L., Tozeto, S. O., and

Barbosa, J. M. N. (1987). Juvenile hormone titers in Africanized and European honey bees in Brazil. *General and Comparative Endocrinology*, **66**, 457–9.

Robinson, G. E., Strambi, C., Strambi, A., and Huang, Zhi-Yong. (1992). Reproduction in worker honey bees is associated with low juvenile hormone titers and rates of biosynthesis. *General and Comparative Endocrinology*, **87**, 471–80.

Röseler, P. F. (1985). Endocrine basis of dominance and reproduction in polistine paper wasps. In *Experimental behavioral ecology and sociobiology*, (ed. B. Hölldobler and M. Lindauer), pp. 259–72. Sinauer, Sunderland.

Röseler, P. F. (1991). Reproductive competition during colony establishment. In *The social biology of wasps* (ed. K. G. Ross and R. W. Matthews), pp. 309–35. Comstock, Ithaca.

Röseler, P. F. and Röseler, I. (1977). Dominance in bumblebees. In *Proceedings of the VII International Congress of the IUSSI, Wageningen*, 232–5.

Röseler, P. F., Röseler, I., and Strambi, A. (1980). The activity of corpora allata in dominant and subordinated females of the wasp *Polistes gallicus*. *Insectes Sociaux*, **27**, 97–107.

Röseler, P. F., Röseler, I., Strambi, A., and Augier, R., (1984). Influence of insect hormones on the establishment of dominance hierarchies among foundresses of the paper wasp, *Polistes gallicus*. *Behavioral Ecology and Sociobiology*, **15**, 133–42.

Röseler, P. F., Röseler, I., and Strambi, A. (1985). Role of ovaries and ecdysteroids in dominance hierarchy establishment among foundresses of the primitively social wasp, *Polistes gallicus*. *Behavioral Ecology and Sociobiology*, **18**, 9–13.

Ross, K. G. (1983). Laboratory studies of the mating biology of the eastern yellow jacket, *Vespula maculifrons* (Hymenoptera: Vespidae). *Journal of the Kansas Entomological Society*, **56**, 523–37.

Ross, K. G. and Carpenter, J. M. (1991). Population genetic structure, relatedness, and breeding systems. In *The social biology of wasps* (ed. K. G. Ross and R. W. Matthews), pp. 451–79. Comstock, Ithaca.

Ross, N. M. and Gamboa, G. J. (1981). Nestmate discrimination in social wasps (*Polistes metricus*, Hymenoptera: Vespidae). *Behavioral Ecology and Sociobiology*, **9**, 163–5.

Rossel, S. and Wehner, R. (1986). Polarization vision in bees. *Nature*, **323**, 128–31.

Roubaud, E. (1910). Evolution de l'instinct chez les Vespides, aperçus biologiques sur les Guêpes sociales d'Afrique des genres *Belonogaster* Sauss *Comptes Rendus de l'Academie des Sciences, Paris*, **151**, 553–6.

Roubaud, E. (1916). Recherches biologiques sur le Guêpes solitaires et sociales d'Afrique. La genèse de la vie sociale et l'évolution de l'instinct maternel chez les Vespides. *Annales des Sciences Naturelles*, **1**, 1–160.

Roubik, D. W. (1989). *Ecology and natural history of tropical bees*. Cambridge University Press.

Ryan, M. J. (1988). Energy, calling, and selection. *American Zoologist*, **28**, 885–98.

Ryan, R. E. and Gamboa, G. J. (1986). Nestmate recognition between males and gynes of the social wasp *Polistes fuscatus* (Hymenoptera: Vespidae). *Annals of the Entomological Society of America*, **79**, 572–5.

Ryan, R. E., Cornell, T. C., and Gamboa, G. J. (1985). Nestmate recognition in the bald-faced hornet, *Dolichovespula maculata* (Hymenoptera: Vespidae). *Zeitschrift für Tierpsychologie*, **69**, 19–26.

Sakagami, S. F. (1982). Stingless bees. In *Social insects*, Vol III, (ed. H. R. Hermann), pp. 361–423. Academic Press, New York.

Sakagami, S. F. (1987). Observations on the rock stenogastrine wasp *Liostenogaster* sp. *Insectaryûmu*, **24**, 16–28.

Sakagami, S. F. and Fukushima, K. (1957a). Some biological observations on a hornet, *Vespa tropica* var. *pulchra* (Du Buysson), with special reference to its dependence on *Polistes* wasps (Hymenoptera). *Treubia*, **24**, 73–82.

Sakagami, S. F. and Fukushima, K. (1957b). *Vespa dybowskii* André as a facultative temporary social parasite. *Insectes Sociaux*, **4**, 1–12.

Sakagami, S. F. and Maeta, Y. (1987a). Sociality, induced and/or natural, in the basically solitary small carpenter bees (*Ceratina*). In *Animal societies: Theories and facts* (ed. Y. Itô, J. Brown, and J. Kikkawa), pp. 1–16. Japan Scientific Societies Press, Tokyo.

Sakagami, S. F. and Maeta, Y. (1987b). Multifemale nests and rudimentary castes of an 'almost' solitary bee *Ceratina flavipes*, with additional observation on multifemale nests of *Ceratina japonica* (Hymenoptera, Apoidea). *Kontyû*, **55**, 391–409.

Sakagami, S. F. and Munakata, M. (1972). Distribution and bionomics of a transpalaearctic eusocial halictine bee, *Lasioglossum (Evylaeus) calceatum*, in northern Japan, with reference to its solitary life cycle at high altitude. *Journal of the Faculty of Science, Hokkaido University, VI, Zoology*, **18**, 411–39.

Sakagami, S. F., Maeta, Y., Nagamori, S., and Saito, K. (1993). Diapause and non-delayed eusociality in a univoltine and basically solitary bee, *Ceratina japonica* (Hymenoptera, Anthophoridae). *Japanese Journal of Entomology*, **61**, (3), 443–57.

Saleh-Mghir, P. E. (1989). La reconnaissance interindividuelle chez l'abeille *Apis mellifica* L. *Année Biologique*, **28**, 187–200.

Samuel, C. T. (1987). Factors affecting colony size in the stenogastrine wasp *Liostenogaster flavolineata*. Unpublished D. Phil. thesis, University of Malaya, Kuala Lumpur.

Santschi, F. (1911). Observations et remarques critiques sur le mécanisme de l'orientation chez le fourmis. *Revue Suisse de Zoologie*, **19**, 303–38.

Sasagawa, H. and Kuwahara, Y. (1990). Comparative study of hydrocarbon profiles on the Japanese and the European honeybee. In *Proceedings of the 11th International Congress IUSSI, India*, pp. 421–2.

Saussure, H. de (1853–1858). *Études sur la famille des Véspides*, Vol. 2. *Monographie des Guêpes sociales, ou de la tribu des Vespiens*. Masson, Paris, and Kessmann, Genève.

Saussure, H. de (1857). Note sur les *Polistes* américains. *Annales de la Société Entomologique de France* (3), **5**, 309–14.

Schaffner, K. (1993). *Discovery and explanation in biology and medicine*. University of Chicago Press.

Scheven, J. (1958). Beitrag zur Biologie der Schmarotzerfeldenwespen *Sulcopolistes atrimandibularis* Zimm., *S. semenowi* F. Morawitz und *S. sulcifer* Zimm. *Insectes Sociaux*, **5**, 409–37.

Schneirla, T. C. (1949). Army-ant life and behavior under dry-season conditions. 3. The course of reproduction and colony behavior. *Bulletin of the American Museum of Natural History*, **94**, 3–81.

Schneirla, T. C. (1971). *Army ants: a study in social organization*. W. H. Freeman and Company, San Francisco.

Schöne, H. and Tengö, J. (1991). Homing in the digger wasp *Bembix rostrata* (Hymenoptera, Sphecidae) – Release direction and weather conditions. *Ethology*, **87**, 160–4.

Schremmer, F. (1972). Beobachtungen zur Biologie von *Apoica pallida* (Olivier, 1791), einer neotropischen sozialen Faltenwespe (Hymenoptera, Vespidae). *Insectes Sociaux*, **19**, 343–57.

Schremmer, F. (1977). Das Baumrinden-Nest der neotropischen Faltenwespen *Nectarinella championi*, umgeben von einem Leimring als Ameisen-Abwehr (Hymenoptera: Vespidae). *Entomologica Germanica*, **3**, 344–55.

Schremmer, F. (1983). Das Nest der neotropischen Faltenwespe *Leipomeles dorsata*. Ein Beitrag zur Kenntnis der Nestarchitektur der sozialen Faltenwespen (Vespidae, Polistinae, Polybiini). *Zoologischer Anzeiger*, **211**, 95–107.

Schremmer, F., März, L., and Simonsberger, P. (1985). Chitin im Speichel der Papierwespen (soziale Faltenwespen, Vespidae): Biologie, Chemismus, Feinstruktur.

Mikroskopie (Wien), **42**, 52–6.

Schwammberger, K. H. (1993). Freilandbeobachtungen zur Nestübernahme bei *Polistes biglumis bimaculatus* (Geoffroy) durch den Sozialparasiten *Sulcopolistes atrimandibularis* (Zimmermann) (Hymenoptera, Vespidae). *Zeitschrift für Angewandte Zoologie*, **79**, 291–7.

Seger, J. (1983). Partial bivoltinism may cause alternating sex ratio biases that favour eusociality. *Nature*, **301**, 59–62.

Sekijima, M., Sugiura, M., and Matsuura, M. (1980). Nesting habits and brood development of *Parapolybia indica* Saussurre (Hymenoptera: Vespidae). *Bulletin of the Faculty of Agriculture, Mie University*, **61**, 11–23.

Shakarad, M. and Gadagkar, R. (1993). Why are there multiple-foundress colonies in *Ropalidia marginata*. In *Proceedings of the XXI Annual Conference of the Ethological Society of India, Tirupati* (In press).

Shellman, J. S. and Gamboa, G. J. (1982). Nestmate discrimination in social wasps: the role of exposure to nest and nestmates (*Polistes fuscatus*, Hymenoptera: Vespidae). *Behavioral Ecology and Sociobiology*, **11**, 51–3.

Shellman-Reeve, J. S. and Gamboa, G. J. (1985). Male social wasps (*Polistes fuscatus*) recognize their male nestmates. *Animal Behaviour*, **33**, 331–2.

Sherman, P. W., Jarvis, J. U. M., and Alexander, R. D. (1991). *The biology of the naked mole-rat*. Princeton University Press.

Sibley, C. G. and Monroe, B. L. (1990). *Distribution and taxonomy of birds of the world*. Yale University Press, New Haven.

Siebold, K. T. von (1871). *Beiträge zur Parthenogenesis der Arthropoden*. W. Engelmann, Leipzig.

Silva, M. N. da (1981). Ciclo de desenvolvimento das colônias de *Mischocyttarus (Kappa) atramentarius* Zikán, 1949 (Hymenoptera, Vespidae): fase de pré-emergência. Unpublished M. S. thesis. University of São Paulo, Rio Claro, Brasil.

Singer, T. L. and Espelie, K. E. (1992). Social wasps use nest paper hydrocarbons for nestmate recognition. *Animal Behaviour*, **44**, 63–8.

Singer, T. L., Camann, M. A., and Espelie, K. E. (1992*a*). Discriminant analysis of cuticular hydrocarbons of social wasp *Polistes exclamans* Viereck and surface hydrocarbons of its nest paper and pedicel. *Journal of Chemical Ecology*, **18**, 785–97.

Singer, T. L., Espelie, K. E., and Himmelsbach, D. S. (1992*b*). Ultrastructural and chemical examination of paper and pedicel from laboratory and field nests of the social wasp *Polistes metricus* Say. *Journal of Chemical Ecology*, **18**, 77–86.

Snelling, R. R. (1952). Notes on nesting and hibernation of *Polistes*. *Pan-Pacific Entomologist*, **28**, 177.

Snelling, R. R. (1983). Taxonomic and nomenclatural studies on American polistine wasps (Hymenoptera: Vespidae). *Pan-Pacific Entomologist*, **59**, 267–80.

Snodgrass, R. E. (1956). *Anatomy of the honey bee*. Cornell University Press, Ithaca.

Solìs, C. R., and Strassmann, J. E. (1990). Presence of brood affects caste differentiation in the social wasp, *Polistes exclamans* Viereck (Hymenoptera: Vespidae). *Functional Ecology*, **4**, 531–41.

Spradbery, J. P. (1973). *Wasps. An account of the biology and natural history of solitary and social wasps*. University of Washington Press, Seattle.

Spradbery, J. P. (1975). The biology of *Stenogaster concinna* Van der Vecht with comments on the phylogeny of the Stenogastrinae (Hymenoptera Vespidae). *Journal of the Australian Entomological Society*, **14**, 309–18.

Spradbery, J. P. (1989). The nesting of *Anischnogaster iridipennis* (Smith) (Hymenoptera: Vespidae) in New Guinea. *Journal of the Australian Entomological Society*, **28**, 225–8.

Spradbery, J. P. (1991). Evolution of queen number and queen control. In *The social biology of wasps* (ed. K. G. Ross and R. W. Matthews), pp. 336–88. Comstock, Ithaca.

Stacey, P. B. and König, W. D. (1990). *Cooperative breeding in birds: long-term studies of ecology and behavior*. Cambridge University Press, Cambridge.

Starr, C. K. (1991). The nest as the locus of social life. In *The social biology of wasps* (ed. K. G. Ross and R. W. Matthews), pp. 480–509. Comstock, Ithaca.

Starr, C. K. (1992). The social wasps (Hymenoptera: Vespidae) of Taiwan. *Bulletin of the National Museum of Natural Sciences*, **3**, 93–138.

Steiner, A. (1930). Die Temperaturregulierung im Nest der Feldwespe (*Polistes gallica* var. *biglumis* L.). *Zeitschrift für Vergleichende Physiologie*, **11**, 452–61.

Steward, O., Torre, E. R., Tomasulo, R., and Lothman, E. (1991). Neuronal activity up-regulates astroglial gene expression. *Proceedings of the National Academy of Science, USA*, **88**, 6819–23.

Strambi, A. (1967). Physiologie des insectes. – Quelques effets de la castration sur la neurosécrétion protocérébrale des femelles de *Polistes* (Hyménoptères Vespides). *Compte Rendu de l'Academie des Sciences, Paris*, **264**, 2031–4.

Strambi, A. (1985). Physiological aspects of caste differentiation in social wasps. In *Caste differences in social insects* (ed. J. A. L. Watson, B. M. Okot-Kotber, Ch. Noirot), pp. 371–84. Pergamon Press, New York.

Strambi, A. and Girardie, A. (1973). Effet de l'implantation de *corpora allata* actifs de *Locusta migratoria* (Orthoptère) dans des femelles de *Polistes gallicus* L. (Hyménoptère) saines et parasitées par *Xenos vesparum* Rossi (Insecte Strepsiptère). *Compte Rendu de l'Academie des Sciences, Paris, D*, **276**, 3319–22.

Strambi, A. and Strambi, C. (1973a). Influence du développement du parasite *Xenos vesparum* Rossi (Insecte, Strepsitère) sur le système neuroendocrinien des femelles de *Polistes* (Hyménoptère, Vespide) au début de leur vie imaginale. *Archives d'Anatomie Microscopique et de Morphologie Experimentale*, **62**, 39–54.

Strambi, A. and Strambi, C. (1973b). Étude histochimique et ultrastructurale des sécrétions élaborées par les péricaryones neurosécréteurs de la pars intercerebralis chez la guêpe *Polistes*. *Acta Histochemique*, **46**, 101–9.

Strassmann, J. E. (1979). Honey caches help female paper wasps (*Polistes annularis*) survive Texas winters. *Science*, **204**, 207–9.

Strassmann, J. E. (1981a). Evolutionary implications of early male and satellite nest production in *Polistes exclamans* colony cycles. *Behavioral Ecology and Sociobiology*, **8**, 55–64.

Strassmann, J. E. (1981b). Kin selection and satellite nests in *Polistes exclamans*. In *Natural selection and social behavior*, (ed. R. D. Alexander and D. W. Tinkle), pp. 45–58. Chiron Press, New York.

Strassmann, J. E. (1981c). Parasitoids, predators and group size in the paper wasp, *Polistes exclamans*. *Ecology*, **62**, 1225–33.

Strassmann, J. E. (1981d). Wasp reproduction and kin selection: reproductive competition and dominance hierarchies among *Polistes annularis* foundresses. *The Florida Entomologist*, **64**, 74–88.

Strassmann, J. E. (1983). Nest fidelity and group size among foundresses of *Polistes annularis* (Hymenoptera: Vespidae). *Journal of the Kansas Entomological Society*, **56**, 621–34.

Strassmann, J. E. (1984). Female-biased sex ratios in social insects lacking morphological castes. *Evolution*, **38**, 256–66.

Strassmann, J. E. (1985a). Relatedness of workers to brood in the social wasp, *Polistes exclamans* (Hymenoptera: Vespidae). *Zeitschrift für Tierpsychologie*, **69**, 141–8.

Strassmann, J. E. (1985b). Worker mortality and the evolution of castes in the social wasp *Polistes exclamans*. *Insectes Sociaux*, **32**, 275–85.

Strassmann, J. E. (1989a). Altruism and relatedness at colony foundation in social insects.

Trends in Ecology and Evolution, **4**, 371–4.

Strassmann, J. E. (1989*b*). Early termination of brood rearing in the social wasp, *Polistes annularis* (Hymenoptera: Vespidae). *Journal of the Kansas Entomological Society*, **62**, 353–62.

Strassmann, J. E. (1991). Costs and benefits of colony aggregation in the social wasp *Polistes annularis*. *Behavioral Ecology*, **2**, 204–9.

Strassmann, J. E. and Hughes, C. R. (1988). Age is more important than size in determining dominance among workers in the primitively eusocial wasp, *Polistes instabilis*. *Behaviour*, **107**, 1–14.

Strassmann, J. E. and Meyer, C. D. (1983). Gerontocracy in the social wasp, *Polistes exclamans*. *Animal Behaviour*, **31**, 431–8.

Strassmann, J. E. and Orgren, M. C. (1983). Nest architecture and brood development times in the paper wasp, *Polistes exclamans* (Hymenoptera: Vespidae). *Psyche*, **90**, 237–48.

Strassmann, J. E. and Queller, D. C. (1989). Ecological determinants of social evolution. In *The genetics of social evolution*. (ed. M. D. Breed and R. E. Page Jr.), pp. 81–101. Westview Press, Boulder.

Strassmann, J. E., Meyer, D. C., and Matlock, R. L. (1984). Behavioral castes in the social wasp, *Polistes exclamans* (Hymenoptera: Vespidae). *Sociobiology*, **8**, (3), 211–24.

Strassmann, J. E., Queller, D. C., and Hughes, C. R. (1988). Predation and the evolution of sociality in the paper wasp *Polistes bellicosus*. *Ecology*, **69**, 1497–505.

Strassmann, J. E., Hughes, C. R., Queller, D. C., Turillazzi, S., Cervo, R., Davis, S. K., and Goodnight, K. F. (1989). Genetic relatedness in primitively eusocial wasps. *Nature*, **342**, 268–70.

Strassmann, J. E., Queller, D. C., Solìs, C. R., and Hughes, C. R. (1991). Relatedness and queen number in the Neotropical wasp, *Parachartergus colobopterus*. *Animal Behaviour*, **42**, 461–70.

Strassmann, J. E., Gastreich, K. R., Queller, D. C., and Hughes, C. R. (1992). Demographic and genetic evidence for cyclical changes in queen number in a Neotropical wasp. *Polybia emaciata*. *American Naturalist*, **140**, (3), 363–72.

Strassmann, J. E., Hughes, C. R., Turillazzi, S., Solìs, C. R., and Queller, D. C. (1994). Genetic relatedness and incipient eusociality in stenogastrine wasps. *Animal Behaviour*, **48**, 813–21.

Stubblefield, J. W. and Charnov, E. L. (1986). Some conceptual issues in the origin of eusociality. *Heredity*, **57**, 181–7.

Sturtevant, A. H. (1971). On the choice of material for genetic studies. *Stadler Symposia*, **1** and **2**, 51–7.

Sullivan, J. D. and Strassmann, J. E. (1984). Physical variability among nest foundresses in the polygynous social wasp, *Polistes annularis*. *Behavioral Ecology and Sociobiology*, **15**, 249–56.

Suzuki, H. and Murai, M. (1980). Ecological studies of *Ropalidia fasciata* in Okinawa island I. Distribution of single- and multiple-foundress colonies. *Researches in Population Ecology*, **22**, 184–95.

Suzuki, T. (1978). Area, efficiency and time of foraging in *Polistes chinensis antennalis* Perez (Hymenoptera, Vespidae). *Japanese Journal of Ecology*, **28**, 179–89.

Suzuki, T. (1985). Mating and laying of female-producing eggs by orphaned workers of a paper wasp, *Polistes snelleni* (Hymenoptera: Vespidae). *Annals of the Entomological Society of America*, **78**, 736–9.

Suzuki, T. and Ramesh, M. (1992). Colony founding in the social wasp, *Polistes stigma* (Hymenoptera Vespidae), in India. *Ethology Ecology & Evolution*, **4**, 333–41.

Svensson, B. G. and Petersson, E. (1992). Why insect swarm: testing the models for lek mating systems on swarming *Empis borealis* females. *Behavioral Ecology and*

Sociobiology, **31**, 253–61.

Swofford, D. L. (1993). *PAUP*, version 3.1. Program and documentation. Illinois Natural History Survey, Urbana.

Tautz, D. (1989). Hypervariability of simple sequences as a general source for polymorphic DNA markers. *Nucleic Acids Research*, **17**, 6463–71.

Taylor, L. H. (1939). Observations on social parasitism in the genus *Vespula* Thomson. *Annals of the Entomological Society of America*, **32**, 304–15.

Ten Cate, C. and Bateson, P. (1988). Sexual selection: The evolution of conspicuous characteristics in birds by means of imprinting. *Evolution*, **42**, 1355–8.

Tengö, J., Schöne, H., and Chmurzynski, J. (1990). Homing in the digger wasp *Bembix rostrata* (Hymenoptera, Sphecidae) in relation to sex and stage. *Ethology*, **86**, 47–56.

Tevis, L. (1958). Interrelations between *Veromessor pergandei* (Mayr) and some desert ephemerals. *Ecology*, **39**, 695–704.

Thiollay, J. M. (1991). Foraging, home range use and social behaviour of a group-living rainforest raptor, the Red-throated Caracara, *Daptrius americanus*. *Ibis*, **133**, 382–93.

Thomsen, M. (1954). Neurosecretion in some hymenopters. *Kongelige Danske Videnskabernes Selskabs Biologiske Skrifter*, **7**, 1–24.

Thornhill, R. and Alcock, J. (1983). *The evolution of insect mating systems*. Harvard University Press, Cambridge.

Tierney, A. J. (1986). The evolution of learned and innate behavior: contributions from genetics and neurobiology to a theory of behavioral evolution. *Animal Learning and Behavior*, **14**, 339–48.

Tillyard, R. J. (1926). *The insects of Australia and New Zealand*. Angus and Robertson, Sydney.

Tinbergen, N. (1958). *Curious naturalists*. Basic, New York.

Tindo, M. (1991). Contribution a l'étude de la biologie de *Belonogaster junceus junceus* (Hymenoptera: Vespidae, Polistinae). Unpublished Maitrise thesis, Université de Yaoundé, Cameroun.

Tomialojc, L. (1992). Colonisation of dry habitats by the Song Thrush *Turdus philomelos*: is the type of nest material an important constraint? *Bulletin of the British Ornithologists' Club*, **112**, 27–34.

Topoff, H. and Zimmerli, E. (1993). Colony takeover by a socially parasitic ant, *Polyergus breviceps*: the role of chemicals obtained during host-queen killing. *Animal Behaviour*, **46**, 479–86.

Townes, H., Momoi, S., and Townes, M. (1965). A catalogue and reclassification of the eastern Palearctic Ichneumonidae. *Memoirs of the American Entomological Institute* No. 5.

Trivers, R. L. (1972). Parental investment and sexual selection. In *Sexual selection and the descent of man* (ed. B. Campbell), pp. 136–79. Heinemann, London.

Trivers, R. L. and Hare, H. (1976). Haplodiploidy and the evolution of social insects. *Science*, **191**, 249–63.

Truman, J. W. and Riddiford, L. M. (1974). Hormonal mechanisms underlying insect behaviour. *Advances in Insect Physiology*, **10**, 297–352.

Tsuchida, K. (1991). Temporal behavioral variation and division of labor among workers in the primitively eusocial wasp, *Polistes jadwigae* Dalla Torre. *Journal of Ethology*, **9**, 129–34.

Tsuchida, K. and Itô, Y. (1991). Negative correlation between dominance and frequency of oviposition and oophagy in a foundress association of the Japanese paper wasp *Polistes jadwigae* Dalla Torre. *Applied Entomology and Zoology*, **26**, 443–8.

Tsuji, K. (1990). Reproductive division of labour related to age in the Japanese queenless ant, *Pristomyrmex pungens*. *Animal Behaviour*, **39**, 843–9.

Tsuji, K. (1992). Sterility for life: applying the concept of eusociality. *Animal Behaviour*,

44, 572–3.

Tsuneki, K. (1957). Ethological studies on *Bembix niponica* Smith, with emphasis on the psychobiological analysis of behavior inside the nest. II. Experimental part. *Memoirs of the Faculty of Liberal Arts, Fukui University, Series II, Natural Science*, **7**, 1–116.

Turillazzi, S. (1979). Tegumental glands in the abdomen of some European *Polistes* (Hymenopetra Vespidae). *Monitore Zoologico Italiano (Nuova Serie)*, **13**, 67–70.

Turillazzi, S. (1980). Seasonal variations in the size and anatomy of *Polistes gallicus* (L.) (Hymenoptera: Vespidae). *Monitore Zoologico Italiano (Nuova Serie)*, **14**, 63–75.

Turillazzi, S. (1981). Use of artificial nests for rearing and studying *Polistes* wasps. *Psyche*, **87**, 131–40.

Turillazzi, S. (1983*a*). Extranidal behaviour in *Parischnogaster nigricans serrei* (Du Buysson) (Hymenoptera Stenogastrinae). *Zeitschrift für Tierpsychologie*. **63**, 27–36.

Turillazzi, S. (1983*b*). Patrolling behaviour in males of *Parischnogaster nigricans serrei* (Du Buysson) and *P. mellyi* (Saussure) (Hymenoptera Stenogastrinae). *Accademia Nazionale dei Lincei, Rendiconti Classe di Scienze Fisiche Matematiche e Naturali*, **72**, 153–7.

Turillazzi, S. (1985*a*). Colonial cycle in *Parischnogaster nigricans serrei* (Du Buysson) in West Java (Hymenoptera Stenogastrinae). *Insectes Sociaux*, **32**, 43–60.

Turillazzi, S. (1985*b*). Brood rearing behaviour and larval development in *Parischnogaster nigricans serrei* (Du Buysson) (Hymenoptera Stenogastrinae). *Insectes Sociaux*, **32**, 117–27.

Turillazzi, S. (1985*c*). Associative nest foundation in the wasp *Parischnogaster alternata*. *Naturwissenschaften*, **72**, 100–2.

Turillazzi, S. (1985*d*). Function and characteristics of the abdominal substance secreted by wasps of the genus *Parischnogaster* (Hymenoptera Stenogastrinae). *Monitore Zoologico Italiano (Nuova Serie)*, **19**, 91–9.

Turillazzi, S. (1986). Colony composition and social behaviour of *Parischnogaster alternata* Sakagami (Hymenoptera Stenogastrinae). *Monitore Zoologico Italiano (Nuova Serie)*, **20**, 333–47.

Turillazzi, S. (1987). Distinguished features of the social behaviour of stenogastrine wasps. In *Chemistry and biology of social insects* (ed. J. Eder and H. Rembold), pp. 492–5. Verlag J. Peperny, München.

Turillazzi, S. (1988). Social biology of *Parischnogaster jacobsoni* (Du Buysson) (Hymenoptera, Stenogastrinae). *Insectes Sociaux*, **35**, 133–43.

Turillazzi, S. (1989). The origin and evolution of social life in the Stenogastrinae (Hymenoptera, Vespidae). *Journal of Insect Behavior*, **2**, 649–61.

Turillazzi, S. (1990*a*). Social biology of *Liostenogaster vechti* Turillazzi 1988 (Hymenoptera Stenogastrinae). *Tropical Zoology*, **3**, 68–87.

Turillazzi, S. (1990*b*). Notes on the biology, social behaviour and nest architecture of *Metischnogaster drewseni* (Saussure) (Hymenoptera Stenogastrinae). *Bollettino di Zoologia*, **57**, 331–9.

Turillazzi, S. (1991). The Stenogastrinae. In *The social biology of wasps* (ed. K. G. Ross and R. W. Matthews), pp. 74–98. Comstock, Ithaca.

Turillazzi, S. (1992). Nest usurpation and social parasitism in *Polistes* wasps: new acquisitions and current problems. In *Biology and evolution of social insects* (ed. J. Billen), pp. 263–72. Leuven University Press, Leuven, Belgium.

Turillazzi, S. (1994). Protection of brood by means of Dufour's gland secretion in two species of stenogastrine wasps. *Ethology Ecology & Evolution (Special Issue)*, **3**, 37–41.

Turillazzi, S. and Calloni, C. (1983). Tegumental glands in the third gastral tergite of male *Parischnogaster nigricans serrei* (du Buysson) and *P. mellyi* (Saussure) (Hymenoptera Stenogastrinae). *Insectes Sociaux*, **3**, 455–60.

Turillazzi, S. and Cervo, R. (1982). Territorial behaviour in males of *Polistes nimpha* (Christ) (Hymenoptera, Vespidae). *Zeitschrift für Tierpsychologie*, **58**, 174–80.

Turillazzi, S. and Francescato, E. (1989). Observations on the behaviour of male stenogastrine wasps (Hymenoptera, Vespidae, Stenogastrinae). *Actes des Colloques Insectes Sociaux*, **5**, 181–7.

Turillazzi, S. and Francescato, E. (1990). Patrolling behaviour and related secretory structures in the males of some stenogastrine wasps (Hymenoptera Vespidae). *Insectes Sociaux*, **37**, 146–57.

Turillazzi, S. and Hansell, M. H. (1991). Biology and social behaviour of three species of *Anischnogaster* (Vespidae, Stenogastrinae) in Papua New Guinea. *Insectes Sociaux*, **38**, 423–37.

Turillazzi, S. and Pardi, L. (1977). Body size and hierarchy in polygynic nests of *Polistes gallicus* (L.) (Hymenoptera Vespidae). *Monitore Zoologico Italiano (Nuova Serie)*, **15**, 1–17.

Turillazzi, S. and Pardi, L. (1981). Ant guards on nests of *Parischnogaster nigricans serrei* (Du Buysson) (Stenogastrinae). *Monitore Zoologico Italiano (Nuova Serie)*, **15**, 1–7.

Turillazzi, S. and Pardi, L. (1982). Social behavior of *Parischnogaster nigricans serrei* (Du Buysson) (Hymenoptera Vespidae) in Java. *Annals of the Entomological Society of America*, **75**: 657–64.

Turillazzi, S. and Ugolini, A. (1978). Osservazioni su colonie miste sperimentali di *Polistes gallicus* (L.) e *Polistes foederatus* (Kohl) (Hymenoptera, Vespidae). *Redia*, **61**, 233–49.

Turillazzi, S. and Ugolini, A. (1979). Rubbing behaviour in some European *Polistes* (Hymenoptera Vespidae). *Monitore Zoologico Italiano (Nuova Serie)*, **13**, 129–41.

Turillazzi, S., Marino Piccioli, M. T., Hervatin, L., and Pardi, L. (1982). Reproductive capacity of single foundress and associated foundress females of *Polistes gallicus* (L.) (Hymenoptera Vespidae). *Monitore Zoologico Italiano (Nuova Serie)*, **16**, 75–88.

Turillazzi, S., Cervo, R., and Cavallari, I. (1990). Invasion of the nest of *Polistes dominulus* by the social parasite *Sulcopolistes sulcifer* (Hymenoptera, Vespidae). *Ethology*, **84**, 47–59.

Turillazzi, S., Cervo, R., and Zanobetti, L. (1991). Control of host reproduction by the social parasite *Sulcopolistes sulcifer* (Hymenoptera, Vespidae). *Actes des Colloques Insectes Sociaux*, **7**, 97–102.

Turillazzi, S., Cervo, R., and Dani, F. R. (1993). Dati preliminari sulle migrazioni altitudinali di maschi e femmine di vespe del genere *Polistes* (Hymenoptera, Vespidae). In *Abstracts del LV Congresso Nazionale dell'Unione Zoologica Italiana, Torino, Italy*, 219.

Turillazzi, S., Francescato, E., Baldini Tosi, A., and Carpenter, J. M. (1994). A distinct caste difference in *Polybioides tabidus* (Fabricius) (Hymenoptera, Vespidae). *Insectes Sociaux*, **41**, 327–30.

Ugolini, A. (1981). Initial homeward orientation after displacement in *Polistes gallicus*. *Naturwissenschaften*, **68**, 487.

Ugolini, A. (1983). Homing in the wasp *Polistes gallicus* (L.): information gained en route and initial orientation. *Monitore Zoologico Italiano (Nuova Serie)*, **17**, 212–13.

Ugolini, A. (1985). Initial orientation and homing in workers of *Polistes gallicus* (L.) (Hymenoptera; Vespidae). *Zeitschrift für Tierpsychologie*, **69**, 133–40.

Ugolini, A. (1986a). Homing ability in *Polistes gallicus* (L.) (Hymenoptera Vespidae). *Monitore Zoologico Italiano (Nuova Serie)*, **20**, 1–15.

Ugolini, A. (1986b). Homing in female *Polistes gallicus* (L.) (Hymenoptera, Vespidae). In *Orientation in space* (ed. G. Beugnon), pp. 57–62. Privat, Toulouse.

Ugolini, A. (1987). Visual information acquired during displacement and initial orientation in *Polistes gallicus* (L.) (Hymenoptera, Vespidae). *Animal Behaviour*, **35**, 590–5.

Ugolini, A. and Cannicci, S. (1991). Homing in *Polistes dominulus* and *Polistes gallicus*. In

Abstracts of the 1st European Congress Social Insects, Leuven, 68.

Ugolini, A. and Samoggia, M. (1991). Workers of *Polistes dominulus* (Christ): infuence of the landscape on initial orientation. *Ethology Ecology & Evolution*, **3**, 247–55.

Ugolini, A., Kesller, A., and Ishay, J. S. (1987). Initial orientation and homing by oriental hornets, *Vespa orientalis* L. (Hymenoptera Vespidae). *Monitore Zoologico Italiano (Nuova Serie)*, **21**, 157–64.

Vander Meer, R. K. and Wojcik, D. P. (1982). Chemical mimicry in the myrmecophilous beetle, *Myrmecophodius excavaticollis*. *Science*, **218**, 806–8.

Vander Meer, R. K., Lofgren, C. L., and Alvarez, F. M. (1985). Biochemical evidence for hybridization in fire ants. *The Florida Entomologist*, **68**, 501–6.

Vander Meer, R. K., Saliwanchik, D., and Lavine, B. (1989*a*). Temporal changes in colony cuticular hydrocarbons patterns of *Solenopsis invicta*. Implications for nestmate recognition. *Journal of Chemical Ecology*, **15**, 2115–25.

Vander Meer, R. K., Jouvenaz, D. P., and Wojcik, D. P. (1989*b*). Chemical mimicry in a parasitoid (Hymenoptera: Eucharitidae) of fire ants (Hymenoptera: Formicidae). *Journal of Chemical Ecology*, **15**, 2247–61.

Vavilov, N. I. (1922). The law of homologous series in variation. *Journal of Genetics*, **12**, (1), 47–89.

Vecht, J. van der (1962). The Indo-australian species of the genus *Ropalidia (Icaria)* (Hymenoptera, Vespidae) (2nd part). *Zoologische Verhandelingen*, **57**, 1–71.

Vecht, J. van der (1965). The geographical distribution of the social wasps (Hymenoptera, Vespidae). In *Proceedings of the XII International Congress of Entomology, London*, 440–1.

Vecht, J. van der (1966). The East-Asiatic and Indo-australian species of *Polybioides* Buysson and *Parapolybia* Saussure (Hymenoptera, Vespidae). *Zoologische Verhandelingen*, **82**, 1–42.

Vecht, J. van der (1968*a*). The geographic variation of *Polistes* (*Megapolistes* subg. n.) *rothneyi* Cameron. *Bijdragen tot de Dierkunde*, **38**, 97–109.

Vecht, J. van der (1968*b*). The terminal gastral sternite of female and worker social wasps (Hymenoptera, Vespidae). *Proceedings of the Koninklijke Nederlandse Akademie van Wetenschappen, Series C*, **71**, 411–22.

Vecht, J. van der (1972). The subgenera *Megapolistes* and *Stenopolistes* in the Solomon Islands (Hymenoptera, Vespidae, *Polistes* Latreille). In *Entomological Essays to Commemorate the Retirement of Professor K. Yasumatsu*, pp. 87–106 (1971). Hokuryukan, Tokyo.

Vecht, J. van der (1984). New species of *Polistes* from Rennell and San Cristobal Islands. In *The natural history of Rennell Island, British Solomon Islands* 8, pp. 171–3. Copenhagen.

Vecht, J. van der and Carpenter, J. M. (1990). A catalog of the genera of the Vespidae (Hymenoptera). *Zoologische Verhandelingen, Leiden*, **260**, 3–62.

Veith, H. J. and Koeniger, N. (1978). Identifizierung von cis-9-Pentacosen als Auslöser für das Wärmen der Brut by der Hornisse. *Naturwissenschaften*, **65**, 263.

Veith, H. J., Koeniger, N., and Maschwitz, U. (1984). 2-methyl-3-butene-2-ol, a major component of the alarm pheromone of the hornet *Vespa crabro*. *Naturwissenschaften*, **71**, 328–9.

Velthuis, H. H. W. (1990). Food and the evolution of sociality in bees. In *Social insects and the environment* (ed. G. K. Veeresh, B. Mallik, and C. A. Viraktamath), pp. 342–3. *Proceedings, 11th International Congress IUSSI, Bangalore, India*. Oxford and IBH, New Delhi.

Velthuis, H. H. W., Ruttner, F. and Crewe, R. M. (1990). Differentiation in reproductive physiology and behaviour during the development of laying worker honey bees. In *Social insects* (ed. W. Engels), pp. 231–43. Springer-Verlag, New York.

Venkataraman, A. B., Swarnalatha, V. B., Nair, P., and Gadagkar, R. (1988). The mechanism of nestmate discrimination in the tropical social wasp *Ropalidia marginata* and its implications for the evolution of sociality. *Behavioral Ecology and Sociobiology*, **23**, 271–9.

Verstraeten, C. (1976). Nidification aberrante de *Polistes* sp. (Hymenoptera, Vespidae) sur *Adesmia variolaris* Oliv. (Coleoptera, Tenebrionidae). *Bulletin et Annales de la Société Royale Belge d'Entomologie*, **112**, 163–4.

Vigilant, L., Stoneking, M., Harpending, H., Kocher, T. D., Hawkes, K., and Wilson, A. C. (1991). African populations and the evolution of human mitochondrial DNA. *Science*, **253**, 1503–7.

Vollrath, F. (1986). Eusociality and extraordinary sex ratios in the spider *Anelosimus eximius* (Araneae: Theridiidae). *Behavioral Ecology and Sociobiology*, **18**, 283–7.

Waddington, C. H. (1961). Genetic assimilation. *Advances in Genetics*, **10**, 257–90.

Waku, Y. and Foldi, I. (1984). The fine structure of insect glands secreting waxy substances. In *Insect ultrastructure*, Vol. 2 (1st edn) (ed. R. C. King and H. Akai), pp. 303–22. Plenum Press, New York.

Waldman, B., Frumhoff, P. C., and Sherman, P. W. (1988). Problems of kin recognition. *Trends in Ecology and Evolution*, **3**, 8–13.

Wallace, A. R. (1876). *The geographical distribution of animals*. Macmillan, London.

Warburg, O. (1919). Über die Geschwindigkeit der photochemischen Kohlensäure-zersetzung in lebenden Zellen. *Biochemische Zeitschrift*, **100**, 230–70.

Waterman, T. H. (1981). Polarization sensitivity. In *Handbook of sensory physiology. Vision in invertebrates*, Vol. VII/6B, (ed. H. Autrum), pp. 281–469. Springer, Berlin.

Wcislo, W. T. (1987). The roles of seasonality, host synchrony, and behaviour in the evolutions and distributions of nest parasites in Hymenoptera (Insecta), with special references to bees (Apoidea). *Biological Reviews*, **62**, 515–43.

Wcislo, W. T. (1989). Behavioural environments and evolutionary change. *Annual Review of Ecology and Systematics*, **20**, 137–69.

Wcislo, W. T., West-Eberhard, M. J., and Eberhard, W. G. (1988). Natural history and behavior of a primitively social wasp, *Auplopus semialatus*, and its parasite, *Irenangelus eberhardi*, (Hymenoptera: Pompilidae). *Journal of Insect Behavior*, **1**, 247–60.

Weber, J. and May, P. (1989). Abundant class of human DNA polymorphisms which can be typed using the polymerase chain reaction. *American Journal of Human Genetics*, **44**, 388–96.

Wehner, R. (1981). Spatial vision in arthropods. In *Handbook of sensory physiology. Vision in invertebrates*, Vol. VII/6C, (ed. H. Autrum), pp. 287–616. Springer, Berlin.

Wehner, R. (1983). Celestial and terrestrial navigation: Human strategies-insect strategies. In *Behavioural physiology and neuroethology* (ed. F. Huber and H. Markl), pp. 366–81. Springer, Berlin.

Wehner, R. (1984). Astronavigation in Insects. *Annual Review of Entomology*, **29**, 277–98.

Wehner, R. (1992). Arthropods. In *Animal homing* (ed. F. Papi), pp. 45–144. Chapman and Hall, London.

Wehner, R. (1993). Les yeux "boussole": un instrument de navigation des insectes sociaux. *Actes des Colloques des Insectes Sociaux*, **8**, 1–8.

Wehner, R. and Menzel, R. (1990). Do insects have cognitive maps? *Annual Revue of Neuroscience*, **13**, 403–14.

Wenzel, J. W. (1987a). Male reproductive behavior and mandibular glands in *Polistes major* (Hymenoptera: Vespidae). *Insectes Sociaux*, **34**, 44–57.

Wenzel, J. W. (1987b). *Ropalidia formosa*, a nearly solitary paper wasp from Madagascar (Hymenoptera: Vespidae). *Journal of the Kansas Entomological Society*, **60**, 549–56.

Wenzel, J. W. (1988). Architecture and the evolution of nest design in paper wasps

(Hymenoptera: Vespidae). Unpublished D. Phil. thesis. University of Kansas.

Wenzel, J. W. (1989). Endogenous factors, external cues and eccentric construction in *Polistes annularis* (Hymenoptera: Vespidae). *Journal of Insect Behavior*, **2**, 679–99.

Wenzel, J. W. (1990). A social wasp's nest from the Cretaceous Period, Utah, USA, and its biogeographical significance. *Psyche*, **97**, 21–9.

Wenzel, J. W. (1991). Evolution of nest architecture. In *The social biology of wasps* (ed. K. G. Ross and R. W. Matthews), pp. 480–519. Comstock, Ithaca.

Wenzel, J. W. (1992a). Behavioral homology and phylogeny. *Annual Review of Ecology and Systematics*, **23**, 361–81.

Wenzel, J. W. (1992b). Extreme queen-worker dimorphism in *Ropalidia ignobilis*, a small-colony wasp (Hymenoptera: Vespidae). *Insectes Sociaux*, **39**, 31–43.

Wenzel, J. W. (1993). Application of the biogenetic law to behavioural ontogeny: a test using nest architecture in paper wasps. *Journal of Evolutionary Biology*, **6**, 229–47.

Wenzel, J. W. and Pickering, J. (1991). Cooperative foraging, productivity and the central limit theorem. *Proceedings of the National Academy of Sciences of USA*, **88**, 36–8.

West, M. J. (1967). Foundress associations in polistine wasps: Dominance hierarchies and the evolution of social behavior. *Science*, **157**, 1584–5.

West-Eberhard, M. J. (1969a) The social biology of polistine wasps. Unpublished D. Phil. thesis. University of Michigan.

West-Eberhard, M. J. (1969b). The social biology of polistine wasps. *Miscellaneous Publications of the Museum of Zoology, University of Michigan*, **140**, 1–101.

West-Eberhard, M. J. (1975). The evolution of social behavior by kin selection. *Quarterly Review of Biology*, **50**, 1–33.

West-Eberhard, M. J. (1977). The establishment of reproductive dominance in social wasp colonies. In *Abstracts of the 8th International Ethological Conference*, Wagenigen, pp. 223–7.

West-Eberhard, M. J. (1978a). Temporary queens in *Metapolybia* wasps: nonreproductive helpers without altruism? *Science*, **200**, 441–3.

West-Eberhard, M. J. (1978b). Polygyny and the evolution of social behavior in wasps. *Journal of the Kansas Entomological Society*, **51**, 832–56.

West-Eberhard, M. J. (1979). Sexual selection, social competition and evolution. *Proceedings of the American Philosophical Society*, **123**, 222–34.

West-Eberhard, M. J. (1981). Intragroup selection and the evolution of insect societies. In *Natural selection and social behavior: recent research and new theory* (ed. R. D. Alexander and D. W. Tinkle), pp. 3–17. Chiron Press, New York.

West-Eberhard, M. J. (1982a). Communication in social wasps: predicted and observed patterns, with a note on the significance of behavioral and ontogenetic flexibility for theories of worker "altruism". In *La communication chez les sociétés d'insectes* (ed. A. de Haro and X. Espadaler), pp. 13–36. Universidad Autonoma de Barcelona, Bellaterra.

West-Eberhard, M. J. (1982b). The nature and evolution of swarming in tropical social wasps (Vespidae, Polistinae, Polybiini). In *Social insects in the tropics*, Vol. 1, (ed. P. Jaisson), pp. 97–128. University of Paris XIII Press, Paris.

West-Eberhard, M. J. (1983). Sexual selection, social competition, and speciation. *Quarterly Review of Biology*, **58**, 155–83.

West-Eberhard, M. J. (1986a). Dominance relations in *Polistes canadensis* (L.): a tropical social wasp. *Monitore Zoologico Italiano (Nuova Serie)*, **20**, 263–81.

West-Eberhard, M. J. (1986b). Alternative adaptations, speciation and phylogeny (A review). *Proceedings of the National Academy of Sciences of the USA*, **83**, 1388–92.

West-Eberhard, M. J. (1987a). Flexible strategy and social evolution. In *Animal societies: theories and facts* (ed. Y. Itô, J. L. Brown, and J. Kikkawa), pp. 35–51. Scientific Societies Press Ltd., Tokyo.

West-Eberhard, M. J. (1987*b*). The epigenetic origins of insect sociality. In *Chemistry and biology of social insects* (ed. J. Eder and H. Rembold), pp. 369–72. Verlag J. Peperny, München.

West-Eberhard, M. J. (1988*a*). Observations of *Xenorhynchium nitidulum* (Fabricius) (Hymenoptera, Eumeninae), a primitively social wasp. *Psyche*, **94**, 317–23.

West-Eberhard, M. J. (1988*b*). Phenotypic plasticity and "genetic" theories of insect sociality. In *Evolution of social behaviour and integrative levels* (ed. E. Tobach and G. Greenberg), pp. 123–33. Erlbanm Press, Hillsdale, New Jersey.

West-Eberhard, M. J. (1989). Phenotypic plasticity and the origins of diversity. *Annual Review of Ecology and Systematics*, **20**, 249–78.

West-Eberhard, M. J. (1991). Introduction. In *The social biology of wasps* (ed. K. G. Ross and R. W. Matthews), pp. 1–4. Comstock, Ithaca.

West-Eberhard, M. J. (1992). Behaviour and evolution. In *Molds, molecules, and Metazoa, growing points in evolutionary biology* (ed. P. R. Grant and H. S. Horn), pp. 56–79. Princeton University Press.

Weyrauch, W. K. (1929). Beitrag zur Biologie von *Polistes. Biologisches Zentralblatt*, **48**, 407–27.

Weyrauch, W. K. (1937). Zur systematik und Biologie der Kuckuckswespen *Pseudovespa, Pseudovespula*, und *Pseudopolistes. Zoologische Jahrbücher, Abteilung Systematik, Ökologie und Geographie der Tiere*, **70**, 243–90.

Weyrauch, W. K. (1938). Zur Systematik und Biologie der palearktischen Polistinen. *Arbeiten über Physiologische Angewandte Entomologie aus Berlin-Dahlem*, **5**, 273–8.

Weyrauch, W. K. (1939). Zur Systematik der paläarktischen Polistinen auf biologischer Grundlage. *Archiv für Naturgeschichte, N. F.*, **8**, 145–97.

Wheeler, D. E. (1986). Developmental and physiological determinants of caste in social Hymenoptera: evolutionary implications. *American Naturalist*, **128**, 13–34.

Wheeler, D. E. (1991). The developmental basis of worker caste polymorphism in ants. *The American Naturalist*, **138**, 1218–38.

Wheeler, D. E. (1994). Nourishment in ants: Patterns in individuals and societies. In *Nourishment and evolution in insect societies*, (ed. J. H. Hunt and C. A. Nalepa), pp. 245–278. Westview Press, Boulder.

Wheeler, W. M. (1892). On the appendages of the first abdominal segment of embryos of insects. *Transactions of the Wisconsin Academy of Sciences*, **8**, 87–145.

Wheeler, W. M. (1913). *The ants – Their structure, development and behavior*. Columbia University Press. New York.

Wheeler, W. M. (1922). Social life among the insects. Lecture II. Part 2. Wasps solitary and social. *The Scientific Monthly*, **15**, 119–31.

Wheeler, W. M. (1923). *Social life among the insects*. Harcourt, Brace and Co., New York.

Wheeler, W. M. (1928). *The social insects: their origin and evolution*. Kegan Paul, Trench, Trubner and Co., London.

Wiley, E. O. (1979). An annotated Linnaean hierarchy, with comments on natural taxa and competing systems. *Systematic Zoology*, **28**, 308–37.

Wiley, E. O. (1987). Methods in vicariance biogeography. In *Systematics and evolution: a matter of diversity* (ed. P. Hovenkamp), pp. 283–306. Institute of Systematic Botany, Utrecht University.

Williams, F. X. (1919). Philippine wasp studies. II. Descriptions of new species and life history studies. *Bulletin of the Experimental Station of the Hawaii Sugar Planters' Association (Entomological Series)*, **14**, 19–186.

Williams, F. X. (1928). Studies in tropical wasps – their hosts and associates (with descriptions of new species). *Bulletin of the Experiment Station of the Hawaiian Sugar Planters' Association*, **19**, 1–179.

Wilson, D. S. and Colwell, R. K. (1981). Evolution of sex ratio in structured demes. *Evolution*, **35**, 882–97.

Wilson, E. O. (1971). *The insect societies*. Harvard University Press, Cambridge.

Winston, M. L. (1987). *The biology of the honey bee*. Harvard University Press, Cambridge.

Yablokov, A. V. (1986). *Phenetics: evolution, population trait*. Columbia Universiy Press, New York.

Yamane, Sk. (1973). Discovery of a pleometrotic association in *Polistes chinensis antennalis* Perez (Hymenoptera, Vespidae). *Life Study*, **17**, 3–4.

Yamane, Sk. (1978). Evolution of social parasitism among the Vespinae. *Panmixia*, **3**, 1–19.

Yamane, Sk. and Kusigemati, K. (1985). Vespoidea and Scoliidae from the Fiji and Solomon Islands (Insecta, Hymenoptera). *Occasional Papers of the Kagoshima University Research Center for the South Pacific*, **5**, 75–9.

Yamane, Sk. and Okazawa, T. (1981). Mature larvae of some polistine wasps from Papua New Guinea and Fiji, with notes on larval characters of the Old World and Oceanian Polistinae (Hymenoptera: Vespidae). *Reports of the Faculty of Science, Kagoshima University (Earth Sciences and Biology)*, **14**, 65–75.

Yamane, Sô. (1969). Preliminary observations on the life history of two polistine wasps, *Polistes snelleni* and *P. biglumis* in Sapporo, northern Japan. *Journal of the Faculty of Science, Hokkaido University, VI, Zoology*, **17**, 78–105.

Yamane, Sô. (1972). Life cycle and nest architecture of *Polistes* wasps in the Okushiri Island, northern Japan (Hymenoptera, Vespidae). *Journal of the Faculty of Science, Hokkaido University, VI, Zoology*, **18**, 440–59.

Yamane, Sô. (1980). Social biology of *Parapolybia varia* in Taiwan. Unpublished D. Phil. thesis, Hokkaido University, Japan.

Yamane, Sô. (1984). Nest architecture of two oriental paper wasps. *Parapolybia varia* and *P. nodosa*, with notes on its adaptive significance (Vespidae, Polistinae). *Zoologische Jahrbücher, Abteilung Systematik, Ökologie und Geographie der Tiere*, **111**, 119–41.

Yamane, Sô. (1986). The colony cycle of the Sumatran paper wasp *Ropalidia (Icariola) variegata jacobsoni* (Buysson), with reference to the possible occurrence of serial polygyny (Hymenoptera Vespidae). *Monitore Zoologico Italiano (Nuova Serie)*, **20**, 135–61.

Yamane, Sô. (1992). A huge nest of *Vespa basalis* collected in Taiwan (Hymenoptera, Vespidae). *Chinese Journal of Entomology*, **12**, 1–11.

Yamane, Sô. and Kawamichi, T. (1975). Bionomic comparison of *Polistes biglumis* (Hymenoptera, Vespidae) at two different localities in Hokkaido, northern Japan, with reference to its probable adaptation to cold climate. *Kontyû*, **43**, 214–32.

Yamane, Sô. and Okazawa, T. (1977). Some biological observations on a paper wasp, *Polistes (Megapolistes) tepidus malayanus* Cameron (Hymenoptera, Vespidae) in New Guinea. *Kontyû*, **45**, 283–99.

Yamane, Sô. and Yamane, Sk. (1979). Polistine wasps from Nepal (Hymenoptera, Vespidae). *Insecta Matsumurana, (N. S.)*, **15**, 1–37.

Yamane, Sô., Sakagami, S. F., and Ohgushi, R. (1983*a*). Multiple behavioural options in a primitively social wasp, *Parischnogaster mellyi*. *Insectes Sociaux*, **30**, 412–15.

Yamane, Sô., Kojima, J., and Yamane, Sk. (1983*b*). Queen/worker size dymorphism in an oriental polistine wasp, *Ropalidia montana* Carl (Hymenoptera: Vespidae). *Insectes Sociaux*, **30**, 416–22.

Yamane, Sô., Abbas, N. D., and Matsuura, M. (1989). Nest architecture of three species of genus *Polistes* with biological notes on *P. tenebricosus hoplites* in Sumatera Barat (Hymenoptera, Vespidae). *Bulletin of the Faculty of Education, Ibaraki University (Natural Sciences)*, **38**, 69–83.

Yamane, Sô., Sakagami, S. F., and Ohgushi, R. (1990). Social behavior in the stenogastrine wasp *Parishnogaster mellyi* (Hymenoptera, Vespidae). In *Natural history of social wasps and bees in Equatorial Sumatra* (ed. S. F. Sakagami, R. Ohgushi, and D. W. Roubik), pp. 97–111. Hokkaido University Press, Sapporo.

Yamane, Sô., Itô, Y., and Spradbery, J. P. (1991). Comb cutting in *Ropalidia plebeiana*: a new process of colony fission in social wasps (Hymenoptera: Vespidae). *Insectes Sociaux*, **38**, 105–10.

Yoshikawa, K. (1955). A polistine colony usurped by a foreign queen. Ecological studies of *Polistes* wasps, II. *Insectes Sociaux*, **2**, 255–60.

Yoshikawa, K. (1956). Compound nest experiments in *Polistes fadwigae* Dalla Torre. Ecological studies of *Polistes* wasps. IV. *Journal of the Institute of Polytechnic, Osaka City University (D)*, **7**, 229–43.

Yoshikawa, K. (1957). A brief note on the temporary polygyny in *Polistes fadwigae* Dalla Torre, the first discovery in Japan. Ecological studies of *Polistes* wasps, III. *Mushi*, **30**, 37–9.

Yoshikawa, K. (1962*a*). Introductory studies on the life economy of polistine wasps. VI. Geographical distribution and its ecological significance. *Journal of Biology, Osaka City University*, **13**, 19–44.

Yoshikawa, K. (1962*b*). Introductory studies on the life economy of polistine wasps. VII. Comparative consideration and phylogeny. *Journal of Biology Osaka City University*. **13**, 45–64.

Yoshikawa, K. (1962*c*). Introductory studies on the life economy of polistine wasps. I. Scope of problems and consideration of the solitary stage. *Bulletin of the Osaka Museum of Natural History*, **15**, 3–27.

Yoshikawa, K. (1963*a*). Introductory studies on the life economy of polistine wasps. III. Social stage. *Journal of Biology, Osaka City University*, **14**, 63–6.

Yoshikawa, K. (1963*b*). Introductory studies on the life economy of polistine wasps. V. Three stages relating to hibernation. *Journal of Biology, Osaka City University*, **14**, 87–96.

Yoshikawa, K., Ohgushi, R., and Sakagami, S. F. (1969). Preliminary report on entomology of the Osaka City University 5th Scientific Expedition to Southeast Asia 1966. With description of two new genera of stenogasterine [sic] wasps by J. van der Vecht. *Nature and Life in South East Asia (Tokyo)*, **6**, 153–82.

Zallen, D. (1993). The 'light' organism for the job: Green algae and photosynthesis research. *Journal of the History of Biology*, **26**, 269–80.

Zandee, M. and Roos, M. C. (1987). Component-compatibility in historical biogeography. *Cladistics*, **3**, 305–32.

Zikan, J. F. (1951). Polymorphismus und Ethologie der sozialen Faltenwespen (Vespidae, Diploptera). *Acta Zoologica Lilloana*, **11**, 5–51.

Zimmermann, K. (1930). Zur Systematik der palearktischen *Polistes*. *Mitteilungen aus dem Zoologische Museum in Berlin*, **15**, 609–21.

Index

Page numbers in *italic* refer to figures and tables.